VLSI DESIGN

A. ALBERT RAJ

Principal
DMI Engineering College
Aralvaimozhi, Kanyakumari

T. LATHA

Associate Professor
Department of Electronics and Communication Engineering
St. Xavier's Catholic College of Engineering
Kanyakumari

PHI Learning Private Limited
Delhi-110092
2014

₹ 350.00

VLSI DESIGN
A. Albert Raj and T. Latha

ISBN-978-81-203-3431-1

The export rights of this book are vested solely with the publisher.

Fifth Printing ... ... **July, 2014**

Published by Asoke K. Ghosh, PHI Learning Private Limited, Rimjhim House, 111, Patparganj Industrial Estate, Delhi-110092 and Printed by Mudrak, 30-A, Patparganj, Delhi-110091.

Contents

Preface xiii

1. INTRODUCTION 1–4

1.1 Evolution of VLSI Device Technology *1*
1.2 Metal Oxide Semiconductor (MOS) and VLSI Technology *3*
Summary 4

2. BASIC MOS STRUCTURE 5–33

2.1 Introduction *5*
2.2 Basic MOS Transistor Operation *6*
2.2.1 Enhancement Mode Transistor Action *7*
2.2.2 Depletion Mode Transistor Action *10*
2.3 MOS Transistor Switches *10*
2.3.1 Complementary CMOS Switch *11*
2.4 NMOS Fabrication *12*
2.5 Basic CMOS Technology *15*
2.5.1 The *p*-well CMOS Process *15*
2.5.2 The *n*-well CMOS Process *16*
2.5.3 The Twin-Well Process *19*
2.5.4 Silicon-On-Insulator Process *19*
2.6 CMOS Process Enhancements *21*
2.6.1 Interconnect *22*
2.6.2 Circuit Elements *25*
2.7 BiCMOS Technology *29*
2.7.1 BiCMOS Fabrication in an n-well Process *32*
2.7.2 Some Aspects of Bipolar and CMOS Devices *32*
Summary 32
Review Questions 33
Short Answer Questions 33

3. MOS DEVICE CHARACTERISTICS 34–62

3.1 Introduction 34
3.2 Static Behaviour of the MOS Transistor 35
3.2.1 The Threshold Voltage 35
3.2.2 Current–Voltage Relations 39
3.2.3 A Model for Manual Analysis 42
3.2.4 MOS Transistor Transconductance g_m and Output Conductance g_{ds} 42
3.2.5 MOS Transistor Figure of Merit, ω_0 43
3.3 Dynamic Behaviour of MOS Transistor 43
3.3.1 MOS Structure Capacitances 43
3.3.2 Channel Capacitance 44
3.3.3 Junction Capacitance 45
3.3.4 Capacitive Device Model 46
3.4 The Actual MOS Transistor—Secondary Effects 46
3.4.1 Threshold Variations 46
3.4.2 Source–Drain Resistance 47
3.4.3 Variation in I-V Characteristics 48
3.4.4 Subthreshold Conduction 49
3.4.5 CMOS Latchup 50
3.5 NMOS Inverter 50
3.6 Determination of Pull-up to Pull-down Ratio ($Z_{p.u}/Z_{p.d}$) for an NMOS Inverter Driven by Another NMOS Inverter 52
3.7 Pull-up to Pull-down Ratio for an NMOS Inverter Driven Through One or More Pass Transistors 54
3.8 Device Models for Simulation 56
3.8.1 MOS Models 56
3.8.2 DC MOSFET Model 56
3.8.3 High Frequency MOSFET Model 57
3.8.4 SPICE Models 60

Summary 62
Review Questions 62
Short Answer Questions 62

4. CMOS INVERTER DESIGN 63–89

4.1 Introduction 63
4.2 CMOS Inverter—DC Characteristics 65
4.3 Design Parameters of CMOS Inverter 75
4.3.1 Symmetric CMOS Inverter 76
4.3.2 Noise Margins of CMOS Inverter 77
4.3.3 Temperature Dependence of VTC of CMOS Inverter 78
4.3.4 Supply Voltage Scaling in CMOS Inverters 78
4.3.5 Power and Area Considerations 79
4.4 Switching Characteristics of CMOS Inverter 80
4.4.1 Estimation of CMOS Inverter Delay 81
4.5 CMOS—Gate Transistor Sizing 85
4.6 Stage Ratio 86

4.7 Power Dissipation *86*
4.7.1 Static Dissipation *87*
4.7.2 Dynamic Dissipation *87*
4.7.3 Short-circuit Dissipation *87*
4.7.4 Total Power Dissipation *87*
4.7.5 Power Economy *88*
Summary 88
Review Questions 88
Short Answer Questions 88

5. MOS CIRCUIT DESIGN PROCESSES 90–115
5.1 Introduction *90*
5.2 Why Design Rules *90*
5.3 MOS Layers *91*
5.4 Stick Diagrams *91*
5.4.1 Stick Layout Using NMOS Design *94*
5.4.2 Stick Layout Using CMOS Design *95*
5.5 Design Rules and Layout *96*
5.5.1 Lambda (λ) Based Design Rules *97*
5.5.2 Double Metal MOS Process Rules *101*
5.5.3 CMOS Lambda-based Design Rules *102*
5.6 Elements of Physical Design *109*
5.6.1 Basic Concepts *110*
5.6.2 Design Hierarchies *111*
Summary 114
Review Questions 114
Short Answer Questions 114

6. SPECIAL CIRCUIT LAYOUTS 116–136
6.1 Introduction *116*
6.2 Tally Circuits *117*
6.3 NAND–NAND, NOR–NOR, and AOI Logic *119*
6.4 Exclusive-OR Structures *122*
6.5 Barrel Shifter *127*
6.6 Transmission Gates *130*
6.7 Latches and Flip-flops *131*
6.7.1 CMOS Static Latches *132*
6.7.2 CMOS Dynamic Latches *132*
6.8 Fan-in and Fan-out of CMOS Logic Design *134*
Summary 136
Review Questions 136
Short Answer Questions 136

7. SUPER BUFFERS, BiCMOS AND STEERING LOGIC 137–157
7.1 Introduction *137*
7.2 RC Delay Lines *138*
7.3 Super Buffers *139*
7.3.1 NMOS Super Super Buffer *141*
7.3.2 NMOS Tristate Super Buffers and Pad-Drivers *142*
7.3.3 CMOS Super Buffers *143*
7.3.4 BiCMOS Gates *144*
7.4 Dynamic Ratioless Inverters *146*
7.5 Large Capacitive Loads *147*
7.6 Pass-Transistor Logic *148*
7.7 General Function Blocks *152*
7.7.1 NMOS Function Blocks *153*
7.7.2 CMOS Function Blocks *155*
Summary 156
Review Questions 156
Short Answer Questions 156

8. CMOS COMBINATIONAL LOGIC CIRCUITS 158–178
8.1 Introduction *158*
8.2 Static CMOS Design *159*
8.2.1 Complementary CMOS *159*
8.2.2 Ratioed Logic *163*
8.2.3 Pass-Transistor Logic *164*
8.3 Dynamic CMOS Design *166*
8.3.1 Dynamic Logic: Basic Principles *166*
8.3.2 Speed and Power Dissipation of Dynamic Logic *167*
8.3.3 Signal Integrity Issues in Dynamic Design *168*
8.3.4 Cascading Dynamic Gates *171*
8.4 Complex Logic Gates in CMOS *172*
Summary 177
Review Questions 177
Short Answer Questions 178

9. CMOS SEQUENTIAL LOGIC CIRCUITS 179–197
9.1 Introduction *179*
9.2 Timing Metrics for Sequential Circuits *180*
9.3 Classification of Memory Elements *181*
9.4 Static Latches and Registers *183*
9.4.1 Bistability Principle *183*
9.4.2 Multiplexer-Based Latches *183*
9.4.3 Master–Slave Edge-Triggered Register *185*
9.4.4 Low Voltage Static Latches *187*
9.5 Dynamic Latches and Registers *187*
9.5.1 Dynamic Transmission-Gate Edge-Triggered Registers *188*
9.5.2 C^2MOS—A Clock Skew Insensitive Approach *189*
9.5.3 True Single-Phase Clocked Register (TSPCR) *190*

9.6 Alternative Register Styles 192
9.6.1 Pulse Registers 192
9.6.2 Sense Amplifier-Based Registers 192
9.7 Non-bistable Sequential Circuits 193
9.7.1 The Schmitt Trigger 193
9.7.2 Monostable Sequential Circuits 194
9.7.3 Astable Circuits 195
Summary 196
Review Questions 196
Short Answer Questions 197

10. DESIGN OF ARITHMETIC BUILDING BLOCKS 198–225
10.1 Introduction 198
10.2 Datapaths 199
10.3 The Adder 200
10.3.1 The Binary Adder: Definitions 200
10.3.2 The Full-Adder: Circuit Design Considerations 202
10.3.3 The Binary Adder: Logic Design Considerations 207
10.4 The Multiplier 217
10.4.1 Multiplier: Definitions 218
10.4.2 Partial-Product Generation 219
10.4.3 Partial-Product Accumulation 220
10.4.4 Final Addition 223
Summary 224
Review Questions 224
Short Answer Questions 225

11. PROGRAMMABLE LOGIC DEVICES 226–258
11.1 Introduction 226
11.2 NMOS PLAs 227
11.2.1 NMOS PLA Layouts 227
11.3 Other Programmable Logic Devices 232
11.3.1 Field Programmable Logic Array (FPLA) 232
11.3.2 Programmable Array Logic (PAL) 233
11.3.3 Dynamic Logic Arrays (DLAs) 233
11.4 The Finite-State Machine as a PLA Structure 235
11.5 Complex Programmable Logic Devices (CPLDs) 237
11.5.1 CPLD Packaging and Programming 239
11.6 Field Programmable Gate Arrays (FPGAs) 242
11.6.1 FPGA Packaging and Programming 243
11.6.2 The XILINX Programmable Gate Array 250
11.6.3 Implementation in FPGAs 256
11.6.4 Design Flow 256
Summary 257
Review Questions 258
Short Answer Questions 258

12. CMOS CHIP DESIGN 259–284

12.1 Introduction *259*

12.2 Design Strategies *260*

12.2.1 Structured Design Strategies *260*

12.2.2 Hierarchy *261*

12.2.3 Regularity *261*

12.2.4 Modularity *261*

12.2.5 Locality 261

12.3 CMOS Chip Design Options *262*

12.3.1 Application Specific Integrated Circuits (ASICs) *262*

12.3.2 Types of ASICs *262*

12.3.3 Economics of ASICs *272*

12.3.4 CMOS Chip Design with Programmable Logic *277*

Summary 283

Review Questions 283

Short Answer Questions 284

13. ROUTING PROCEDURES 285–307

13.1 Introduction *285*

13.2 Global Routing *285*

13.2.1 Goals and Objectives *285*

13.2.2 Measurement of Interconnect Delay *286*

13.2.3 Global Routing Methods *289*

13.2.4 Global Routing Between Blocks *289*

13.2.5 Global Routing Inside Flexible Blocks *291*

13.2.6 Timing-Driven Methods *293*

13.2.7 Back-Annotation *294*

13.3 Detailed Routing *294*

13.3.1 Goals and Objectives *298*

13.3.2 Measurement of Channel Density *298*

13.3.3 Algorithms *299*

13.3.4 Left-Edge Algorithm *299*

13.3.5 Constraints and Routing Graphs *299*

13.3.6 Area-Routing Algorithms *302*

13.3.7 Multilevel Routing *303*

13.3.8 Timing-driven Detailed Routing *303*

13.3.9 Final Routing Steps *304*

13.4 Special Routing *304*

13.4.1 Clock Routing *304*

13.4.2 Power Routing *305*

Summary 306

Review Questions 306

Short Answer Questions 306

14. CMOS TESTING 308–351

14.1 Introduction 308
14.2 Need for Testing 308
14.2.1 Functionality Tests 309
14.2.2 Manufacturing Tests 309
14.2.3 Test Process 310
14.3 General Concepts of Testing 310
14.3.1 Reliability 311
14.3.2 Reliability Modelling 312
14.4 Manufacturing Test Principles 314
14.4.1 Fault Models 314
14.4.2 Gate Level Testing 317
14.4.3 Observability 321
14.4.4 Controllability 321
14.4.5 Fault Coverage 321
14.4.6 Automatic Test Pattern Generation (ATPG) 322
14.4.7 Fault Grading and Fault Simulation 326
14.4.8 Delay Fault Testing 327
14.4.9 Statistical Fault Analysis 328
14.4.10 Fault Sampling 329
14.5 Design Strategies for Test 330
14.5.1 Design for Testability 330
14.5.2 Ad hoc Testing 330
14.5.3 Scan-Based Test Techniques 333
14.5.4 Self-Test Techniques 339
14.5.5 I_{DDQ} Testing 342
14.6 Chip-Level Test Techniques 343
14.6.1 Regular Logic Arrays 343
14.6.2 Memories 343
14.6.3 Random Logic 344
14.7 System-Level Test Techniques 344
14.7.1 Boundary Scan 344
14.8 Layout Design for Improved Testability 350

Summary 350
Review Questions 350
Short Answer Questions 350

15. VERILOG HDL 352–365

15.1 Introduction 352
15.2 Basic Concepts 352
15.3 Structural Gate-Level Modelling 354
15.3.1 Verilog by Example 354
15.4 Switch-Level Modelling 359
15.5 Design Hierarchies 363

Summary 365
Review Questions 365
Short Answer Questions 365

16. BEHAVIOURAL MODELLING 366–396

16.1 Behavioural and RTL Modelling *366*
16.1.1 Dataflow Modelling and RTL *371*
16.2 General VLSI System Components *372*
16.2.1 Multiplexers *372*
16.2.2 Binary Decoders *375*
16.2.3 Equality Detectors and Comparators *378*
16.2.4 Priority Encoder *382*
16.2.5 Latches *386*
16.3 Combinational Logic Designs *387*
16.4 CMOS VLSI Latch *388*
16.4.1 D Flip-Flop *390*
Summary *395*
Review Questions *395*
Short Answer Questions *395*

17. ARITHMETIC CIRCUITS IN CMOS VLSI 397–413

17.1 Introduction *397*
17.2 Bit Adder Circuits *397*
17.3 Ripple-Carry Adders *405*
17.4 Carry Look-Ahead Adders *408*
Summary *412*
Review Questions *413*
Short Answer Questions *413*

18. VHDL 414–452

18.1 Introduction to VHDL *414*
18.2 Hardware Abstraction *415*
18.2.1 Basic Terminology *415*
18.2.2 Entity Declaration *416*
18.2.3 Architecture Body *416*
18.2.4 Structural Style of Modelling *417*
18.2.5 Dataflow Style of Modelling *417*
18.2.6 Behavioural Style of Modelling *417*
18.2.7 Mixed Style of Modelling *417*
18.3 VHDL is Like a Programming Language *418*
18.3.1 Lexical Elements *418*
18.3.2 Data Types and Objects *419*
18.4 VHDL Program *422*
18.4.1 Basic Logic Gates *422*
18.4.2 Combinational Logic Design *423*
18.4.3 Typical Combinational Components *425*
18.4.4 Typical Sequential Components *431*

18.5 Finite State Machine 436
18.6 Memories 438
18.6.1 wait Statement 439
18.6.2 Concurrent Signal Assignment Statements 442
18.7 Subprograms 445
18.7.1 Function 446
18.7.2 Procedure 447
18.8 Structural Modelling 448
18.9 Test Benches 449
18.10 VHDL vs. Verilog 451
Summary 451
Review Questions 452
Short Answer Questions 452

INDEX 453–458

Preface

This textbook has been designed for undergraduate-level courses in VLSI Design conducted in disciplines of electrical and electronics engineering, electronics and communication engineering, electronics and instrumentation engineering, and is also appropriate for postgraduate-level courses in Applied Electronics and VLSI Design.

The ability to design VLSI circuits has become increasingly important for those in the fields of electronics and computer engineering. Indeed the design and applications of VLSI extend beyond the boundaries of electronics engineering. This book is written in a simple and lucid manner with the topics arranged systematically to enable the reader to get enough knowledge of the subject. It includes 18 chapters covering designing, manufacturing, listing and programming concepts. The text is organized to give a clear and logical sequence of topics. The introductory chapter gives a brief historical background of VLSI circuits and explains how the design aims of low power consumption and smaller area are achieved. The book then progresses systematically from the description of basic MOS structure to analysis and application of VLSI circuits and ends with descriptive examples of VHDL and Verilog programming aspects.

Constructive suggestions and comments to help us improve this book would be most welcome.

A. Albert Raj

T. Latha

1 Introduction

Since the invention of the bipolar transistor in 1947, there has been an unprecedented growth of the semiconductor industry, with an enormous impact on the way people work and live. In the last twenty years or so by far, the strongest growth area of the semiconductor industry has been in silicon Very-Large Scale Integration (VLSI) technology. The sustained growth in VLSI technology is fuelled by the continued shrinking of transistors to ever smaller dimensions. The benefits of miniaturization—higher packing densities, higher circuit speeds, and lower power dissipation—have been key in the evolutionary progress leading to today's computers and communication systems that offer superior performance, dramatically reduced cost per function, and much reduced physical size, in comparison with their prodecessors. On the economic side, the integrated circuit (IC) business has grown worldwide from $1 billion in 1970 to 20 billion in 1984 and to over $300 billion in 2007. The electronics industry is now among the largest industries in terms of the output as well as employment in many nations. The importance of microelectronics in the economic, social, and even political development throughout the world will no doubt continue to ascend. The large worldwide investment in VLSI technology is the driving force behind the ever new developments in IC integration density and speed.

1.1 EVOLUTION OF VLSI DEVICE TECHNOLOGY

An excellent account of the evolution of the metal-oxide-semiconductor field-effect-transistor (MOSFET), from its initial concept to VLSI applications in the mid-1980s, can be found in the paper by Sah (1988). Figure 1.1 gives a chronology of the major milestone events in the development of VLSI technology. The bipolar transistor technology was developed early on and was applied to the first integrated circuit memory in mainframe computers in the 1960s. Bipolar transistors have been used all along where raw circuit speed is important, for bipolar circuits remain the fastest at the individual-circuit level. However, the large power dissipation of bipolar transistors has severely limited their integration level to about 10^4 circuits per chip. This integration level is quite low by today's VLSI standard.

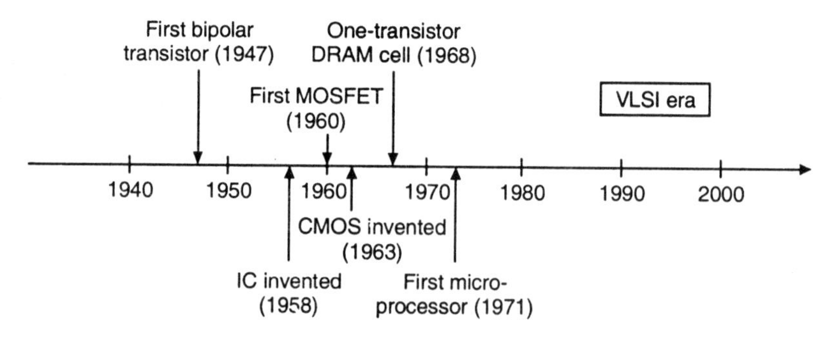

Figure 1.1 A brief chronology of the major milestone events in the development of VLSI technology.

The first MOSFET in the silicon substrate using SiO_2 as the gate insulator was fabricated in 1960. During the 1960s and 1970s, *n*-channel and *p*-channel MOSFETs were widely used, along with bipolar transistors, for implementing circuit functions on a silicon chip. Although the MOSFET devices were slow as compared to the bipolar devices, they had a higher layout density and were relatively simple to fabricate. The major breakthrough in the level of integration came in 1963 with the invention of CMOS (complementary MOS), in which, both *n*-channel and *p*-channel MOSFETs are constructed simultaneously on the same substrate. A CMOS circuit typically consists of a *n*-channel MOSFET and a *p*-channel MOSFET are connected in series between the power supply terminals, so that there is negligible standby power dissipation. Significant power is dissipated only during switching of the circuit. By clever design of the switch activities of the circuits on a chip to minimize active power dissipation, it has become possible to integrate hundreds of millions of CMOS transistors on a single chip. Another advantage of CMOS circuits comes from the ratioless, full rail-to-rail logic swing, which improves the noise margin and makes a CMOS chip easier to design.

As linear dimensions reached the 0.5 μm level in the early 1990s, the performance advantages of bipolar transistors were outweighed by the significantly greater circuit density of CMOS devices. Practically, all the VLSI chips in production today are based on CMOS technology. Bipolar transistors are used only where raw circuit speed makes an important difference. Consequently, bipolar transistors are widely used in small-size bipolar-only chips, or in BiCMOS chips where most functions are implemented using CMOS transistors and only a relatively small number are implemented using bipolar transistors.

Advances in lithography and etching techologies have enabled the industry to scale down transistors in physical dimensions, and to pack more transistors in the same chip area. Such progress, combined with a steady growth in chip size, resulted in an exponential growth in the number of transistors and memory bits per chip. The recent trends and future projections in these are illustrated in Figure 1.2. The dynamic RAMs (DRAMs) have the highest component count for IC chips. This has been so because of the smaller size of the one-transistor memory cell and because of the large and often insatiable demand for more memory in computing systems.

One remarkable feature of silicon devices that fuelled the rapid growth of the IT industry is their decreasing size and increasing speed. The transistors manufactured today are 20 times faster and occupy less than 1% of the area of those built 20 years ago. This is illustrated in Figure 1.2.

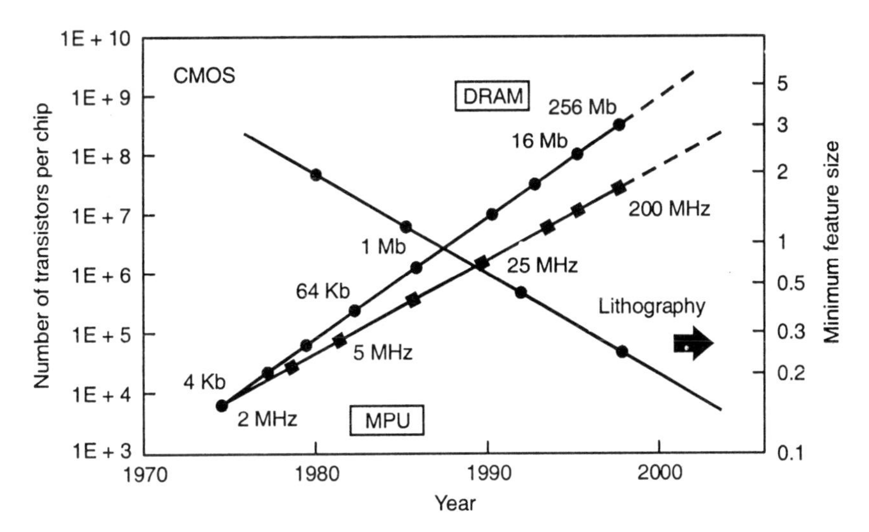

Figure 1.2 Trends in lithographic feature size, and the number of transistors per chip for DRAM and microprocessor chips.

1.2 METAL OXIDE SEMICONDUCTOR (MOS) AND VLSI TECHNOLOGY

Within the bounds of MOS technology, the possible circuit realizations may be based on PMOS, NMOS, CMOS and now BiCMOS devices. Although CMOS is the dominant technology, some of the examples used to illustrate the design processes will be presented in NMOS form for the following reasons:

- For NMOS technology, the design methodology and the design rules are easily learned, thus providing a simple but excellent introduction to structured design for VLSI.
- NMOS technology and design processes provide an excellent background for other technologies. In particular, some familiarity with NMOS allows a relatively easy transition to CMOS technology and design.

VLSI technology provides the user with a new and more complex range of 'off the shelf' circuits, and VLSI design processes are such that system designers can readily design their own special circuits of considerable complexity. This provides a new degree of freedom for designers and it is probable that some very significant advances will result. As advances in technology shrink the feature size for circuits integrated in silicon, the integration density is increasing rapidly.

Figure 1.3 illustrates clearly the approximate minimum line width of commercial products. Simultaneously, the effectiveness of the circuits produced has increased with scaling down. A common measure of effectiveness is the speed power product of the basic logic gate circuit of the technology. The speed–power product is measured in picojoules (pJ) and is the product of the gate switching delay in nanoseconds and the gate power dissipation in milliwatts.

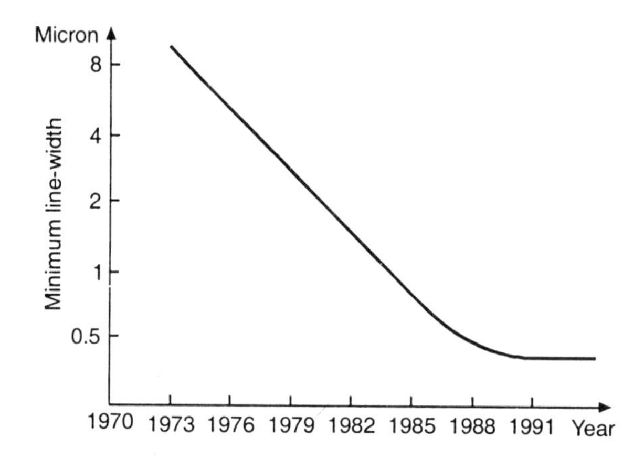

Figure 1.3 Approximate minimum line-width of commercial product vs. year.

SUMMARY

VLSI circuits have found a lot of applications in the recent developments of ICs. The dimensions of transistors have shrunk enormously, which has a great positive impact on VLSI technology. This introductory chapter describes briefly the evolution of VLSI device technology from the invention of first bipolar transistor, MOSFET, CMOS up to the VLSI era. It also gives a clear view of the power dissipation aspects of various types of transistor technologies. The CMOS circuits include both *p*-channel and *n*-channel MOSFETs which causes only negligible standby power dissipation. The active power dissipation occurring during switching activities can be minimized by good design of the circuits. By this characteristics, hundreds of millions of CMOS transistors can be integrated on a single chip. Also, the noise margin of CMOS-based circuits is improved leading to easy design of CMOS chip. The speed of silicon-based devices increased and cost decreased with the smaller sized ICs. The reduced feature size of silicon ICs rapidly increases the integration density.

2 Basic MOS Structure

2.1 INTRODUCTION

Silicon, a semiconductor, forms the basic material for a large class of integrated circuits. An MOS (metal-oxide-semiconductor) structure is created by superimposing several layers of conducting, insulating and transistor forming materials to create a sandwich like structure. These structures are created by a series of chemical processing steps involving oxidation of the silicon, diffusion of impurities into the silicon to give it certain conduction characteristics, and depositing and etching of aluminium on the silicon to provide interconnection. This construction process is carried out on a single crystal of silicon, available in the form of thin, flat circular wafers.

Typical physical structures for the two types of MOS transistors are shown in Figure 2.1. The NMOS transistors are formed in a p-type substrate of moderate doping level. The source and drain regions are formed by diffusing n-type impurities through suitable masks into these areas to give the desired n-impurity concentration and give rise to depletion regions. Thus, the

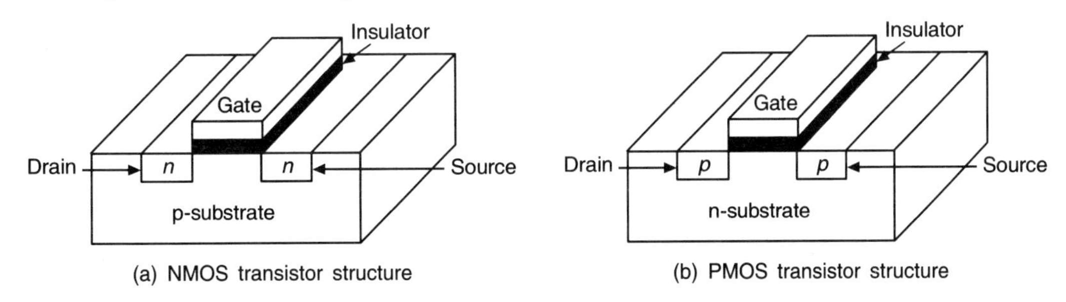

(a) NMOS transistor structure

(b) PMOS transistor structure

Figure 2.1 Basic MOS (BiCMOS) transistor physical structure.

source and the drain are isolated from one another by two diodes. The area separating the n-regions is capped with a sandwich consisting of SiO_2 (insulator) and a conducting electrode (polycrystalline silicon) called the gate. Similarly, the structure of PMOS transistor consists of n-type silicon substrate separating two p-type regions called source and drain. The p-type transistor also has a gate electrode in common with the n-transistor.

The circuit symbols for various MOS transistors are shown in Figure 2.2. The MOS transistor is a four-terminal device with gate, source, drain and body terminals. Since the body is generally connected to dc supply that is identical for all devices of the same type (GND for NMOS, V_{DD} for PMOS), most often it is not shown on the schematics. If the fourth terminal is not shown, it is assumed that the body is connected to the appropriate supply.

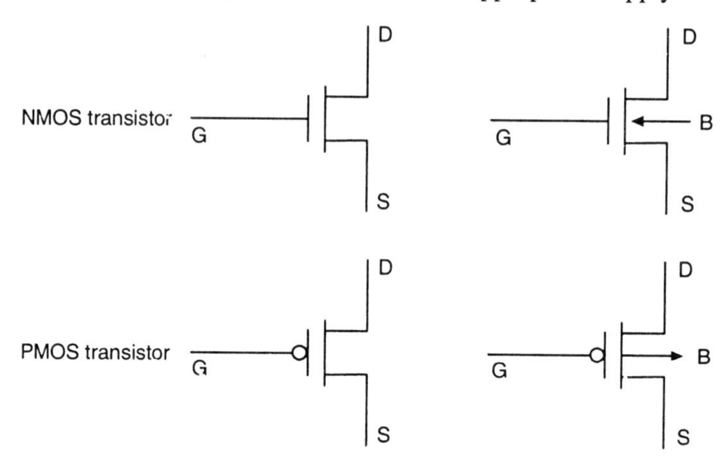

Figure 2.2 MOS transistor—circuit symbols.

2.2 BASIC MOS TRANSISTOR OPERATION

An MOS transistor is termed a majority-carrier device in which the current in a conducting channel between the source and drain is modulated by a voltage applied to the gate. In an NMOS transistor, the majority carriers are electrons. A positive voltage applied on the gate with respect to substrate enhances the number of electrons in the channel (the region immediately under the gate) and hence increases the conductivity of the channel. For gate voltages less than a threshold voltage, V_T, the channel is cut-off, thus causing a very low drain-to-source current. The operation of a *p*-type transistor is analogous to the NMOS transistor, with the exception that the majority carriers are holes and the voltages are negative with respect to the substrate.

The first parameter of interest that characterizes the switching behaviour of an MOS device is the threshold voltage V_T. It is defined as the gate-to-source voltage at which an MOS device begins to conduct. The graph shown in Figure 2.3 illustrates the relative conduction against the difference in gate-to-source voltage in terms of source–drain current (I_{DS}) and the gate-to-source voltage (V_{GS}). The *n*-channel devices that are normally cut-off (i.e., non-conducting) with zero gate bias are called Enhancement Mode Devices (EMD) whereas those devices that conduct with zero gate bias are called Depletion-Mode Devices (DMD). The *n*-channel and *p*-channel devices are duals of each other; i.e., the voltage polarities required for the correct operation are the opposite. The threshold voltages for the *n*-channel and the *p*-channel devices are denoted by V_{Tn} and V_{Tp} respectively. In CMOS technologies, both *n*-channel and *p*-channel transistors are fabricated on the same chip. Furthermore, most CMOS integrated circuits at present use enhancement type transistors.

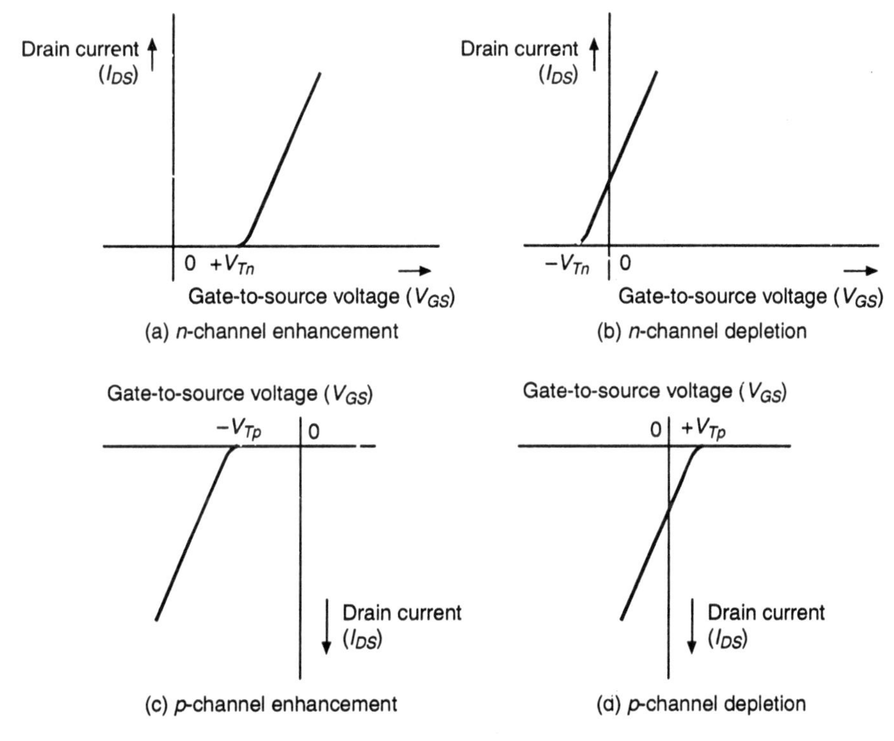

Figure 2.3 Conduction characteristics of EMOS and DEMOS transistors for fixed bias.

2.2.1 Enhancement Mode Transistor Action

NMOS Enhancement transistor. The structure of an *n*-channel enhancement type transistor is shown in Figure 2.4. It consists of a moderately doped silicon substrate onto which two heavily doped *n*+ regions, the source and drain are diffused. Between these two regions, there is a narrow region of *p*-type substrate called the channel, which is covered by a thin insulating layer of SiO_2 called the gate oxide. Over this oxide layer is a polysilicon electrode, referred to as the gate. Since the oxide layer is an insulator, the dc current from the gate-to-channel is essentially zero.

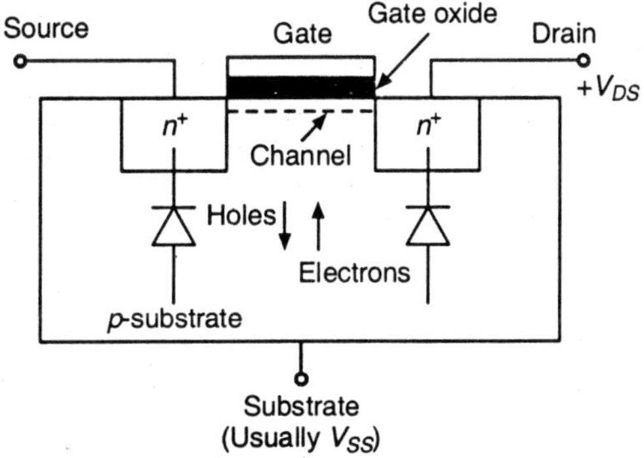

Figure 2.4 NMOS transistor (detailed structure).

Assume that a positive voltage is applied between the drain and source (V_{DS}). With zero gate voltage ($V_{GS} = 0$), no current flows from source to drain because they are effectively insulated from each other by the two reverse biased *pn*-junctions (indicated by the diode symbols in Figure 2.4) and source, an electric field is produced across the substrate, which attracts electrons towards the gate and repels holes. If the gate voltage is sufficiently large, the region under the gate changes from *p*-type to *n*-type due to accumulation of attracted electrons and provides a conduction path between the source and drain. Under such conditions, the surface of the underlying *p*-type silicon is said to be inverted.

Figure 2.5 shows the initial distribution of mobile positive holes in a *p*-type silicon substrate of an MOS structure for a voltage V_{GS} much less than the threshold voltage, V_T. This is termed as the accumulation mode [Figure 2.5(a)]. As V_{GS} is raised above V_T, the holes are repelled causing a depletion region under the gate. Now, the structure is in the depletion mode [Figure 2.5(b)].

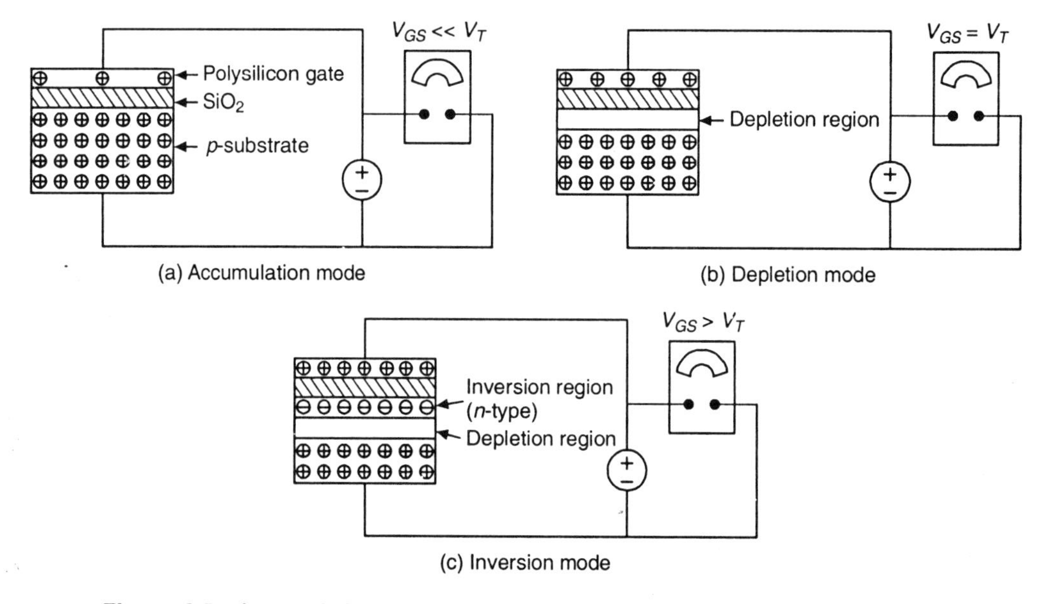

Figure 2.5 Accumulation, depletion and inversion modes in an MOS structure.

Raising V_{GS} above V_T results in electrons being attracted to the region of the substrate under the gate. A conductive layer of the electrons in the *p*-substrate gives rise to the name inversion mode [Figure 2.5(c)]. In an inversion layer substrate junction, the *n*-type layer is induced by the electric field *E* applied to the gate. Thus, this junction is a field-induced junction.

Electrically, an MOS device acts as a voltage-controlled switch that conducts initially when the V_{GS} voltage is equal to the threshold voltage V_T. When a voltage V_{DS} is applied between source and drain, with $V_{GS} = V_T$, the horizontal and vertical components of the electric field due to the source–drain voltage and gate-to-substrate voltage interact, causing conduction to occur along the channel. As the voltage from drain-to-source is increased, the resistive drop along the channel begins to change the shape of the channel characteristics [Figure 2.6(a)]. At the source-end of the channel, the full gate voltage is effective in inverting the channel. However, at the drain-end of the channel, only the difference between the gate and drain voltages is effective.

When the effective gate voltage ($V_{GS} - V_T$) is greater than the drain voltage, the channel becomes deeper as V_{GS} is increased. This is termed as linear, resistive, non-saturated or ohmic region where the channel current I_{DS} is a function of both gate and drain voltages [Figure 2.6(b)]. If $V_{DS} > V_{GS} - V_T$, then channel becomes pinched-off, i.e., the channel no longer reaches the drain (saturation region). This is illustrated in Figure 2.6(c).

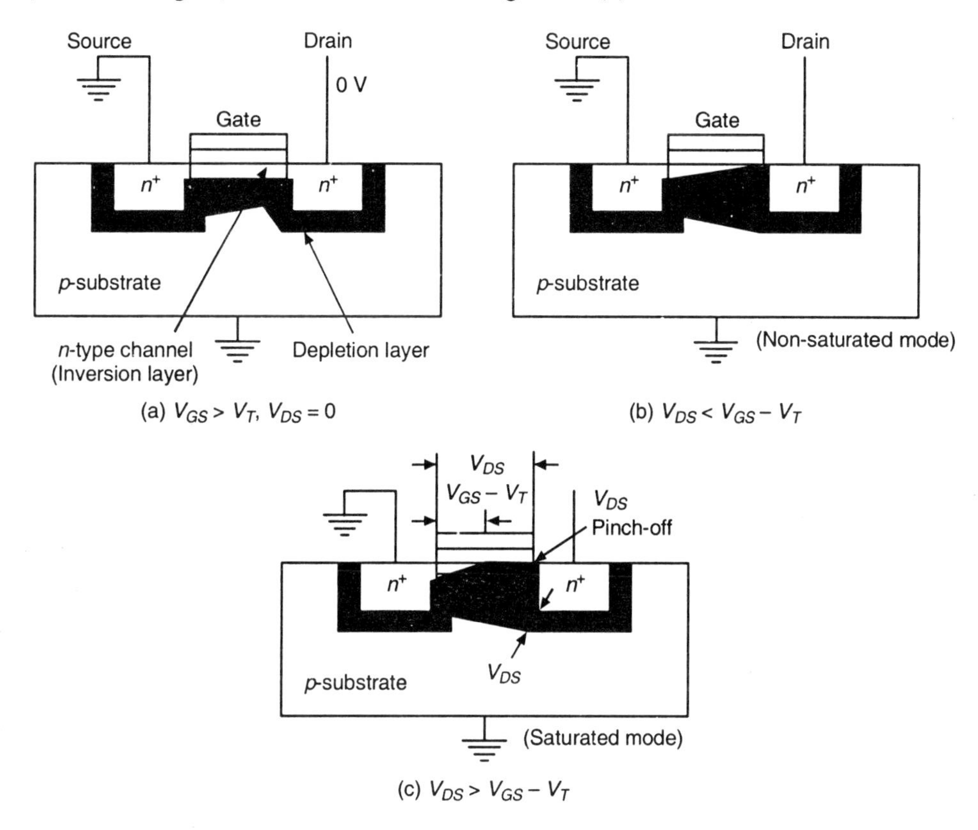

Figure 2.6 NMOS device behaviour under the influence of different terminal voltages.

The normal conduction characteristics of an MOS transistor can be given below:

- **Cut-off region:** Where the current flow is essentially zero (accumulation mode).
- **Non-saturation region:** Weak inversion region where the drain current is dependent on the gate and the drain voltage with reference to the substrate.
- **Saturation region:** Channel is strongly inverted and the drain current flow is ideally independent of the drain–source voltage (strong-inversion region).

An abnormal conduction condition called avalanche breakdown or punch-through can occur if very high voltages are applied to the drain. Under these circumstances, the gate has no control over the drain current.

PMOS Enhancement Transistor. A reversal of *n*-type and *p*-type regions of an NMOS transistor yields a PMOS transistor. This is illustrated in Figure 2.7. The application of negative gate voltage with respect to source draws holes into the region below the gate, resulting in the

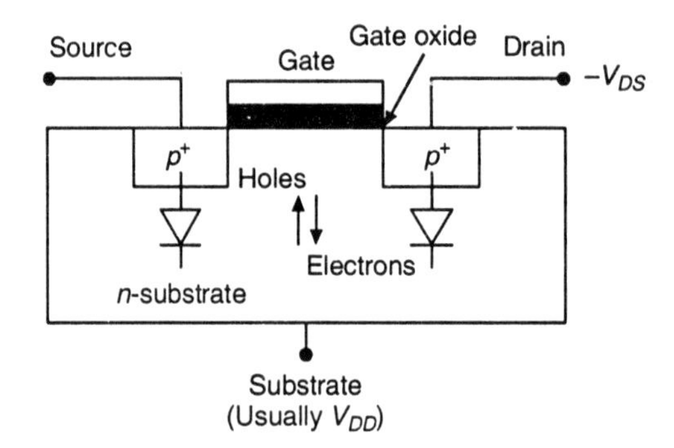

Figure 2.7 Physical structure of a PMOS transistor (detailed view).

channel changing from n-type to p-type. Thus, a conduction path is created between the source and drain. In this case, conduction results from the movement of holes in the channel. A negative gate voltage sweeps the holes from the source through the channel to the drain.

2.2.2 Depletion Mode Transistor Action

For depletion mode devices, the channel is established because of the implant, even when $V_{GS} = 0$, and to cause the channel to cease to exist, a negative voltage V_{TD} must be applied between gate and source. The value of V_{TD} is typically $< -0.8\ V_{DD}$ depending on the implant and substrate bias.

2.3 MOS TRANSISTOR SWITCHES

The gate controls the passage of current between the source and drain. This allows the MOS transistor to be viewed as simple on/off switch. The NMOS and PMOS switches are shown in Figure 2.8. In this, the gate has been labelled with the signal s, the drain 'a', and the source 'b'. In an N-switch, the switch is closed or ON if the drain and source are connected. This occurs

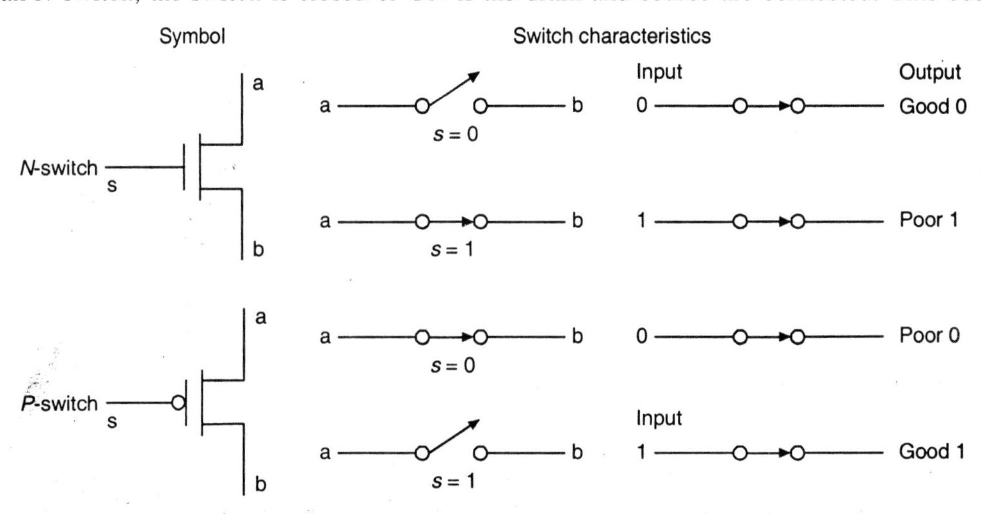

Figure 2.8 NMOS and PMOS switch symbols and their characteristics.

when there is a 1 on the gate. The switch is open or OFF if the drain and source are disconnected. A 0 on the gate ensures this condition. As it can be seen, an *N*-switch is almost a perfect switch when a 0 is to be passed from an input to output (say *a* to *b*). However, an *N*-switch is an imperfect switch when passing a 1.

It can be clearly noticed that PMOS switch has different properties from the *N*-switch. The *P*-switch is closed or ON when there is a 0 on the gate. The switch is open or OFF when there is a 1 on the gate. The *P*-switch is almost a perfect switch for passing 1 signals but imperfect when passing 0 signals. The output logic levels of an *N*-switch or a *P*-switch are summarized in Table 2.1.

Table 2.1 The output logic levels of *N*-switches and *P*-switches

Level	*Symbol*	*Switch condition*
Strong 1	1	*P*-switch gate = 0, Source = V_{DD}
Weak 1	1	*N*-switch gate = 1, Source = V_{DD} or *P*-switch connected V_{DD}
Strong 0	0	*N*-switch gate = 1, Source = V_{SS}
Weak 0	0	*P*-switch gate = 0, Source = V_{SS} or *N*-switch connected V_{SS}
High impedence	*Z*	*N*-switch gate = 0 or *P*-switch gate = 1

2.3.1 Complementary CMOS Switch

By combining an *N*-switch and a *P*-switch in parallel (Figure 2.9), we can obtain a complementary CMOS switch or *C*-switch in which 0s and 1s are passed in an acceptable fashion. In a circuit where only a 0 or a 1 has to be passed, the appropriate sub-switch (*n* or *p*) may be deleted, reverting to a *P*-switch or an *N*-switch. The control signal is applied to the *n*-transistor and the complement of the *p*-transistor. The complementary switch is also called a transmission gate or pass gate (complementary). Figures 2.9(a), (b) and (c) show the CMOS switch (or *C*-switch), the switch characteristics and different symbols of the switch.

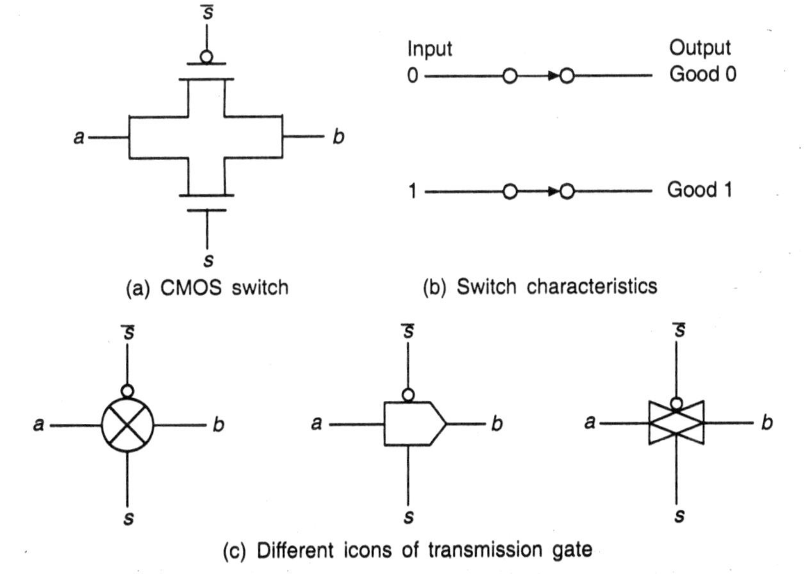

(a) CMOS switch

(b) Switch characteristics

(c) Different icons of transmission gate

Figure 2.9 A CMOS switch.

2.4 NMOS FABRICATION

The fabrication of NMOS can be considered a 'standard' process. The advantages of this process over the other processes are that it is conceptually and physically simpler than other modern processes because it requires less photolithographic steps. It has high functional density, good speed/power performance, and has tremendous future potential due to its scalability.

The major drawbacks of NMOS process are its high absolute power consumption and its electrical asymmetry. CMOS is replacing NMOS as the 'standard' process because it minimizes both of the above disadvantages.

A brief introduction to the general aspects of the polysilicon gate self-aligning NMOS fabrication process will now be given. The fabrication process used for NMOS are relevant to CMOS and BiCMOS which may be viewed as involving additional fabrication steps. Figure 2.10 shows the NMOS fabrication sequence.

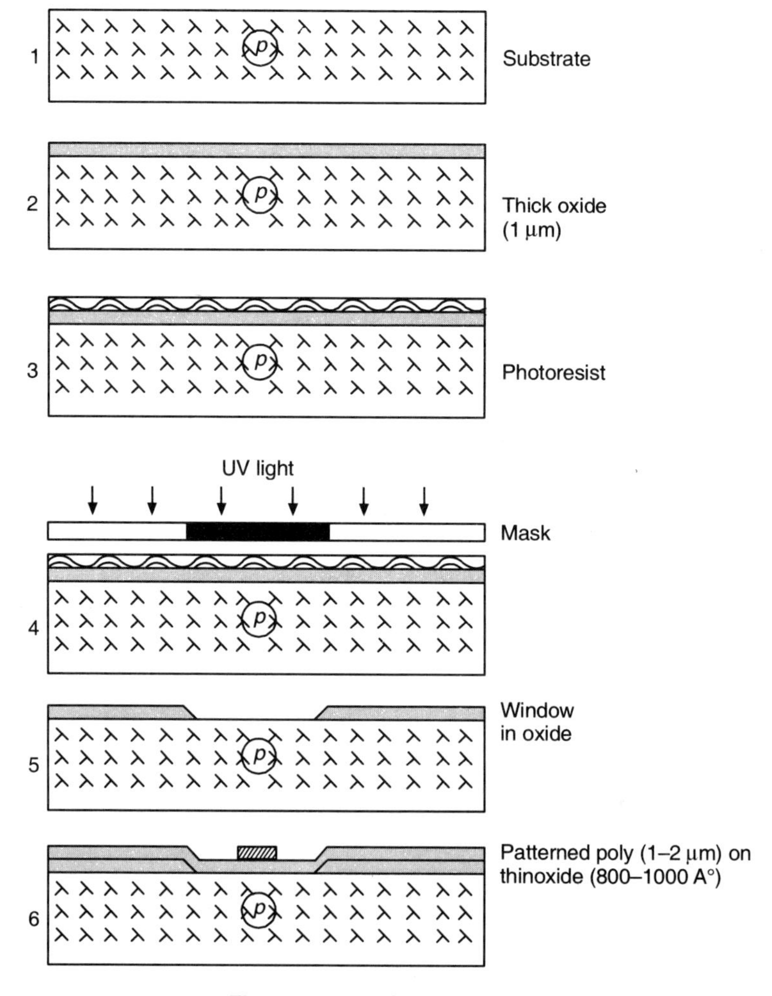

Figure 2.10 (Contd).

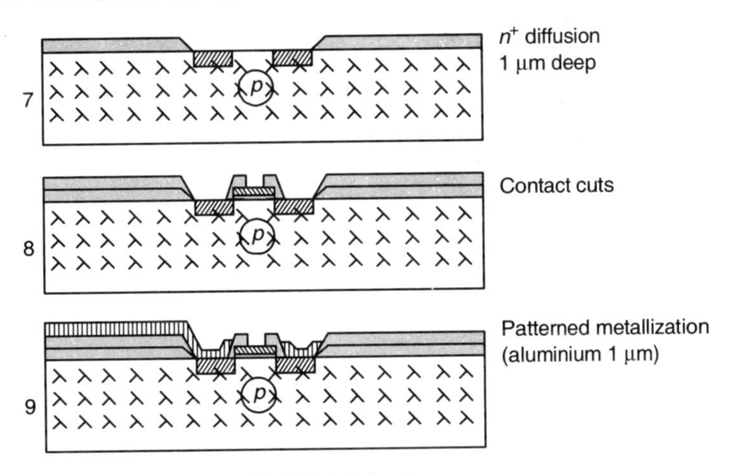

Figure 2.10 NMOS fabrication process.

Step 1: A thin wafer cut from a single crystal of silicon of high purity, into which the required *p*-impurities are introduced as the crystal is grown, is used for NMOS processing. The processing is performed on this wafer. Such wafers are typically 75 to 150 mm in diameter and 0.4 mm thick and are doped with, e.g., boron to impurity concentrations of 10^{15} cm^{-3} to 10^{16} cm^{-3}. At such doping concentrations, the wafer has resistivity in the range of 25 Ω-cm to 2 Ω-cm.

Step 2: By total or overall thermal oxidation process, 1 μm thick silicon dioxide layer is deposited throughout the wafer surface. The layer used as protective layer, acts as a barrier to dopants during processing, and to provide a common insulating substrate onto which other layers may be grown.

Step 3: Next, the surface is covered with a photoresist, which is deposited onto the wafer and spun to achieve an even distribution of the required thickness.

Step 4: The photoresist is then exposed to UV light through a mask, which defines those regions into which diffusion is to take place together with transistor channels. Here, the unexposed areas are left unaffected and may be dissolved in proper etchant, whereas exposed areas are polymerized and become hardened.

Step 5: The unexposed portions are subsequently etched away together with the underlying silicon dioxide so that the wafer surface is exposed in the window defined by the mask.

Step 6: The remaining photoresist is removed and a thin layer of SiO_2 (0.1 μm typical) is grown over the entire chip surface and then polysilicon is deposited on top of this to form the gate structure. The polysilicon layer consists of heavily doped polysilicon deposited by chemical vapour deposition (CVD). In the fabrication of fine pattern devices, precise control of thickness, impurity concentration, and resistivity are necessary.

Step 7: Further photoresist coating and masking allows the polysilicon to be patterned (as shown in Step 6), and then the thin oxide is removed to expose areas into which *n*-type impurities are to be diffused to form the source and drain as shown. Diffusion is achieved by heating the wafer to a high temperature and passing a gas containing the desired n-type impurity (for example, phosphorus) as shown in Figure 2.11. The polysilicon with underlying thin oxide and the thick oxide act as masks during diffusion—the process is known as self-aligning.

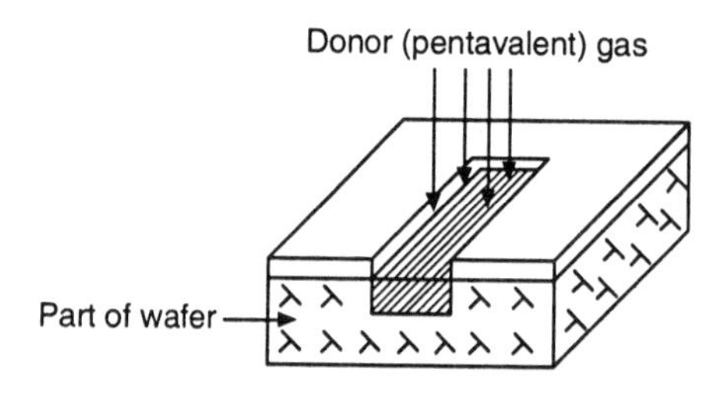

Figure 2.11 Impurity doping by diffusion process.

Step 8: Once again, thick oxide (SiO_2) is deposited over all and is then masked with photoresist and etched to expose selected areas of the polysilicon gate and the drain and source areas where connections (i.e., contact cuts) are to be made.

Step 9: The whole chip then has metal (aluminium) deposited over its surface to a thickness typically of 1 μm. This metal layer is then masked and etched to form the required interconnection pattern.

To the fabrication process of depletion mode devices, one more masking step for ion-implantation is added between Steps 5 and 6 in Figure 2.10. All other steps will be similar to that of fabricating NMOS devices. Here, once again the thick oxide acts as mask and this process stage is also self-aligning.

The formation of an NMOS device can be summarized as below:

- Processing takes place on a p-doped wafer on which a thick layer of silicon dioxide is grown.
- Mask 1: Pattern SiO_2 to expose the silicon surface in areas where paths in the areas where paths in the diffusion layer or source, drain or gate areas of transistors are required. Deposit thin oxide over all. Due to this, this mask is often known as 'thinox' mask but sometimes it is called as diffusion mask.
- Mask 2: Thinox region is patterned by ion-implantation process, where depletion mode devices are required to be formed—self-aligning.
- Mask 3: A polysilicon layer of typically 1.5 μm thick is deposited on entire area of the wafer, then pattern it by using Mask 3. Using the same mask, the thin oxide layer, not covering the polysilicon is removed.
- Mask 4: Diffuse n^+ regions into areas where thinox has been removed. Thus, transistor drains and sources are self-aligned with respect to the gate structures.
- Mask 5: Growing thick oxide layer over the entire substrate and then etching for contact cuts.
- Mask 6: The metal is deposited and then patterned with Mask 6.
- Mask 7: This will be required for over glassing processing step.

2.5 BASIC CMOS TECHNOLOGY

The CMOS (complementary metal oxide silicon) technology is recognized as the leading VLSI systems technology. CMOS provides an inherently low power static circuit technology that has the capability of providing lower power-delay product than bipolar, NMOS or GaAs technologies.

The four main CMOS technologies are:

1. *p*-well process
2. *n*-well process
3. Twin-tub process
4. Silicon On Insulator (SOI)/Silicon On Sapphire (SOS) process

In addition, by adding bipolar transistors, a range of BiCMOS processes are available.

2.5.1 The *p*-well CMOS Process

In this, the structure consists of an *n*-type substrate in which *p*-type devices may be formed by suitable masking and diffusion. In order to accommodate *n*-type devices, a deep *p*-well is diffused into the *n*-type substrate as shown in Figure 2.12.

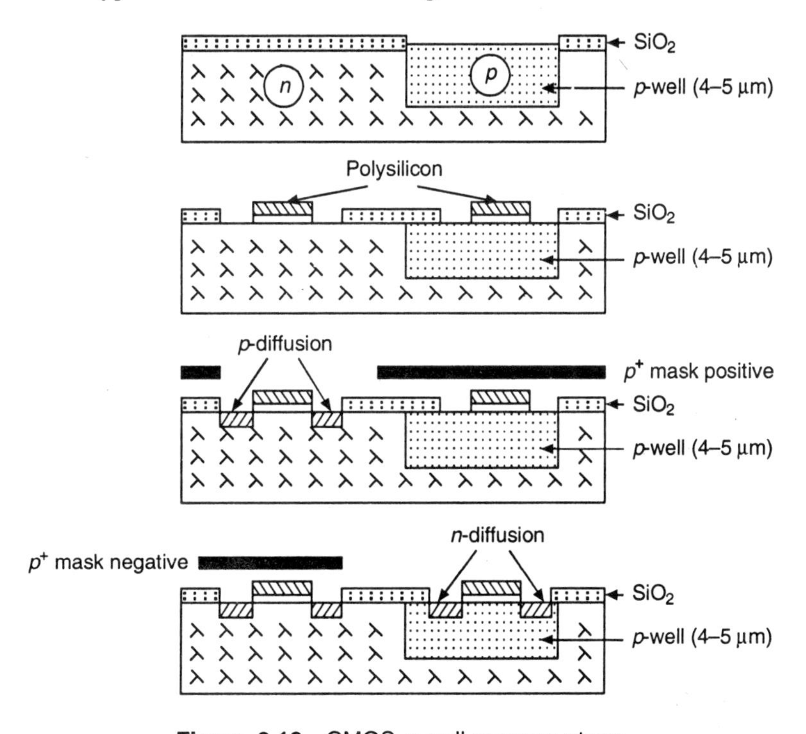

Figure 2.12 CMOS *p*-well process steps.

The diffusion must be carried out with special care since the p-well doping concentration and depth will affect the threshold voltages as well as the breakdown voltages of the n-transistors. To achieve low threshold voltages (0.6 to 1.0 V), we need either deep well diffusion or high well resistivity. However, deep wells require larger spacing between the n-type and p-type transistors and wires because of lateral diffusion and hence a large chip area.

The p-wells act as substrates for the n-devices within the parent n-substrate, and the two areas are electrically isolated.

In summary, typical processing steps are:

- Mask 1: Defines the areas in which the deep p-well diffusions are to take place.
- Mask 2: Defines the thinox regions, namely those areas where the thick oxide is to be stripped and thin oxide grown to accommodate p-transistors and n-transistors and diffusion wires.
- Mask 3: Used to pattern the polysilicon layer which is deposited after the thinox.
- Mask 4: A p^+ mask is now used (to be in effect ANDed with Mask 2) to define all areas where p-diffusion is to take place.
- Mask 5: This is usually performed using the negative form of the p^+ mask and, with Mask 2, defines those areas where n-type diffusion is to take place.
- Mask 6: Contact cuts are now defined.
- Mask 7: The metal layer pattern is defined by this mask.
- Mask 8: An overall passivation (overglass) layer is now applied and by this mask, the openings for bonding pads are defined.

The CMOS inverter circuit using p-well technology is shown in Figure 2.13.

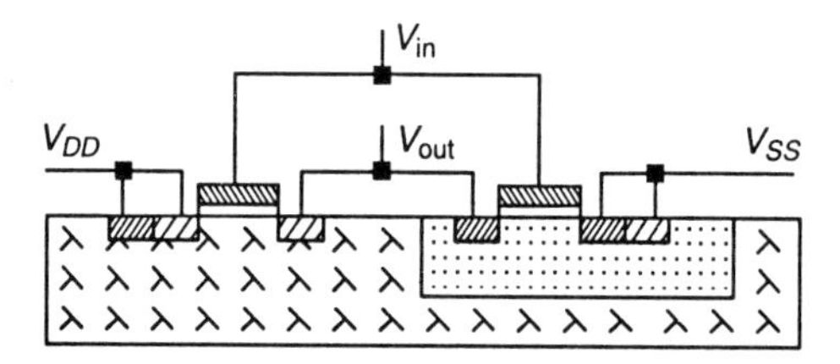

Figure 2.13 CMOS p-well inverter showing V_{DD} and V_{SS} substrate connection.

2.5.2 The n-well CMOS Process

Although the p-well process is widely used, n-well fabrication also gained wide acceptance. n-well CMOS circuits are also superior to p-well because of the lower substrate bias effects on transistor threshold voltage and inherently lower parasitic capacitances associated with source and drain regions. The flow diagram of the fabrication steps for NMOS process are illustrated in Figure 2.14.

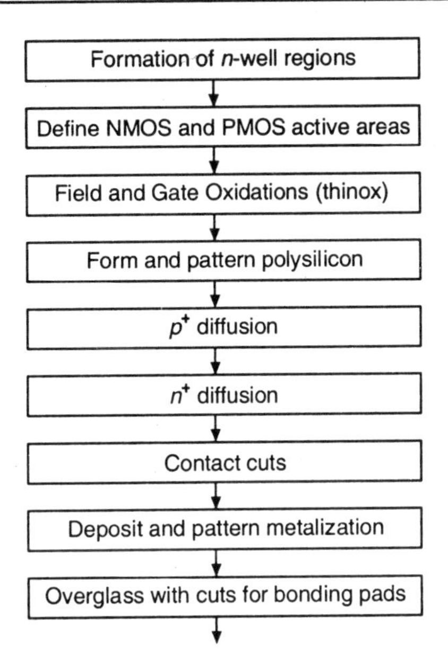

Figure 2.14 Fabrication steps for NMOS devices.

1. The structure comprised of *p*-type substrate in which *n*-devices are formed by suitable masking and diffusion. For accommodating *p*-devices, a deep *n*-well is diffused into the *p*-type substrate.
2. The diffusion step should be made with extra precaution, as the threshold voltage and the breakdown voltage of the transistors is affected by *n*-well doping concentration and depth of *n*-well. For obtaining low threshold voltages typically in the range of 0.6 V to 1.0 V, a deep diffused well or high resistivity well will be required. But the large chip spacing between the *n*-type and *p*-type transistors and wires are required for deep wells. This is due to the lateral diffusion.
3. The *n*-well serves as substrate for *p*-type devices within the parent *n*-substrate/wafer. Here, it has been observed that except the voltage polarity restriction, the *n*- and *p*-areas are isolated from each other.

By applying the abovementioned steps, the *n*-type wells are created in *p*-type substrate. The typical processing steps for fabrication of CMOS devices may be summarized as below:

- Mask 1: It defines the areas in which the deep *n*-well diffusions has to take place.
- Mask 2: It defines the thinoxide regions, i.e., those areas in which the thickoxide layer is to be stripped and thinoxide is to be grown for accommodating *n*- and *p*-type transistors and diffusion wires.
- Mask 3: Mask 3 is used to pattern the polysilicon layer, which is to be deposited after thinoxide.
- Mask 4: In this a n^+ mask is used (Mask 4 is added, i.e., in effect with Mask 2) to define the areas for all types of diffusion of *p*-type impurity.

- Mask 5: It is generally performed using negative form of n^+ mask and with Mask 2. It is used to define the areas for p-type diffusion.
- Mask 6: At this stage, contact cuts are defined.
- Mask 7: By this mask, the metal layer pattern is defined.
- Mask 8: At this stage, the overall passivation layer is applied and by this mask, the openings for bonding pads are defined.

These steps are clearly illustrated in Figure 2.15. And by way of illustration, Figure 2.16 shows an inverter circuit fabricated by the n-well process.

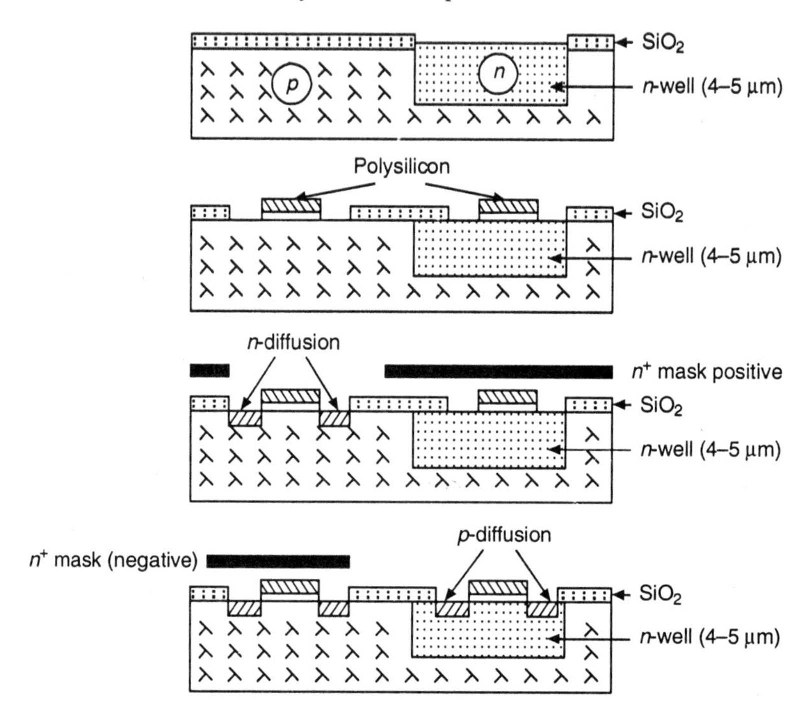

Figure 2.15 Steps for CMOS n-well process.

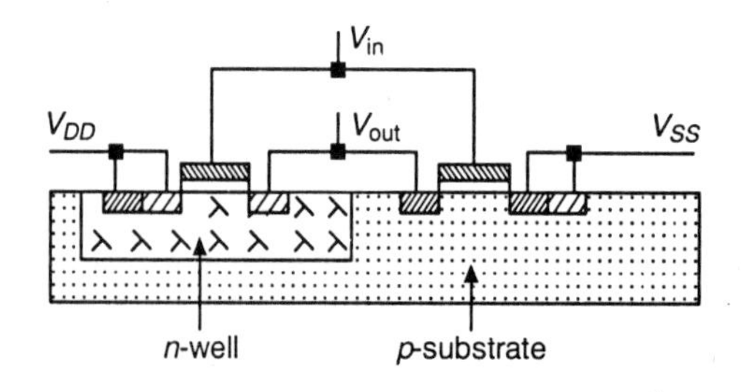

Figure 2.16 CMOS n-well inverter showing V_{DD} and V_{SS} substrate connections.

Using a low-resistivity epitaxial *p*-type substrate as the starting material can considerably reduce latchup problems. However, a factor of the *n*-well process is that the performance of the already poorly performing *p*-transistor is even further degraded.

2.5.3 The Twin-Well Process

A logical extension of the *p*-well and the *n*-well approaches is the twin-tub process. In this process, we start with a substrate of high resistivity *n*-type material and then create both *n*-well and *p*-well regions. Through this process, it is possible to preserve the performance of the *n*-transistors without compromising *p*-transistors. Doping control is more readily achieved and some relaxation in manufacturing tolerance results. This is particularly important as far as the latchup is concerned. The arrangement of an inverter using the twin-tub process is shown in Figure 2.17 which in turn may be compared with Figures 2.13 and 2.16. The twin-tub CMOS technology provides the basis for separate optimization of the *p*-type and *n*-type transistors, thus making it possible for threshold voltage, body effect, and the gain associated with *n*- and *p*-devices to be independently optimized.

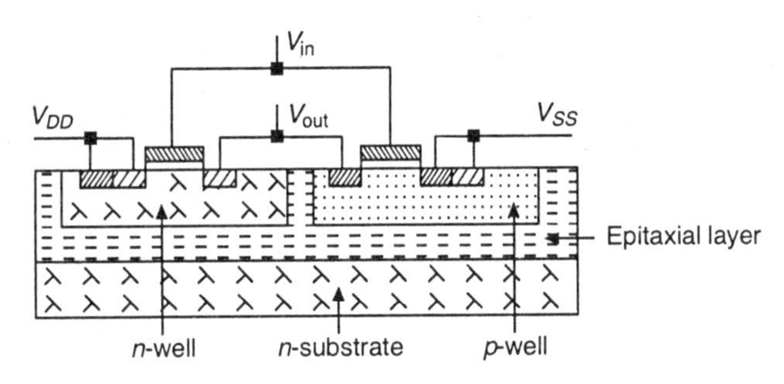

Figure 2.17 Twin-tub structure.

2.5.4 Silicon-On-Insulator Process

Rather than using silicon as the substrate, technologies have sought to use an insulating substrate to improve process characteristics such as latchup and speed. Hence is the emergence of Silicon-On-Insulator (SOI) technologies. Silicon-On-Sapphire (SOS) is the highest performance SOI technology today. In this approach, silicon is grown on a sapphire substrate, and islands are formed by implant or diffusion. *n*-channel and *p*-channel transistors are built on the islands. High performance is achieved due to a significant reduction in parasitic capacitance, and high gate density is achieved because no guard rings are needed.

Sapphire (Al_2O_3) is a good insulator and the lattice constants of silicon and sapphire match well. When sapphire is used as the substrate, the epitaxial growth of silicon yields monocrystalline material. Sapphire is also not affected by radiation as bulk silicon is, which makes it a preferred material for military applications which require radiation hardened devices.

The steps used in typical SOI CMOS processes are as follows:

- A thin film (7–8 μm) of very lightly doped *n*-type silicon is grown over an insulator. Sapphire or silicon dioxide is a commonly used insulator [Figure 2.18(a)].

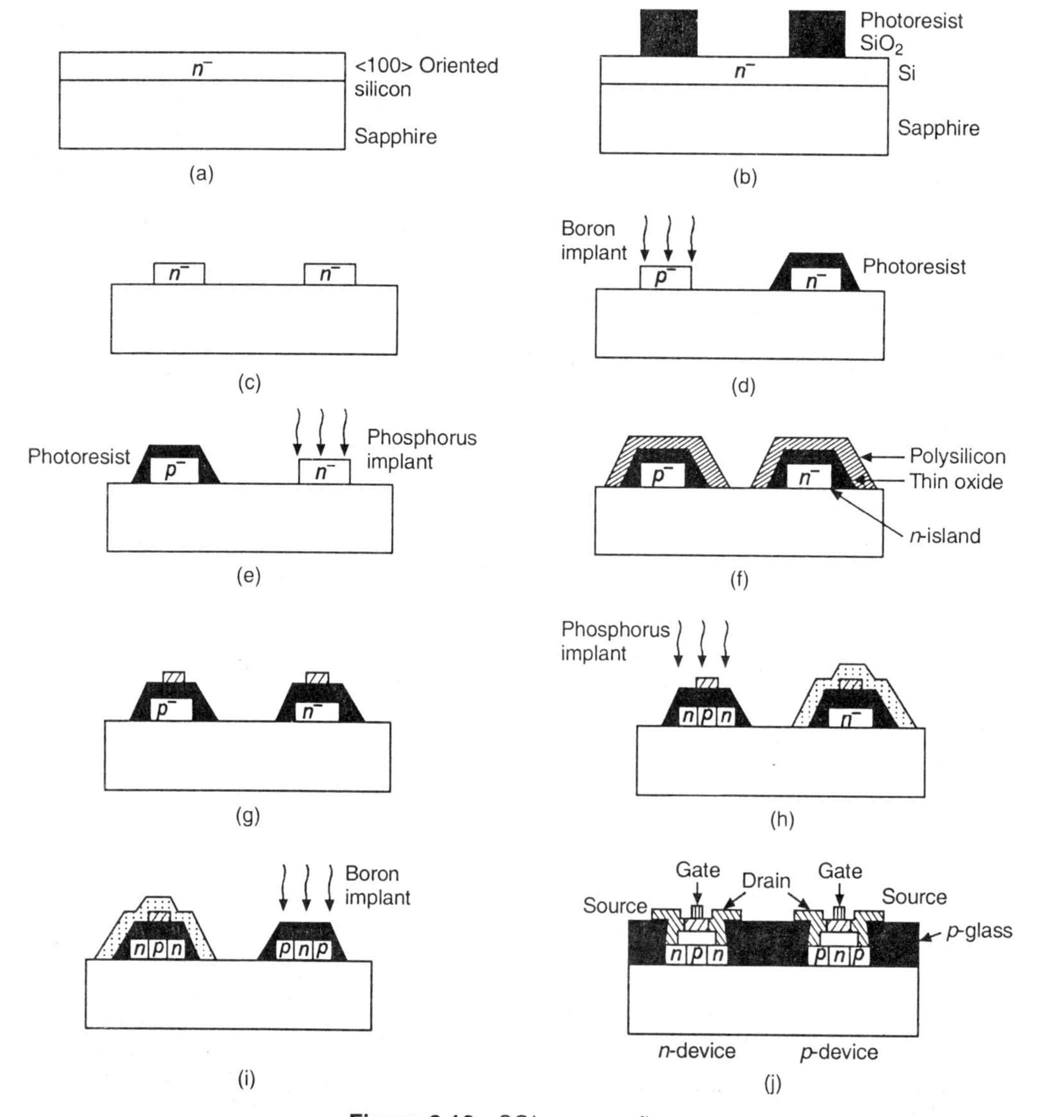

Figure 2.18 SOI process flow.

- An isotropic etch is used to etch away the silicon except where a diffusion area (*n* or *p*) will be needed [Figures 2.18(b) and (c)].
- The *p*-islands are formed next by masking the *n*-islands with a photoresist. A *p*-type dopant, e.g., boron is then implanted. The *p*-islands will become the *n*-channel devices [Figure 2.18(d)].
- The *p*-islands are then covered with a photoresist and an *n*-type dopant, e.g., phosphorus is completed to form the *n*-islands. The *n*-islands will become the *p*-channel devices [Figure 2.18(e)].

- A thin gate oxide (around 100–250 Å) is grown over all of the structures. This is normally done by thermal oxidation.
- A polysilicon film is deposited over the oxide. Often the polysilicon is doped with phosphorus to reduce its resistivity [Figure 2.18(f)].
- The polysilicon is then patterned by photomasking and is etched. This defines the polysilicon layer in the structure [Figure 2.18(g)].
- The next step is to form the *n*-doped source and drain of the *n*-channel devices in the *p*-islands. The *n*-islands are covered with a photoresist and an *n*-type dopant, normally phosphorus, is implanted.
- The dopant will be blocked at the *n*-islands by the photoresist, it will be blocked from the gate region of the *p*-islands by the polysilicon. After this step, the *n*-channel devices are complete [Figure 2.18(h)].
- The *p*-channel devices are formed next by masking the *p*-islands and implanting a *p*-type dopant such as boron. The polysilicon over the gate of the *n*-islands will block the dopant from the gate, thus forming the *p*-channel devices [Figure 2.18(i)].
- A layer of phosphorus glass or some other insulator such as SiO_2 is then deposited over the entire structure.
- The glass is etched at contact-cut locations. The metallization layer is formed next by evaporating aluminium over the entire surface and etching it to leave only the desired metal wires. The aluminium will flow through the contact cuts to make contact with the diffusion or polysilicon regions [Figure 2.18(j)].
- A final passivation layer of phosphorus glass is deposited and etched over bonding pad locations.

Advantages of SOI technology

1. Due to the absence of wells, transistor structures denser than bulk region are feasible.
2. Lower substrate capacitances provide the possibility for faster circuits.
3. No field-inversion problems exist.
4. There is no latchup because of the isolation of *n*- and *p*-transistors by the insulating material.
5. As there is no conducting substrate, there are no body effect problems. However, the absence of a backside substrate contact could lead to add device characteristics, such as the 'kink' effect in which the drain current increases abruptly at around 2 to 3 V.
6. There is enhanced radiation tolerance.

Limitation. Manufacturing difficulty is a major disadvantage of SOS technology. Cost is also a serious problem primarily due to the high cost of sapphire wafers.

2.6 CMOS PROCESS ENHANCEMENTS

A number of enhancements may be added to the CMOS processes, primarily to increase routability of circuits, provide high quality capacitors for analog circuits and memories, or provide resistors of variable characteristics.

These enhancements include

Double- or triple- or quadruple-level metal (or more)
Double- or triple-level poly (or more)
Combinations of the above

We will examine these additions in terms of the additional functionality that they bring to a basic CMOS process.

2.6.1 Interconnect

Probably, the most important additions for CMOS logic processes are additional signal- and power-routing layers. This eases the routing of logic signals between modules and improves the power and clock distribution to modules. Improved routability is achieved through additional layers of metal or by improving the existing polysilicon interconnection layer. The various types of interconnects like metal interconnect, polysilicon/refractory metal interconnect and local interconnect are described here.

2.6.1.1 *Metal interconnect*

A second level of metal is essential for modern CMOS digital design. A third layer is becoming common and is certainly required for leading-edge high density high-speed chips. Normally, aluminium is used for the metal layers. If some form of planarization is employed, the second-level metal pitch can be the same as the first. As the vertical topology becomes more varied, the width and spacing of metal conductors has to increase so that the conductors do not become thin and hence break at vertical topology jumps (step coverage).

Contact from the second-layer metal to the first-layer metal is achieved by a via, as shown in Figure 2.19. If further contact to diffusion or polysilicon is required, a separation between the via and the contact cut is usually required. This requires a first-level metal tab to bridge between Metal 2 and the lower level conductor. It is important to realize that in contemporary processes,

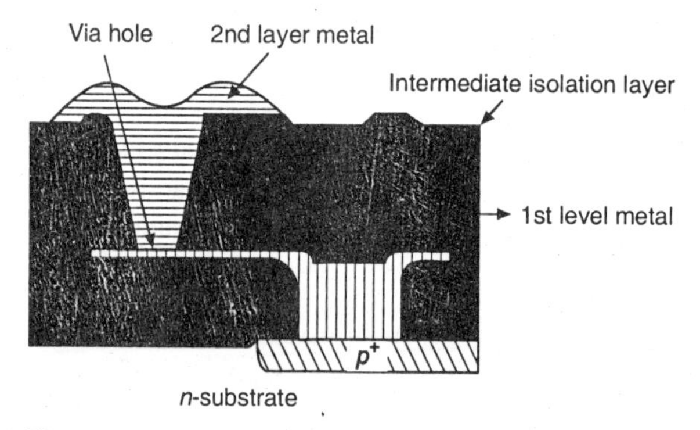

Figure 2.19 Two-level metal process cross section.

first-level metal must be involved in any contact to underlying areas. A number of contact geometries are shown in Figure 2.20. Processes usually require metal borders around the via on both levels of metal although some processes require none. Processes may have no restrictions on the placement of the via with respect to underlying layers [Figure 2.20(a)] or they may have to be placed inside [Figure 2.20(b)] or outside [Figure 2.20(c)] the underlying polysilicon or diffusion areas. Aggressive processes allow the seeking of vias on top of contacts, as shown in Figure 2.20(d).

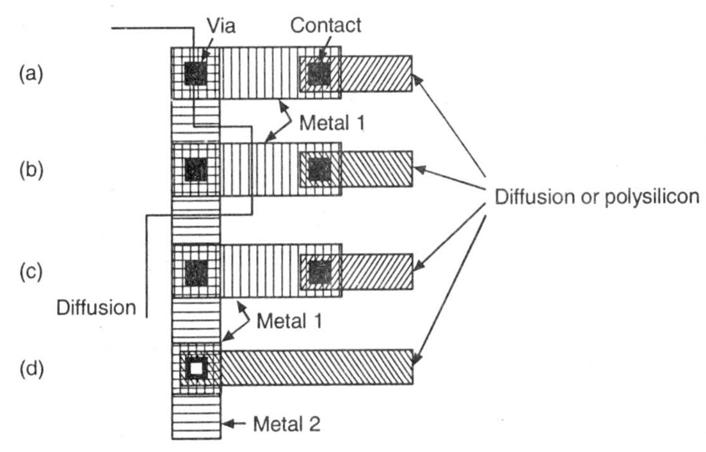

Figure 2.20 Two-level metal via/contact geometries.

Consistent with the relatively large thickness of the intermediate isolation layer, the vias might be larger than contact cuts and second-layer metal may need to be thicker and require a larger via overlap although modern processes strive for uniform pitches on Metal 1 and Metal 2.

The process steps for a two-metal process are briefly as follows:

The oxide below the first metal layer is deposited by atmospheric chemical vapour deposition (CVD).

The second oxide layer between the two metal layers is applied in a similar manner.

Depending on the process, removal of the oxide is accomplished using a plasma etch designed to have a high rate of vertical ion bombardment. This allows fast and uniform etch rates. The structure of a via etched using such a method is shown in Figure 2.19.

2.6.1.2 Polysilicon/Refractory metal interconnect

The polysilicon layer used for the gates of transistors is commonly used as an interconnect layer. However, the sheet resistance of doped polysilicon is between 20 and 40 Ω/square. If used as a long-distance conductor, a polysilicon wire can represent a significant delay. The polysilicon resistance can be reduced by combining it with a refractory metal. This improvement is achieved with a method which requires no extra mask levels. There are three approaches to achieve this improvement in resistance of polysilicon, namely,

1. Silicide gate approach
2. Polycide approach
3. Salicide process approach

In silicide gate approach, a silicide (e.g., silicon and tantalum) is used as the gate material. Sheet resistances of the order of 1 to 5 Ω/square may be obtained. Silicides are mechanically strong and may be dry etched in plasma reactors. Tantalum silicide is stable throughout standard processing and has the advantage that it may be retrofitted into the existing process lines [Figure 2.21(a)].

In polycide approach, a sandwich of silicide upon polysilicon is used [Figure 2.21(b)]. The third process called the salicide process (self-aligned Silicide) approach is the extension of the silicide/polysilicon approach to include the formation of source and drain regions using the silicide [Figure 2.21(c)].

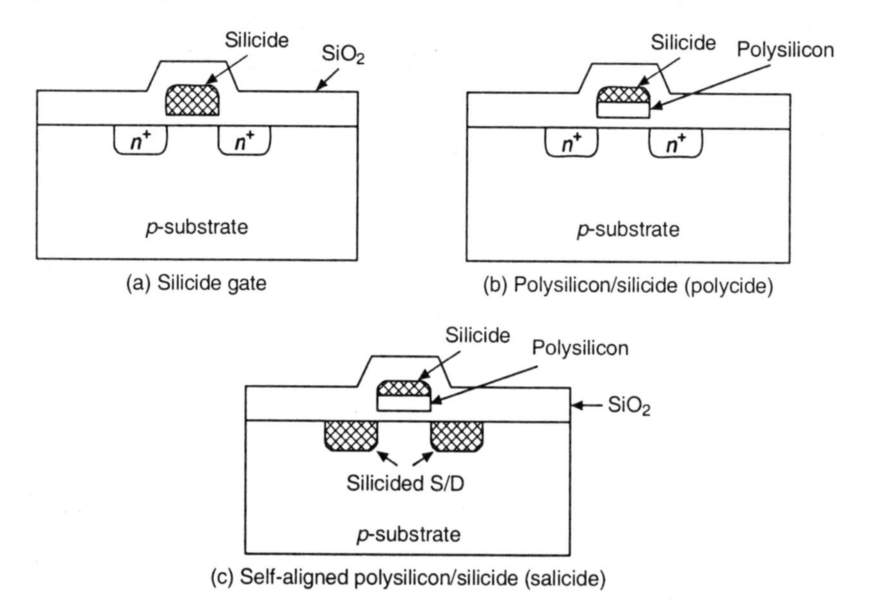

Figure 2.21 Refractory metal interconnect.

The effect of all these processes is to reduce the 'second layer' interconnect resistance, allowing the gate material to be used as a moderate long-distance interconnect. This is achieved by minimum perturbation of an existing process. An increasing trend is to use the salicide approach in the processes to reduce the resistance of both the gate and source/drain conductors.

2.6.1.3 *Local interconnect*

The silicide itself may be used as a 'local interconnect' layer for connection within cells (e.g., TiN). Local interconnect allows a direct connection between polysilicon and diffusion, thus alleviating the need for area-intensive contacts and metal.

Figure 2.22 shows a portion (*p*-devices only) of a six-transistor SRAM cell that uses local interconnect. The local interconnect makes the polysilicon-to-diffusion connections within the cell, thereby alleviating the need to use metal (and contacts). Metal 2 (not shown) bit lines run over the cell vertically. The use of local interconnect in this RAM reduced the cell area by 25%. In general, local interconnect if available can be used to complete intracell routing, leaving the remaining metal layers for global wiring.

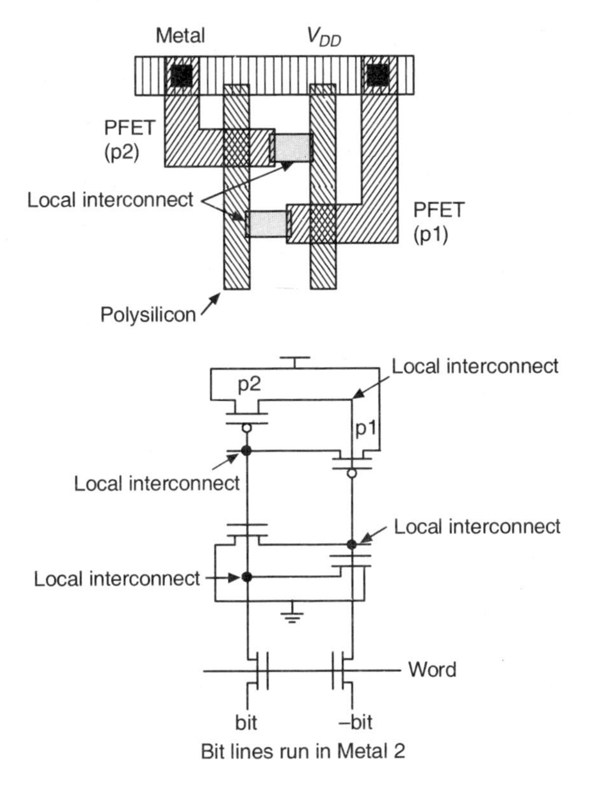

Figure 2.22 Local interconnect used in a RAM cell.

2.6.2 Circuit Elements

2.6.2.1 *Resistors*

Polysilicon, if left undoped, is highly resistive. This property is used to build resistors that are used in static memory cells. The process step is achieved by preventing the resistor areas from being implanted during normal processing. Resistors in the tera-Ω (10^{12} Ω) region are used. A value of 3 TΩ results in a standby current of 2 μA for a 1 Mbit memory.

For mixed signal CMOS (analog and digital), a resistive metal such as nichrome may be added to produce high value, high-quality resistors. The resistor accuracy might be further improved by laser trimming the resulting resistors on each chip to some predetermined test specification. In this process, a high-powered laser vapourizes areas of the metal resistor until it meets a measurement constraint. Sheet resistance values in the kΩ/square are normal. The resistors have excellent temperature stability and long-term reliability.

2.6.2.2 *Capacitors*

Good quality capacitors are required for switched-capacitor analog circuits while small high-value/area capacitors are required for dynamic memory cells. Both types of capacitors are usually added by using at least one extra layer of polysilicon, although the process techniques are very different.

Polysilicon capacitors for analog applications are the most straight-forward. A second thinoxide layer is required in order to have an oxide sandwich between the two polysilicon layers yielding a high capacitance/unit area. Figure 2.23 shows a typical polysilicon capacitor. The presence of this second oxide can also be used to fabricate transistors. These may differ in characteristics from the primary gate oxide devices.

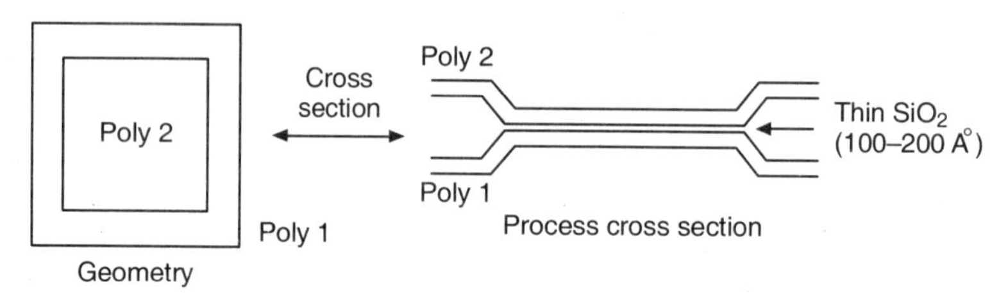

Figure 2.23 Polysilicon capacitor.

For memory capacitors, recent processors have used three dimensions to increase the capacitance/area. One popular structure called the trench capacitor has evolved considerably over the years to push memory densities to 64 Mb and beyond. A typical trench capacitor structure is shown in Figure 2.24(a). The sides of the trench are doped n^+ and coated with a thin 10 nm oxide. Sometimes, oxynitride is used because its high dielectric constant increases the capacitance. The trench is filled with a polysilicon plug, which forms the bottom plate of the cell storage capacitor. This is held at $V_{DD}/2$ via a metal connection at the edge of the array. The side wall n^+ forms the other side of the capacitor and one side of the pass transistor that is used to enable data onto the bit lines. The bottom of the trench has p^+ plug that forms a channel-stop region to isolate adjacent capacitors. The trench is 4 μm deep and has a capacitance of 90 μF. Figure 2.24(b) shows a fin-type capacitor used in a 64 Mb-DRAM. The storage capacitance is 20 to 30 μF. The fins have the additional advantage of reducing the bit capacitance by shielding the bit lines.

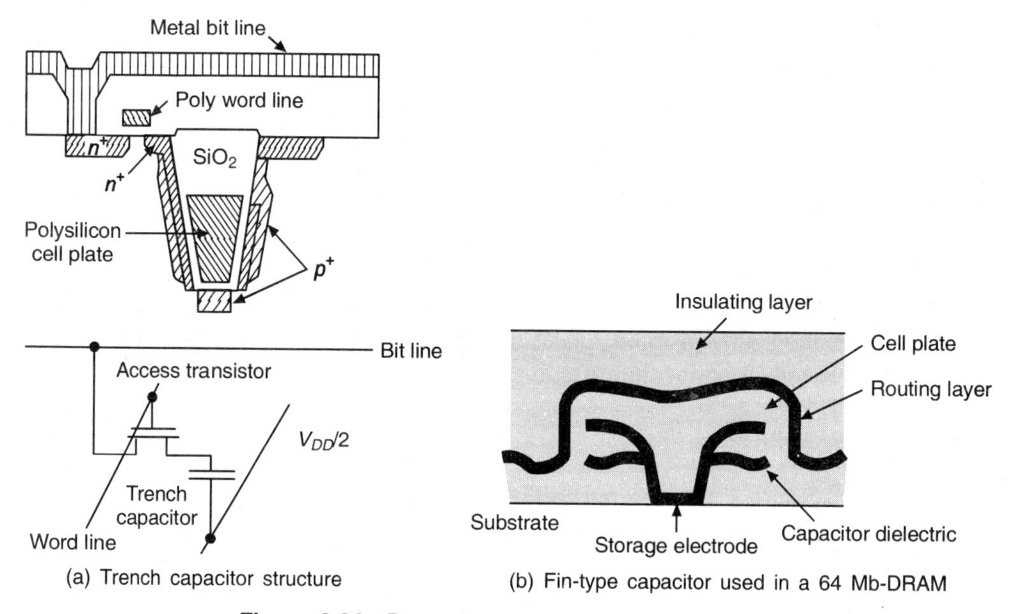

(a) Trench capacitor structure

(b) Fin-type capacitor used in a 64 Mb-DRAM

Figure 2.24 Dynamic memory capacitors.

2.6.2.3 Electrically alterable ROM

Frequently, electrically alterable/erasable ROM (EAROM/EEROM) is added to CMOS processes to yield permanent but reprogrammable storage to a process. This is usually done by adding a polysilicon layer. Figure 2.25 shows a typical memory structure, which consists of a stacked-gate structure. The normal gate is left floating, while a control gate is placed above the floating gate. A very thinoxide called the tunnel oxide separates the floating gate from the source, drain and substrate. This is usually about 10 nm thick. Another thinoxide separates the control gate from the floating gate. By controlling the control gate, source and drain voltages, the very thin tunnel oxide between the floating gate and the drain of the device is used to allow electrons to 'tunnel' to or from the floating gate to turn the cell OFF or ON, respectively, using Fowler–Nordheim tunnelling. Alternatively, by setting the appropriate voltages on the terminals, 'hot electrons' can be induced to charge the floating gate, thereby programming the transistor. In non-electrically alterable versions of the technology, the process can be reversed by illuminating the gate with UV light. In these cases, the chips are usually housed in glass-lidded packages.

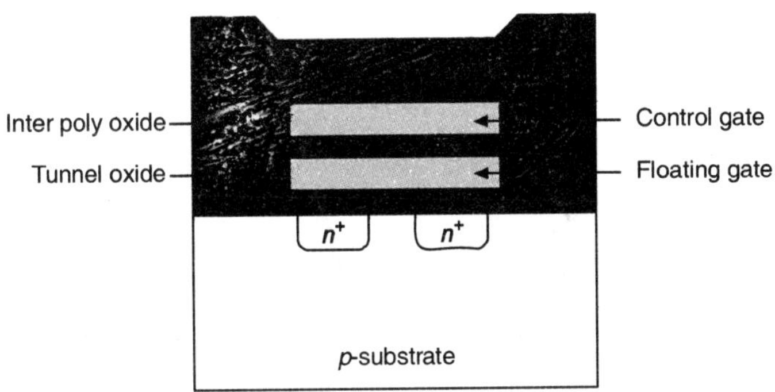

Figure 2.25 EEPROM Technology.

2.6.2.4 Bipolar transistors

The addition of a bipolar transistor to the device structure forms the basis for BiCMOS processes. Adding an *npn*-transistor can markedly aid in reducing the delay times of highly loaded signals, such as memory word lines and microprocessor buses. Additionally, for analog applications, bipolar transistors may be used to provide better performance analog functions than MOS alone.

To get merged bipolar/CMOS functionality, MOS transistors can be added to a bipolar process or vice versa. A BiCMOS process should have excellent gate oxides and also precisely controlled diffusions.

A mixed signal BiCMOS process cross section is shown in Figure 2.26. This process features both *npn*- and *pnp*-transistors in addition to PMOS and NMOS transistors. The major processing steps are summarized in Figure 2.27, showing the particular device to which they correspond. The starting material is a lightly doped *p*-type substrate into which antimony or arsenic are diffused to form an n^+ buried layer. Boron is diffused to form a buried p^+ layer.

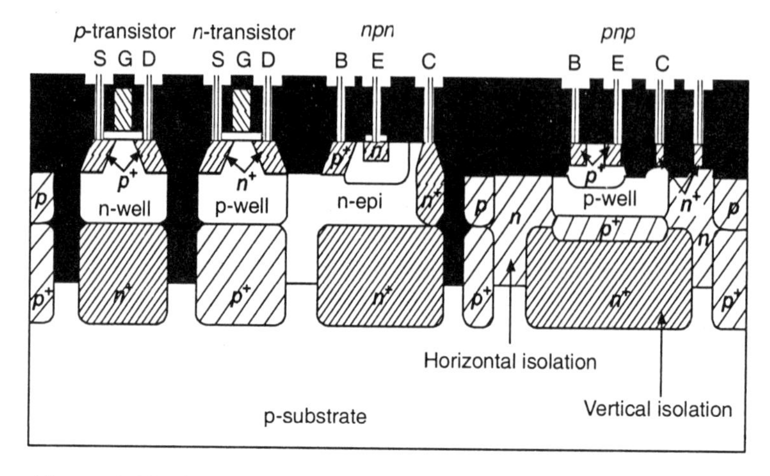

Figure 2.26 Typical mixed signal BiCMOS process cross section.

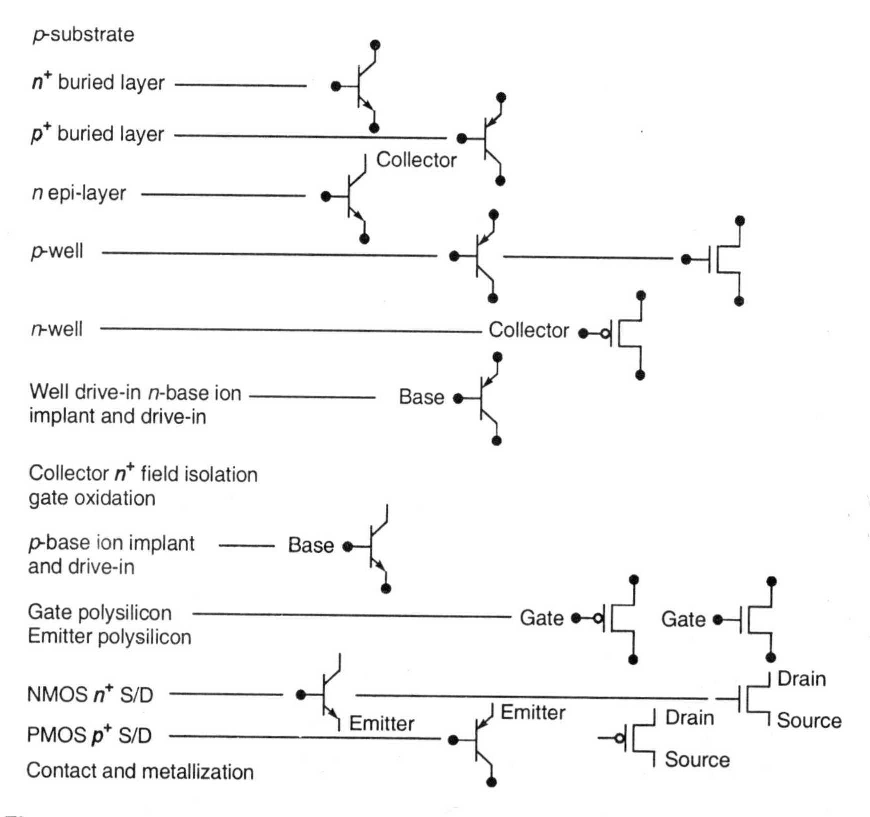

Figure 2.27 BiCMOS process steps for the cross section shown in Figure 2.24.

An n-type epitaxial layer 4 μm thick is then grown. n-wells and p-wells are then diffused so that they join in the middle of the epitaxial layer. This epitaxial layer isolates the *pnp*-transistor in the horizontal direction, while the buried n^+ layer isolates it vertically. The *npn*-transistor is junction isolated. The base for the *pnp* is then ion-implanted using phosphorus. A diffusion step follows this to get the right doping profile. The *npn*-collector is formed by depositing phosphorus before LOCOS. Field oxidation is carried out and the gate oxide is grown. Boron is then used to form the p-type base of the *npn*-transistor. Following the threshold adjustment of the NMOS transistors, the polysilicon gates are defined. The emitters of the *npn*-transistors employ polysilicon rather than a diffusion. These are formed by opening windows and depositing polysilicon. The n^+ and p^+ source/drain implants are then completed. This step also dopes the *npn*-emitter and the extrinsic bases of the *npn*- and *pnp*-transistors. Following the deposition of PSG, the normal two-layer metallization steps are completed.

The cross section of digital BiCMOS process is shown in Figure 2.28. In this, the buried-layer epitaxial layer is a 0.8 μm process. The lightly doped drain (LDD) structures must be constructed for the p-transistors and the n-transistors. The *npn* is formed by a double diffused sequence in which both base and emitter are formed by impurities that diffuse out of covering layer of polysilicon. This process, used for logic applications, has only an *npn*-transistor. The collector of the *npn* is connected to the n-well, which in turn, is connected to the V_{DD} supply. Thus, all *npn*-collectors are commoned.

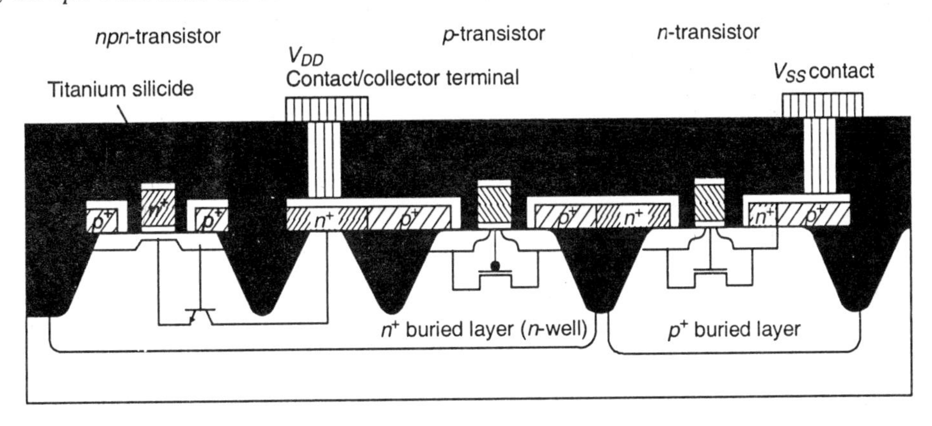

Figure 2.28 Digital BiCMOS process cross section.

2.7 BiCMOS TECHNOLOGY

A deficiency of MOS technology lies in the limited load driving capabilities of MOS transistors. This is due to the limited current sourcing and current sinking abilities associated with both p- and n-transistors. Bipolar transistors also provide higher gain and have generally better noise and high frequency characteristics than MOS transistors (Figure 2.29). From the figure, it may be seen that using BiCMOS gates may be an effective way of speeding up VLSI circuits. However, the application of BiCMOS in subsystems such as ALU, ROM, a register-file or a barrel shifter is not always an effective way of improving speed. This is because most gates in such structures do not have to drive large capacitive loads so that the BiCMOS arrangements

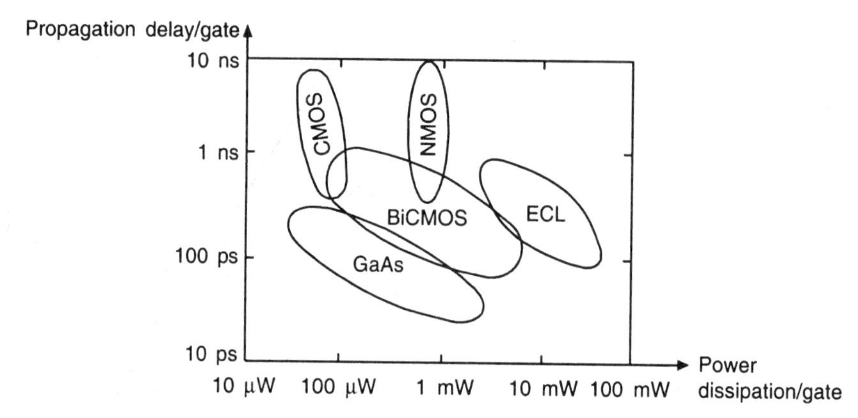

Figure 2.29 Speed/power performance of the available technologies.

give no speed advantage. To take advantage of BiCMOS, the whole functional entity must be considered. A comparison between the characteristics of CMOS and bipolar circuits is given in Table 2.2.

Table 2.2 Comparisons between CMOS and bipolar technologies

CMOS technology	*Bipolar technology*
• Low static power dissipation	• High power dissipation
• High input impedance (low drive current)	• Low input impedance (high drive current)
• High noise margin	• Low voltage swing logic
• High packing density	• Low packing density
• High delay sensitivity to load (fan-out limitations)	• Low delay sensitivity to load
• Low output drive current	• High output drive current
• Low g_m ($g_m \propto V_{in}$)	• High g_m ($g_m \propto e^{V_{in}}$)
• Bidirectional capability	• High f_t at low current
• Scalable threshold voltage	• Essentially unidirectional
• A near ideal switching device	

When considering CMOS technology, it becomes clear that theoretically there should be little difficulty in extending the fabrication processes to include bipolar as well as MOS transistors. In deed, a problem of p-well and n-well CMOS processing is that parasitic bipolar transistors are inadvertently formed as part of outcome of fabrication. The production of *npn* bipolar transistors with good performance characteristics can be achieved, for example, by extending the standard n-well CMOS processing to include further masks to add two additional layers—the n^+ subcollector and p^+ base layers. The *npn* transistor is formed in an n-well and the additional p^+ base region is located in the well to form the p-base region of the transistor. The second additional layer, the buried n^+ subcollector (BCCD), is added to reduce the n-well (collector) resistance and thus improve the quality of the bipolar transistor. The simplified general arrangement of such a bipolar *npn* transistor may be as shown in Figures 2.30(a)

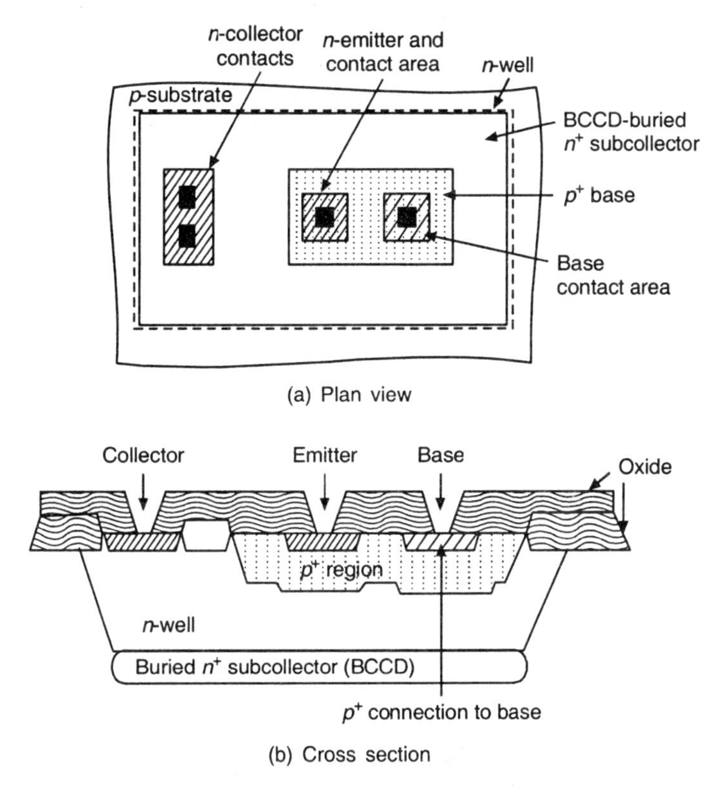

Figure 2.30 Arrangement of BiCMOS *npn* transistor (orbit 2 μm, CMOS).

and (b). Since extra design and processing steps are involved, there is an inevitable increase in cost and it is reflected in Figure 2.31, which also includes ECL and GaAs gates for cost comparison.

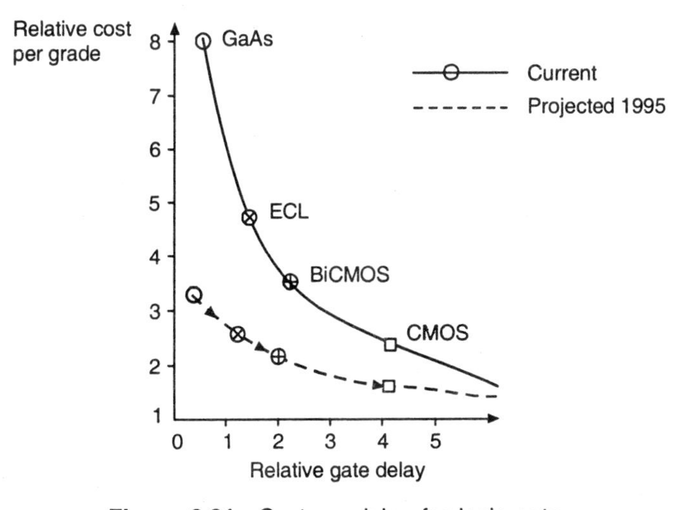

Figure 2.31 Cost vs. delay for logic gate.

2.7.1 BiCMOS Fabrication in an *n*-well Process

The basic process steps are those already outlined for CMOS but with additional process steps and additional masks defining (i) the p^+ base region, (ii) n^+ collector area, and (iii) the buried subcollector (BCCD). Table 2.3 sets out the process steps for a single poly single metal CMOS n-well process, showing the additional process steps for the bipolar devices.

Table 2.3 *n*-well BiCMOS fabrication process steps

Single poly single metal CMOS	*Additional steps for bipolar devices*
Form n-well	Form buried n^+ layer (BCCD)
Delineate active areas	
Channel stop	
Threshold V_T adjustment	Form deep n^+ collector
Delineate poly/gate areas	
Form n^+ active areas	
Form p^+ active areas	
Define contacts	Form p^+ base for bipolars
Delineate the metal areas	

2.7.2 Some Aspects of Bipolar and CMOS Devices

There are several advantages if the properties of CMOS and bipolar technologies could be combined. This is achieved to a significant extent in the BiCMOS technology. The penalty which arises from the additional process steps is the loss of packing density and thus higher cost. A cost comparison of all current high-speed technologies may be assessed from Figure 2.31. A further advantage of using BiCMOS technology is that analog amplifier design is facilitated and improved. High impedance CMOS transistors may be used for the input circuitry while the remaining stages and output drivers are realized using bipolar transistors. To take maximum advantage of available silicon technologies, one might envisage the following mix of technologies in a silicon system:

- CMOS for logic
- BiCMOS for I/O and driver circuits
- ECL for critical high speed parts of the system.

SUMMARY

In Chapter 2, the basic MOS structures of p-channel and n-channel MOSFETs are explained with clear illustrations. The conduction characteristics give the transfer characteristics and threshold voltages for enhancement and depletion mode transistors of n- and p-types. The various modes of operation such as accumulation mode, depletion mode and inversion mode are described in detail with reference to an NMOS enhancement transistor. The operating regions: cutoff, non-saturated and saturated regions are dealt with a normal conduction characteristics of MOS transistor. MOS and CMOS switches with their symbol and characteristics are explained in this chapter. The NMOS fabrication process is then described with illustrations showing the different steps involved in the process. The chapter also provides a detailed description of the CMOS fabrication technologies: *p*-well, *n*-well, twin-tub and SOI or SOS processes. The twin-tub process combines both *p*-well and *n*-well approaches and provides the optimization of *p*-type and *n*-type transistors. Improvement in process characteristics such as CMOS latchup and speed are achieved with Silicon On Insulator (SOI)/Silicon On Sapphire (SOS) processes. Since sapphire is used as the substrate material, it is not affected by radiation as bulk silicon, but higher cost is the serious problem in SOS technology. The CMOS process enhancements are described in detail. Processes are constantly under development with new structures and techniques being introduced to yield smaller, higher speed, less costly and more reliable ICs. Then the CMOS process enhancements such as double- or triple- or quadruple level metal, double- or triple- level poly and their combinations are described in detail. Finally, the design criterion of BiCMOS technology is highlighted, which integrates both bipolar and MOS transistors on the same chip, thus assumes the advantages of both technologies.

REVIEW QUESTIONS

1. Explain NMOS fabrication.
2. Give brief notes on CMOS technologies.
3. Explain diff circuit elements.
4. Explain BiCMOS technology.

SHORT ANSWER QUESTIONS

1. What are Enhancement Mode Devices (EMDs).

   Ans. The *n*-channel devices that are cut off with zero gate bias.

2. Accumulation mode is ______________.

   Ans. $V_{GS} < V_T$.

3. Major drawbacks of NMOS process

   (a) high absolute power consumption.

   (b) electrical asymmetry.

4. The mask is often known as ______________.

   Ans. Thinox.

5. What is twin-well process?

   Ans. A logical extension of the *p*-well and the *n*-well approaches is the twin-tub process.

6. ______________ is the highest performance Silicon on Insulator (SOI) today.

   Ans. Silicon on Sapphire.

3 MOS Device Characteristics

3.1 INTRODUCTION

If the design is to be effectively carried out, or indeed if the performance of the circuits realized in MOS technology is to be properly understood, then the practitioners must have a sound understanding of the MOS active devices. This chapter deals with the basic characteristics of MOS transistor. After a generic overview of the device, an analytical description of the transistor from both a static (steady state) and a dynamic (transient) viewpoint is presented. We conclude the discussion with an enumeration of some second order effects and an introduction of the device models of MOS transistor for simulation purpose.

VLSI designers should have a very good knowledge of the behaviour of the circuits, they are designing. Even if large systems are being designed using computer-aided design processes, it is essential that the designs be based on a sound foundation of understanding if those systems are to meet performance specifications.

A first glance at the device. The MOSFET is a four-terminal device. The voltage applied to the gate terminal determines if and how much current flows between the source and the drain. The body represents the fourth terminal of the transistor.

A cross section of a typical n-channel transistor is shown in Figure 3.1. Heavily-doped n-type source and drain regions are implanted (or diffused) into a lightly doped p-type substrate

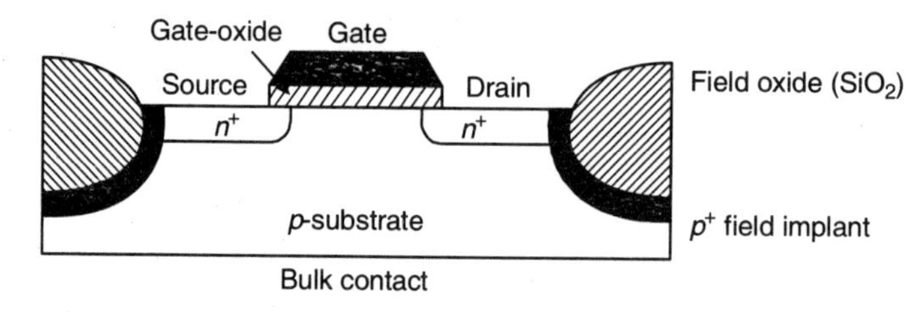

Figure 3.1 Cross section of NMOS transistor.

(or body). A thin layer of silicon dioxide (SiO_2) is grown over the region between the source and drain and is covered by a conductive material, most often polycrystalline silicon (polysilicon). The conductive material forms the gate of the transistor. Neighbouring devices are insulated from each other with the aid of a thick layer of SiO_2 (called the field oxide) and a reverse-biased *np* device formed by adding an extra p^+ region, called the channel-stop implant (or field implant).

At the most superficial level, the transistor can be thought of as a switch. When voltage is applied to the gate larger than a given value called threshold voltage, V_T, a conducting channel is formed between drain and source. If a potential difference exists between them, electrical current flows between the two. The conductivity of the channel is modulated by the gate voltage—the larger the voltage difference between source and gate, the smaller the resistance of the conducting channel and the larger the current. When the gate voltage is lower than the threshold, no channel exists, and the switch is considered open.

Two types of MOSFET devices can be identified. The NMOS transistor consists of n^+ source and drain regions embedded in a *p*-type substrate. The current is carried by electrons moving through an *n*-type channel between the source and drain. But in PMOS transistor made by using an *n*-type substrate and p^+ drain and source regions, the current is carried by holes moving through a *p*-type channel. In a complementary MOS technology (CMOS), both devices are present.

3.2 STATIC BEHAVIOUR OF THE MOS TRANSISTOR

In the derivation of the static model of the MOS transistor, we concentrate on the NMOS device. All of the arguments are valid for PMOS devices as well.

3.2.1 The Threshold Voltage

The threshold voltage V_T, for an MOS transistor can be defined as the voltage applied between the gate and the source of the MOS transistor below which the drain to source current I_{DS}, effectively drops to 0. In general, the threshold voltage is a function of the number of parameters as below:

1. Gate conducting material
2. Gate insulation material
3. Gate insulator thickness-channel doping
4. Impurities at Silicon–Silicon dioxide interface
5. Voltage between source and substrate V_{SB}

In addition, the absolute value of V_T decreases with increase in temperature. This variation is approximately – 4 mV/°C for high substrate doping levels and –2 mV/°C for low doping levels.

Consider first the case in which $V_{GS} = 0$, and drain, source, and bulk are connected to ground (Figure 3.2). The drain and source are connected by back-to-back *pn*-junctions (substrate–source and substrate–drain). Under the mentioned conditions, both the junctions have a 0 V bias and can be considered off which results in an extremely high resistance between drain and source.

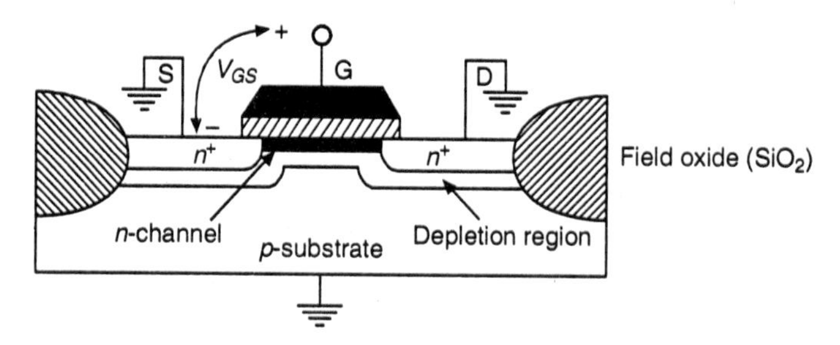

Figure 3.2 NMOS transistor for positive V_{GS} showing depletion region and induced channel.

Assume now that a positive volatge is applied to the gate (with reference to source) as in Figure 3.2. The gate and substrate form the plates of a capacitor with the gate oxide as the dielectric. The positive gate voltage causes positive and negative charge to accumulate on the gate electrode and the substrate side, respectively. The latter manifests itself initially by repelling mobile holes. Hence, a depletion region is formed below the gate. The width and the space charge per unit area of the depletion region are given by

$$W_d = \sqrt{\frac{2\varepsilon_{si}\varphi}{qN_A}} \tag{3.1}$$

and

$$Q_d = \sqrt{2qN_A\varepsilon_{si}\varphi} \tag{3.2}$$

where N_A = substrate doping
φ = voltage across the depletion region, i.e., the potential at the oxide–silicon boundary
q = charge of electron (1.6×10^{-19} C)
ε_{si} = relative permittivity of silicon

As the gate voltage increases, the potential at the silicon surface reaches a critical value at some point, where the semiconductor surface inverts to n-type material. This leads to a phenomenon called strong inversion, which occurs at a voltage equal to twice the Fermi potential (φ_F is approximately equal to –0.3 V for typical p-type silicon substrates).

Further increases in the gate voltage produce no further changes in the depletion layer width, but results in additional electrons in the thin inversion layer directly under the oxide. These are drawn into the inversion layer from the heavily doped n^+ source region. Hence, a continuous n-type channel is formed between the source and the drain regions, whose conductivity is modulated by the gate–source voltage.

In the presence of an inversion layer, the charge stored in the depletion region is fixed and equals to

$$Q_{Bo} = \sqrt{2qN_A\,\varepsilon_{si}\,|-2\varphi_F|} \tag{3.3}$$

When a substrate bias voltage V_{SB} is applied between source and body, the surface potential required for strong inversion increases to $|-2\varphi_F + V_{SB}|$. The charge stored in the depletion region is then expressed as:

$$Q_B = \sqrt{2qN_A\,\varepsilon_{si}\,|-2\varphi_F + V_{SB}|} \tag{3.4}$$

The value of V_{GS} where strong inversion occurs is called the threshold voltage V_T. The expression of V_T consists of several components:

1. A flat-band voltage V_{SB} that represents the built-in voltage offset across the MOS structure. It consists of the work function difference, φ_{MS} existing between the gate polysilicon and the silicon, and some extra components to compensate for the (undesired) fixed charge Q_{ox} at the oxide–silicon interface, and the threshold-adjusting implanted impurities Q_I.
2. A second term V_B represents the voltage drop across the depletion region at inversion and equals $-2\varphi_F$.
3. A final component V_{ox} stands for the potential drop across the gate-oxide and is equal to Q_B/C_{ox}, with C_{ox} representing the gate–oxide capacitance per unit area.

$$V_T = V_{FB} + V_B + V_{ox}$$

$$= \left[\varphi_{MS} - \left(\frac{Q_{ox}}{C_{ox}}\right) - \left(\frac{Q_I}{C_{ox}}\right) - 2\varphi_F - \left(\frac{Q_B}{C_{ox}}\right)\right] \tag{3.5}$$

Here, Q_B is a function of V_{SB}.

The above expression can be reorganized as

$$V_T = V_{To} + \gamma(\sqrt{|-2\varphi_F + V_{SB}|} - \sqrt{|-2\varphi_F|} \tag{3.6}$$

where,
$$V_{To} = \left[\varphi_{MS} - \left(\frac{Q_{ox}}{C_{ox}}\right) - \left(\frac{Q_I}{C_{ox}}\right) - 2\varphi_F - \left(\frac{Q_{B0}}{C_{ox}}\right)\right] \tag{3.7}$$

$$\gamma = \frac{\sqrt{2qN_A\,\varepsilon_{si}}}{C_{ox}} \tag{3.8}$$

V_{T0} represents the threshold voltage for $V_{SB} = 0$ V and γ is called the body-effect coefficient or body factor. The gate capacitance per unit area C_{ox} is given by

$$C_{ox} = \frac{\varepsilon_{ox}}{t_{ox}} \tag{3.9}$$

$\varepsilon_{ox} = 3.97 \times \varepsilon_o = 3.5 \times 10^{-13}$ F/cm is the oxide permittivity, $t_{ox} \approx 200$ Å or less.

The threshold voltage has a positive value for a typical NMOS device while it is negative for PMOS device. The effect of the substrate-bias on the threshold volatge of an NMOS transistor is plotted in Figure 3.3 for typical values of $|-2\varphi_F| = 0.6$ V and $\gamma = 0.4V^{0.5}$. A negative bias on the substrate causes the threshold to increase from 0.45 V to 0.85 V. Note that V_{SB} always has to be larger than -0.6 V in an NMOS device. If not, the source–body diode becomes forward biased, which deteriorates the transistor action.

Threshold voltage–body effect. The threshold voltage V_T is not a constant with respect to the voltage difference between the substrate and the source of MOS transistor. This effect is called substrate-bias effect or body effect.

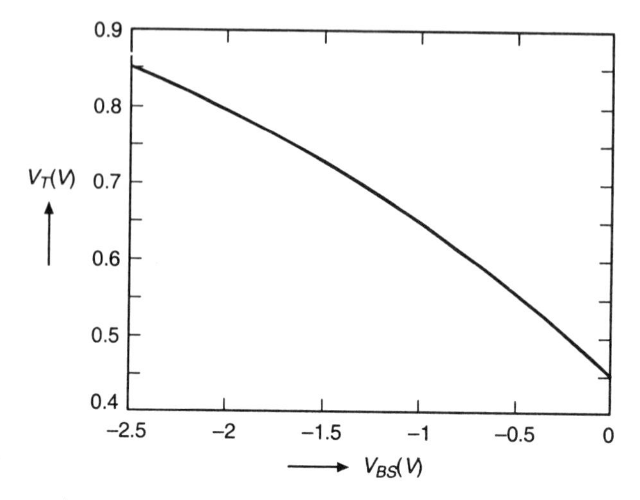

Figure 3.3 The effect of body-bias on threshold.

The effect of substrate bias on series-connected *n*-transistors is shown in Figure 3.4. In this, all devices comprising on MOS devices are made on a common substrate. Hence, the substrate voltage of all devices is normally equal. However, a series connection of several devices may result in an increase in source-to-substrate voltage as we proceed vertically along the series chain (V_{SB1} = 0, V_{SB2} not equals 0).

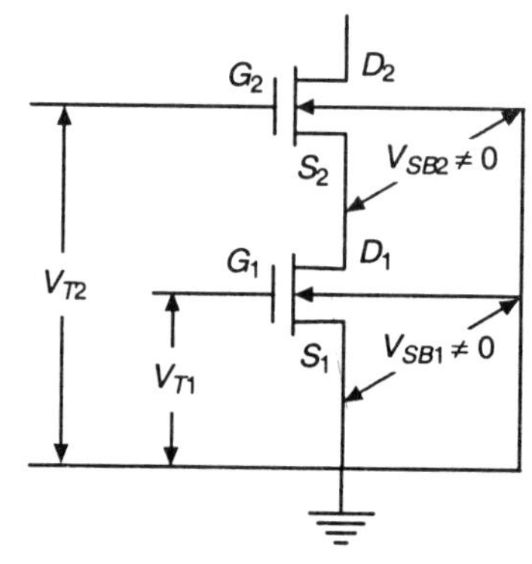

Figure 3.4 The effect of substrate bias on series-connected *n*-transistors.

Under normal conditions, i.e., when $V_{GS} > V_T$, the depletion region width remains constant and charge carriers are pulled into the channel from source. However, as the substrate bias V_{SB} ($V_{\text{source}} - V_{\text{substrate}}$) is increased, the width of the channel–substrate region also increases, resulting in an increase in the density of the trapped carriers in the depletion layer. For charge neutrality to hold, the channel charge must decrease. The resultant effect is that the substrate voltage V_{SB} adds to the channel–substrate junction potential. This increases the gate–channel voltage drop. The overall effect is an increase in the threshold voltage V_T (i.e., $V_{T2} > V_{T1}$).

The silicon substrate is usually connected to our system's circuit ground during packaging. However, a fixed voltage is sometimes applied between circuit ground and substrate as in Figure 3.5, and this bias must be taken into account in estimating the body effect. If the source-to-bulk (substrate) voltage V_{SB}, equals zero, then V_T is at its minimum value, approximately 0.2 V_{DD}. As V_{SB} is increased, V_T increases slightly.

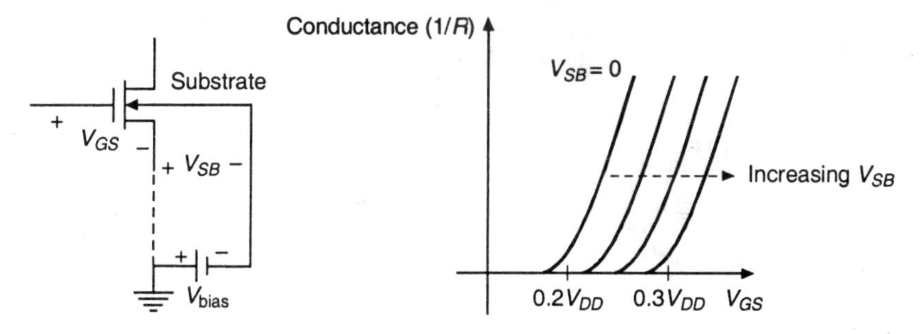

Figure 3.5 The body effect.

Example 3.1 An MOS transistor has a threshold voltage of 0.75 V while the body-effect coefficient equals 0.54. Compute the threshold voltage for V_{SB} = 5 V and $2\varphi_F = -0.6$ V.

We have from Eq. (3.6),

$$\begin{aligned} V_T &= V_{To} + \gamma\,(\sqrt{|-2\varphi_F + V_{SB}|} - \sqrt{|-2\varphi_F|}) \\ &= 0.75 + 0.54\,(2.3564 - 0.77596) \\ &= 1.6 \text{ V} \end{aligned}$$

3.2.2 Current–Voltage Relations

Linear region. Assume $V_{GS} > V_T$. A small voltage V_{DS} is applied between source and drain and causes a current I_D to flow from drain to source (Figure 3.6). Using a simple analysis, a first-order expression of the current as a function of V_{GS} and V_{DS} can be obtained.

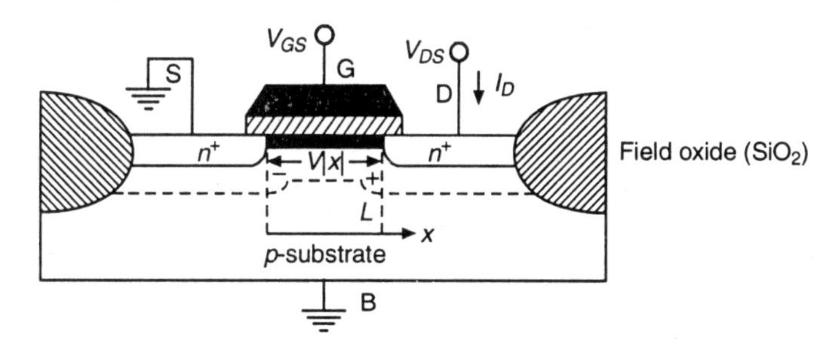

Figure 3.6 NMOS transistor with bias voltages.

At a point x along the channel, the voltage is $V(x)$, and the gate-to-channel voltage at that point equals $V_{GS} - V(x)$. Under the assumption that this voltage exceeds the threshold voltage all along the channel, the induced charge per unit area at point x can be computed by using the following equation:

$$Q_i(x) = -C_{ox}\,[V_{GS} - V(x) - V_T] \tag{3.10}$$

The current is given as the product of the drift velocity of the carriers v_n and the available charge. Due to charge conservation, it is constant over the length of the channel. W is the width of the channel in a direction perpendicular to the current flow. The current equation is

$$I_D = -v_n(x) \cdot Q_i(x) \cdot W \tag{3.11}$$

But, $v_n = -\mu_n E(x) = \mu_n (dV/dx)$ (3.12)

$\mu_n \rightarrow$ mobility in cm^2/V-sec

$E(x) \rightarrow$ electric field strength

Equation (3.12) gives the relationship between electron velocity and electric field strength.

Substituting Eqs. (3.10) and (3.12) in Eq. (3.11),

$$I_D = C_{ox}\,[V_{GS} - V(x) - V_T]\,\mu_n\,(dV/dx)\,W$$

$$\Rightarrow \quad I_D dx = \mu_n C_{ox} W\,[V_{GS} - V(x) - V_T]\,dV$$

$$\Rightarrow \quad I_D dx = k_n' W\,[V_{GS} - V(x) - V_T]\,dV \tag{3.13}$$

where, $k_n' = \mu_n C_{ox} = \mu_n \varepsilon_{ox}/t_{ox}$ = process transconductance parameter

Typical values are t_{ox} = 20 nm, k_n' = 80 μA/V^2.

Integrating Eq. (3.13) over the length L,

$$I_D L = k_n' W\,[(V_{GS} - V_T)\,V_{DS} - (V_{DS}^2/2)]$$

$$\Rightarrow \quad I_D = k_n'(W/L)\,[(V_{GS} - V_T)\,V_{DS} - (V_{DS}^2/2)]$$

$$\Rightarrow \quad I_D = k_n\,[(V_{GS} - V_T)\,V_{DS} - (V_{DS}^2/2)] \tag{3.14}$$

where, $k_n = k'_n(W/L)$ = gain factor, the product of process transconductance, k'_n and the aspect ratio W/L, of the transistor. Equation (3.14) represents the current–voltage relation of an NMOS transistor.

For small values of V_{DS}, the quadratic factor in Eq. (3.14) can be ignored, and we observe a linear dependence between I_D and V_{DS}. The operation region where Eq. (3.14) holds is thus called the resistive or linear region or non-saturated region or triode region. One of its main properties is that it displays a continuous conductive channel between the drain and source regions. This condition exists for $V_{GS} > V_T$ and $V_{DS} \leq V_{GS} - V_T$.

Saturation region. As the value of the drain–source voltage is further increased, the assumption that the channel voltage is larger than the threshold all along the channel ceases to hold. This happens when $V_{GS} - V(x) < V_T$. At that point, the induced charge is zero, and the conducting channel disappears or is pinched off. This is illustrated in Figure 3.7, which shows how the channel thickness is gradually reduced from source-to-drain until pinch-off occurs. No channel exists in the vicinity of the drain region.

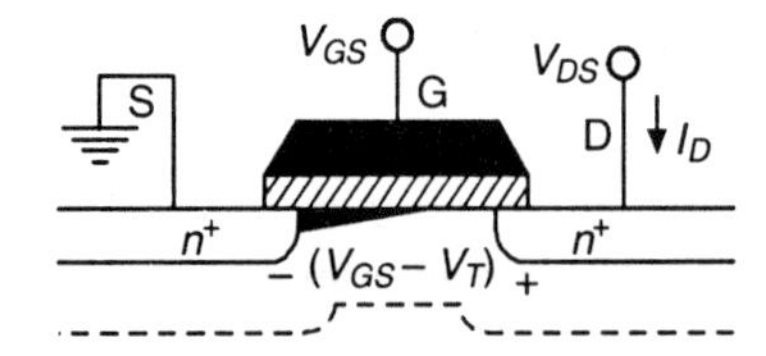

Figure 3.7 NMOS Transistor under pinch-off condition.

Obviously, for this phenomenon to occur, it is essential that the pinch-off condition be met at the drain region, or $V_{GS} - V_{DS} \leq V_T$. Under these circumstances, the transistor is in the saturation region, and Eq. (3.14) no longer holds.

The voltage difference over the induced channel (from the pinch-off point to the source) remains fixed at $V_{GS} - V_T$, and consequently, the current remains constant (or saturates).

Replacing V_{DS} by $V_{GS} - V_T$ in Eq. (3.14), yields the drain current for the saturation mode.

$$\Rightarrow \qquad I_D = \frac{k_n}{2}(V_{GS} - V_T)^2 \qquad (3.15)$$

From Eq. (3.15), it is observed that the current is no longer a function of V_{DS}.

Channel-length modulation. Equation (3.15) suggests that the transistor in the saturation mode acts as a perfect current source—the current between drain and source terminals is constant and independent of the applied voltage over the terminals. This is not entirely correct. The effective length of the conductive channel is actually modulated by the applied V_{DS}, increasing V_{DS} causes the depletion region at the drain junction to grow, reducing the length of the effective channel. From Eq. (3.15), the current increases when the length L is decreased. A more accurate description of the current of the MOS transistor is therefore given by

$$I_D = \frac{k_n}{2}(V_{GS} - V_T)^2(1 + \lambda V_{DS}) \qquad (3.16)$$

where λ is an empirical parameter called the channel-length modulation.

Figure 3.8 plots I_D versus V_{DS} for different values of V_{GS} for an NMOS transistor. In the linear or resistive region, the transistor behaves like a voltage controlled resistor, while in the saturation region, it acts as a voltage-controlled current source (when the channel-length modulation effect is ignored) [Figure 3.8(a)]. Also, a plot of $\sqrt{I_D}$ as a function of V_{GS} (with constant V_{DS}) is shown in Figure 3.8(b).

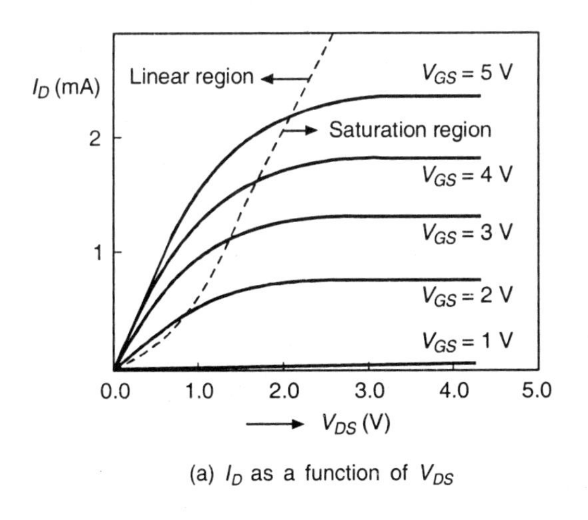

(a) I_D as a function of V_{DS}

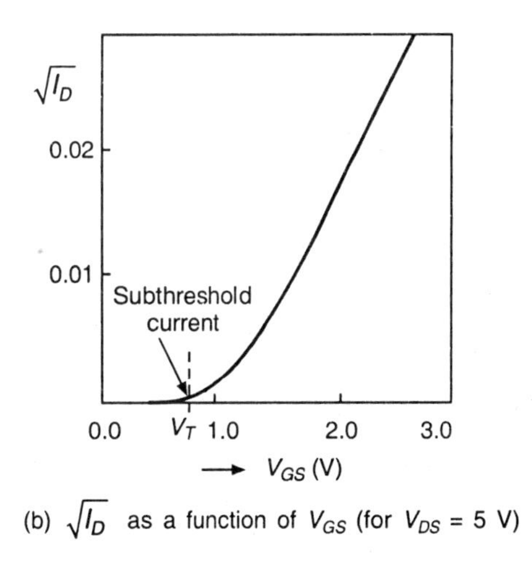

(b) $\sqrt{I_D}$ as a function of V_{GS} (for V_{DS} = 5 V)

Figure 3.8 I-V characteristics of NMOS transistor.

From Figure 3.8(b), it is observed that a linear relationship exists for values of $V_{GS} >> V_T$. Also, at $V_{GS} = V_T$, the current does not abruptly drop to zero, the device goes into subthreshold operation. To turn the device completely off, the gate–source voltage has to be substantially lower than V_T.

Extensions of output charaterictics of NMOS transistor showing the effect of channel-length modulation is shown in Figure 3.9.

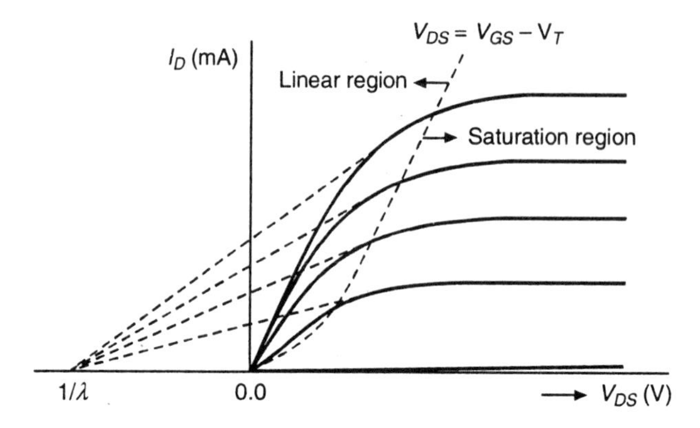

Figure 3.9 Extensions of output characteristics of NMOS transistor showing the effect of channel-length modulation.

All the derived equations hold for the PMOS transistor as well. The only difference is that for PMOS devices, the polarities of all voltages and currents are reversed.

3.2.3 A Model for Manual Analysis

The derived equations can be combined into a simple device model, which will be employed for manual analysis of MOS circuits. It is given in Figure 3.10.

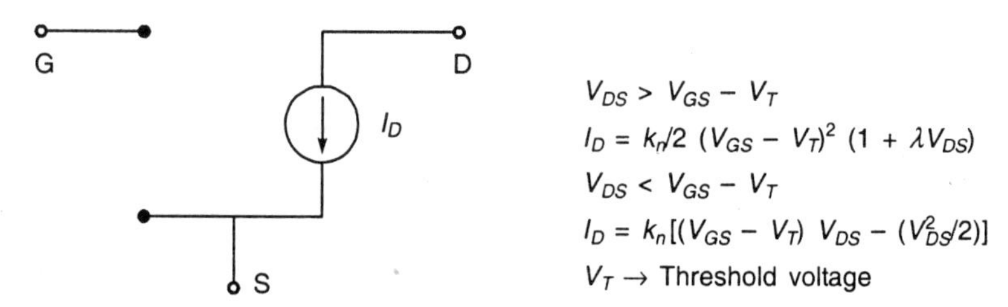

Figure 3.10 An MOS model for manual analysis.

3.2.4 MOS Transistor Transconductance g_m and Output Conductance g_{ds}

Transconductance g_m, expresses the relationship between output current I_D, and input voltage V_{GS}, and is defined as

$$\left(\frac{\delta I_D}{\delta V_{GS}}\right) \Big/ V_{DS} = \text{constant} \tag{3.17}$$

The expression for g_m in terms of circuit and transistor parameters is given by

$$g_m = \frac{k_n' W}{L}(V_{GS} - V_T) \tag{3.18}$$

From Eq. (3.18), we observe that it is possible to increase the g_m of a MOS device by increasing its width. However, this will also increase the input capacitance and area occupied. A reduction in channel length results in an increase in ω_0 owing to the higher g_m. However, the gain of the MOS device decreases owing to the strong degradation of the output resistance.

The output conductance g_{ds}, can be expressed by

$$g_{ds} = \frac{\delta I_D}{\delta V_{DS}} = \lambda I_{DS} \propto \left(\frac{1}{L}\right)^2 \tag{3.19}$$

Here, the strong dependence on the channel length is demonstrated as $\lambda \propto (1/L)$ and $I_D \propto 1/L$ for the MOS device.

3.2.5 MOS Transistor Figure of Merit, ω_0

An indication of frequency response may be obtained from the parameter ω_0 where,

$$\omega_0 = \frac{g_m}{C_g} = \frac{\mu}{L^2}(V_{GS} - V_T) \tag{3.20}$$

Here, C_g is the gate–channel capacitance and μ is the bulk mobility. This shows that switching speed depends on gate voltage above threshold and on carrier mobility and inversely as the square of channel length. A fast circuit requires that g_m must be as high as possible.

3.3 DYNAMIC BEHAVIOUR OF MOS TRANSISTOR

The MOSFET is a majority carrier device. Hence, its dynamic response is determined by the time to (dis)charge the capacitance between the device ports and from the interconnecting lines. An accurate analysis of the nature and behaviour of these capacitances is essential when designing high performance digital circuits. They originate from the sources:

- the basic MOS structure
- the channel charge
- depletion region of reverse-biased *pn*-junctions of drain and source

Aside from the MOS structure capacitances, all capacitors are non-linear and vary with the applied voltage.

3.3.1 MOS Structure Capacitances

The gate of the MOS transistor is isolated from the conducting channel by the gate-oxide that has a capacitance per unit area equal to $C_{ox} = \varepsilon_{ox}/t_{ox}$. From the I-V relations, it is noted that it is useful to have C_{ox} as large as possible, or to keep the oxide thickness very thin. The total capacitance is $C_{ox}WL$ and is called gate capacitance. The gate capacitance can be decomposed into a number of elements, each with a different behaviour. Obviously, one part of C_g contributes to the channel charge, another part is due to the topological structure of the transistor.

Consider the transistor structure of Figure 3.11. Ideally, the source and drain diffusion should end right at the edge of the gate-oxide. In reality, both source and drain tend to extend somewhat below the oxide by an amount x_d, called the lateral diffusion. Hence, the effective channel length of the transistor, L_{eff} becomes shorter than the drawn length by a factor of $2x_d$. It also gives rise to parasitic capacitance between gate and source (drain). That is called the overlap capacitance. This capacitance is strictly linear and has a fixed value

$$C_{gso} = C_{gdo} = C_{ox} x_d W = C_o W \tag{3.21}$$

where, C_o is the overlap capacitance with unit transistor width.

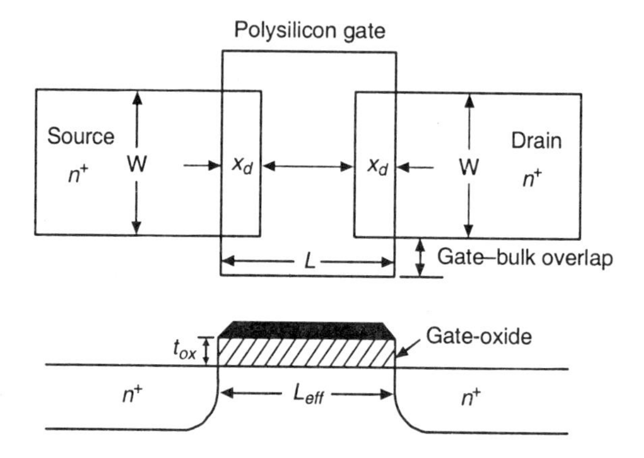

Figure 3.11 MOSFET overlap capacitance.

3.3.2 Channel Capacitance

The gate-to-channel capacitance can be decomposed into three parts:

(i) C_{gs}—capacitance between gate and source

(ii) C_{gd}—capacitance between gate and drain

(iii) C_{gb}—capacitance between gate and bulk regions

All those components are non-linear, and their values depend on the operation regions. To simplify the analysis, estimated and average values are used. For instance, in the cut-off mode, no channel exists, and the total capacitance $C_{ox}WL_{\text{eff}}$ appears between gate and bulk. In the linear/triode region, an inversion layer is formed, which acts as a conductor between source and drain. Consequently, $C_{gb} = 0$ as the bulk electrode is shielded from the gate by the channel. Symmetry dictates that $C_{gs} = C_{dg} = C_{ox}WL_{\text{eff}}/2$. Finally, in the saturation mode, the channel is pinched off. The capacitance between the gate and drain is thus approximately zero, and so is the gate–bulk capacitance. The value of C_{gs} averages to $(2/3)C_{ox}WL_{\text{eff}}$. Table 3.1 gives the average channel capacitance of MOS transistor for different operation regions.

Table 3.1 Average channel capacitances of MOS transistor for different operation regions

Operation region	C_{gb}	C_{gs}	C_{gd}
Cut off	$C_{ox}WL_{eff}$	0	0
Linear	0	$(1/2)C_{ox}WL_{eff}$	$(1/2)C_{ox}WL_{eff}$
Saturation	0	$(2/3)C_{ox}WL_{eff}$	0

3.3.3 Junction Capacitance

It is contributed by the reverse-biased source–bulk and drain–bulk *pn*-junctions. The depletion region capacitance is non-linear and decreases when the reverse bias is raised. It is often called diffusion capacitance. The detailed picture, shown in Figure 3.12, illustrates that the junction consists of two components:

- The bottom-plate junction
- The side-wall junction

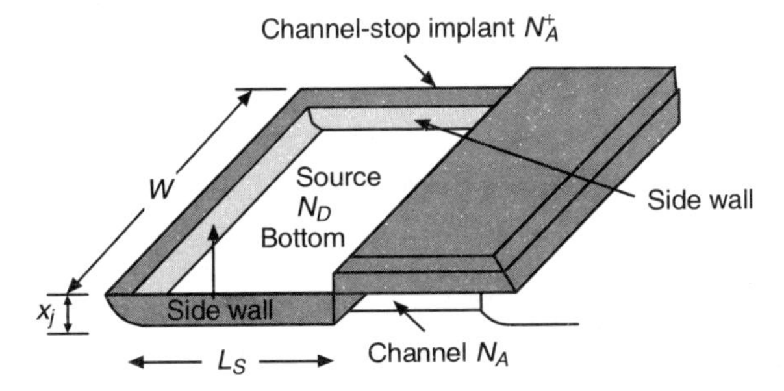

Figure 3.12 Detailed view of the source junction.

The bottom-plate junction is formed by the source region with doping N_D and substrate with doping N_A. The total depletion region capacitance for this components equals $C_{\text{bottom}} = C_jWL_s$, where C_j is the junction capacitance per unit area.

The side-wall junction is formed by the source region with doping N_D and the p^+ channel-stop implant with doping level $N_A{}^+$. The sidewall capacitance is given by

$$\begin{aligned} C_{sw} &= C'_{jsw}\,x_j\,(W + 2L_s) \\ &= C_{jsw}(W + 2L_s) \end{aligned} \tag{3.22}$$

where, x_j is the junction depth.

$$C_{jsw} = C'_{jsw}\,x_j = \text{Capacitance per unit perimeter}$$

Thus, the total junction capacitance is given by

$$\begin{aligned} C_{\text{diff}} &= C_{\text{bottom}} + C_{sw} \\ &= C_j\,(\text{Area}) + C_{jsw}\,(\text{Perimeter}) \\ &= C_jWL_s + C_{jsw}\,(W + 2L_s) \end{aligned} \tag{3.23}$$

3.3.4 Capacitive Device Model

All the above contributions can be combined in a single capacitance model for the MOS transistor as in Figure 3.13. The various values of capacitances are given as:

$$C_{GS} = C_{gs} + C_{gso};$$
$$C_{GD} = C_{gd} + C_{gdo};$$
$$C_{GB} = C_{gb};$$
$$C_{SB} = C_{\text{sdiff}};$$
$$C_{DB} = C_{\text{Ddiff}}; \tag{3.24}$$

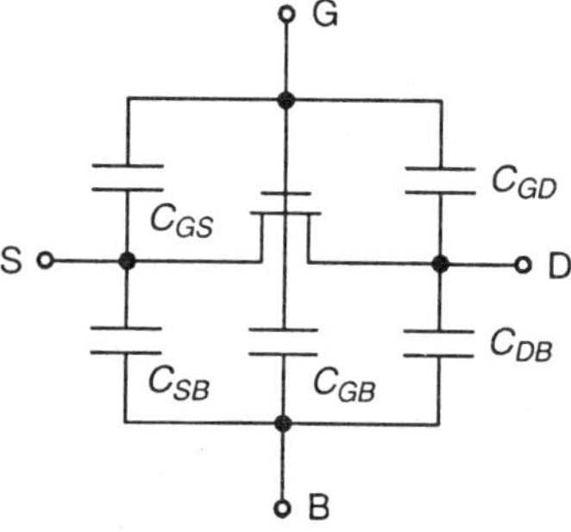

Figure 3.13 MOSFET capacitance model.

3.4 THE ACTUAL MOS TRANSISTOR—SECONDARY EFFECTS

The operation of an actual MOS transistor can deviate substantially from the behaviour of an ideal MOS transistor. This is especially true when the dimensions of the device reach μm range. At that point, the channel-length becomes comparable to other device parameters such as the depth of the drain and source junctions, and the width of their depletion regions. Such a device is called a short-channel transistor. The behaviour of a long-channel device is adequately described by a 1D model, where it is assumed that all current flows on the surface of silicon and the electric fields are oriented along the plane. In short-channel devices, those assumptions are no longer valid and a two-dimensional model is more appropriate.

The understanding of some of these second-order effects and their impact on the device behaviour is essential in the design of contemporary digital circuits.

3.4.1 Threshold Variations

Equation (3.5) states that the threshold voltage is only a function of the manufacturing technology and the applied body bias, V_{SB}. Therefore, the threshold can be considered as a constant over all NMOS (PMOS) transistors in a design. As the device dimensions are increased, this model becomes inaccurate, since the threshold potential becomes a function of L, W and V_{DS}. The value of V_{To} decreases with L for short-channel devices. Figure 3.14(a) shows the threshold as a function of length (for low V_{DS}).

A similar effect can be obtained by raising the drain–source (bulk) voltage, as this increases the width of the drain junction depletion region. Consequently, the threshold decreases with increasing V_{DS}. This effect, called the drain-induced barrier lowering (DIBL) causes the threshold potential to be a function of the operating voltages [Figure 3.14(b)]. For high enough values of the drain voltage, the source and the drain regions can ever be shorted together, and normal transistor action ceases to exist. This effect is called punchthrough, which has a major impact on the system design. For instance, in dynamic memories, where the leakage current of a cell becomes a function of the voltage on the dataline, which is shared with many other cells.

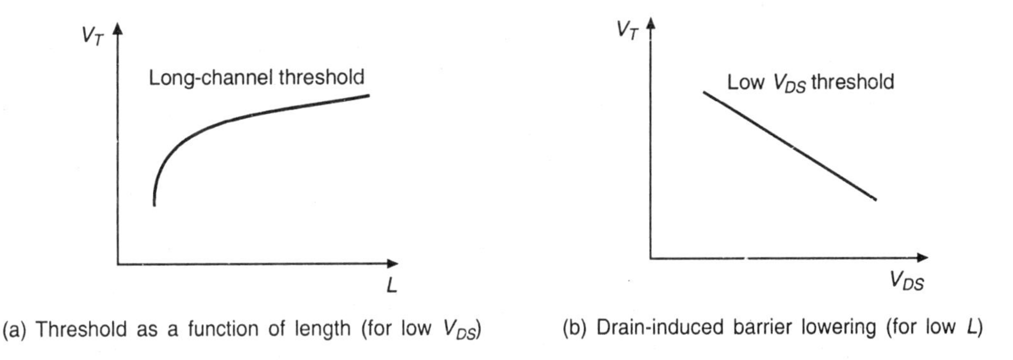

(a) Threshold as a function of length (for low V_{DS})   (b) Drain-induced barrier lowering (for low L)

Figure 3.14 Threshold variations.

The threshold voltages in short-channel devices also have the tendency to drift over time. This is the result of hot-carrier effect. The device dimensions have been scaled down continuously while keeping the power supply and operating voltages constant. This increases the electric field strength which, in turn, increases the velocity of electrons. These electrons can now tunnel into the gate–oxide region reaching a high energy level (hot). Electrons trapped in the oxide change the threshold voltage, typically increasing the threshold of NMOS devices, while decreasing the V_T of PMOS devices. For an electron to become hot, an electrical heat of at least 10^4 V/cm is necessary. This condition is easily met in devices with channel length around or below 1 μm. The hot-electron phenomenon can lead to a long-term reliability problem.

3.4.2 Source–Drain Resistance

When transistors are scaled down, their junctions are shallower, and the contact openings become smaller. This results in an increase in the parasitic resistance in series with the drain and source regions as shown in Figure 3.15. Figure 3.15(a) shows the modelling of the series resistance.

The resistance of the drain (source) region can be expressed as

$$R_{S,D} = \frac{L_{S,D}}{W} R_o + R_C \tag{3.25}$$

where, R_C is the contact resistance W, the width of the transistor, and $L_{S,D}$ the length of the source or drain region [Figure 3.15(b)]. R_o is the sheet resistance per square of the drain–source diffusion.

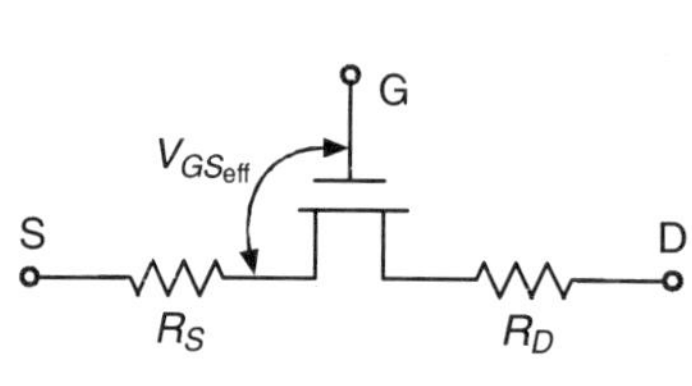

(a) Modelling the series resistance

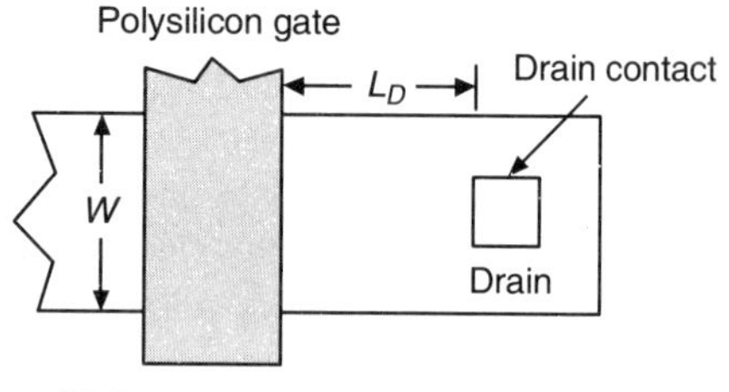

(b) Parameters of the series resistance

Figure 3.15 Series drain and source resistance.

The series resistance causes a deterioration in the device performance, as it reduces the drain current for a given control voltage. Keeping its value as small as possible is thus an important design goal. The two ways to achieve this goal are:

- Silicidation → Covering the drain and source regions with a low resistivity material, such as titanium or tungsten.
  → Effectively reduces the parasitic resistance
  → Also used to reduce the resistance of the polysilicon gate.
- Making the transistor wider than needed.

3.4.3 Variation in I-V Characteristics

The voltage–current relations of a short-channel device deviate considerably from the ideal expressions of Eqs. (3.14) and (3.16). This occurs due to the velocity saturation and the mobility degradation effects.

Velocity Saturation. From Eq. (3.12), it was stated that the velocity of the carriers is proportional to the electric field. However, when the electric field along the channel reaches a critical value, E_{sat}, the velocity of the carriers tends to saturate as shown in Figure 3.16(a).

For *p*-type silicon, the critical field at which electron saturation occurs is 1.5×10^4 V/cm (or 1.5 V/μm) and the saturation velocity, $v_{sat} = 10^7$ cm/sec. The holes in *n*-type silicon saturate at the same velocity but a higher electric field $\geq 10^5$ V/cm is needed to achieve saturation.

This effect has a profound impact on the operation of the transistor. Combining Eqs. (3.9) and (3.10), and setting v_n to v_{sat}, we get,

$$I_{D\text{sat}} = v_{\text{sat}} C_{ox} W (V_{GS} - V_{D\text{sat}} - V_T) \tag{3.26}$$

where $V_{D\text{sat}}$ → drain–source voltage at which velocity saturation comes into play.

From Eq. (3.26), it is observed that $I_{D\text{sat}}$ linearly depends on V_{GS}. Consequently, reducing the operating voltage V_{GS} does not have such a significant effect in submicron devices (short-channel devices) as it would have in a long-channel transistor. Also, I_D is independent of L, the channel length in velocity-saturated devices, so the current drive cannot be further improved by decreasing the channel length as in long-channel devices.

Mobility degradation. Reducing the channel length has another impact on the transistor current: even at normal electric field levels, a reduction in the electron mobility can be observed. This effect is called mobility degradation and can be attributed to the vertical component of the electrical field [Figure 3.16(b)].

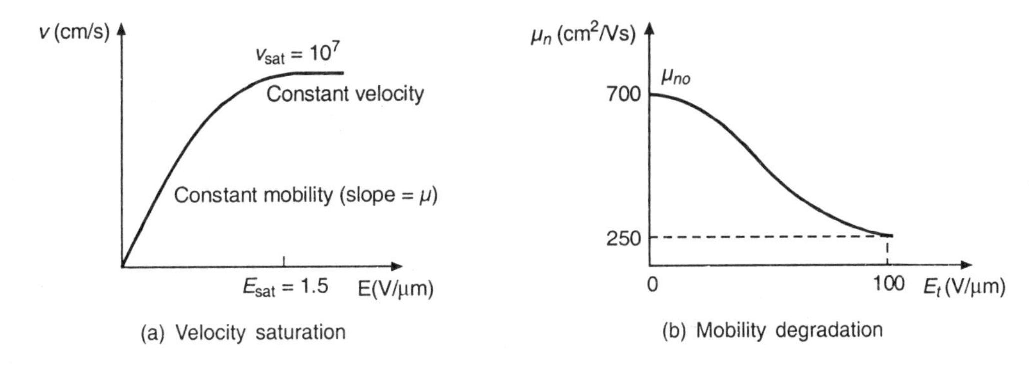

(a) Velocity saturation (b) Mobility degradation

Figure 3.16 Effect of electric field on electron velocity and mobility.

3.4.4 Subthreshold Conduction

The MOS transistor is partially conducting for voltages below the threshold voltage. This effect is called subthreshold or weak-inversion conduction. In Figure 3.17, the curve I_D vs. V_{GS} drawn on a logarithmic scale clearly demonstrates that the current does not drop to zero immediately for $V_{GS} < V_T$, but actually decays in an exponential function, similar to the operation of BJT. In the absence of conducting channel, the n^+ (source)-p (bulk)-n^+ (drain) terminals actually form a parasitic bipolar transistor. In this operation region, the (inverse) rate of decrease of the current with respect to V_{GS} is approximated as

$$\left[\frac{d}{dV_{GS}}(\ln I_D)\right]^{-1} = \frac{kT}{q}\ln_{10}(1+\alpha) \tag{3.27}$$

The expression (kT/q) ln (10) evaluates to 60 mV/decade at room temperature. α is 0 for ideal device and larger than 1 for actual devices. It is a function of transistor capacitances. Reducing it requires advanced and expensive technologies, such as silicon-on-insulator.

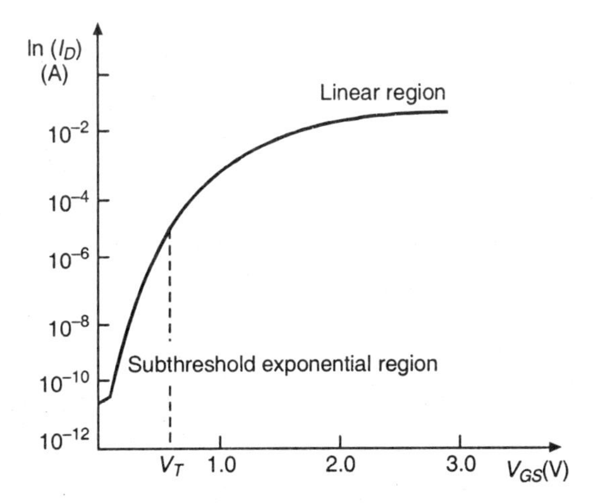

Figure 3.17 I_D vs. V_{GS} (logarithmic scale) showing the exponential characteristics of the subthreshold region.

The presence of subthreshold current detracts from the ideal switch model that we like to assume for the MOS transistor. In general, we want the current to be as close as possible to zero at $V_{GS} = 0$. This is especially important in the so-called dynamic circuits, which rely on the storage of charge on a capacitor and whose operation can be severely degraded by subthreshold leakage.

3.4.5 CMOS Latchup

A problem which is inherent in the *p*-well and *n*-well processes is due to the relatively large number of junctions which are formed in these structures and, the consequent presence of parasitic transistors and diodes. Latchup is a condition in which the parasitic components give rise to the establishment of low resistance conducting paths between V_{DD} and V_{SS} with disastrous results. Careful control during fabrication is necessary to avoid this problem.

Consider the *n*-well structure of Figure 3.18(a). The *n-p-n-p* structure is formed by the source of the NMOS, the *p*-substrate, the *n*-well and the source of the PMOS. An equivalent circuit is shown in Figure 3.18(b). When one of the two bipolar transistors gets forward-biased (e.g., due to current flowing through the well or substrate), it feeds the base of the other transistor. This positive feedback increases the current until the circuit fails or burns out. The remedies for the latchup problem include:

(i) an increase in substrate doping levels with a consequent drop in the value of R_{psubs}.
(ii) reducing R_{nwell} by control of fabrication parameters and ensuring a low contact resistance to V_{DD}.
(iii) by introducing guard rings.

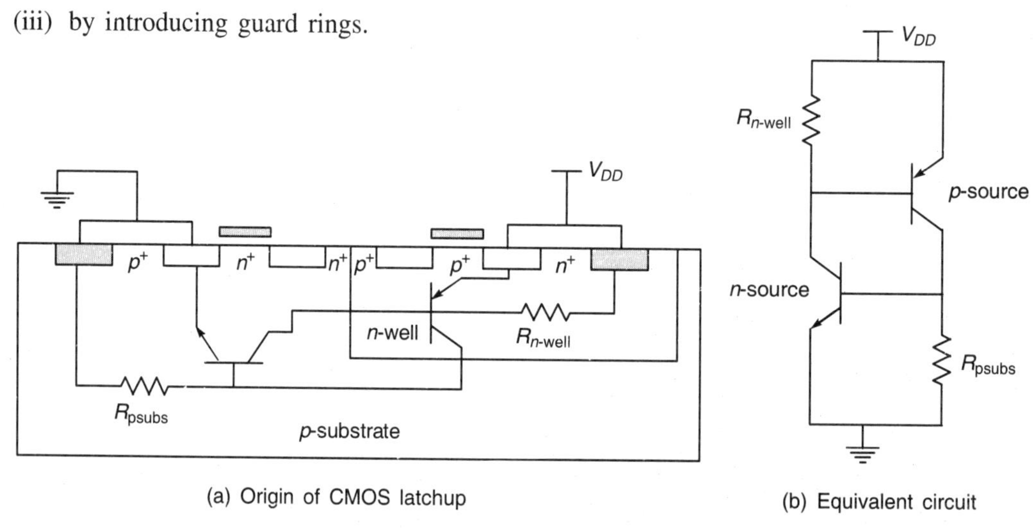

(a) Origin of CMOS latchup

(b) Equivalent circuit

Figure 3.18 CMOS latchup.

3.5 NMOS INVERTER

A basic requirement for producing a complete range of logic circuits is the inverter. This is needed for restoring logic levels for NAND and NOR gates, and for sequential and memory circuits of various forms.

The basic inverter circuit consists of a transistor with source connected to ground and a load resistor connected from the drain to the positive supply rail V_{DD}. The input is applied to the gate and the output is taken from drain terminal. A depletion mode transistor can also be used as the load (Figure 3.19).

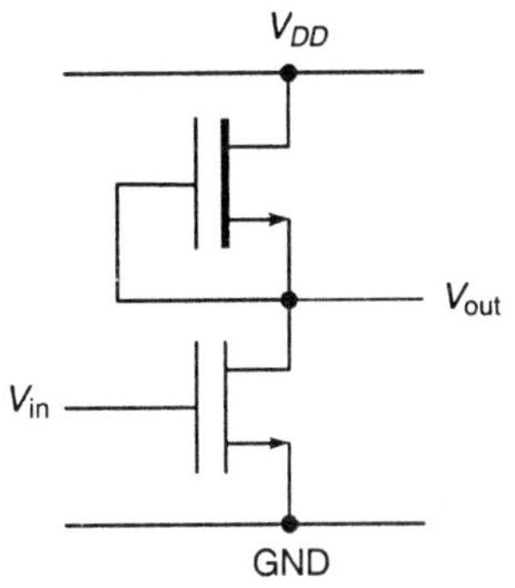

Figure 3.19 NMOS inverter.

Now, the following points can be considered to describe the circuit:

- With no current drawn from the output, the currents I_{DS} for both transistors be equal.
- The gate of depletion mode transistor is connected to the source. Hence, it is always ON and the only characteristic curve $V_{GS} = 0$ is relevant.
- In this configuration, the depletion mode device is called the pull-up (p.u) transistor and the enhancement mode device is the pull-down (p.d) transistor.
- The inverter transfer characteristic is obtained by superimposing the $V_{GS} = 0$ depletion-mode characteristic curve on the family of curves for the enhancement mode device. The maximum voltage across the enhancement mode device corresponds to minimum voltage across the depletion mode transistor (Figure 3.20).

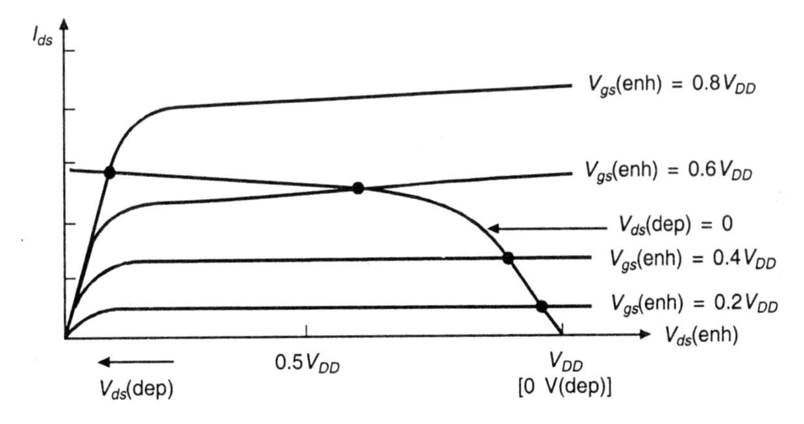

Figure 3.20 Derivation of NMOS inverter transfer characteristic.

- The points of intersection of the curves give points on the transfer characteristic which has the form of Figure 3.21.

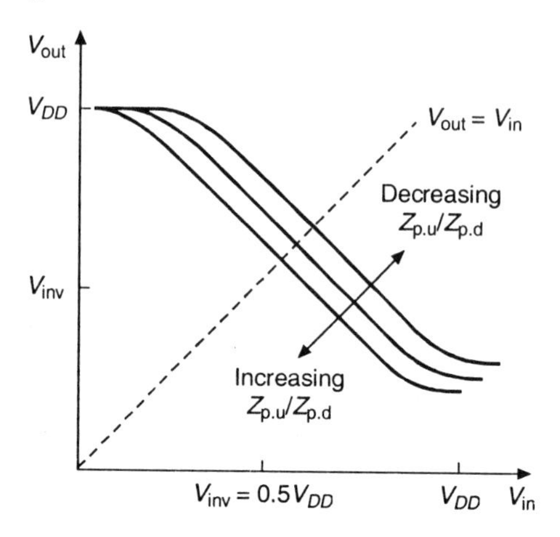

Figure 3.21 NMOS inverter transfer characteristic.

- When V_{in} exceeds the threshold voltage of the pull-down transistor, current begins to flow. The output voltage V_{out} thus decreases and the subsequent increases in V_{in} will cause the pull-down transistor to come out of saturation and become resistive. The pull-up transistor is initially resistive as the pull-down transistor turns ON.
- During transition, the slope of the transfer characteristic determines the gain:

$$\text{Gain} = \frac{\delta V_{out}}{\delta V_{in}}$$

- The point at which $V_{out} = V_{in}$ is denoted as V_{inv} and the transfer characteristic and V_{inv} can be shifted by variations of the ratio of pull-up to pull-down resistances denoted as $(Z_{p.u}/Z_{p.d})$ where Z is determined by the length to width ratio of the transistor.

3.6 DETERMINATION OF PULL-UP TO PULL-DOWN RATIO $(Z_{p.u}/Z_{p.d})$ FOR AN NMOS INVERTER DRIVEN BY ANOTHER NMOS INVERTER

Consider the arrangement shown in Figure 3.22 in which an inverter is driven from the output of another similar inverter. Consider the depletion mode transistor for which $V_{GS} = 0$ under all conditions, and further assume that in order to cascade inverters without degradation of levels, we aim to meet the requirement:

$$V_{in} = V_{out} = V_{inv}$$

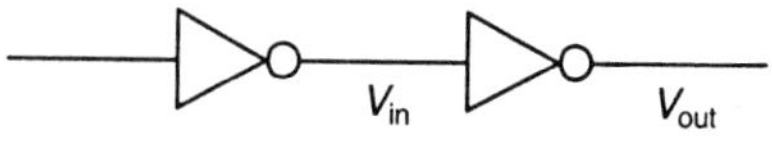

V_{in} V_{out}

Figure 3.22 NMOS inverter driven directly by another inverter.

For equal margins around the inverter threshold, we set $V_{inv} = 0.5V_{DD}$. At this point, both transistors are in saturation and

$$I_D = \frac{k_n^1}{2}\frac{W}{L}(V_{GS} - V_T)^2$$

In the depletion mode,

$$I_D = \frac{k_n^1}{2}\frac{W_{p.u}}{L_{p.u}}(-V_{Td}^2) \qquad \text{(since } V_{GS} = 0\text{)}$$

and in the enhancement mode,

$$I_D = \frac{k_n^1}{2}\frac{W_{p.d}}{L_{p.d}}(V_{inv} - V_T)^2 \qquad \text{(since } V_{GS} = V_{inv}\text{)}$$

Equating, we have,

$$\frac{W_{p.d}}{L_{p.d}}(V_{inv} - V_T)^2 = \frac{W_{p.u}}{L_{p.u}}(-V_{Td})^2$$

where, $W_{p.d}$, $L_{p.d}$, $W_{p.u}$, and $L_{p.u}$ are the widths and lengths of the pull-down and pull-up transistors respectively.
Now,

$$Z_{p.d} = \frac{L_{p.d}}{W_{p.d}}; \quad Z_{p.u} = \frac{L_{p.u}}{W_{p.u}}$$

$$\therefore \qquad \frac{1}{Z_{p.d}}(V_{inv} - V_T)^2 = \frac{1}{Z_{p.u}}(-V_{Td})^2$$

Simplifying,

$$V_{inv} = V_T - \frac{V_{Td}}{\sqrt{(Z_{p.u}/Z_{p.d})}}$$

Typical values are as follows:

$$V_T = 0.2V_{DD}; \quad V_{Td} = -0.6V_{DD};$$

$$V_{inv} = 0.5V_{DD}$$

$$\therefore \qquad 0.5 = 0.2 + \frac{0.6}{\sqrt{Z_{p.u}/Z_{p.d}}}$$

$$\Rightarrow \sqrt{\frac{Z_{p.u}}{Z_{p.d}}} = 2$$

$$\Rightarrow \frac{Z_{p.u}}{Z_{p.d}} = \frac{4}{1}$$

for an inverter directly driven by another inverter.

3.7 PULL-UP TO PULL-DOWN RATIO FOR AN NMOS INVERTER DRIVEN THROUGH ONE OR MORE PASS TRANSISTORS

Now, consider the arrangement of Figure 3.23 in which the input to the inverter 2 comes from the output of inverter 1 but passes through one or more NMOS pass transistors. The connection of pass transistors in series will degrade the logic 1 level into inverter 2 so that the output will not be a proper logic 0 level. If the point A is at 0 V, and point B is thus at V_{DD}, but the voltage into inverter 2 at point C is now reduced from V_{DD} by the threshold voltage of the series pass transistor. However, there can be no voltage drop in the channels, since no static current flows through the pass transistors. With all pass transistors connected to V_{DD}, there is a loss of V_{Tn}.

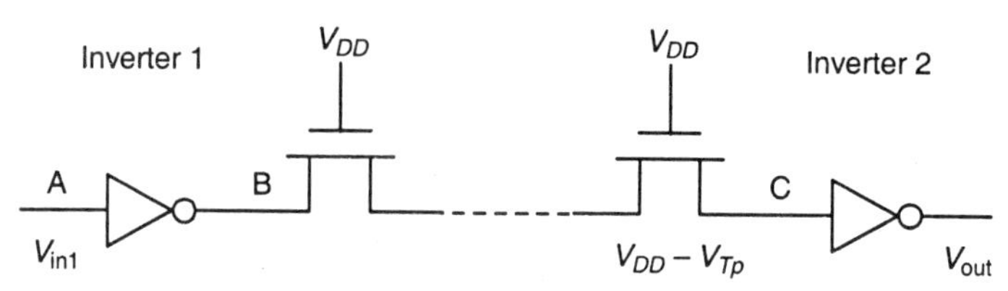

Figure 3.23 Pull-up to pull-down ratios for inverting logic coupled bypass transistors.

Therefore, the input voltage to inverter 2 is

$$V_{in2} = V_{DD} - V_{Tn}$$

where, V_{Tn} is the threshold voltage of the pass transistor.

We must ensure that for this input voltage, we get out the same voltage as would be the case for inverter 1 driven with input = V_{DD}

Consider inverter 1 [Figure 3.24(a)] with input = V_{DD}. If the input is at V_{DD}, then the pull-down transistor T_2 is conducting but with a low voltage across it; therefore, it is in its resistive region represented by R_1. Meanwhile, the pull-up transistor T_1 is in saturation and is represented as a current source.

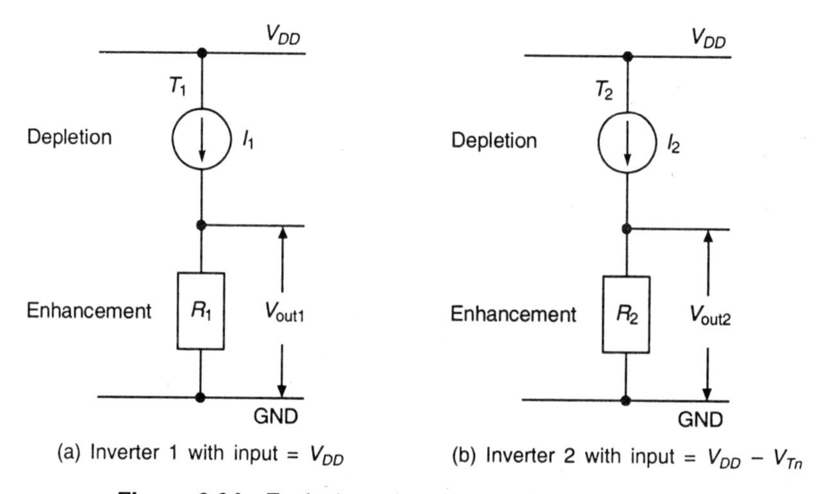

(a) Inverter 1 with input = V_{DD} (b) Inverter 2 with input = $V_{DD} - V_{Tn}$

Figure 3.24 Equivalent circuits of inverters 1 and 2.

For the pull-down transistor,

$$I_{DS} = k_n \frac{W_{\text{p.d.1}}}{L_{\text{p.d.1}}}\left((V_{DD} - V_T)V_{DSI} - \frac{V_{DS1}^2}{2}\right)$$

$$\therefore \qquad R_1 = \frac{V_{DS1}}{I_{DS}} = \frac{1}{k_n}\frac{L_{\text{p.d.1}}}{W_{\text{p.d.1}}}\left(\frac{1}{V_{DD} - V_T - (V_{DS1}/2)}\right)$$

V_{DS1} is small and $(V_{DS1}/2)$ may be neglected.
Thus,

$$R_1 = \frac{1}{k_n} Z_{\text{p.d.1}}\left(\frac{1}{V_{DD} - V_T}\right)$$

Now, for the depletion mode pull-up transistor in saturation, with $V_{GS} = 0$,

$$I_1 = I_{DS} = k_n \frac{W_{\text{p.u.1}}}{L_{\text{p.u.1}}}\frac{(-V_{Td})^2}{2}$$

where V_{Td} represents the threshold voltage for the depletion mode transistor.

$$\therefore \qquad V_{\text{out1}} = I_1 R_1 = \frac{Z_{\text{p.d.1}}}{Z_{\text{p.u.1}}}\left(\frac{1}{V_{DD} - V_T}\right)\frac{(V_{Td})^2}{2}$$

Consider inverter 2 [Figure 3.24(b)] when input = $V_{DD} - V_{Tn}$
As for inverter 1,

$$R_2 = \frac{1}{k_n} Z_{\text{p.d.2}} \frac{1}{[(V_{DD} - V_{Tn}) - V_T]}$$

$$I_2 = k_n \frac{1}{Z_{\text{p.u.2}}}\frac{(-V_{Td})^2}{2}$$

$$\Rightarrow \qquad V_{\text{out2}} = I_2 R_2$$

$$= \frac{Z_{\text{p.d.2}}}{Z_{\text{p.u.2}}}\left(\frac{1}{V_{DD} - V_{Tn} - V_T}\right)\frac{(-V_{Td}^2)}{2}$$

If inverter 2 is to have the same output voltage under these conditions,

$$V_{\text{out 1}} = V_{\text{out 2}}$$

$$\Rightarrow \qquad \frac{Z_{\text{p.u.2}}}{Z_{\text{p.d.2}}} = \frac{Z_{\text{p.u.1}}}{Z_{\text{p.d.1}}}\frac{(V_{DD} - V_T)}{(V_{DD} - V_{Tn} - V_T)}$$

Substituting typical values as below:

$$V_T = 0.2V_{DD};$$

$$V_{Tn} = 0.3V_{DD};$$

$$\frac{Z_{\text{p.u.2}}}{Z_{\text{p.d.2}}} = \frac{Z_{\text{p.u.1}}}{Z_{\text{p.d.1}}} \frac{0.8}{0.5}$$

$$\therefore \quad \frac{Z_{\text{p.u.2}}}{Z_{\text{p.d.2}}} = 2\frac{Z_{\text{p.u.1}}}{Z_{\text{p.d.1}}} = \frac{8}{1}$$

Summarizing for an NMOS transistor:

- An inverter driven directly from the output of another should have a $Z_{p.u}/Z_{p.d}$ ratio $\geq 4/1$.
- An inverter driven through one or more pass transistors should have a $Z_{p.u}/Z_{p.d}$ ratio $\geq 8/1$.

3.8 DEVICE MODELS FOR SIMULATION

The fundamental goal in device modelling is to obtain the functional relationship among the terminal electrical variables of the device that is to be modelled. These electrical characteristics depend upon a set of parameters including both geometric variables and variables dependent upon the device physics. The device models have electrical characteristics identical to those of relatively simple circuits composed of basic circuit components. The models are initially developed by analytically applying basic physical principles and then empirically modifying the resulting mathematical expressions to improve agreement between theoretical and experimental results.

3.8.1 MOS Models

The *n*-channel and *p*-channel enhancement MOSFET devices along with the conventions for the electrical variables are shown in Figures 3.25(a) and (b). The same electrical conventions and geometric gate definitions are followed for depletion devices.

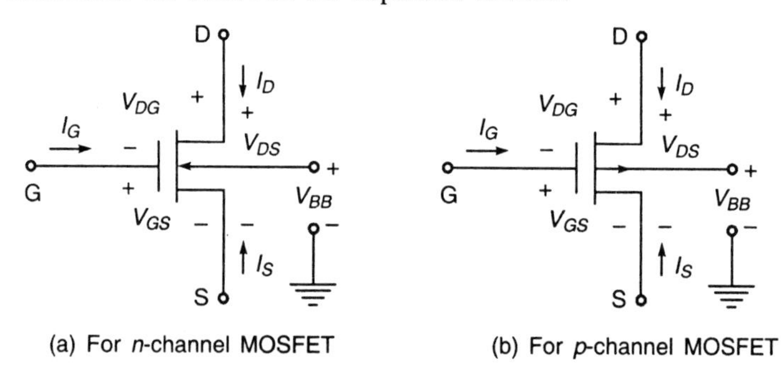

(a) For *n*-channel MOSFET (b) For *p*-channel MOSFET

Figure 3.25 Convention for electrical variables of MOSFETs.

3.8.2 DC MOSFET Model

The dc model of a device is a mathematical or numerical relationship that relates the actual terminal voltages and currents of the device at dc and low frequencies. The dc model should be valid over a large range of terminal voltages and currents.

For MOSFET, the dc model introduced by Sah in 1964 is given by

$$I_D = k_n [(V_{GS} - V_T) V_{DS} - (V_{DS}^2/2)], \quad V_{GS} > V_T$$

$$I_D = 0, \quad V_{GS} < V_T \tag{3.28}$$

The low frequency MOSFET model for both n-channel and p-channel devices adequate for most dc hand calculations is shown in Table 3.2.

Table 3.2 Low-frequency MOSFET model

n-channel MOSFET		
$I_G = 0$		(1)
$I_D = 0,$	$V_{GS} < V_T$ (cutoff), $V_{DS} \geq 0$	(2)
$I_D = k_n' \frac{W}{L}\left[(V_{GS} - V_T) V_{DS} - \frac{V_{DS}^2}{2}\right],$	$V_{GS} > V_T$, $0 < V_{DS} < V_{GS} - V_T$ (Ohmic or linear region)	(3)
$I_D = k_n' \frac{W}{2L}(V_{GS} - V_T)^2,$	$V_{GS} > V_T$, $V_{DS} > V_{GS} - V_T$ (Saturation)	(4)
	$V_T \rightarrow$ Threshold voltage	
p-channel MOSFET		
$I_G = 0$		(5)
$I_D = 0,$	$V_{GS} > V_T$ (cutoff), $V_{DS} \leq 0$	(6)
$I_D = -k_n' \frac{W}{L}\left[(V_{GS} - V_T) V_{DS} - \frac{V_{DS}^2}{2}\right],$	$V_{GS} < V_T$, $0 > V_{DS} > V_{GS} - V_T$ (Ohmic or linear region)	(7)
$I_D = -k_n' \frac{W}{2L}(V_{GS} - V_T)^2,$	$V_{GS} < V_T$, $V_{DS} < V_{GS} - V_T$ (Saturation)	(8)
	$V_T \rightarrow$ Threshold voltage	

Design parameters: W = channel width
L = channel length

Process parameters: k_n' = transconductance parameter
V_{To} = Threshold voltage for VSB = 0
γ = bulk threshold parameter
λ = strong inversion surface potential
φ_F = channel length modulation parameter

3.8.3 High Frequency MOSFET Model

At high frequencies, dc and small signal models of the MOSFET are generally considered inadequate, due to the parasitic capacitances existing in MOS structures. These parasitic capacitances can be divided into two groups. The first group is composed of those parasitic capacitors formed by sandwiching an insulating dielectric of fixed geometric dimensions between two conductive regions. This capacitance is a constant irrespective of the local changes in the voltage applied to the plates of the capacitor.

Let A be the area of the capacitor plates, and d the distance between the plates. Then, the capacitance is given by

$$C = \frac{\varepsilon A}{d} \tag{3.29}$$

where ε is the permittivity of the dielectric material separating the plates.

Taking $\varepsilon/d = C_d$ = capacitance density,

$$C = C_d A \tag{3.30}$$

C_d is called process parameter and A is a design parameter.

The second group is composed of the capacitors formed by the separation of charge associated with a *pn*-junction. The depletion region associated with the semiconductor junction serves as the dielectric. These junction capacitors are quite voltage dependent. This capacitance can be approximated by

$$C = \frac{C_{jo} A}{1 - (V_F/\varphi_B)} \tag{3.31}$$

where, $C_{j0} \rightarrow$ junction capacitance density at zero volts bias

$A \rightarrow$ junction area

$V_F \rightarrow$ dc forward bias voltage of the *p-n* junction

$\Phi_B \rightarrow$ barrier potential

Equation (3.31) is reasonably good for $-\alpha < V_F < \varphi_B/2$.

The parasitic capacitors that dominate high frequency behaviour of MOS transistors is shown in Figure 3.26.

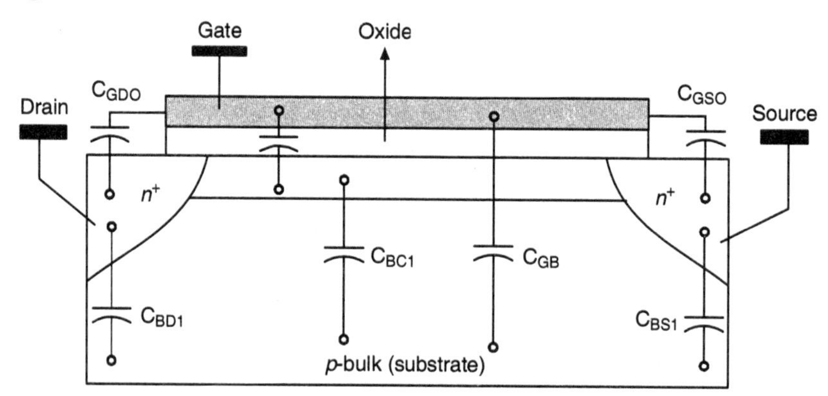

Figure 3.26 Parasitic capacitors in MOS transistors shown for *n*-channel devices.

The capacitors C_{BD1}, C_{BC1} and C_{BS1} are all voltage-dependent capacitors governed by Eq. (3.31). The remaining capacitors are parallel plate capacitors governed by Eq. (3.30).

$C_{GDO} \rightarrow$ gate–drain overlap capacitor

$C_{GSO} \rightarrow$ gate–source overlap capacitor

C_{GC} → gate–channel capacitance

C_{GB} → gate–bulk capacitance

C_{BC1} → bulk–channel junction capacitance

C_{BS1} → capacitance of bulk–source junction

C_{BD1} → capacitance of bulk–drain junction

The lumped model of parasitic capacitors considered in Figure 3.26 is shown in Figure 3.27(a). The values for these capacitors in the three regions of operation is listed in Figure 3.27(b). Here, C_{ox} represents the capacitance density of the gate/oxide/channel capacitor and is given by

$$C_{ox} = \frac{\varepsilon_0 \varepsilon_{SiO2}}{t_{ox}} \tag{3.32}$$

where, ε_0 → permittivity of free space

ε_{SiO2} → relative permittivity of SiO_2 dielectric

t_{ox} → thickness of the dielectric

and L_D represents the distance of the lateral moat diffusion under the gate.

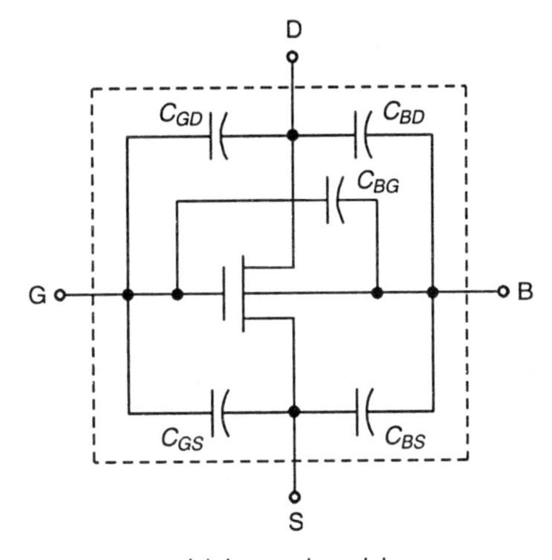

(a) Lumped model

		Region	
	Cut-off	Ohmic	Saturation
C_{GD}	$C_{ox}WL_D$	$C_{ox}WL_D + (1/2)C_{ox}WL$	$C_{ox}WL_D$
C_{GS}	$C_{ox}WL_D$	$C_{ox}WL_D + (1/2)C_{ox}WL$	$C_{ox}WL_D + (2/3)C_{ox}WL$
C_{BG}	$C_{ox}WL$	0	0
C_{BD}	C_{BD1}	$C_{BD1} + (1/2)C_{BC1}$	C_{BD1}
C_{BS}	C_{BS1}	$C_{BS1} + (1/2)C_{BC1}$	$C_{BS1} + (2/3)C_{BC1}$

(b) Capacitor values in the three regions of operation

Figure 3.27 (Contd.)

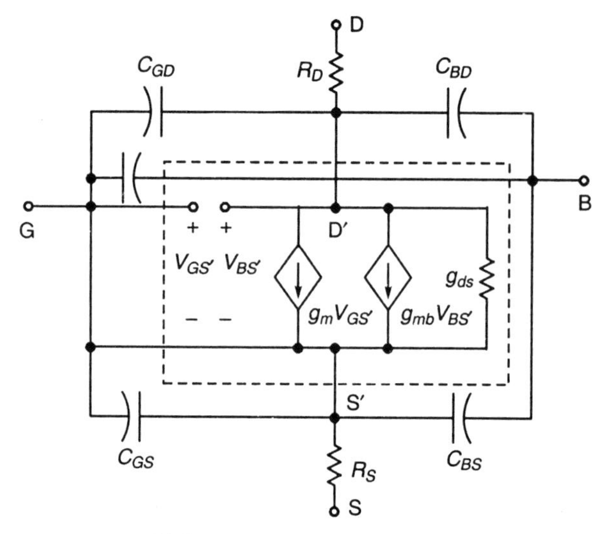

(c) Small signal equivalent circuit

Figure 3.27 Parasitic capacitors.

The small signal equivalent circuit of the MOSFET including the capacitive parasitics is shown in Figure 3.27(c). The resistors R_S and R_D have been included to account for the ohmic resistance in the drain and source regions from the physical source and the drain connections to the actual drain (D') and source of the device. For large physical drain and source areas, these resistors become quite large.

3.8.4 SPICE Models

A circuit simulator, of which SPICE is the prototype example, provides the most accurate description of system behaviour by solving for voltages currents over time. The basis for circuit simulation is Kirchoff's laws, which describe the relationship between voltages and currents. Linear elements like resistors and capacitors have constant values in Kirchoff's laws, so the equations can be solved by standard linear algebra techniques. However, transistors are non-linear greatly complicating the solution of the circuit equations. The circuit simulator uses a model — an equivalent circuit whose parameters may vary with values of other circuits voltages and currents — to represent a transistor. Unlike linear circuits, which can be solved analytically, numerical solution techniques must be used to solve non-linear circuits. The solution is generated as a sequence of points in time.

A circuit simulation is only as accurate as the model for the transistor. Most versions of SPICE offer three MOS transistor models, called naturally enough level 1, level 2, and level 3. The level 1 SPICE model gives the device equations roughly. The level 2 SPICE model provides more accurate determination of effective channel length and the transition between the linear and saturation regions, but is less frequently used today. The level 3 SPICE model uses empirical parameters to provide a better fit to the measured device characteristics. Each model requires a number of parameters, which should be supplied by the fabrication vendor. The level 4 model is also known as BSIM model and uses some extracted parameters but is smaller and more efficient. Even more recent models, including level 28 (BSIM2) and level 47 (BSIM3) have been developed recently to more accurately model deep sub-micron transistors. Table 3.3 gives the SPICE names for some common parameters of SPICE models.

Table 3.3 Names of some SPICE parameters

Parameter	*Symbol*	*SPICE name*
Channel drawn length	L	L
Channel width	W	W
Source, drain areas		AS, AD
Source, drain perimeters		PS, PD
Source, drain resistances	R_s, R_d	RS, RD
Source/Drain sheet resistance		RSH
Zero-bias bulk junction capacitance	C_{j0}	CJ
Bulk junction grading co-efficient	m	MJ
Zero-bias sidewall capacitance	C_{jsw0}	CJSW
Side wall grading co-efficient	m_{sw}	MJSW
Gate–bulk/source/drain overlap capacitances	$C_{gb0}/C_{gs0}/\ C_{gd0}$	CGBO, CGSO, CGDO
Bulk junction leakage current	I_s	IS
Bulk junction leakage current density	J_s	JS
Bulk junction potential	Φ_0	PB
Zero-bias threshold voltage	V_{T0}	VT0
Transconductance	K'	KP
Body bias factor	γ	GAMMA
Channel modulation	λ	LAMBDA
Oxide thickness	t_{ox}	TOX
Lateral diffusion	x_d	LD
Metallurgical junction depth	x_j	XJ
Surface inversion potential	$2\varphi_F$	PHI
Substrate doping	N_A, N_D	NSUB
Surface state density	Q_{ss}/q	NSS
Surface mobility	μ_0	U0
Maximum drift velocity	v_{max}	VMAX
Mobility critical field	E_{crit}	UCRIT
Critical field exponent in mobility degradation		UEXP
Type of gate material		TPG

SUMMARY

Chapter 3 is concerned with MOS transistor characteristics: static and dynamic behaviours. The static model includes threshold voltage, drain current relations for the linear and saturation modes of operations. An accurate analysis for drain current for saturated region depends on the channel-length modulation. In the linear mode, an NMOS transistor acts as a voltage-controlled resistor, while in the saturation region, it behaves like a voltage-controlled current source, if the channel-length modulation effect is ignored. The chapter describes the dynamic behaviour of the MOS transistor by the accurate analysis of the nature and behaviour of the capacitance between the device ports and from interconnecting lines. The various capacitance values of MOS structure capacitance, channel capacitance and junction capacitance in a MOS transistor are derived and explained in the dynamic behaviour. Then the chapter deals in detail with the actual MOS transistor, stating some of the secondary effects existing in it. The threshold voltage variations, the current–voltage characteristics variations, subthreshold conduction effects, CMOS latchup conditions all these effects cause the characteristics of an actual MOS transistor to deviate from an ideal transistor. The NMOS inverter circuit, its transfer characteristics, the pull-up to pull-down ratio driven by another NMOS inverter and through one or more pass transistors are derived. Finally, the chapter describes various device models such as the DC MOSFET model, high frequency MOSFET model, etc. developed to provide a functional relationship among the terminal electrical variables for the MOS transistor.

REVIEW QUESTIONS

1. Explain threshold voltage for MOS transistor.
2. Explain the dynamic behaviour of MOS transistor.
3. Explain NMOS inverter and draw the transfer characteristics of NMOS inverter.
4. Determine pull-up to pull-down ratio for an NMOS inverter.
5. Explain the device models for simulation.

SHORT ANSWER QUESTIONS

1. What is the fermi potential value for typical *p*-type silicon substrates?

   Ans. – 0.3 V.

2. What is body effect?

   Ans. The threshold voltage is not a constant (varies) with respect to the voltage difference between the substrate and the source of MOS transistor.

3. Name the part into which gate-to-channel capacitance can be decomposed.

   Ans. (a) Capacitance between gate and channel

   (b) Capacitance between gate and source

   (c) Capacitance between gate and bulk regions.

4. The threshold voltages in short-channel devices also have the tendency to ________ over time.

   Ans. drift

5. What is mobility degradation?

   Ans. The reduction in electron mobility.

4 CMOS Inverter Design

4.1 INTRODUCTION

The inverter is the most fundamental logic gate that performs a Boolean operation on a single input variable. The inverter design, therefore, forms a significant basis for digital circuit design. In complementary MOS (CMOS), both *p*- and *n*-channel transistors are used. The silicon substrate is an *n*-type, in which a *p*-type well or tub is created by diffusion. The *n*-channel transistors are created in the *n*-substrate under an ion-implanted layer called the p^+-layer. The basic structure of the *p*-well CMOS inverter is shown in Figure 4.1. The special *p*-plugs are used to connect the *p*-well substrate and the source of the *n*-channel transistor to ground. Similarly, an *n*-plug connects the *n*-substrate and the source of the *p*-type transistor to V_{DD}, a positive supply voltage. Other structures, such as guard rings to prevent formation of parasitic transistors and contact cuts, are not described in this premilinary sketch of Figure 4.1.

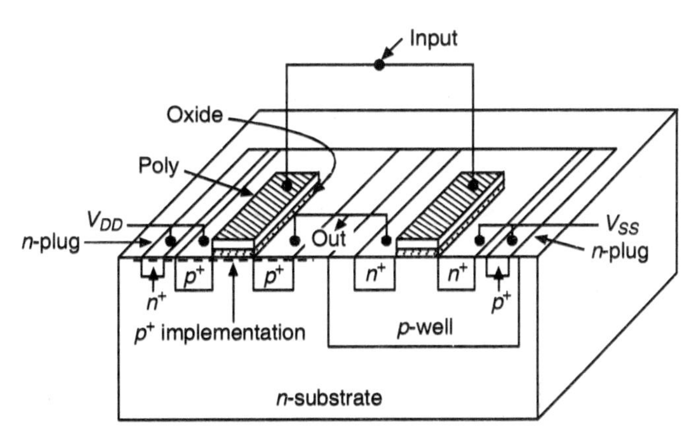

Figure 4.1 Structure of a CMOS inverter.

A complementary MOS (CMOS) inverter is realized by the series connection of *p*- and *n*-devices, as shown in Figure 4.2(a). The simplified view of the CMOS inverter, consisting of two complementary non-ideal switches is shown in Figure 4.2(b). Thus, its operation is readily understood with the aid of the simple switch model of the MOS transistor. The transistor is nothing but a switch with an infinite OFF—resistance (for $|V_{GS}| < |V_T|$) and a finite ON—resistance (for $|V_{GS}| > |V_T|$). This leads to the following interpretation of the inverter.

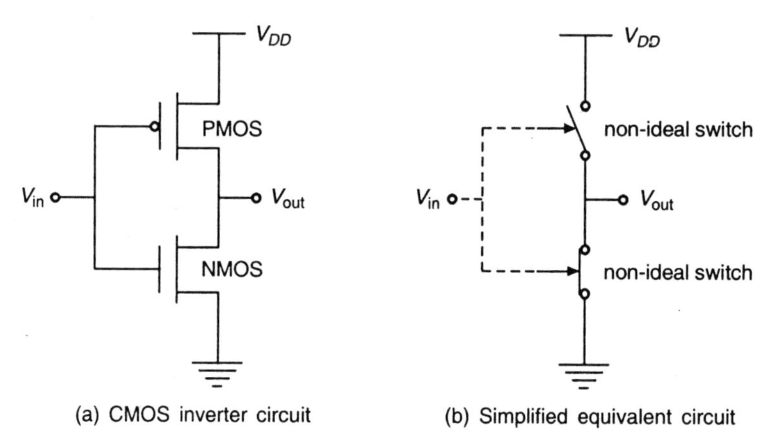

Figure 4.2 CMOS inverter.

When V_{in} is high and equal to V_{DD}, the NMOS transistor is ON and the PMOS transistor is OFF. This yields the equivalent circuit of Figure 4.3(a). A direct path exists between V_{out} and the ground node, resulting in a steady-state value of 0 V. On the other hand, when the input V_{in} in low (0 V), NMOS and PMOS transistors are OFF and ON respectively. The equivalent circuit of Figure 4.3(b) shows that a path exists between V_{DD} and V_{out}, resulting in a high output voltage. Thus, the gate functions as an inverter. It can be noticed that the NMOS transistor is ON for high input and OFF for low input, thereby operating under positive logic, while PMOS transistor is OFF for high input and ON for low input with negative logic operation.

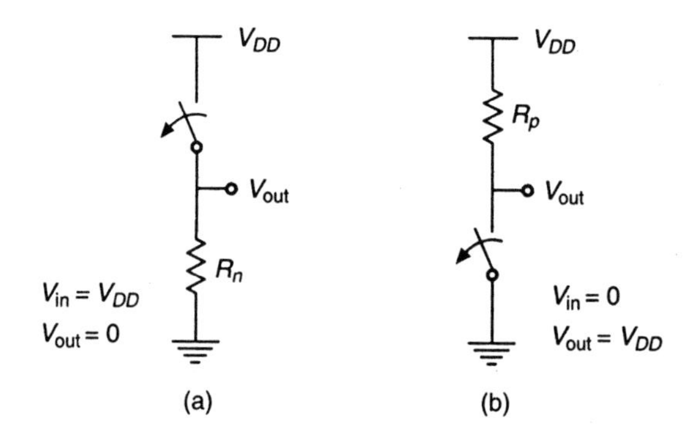

Figure 4.3 Switch models of CMOS inverter.

The CMOS inverter has two important advantages over the other inverter configurations:

- The steady state power dissipation of the CMOS inverter circuit is negligible.
- The voltage transfer characteristic (VTC) exhibits a full output voltage swing between 0 V and V_{DD}. This results in high noise margin. Also, the VTC transition is usually very sharp and hence, the CMOS inverter resembles an ideal inverter characteristic.

The CMOS inverter circuit has its own limitations also:

1. Since NMOS and PMOS transistors must be fabricated on the same chip side-by-side, the CMOS process is more complex than the standard NMOS—only process.
2. Formation of parasitic bipolar transistors due to the close proximity of NMOS and PMOS transistors causes CMOS latchup condition. Additional guard rings must be built around the NMOS and PMOS transistors to prevent this undesirable effect.

Thus, the increased process complexity of CMOS fabrication may be considered as the price being paid for the improvements achieved in power consumption and noise margins.

4.2 CMOS INVERTER—DC CHARACTERISTICS

In Figure 4.2(a), the input voltage is connected to the gate terminals of both the NMOS and the PMOS transistors. Thus, both the transistors are driven directly by the input signal, V_{in}. The substrate of the NMOS transistor should be connected to the ground, while the substrate of the PMOS transistor is to be connected to the power supply voltage, V_{DD} in order to reverse bias the source and drain junctions. Since $V_{SB} = 0$ V for both devices, there will be no substrate-bias effect for either device.

It can be seen from Figure 4.2(a) that

$$\begin{aligned} I_{DSp} &= -I_{DSn} \\ V_{GSn} &= V_{\text{in}} \\ V_{GSp} &= V_{\text{in}} - V_{DD} \\ V_{DSn} &= V_{\text{out}} \\ V_{DSp} &= V_{\text{out}} - V_{DD} \end{aligned} \tag{4.1}$$

We have seen that the current–voltage relations for the MOS transistor may be written as

$$I_D = k_n\left[(V_{GS} - V_T)\,V_{DS} - \frac{V_{DS}^2}{2}\right] \quad \text{in the linear region}$$

and

$$I_D = \frac{k_n}{2}(V_{GS} - V_T)^2 \quad \text{in the saturation region}$$

Let V_{Tn} be the threshold voltage of the n-channel device.

Let V_{Tp} be the threshold voltage of the p-channel device.

We begin with the graphical representation of the simple algebraic equations of drain currents for the two inverter transistors as shown in Figure 4.4(a). The absolute value of the p-transistor drain current, I_D inverts this characteristic. This allows the I-V characteristics for the device to be reflected about the X-axis [Figure 4.4(b)]. This is followed by taking the absolute value of the p-device voltage, V_{DS}, and superimposing the two characteristics, yielding resultant curves shown in Figure 4.4(c). The steady state input–output voltage characteristics of the CMOS inverter can be better visualized by considering the intersection of individual NMOS and PMOS transistor characteristics in the current–voltage space (Figure 4.5).

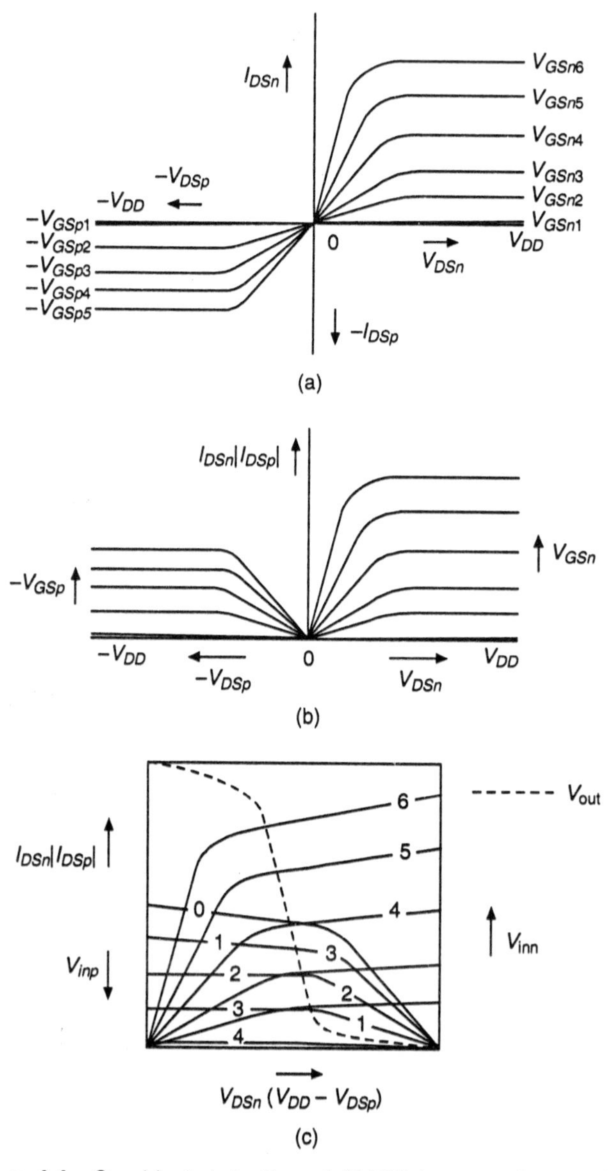

Figure 4.4 Graphical derivation of CMOS inverter characteristics.

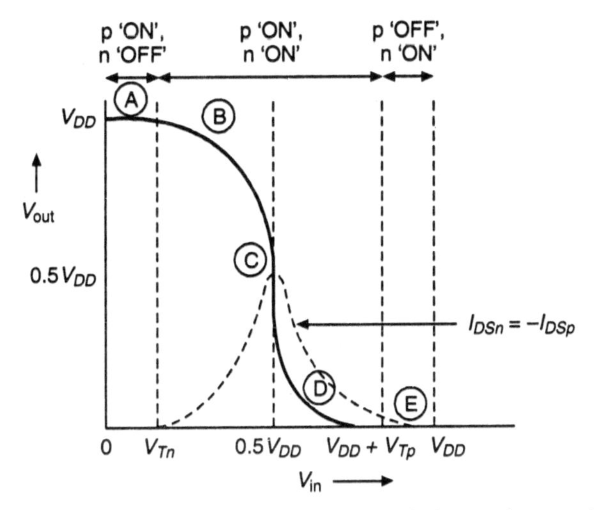

Figure 4.5 CMOS inverter dc transfer characteristics and operating regions.

The operation of the CMOS inverter can be divided into five regions (Figure 4.5). The behaviour of *n*- and *p*-devices in each of the regions may be found by using Table 4.1.

Table 4.1 Behaviour of *n*- and *p*-transistors in the five regions of CMOS inverter.

Region	V_{in}	V_{out}	*NMOS*	*PMOS*
A	$< V_{Tn}$	V_{OH}	Cut-off	Linear
B	V_{IL}	High $\simeq V_{OH}$	Saturation	Linear
C	V_{TH}	V_{TH}	Saturation	Saturation
D	V_{IH}	Low $\simeq V_{OL}$	Linear	Saturation
E	$> V_{DD} + V_{Tp}$	V_{OL}	Linear	Cut-off

The expression for output voltage (V_{out}) in terms of the input voltage (V_{in}) can be derived for the different operating regions as below:

Region A. This region is defined by $0 \leq V_{in} \leq V_{Tn}$ in which the *n*-device is cut-off ($I_{DSn} = 0$), and the *p*-device is in the linear region.

We have, from Eq. (4.1),

$$I_{DSn} = -I_{DSp};$$

Since

$$I_{DSn} = 0,$$

$$\Rightarrow \qquad I_{DSp} = 0 \tag{4.2}$$

Also, $V_{DSp} = V_{out} - V_{DD}$,

From Eq. (4.2), the drain-to-source current for *p*-device is zero and hence the voltage drop from drain to source for *p*-device is also zero ($V_{GSp} = 0$).

Therefore, the above expression becomes

$$V_{out} - V_{DD} = 0$$

$$\Rightarrow \qquad V_{out} = V_{DD} \tag{4.3}$$

Region B. This region is characterized by $V_{Tn} \leq V_{in} < (V_{DD}/2)$ in which the *p*-device is in its non-saturated or linear region ($V_{DS} \neq 0$) while the *n*-device is in saturation. The equivalent circuit for the inverter in this region can be represented by a resistor for the *p*-transistor and a current source for the *n*-transistor as shown by Figure 4.6.

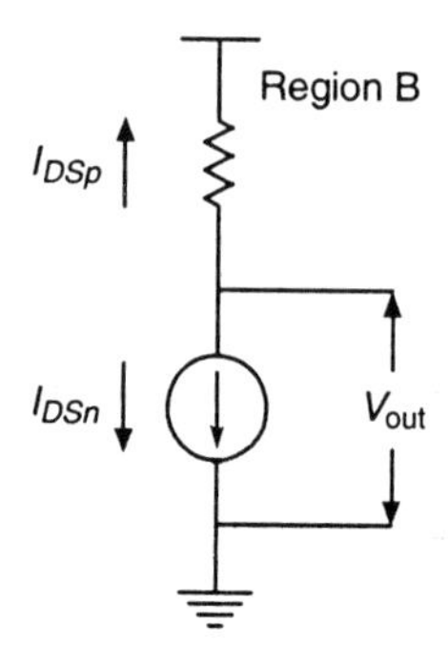

Figure 4.6 Equivalent circuit for region B of a CMOS inverter.

The saturation current for the *n*-device, I_{DSn} is obtained by setting $V_{GSn} = V_{in}$ in Eq. (3.15).

$$I_{DSn} = \frac{k_n}{2}(V_{in} - V_{Tn})^2 \tag{4.4}$$

where, $k_n = \mu_n C_{ox} W_n/L_n$

$V_{Tn} \rightarrow$ threshold voltage for the *n*-device
$\mu_n \rightarrow$ mobility of electrons
$W_n \rightarrow$ channel width of *n*-device
$L_n \rightarrow$ channel length of *n*-device

The drain current for the *p*-device can be obtained by substituting in Eq. (3.13)

$$V_{GSp} = V_{in} - V_{DD}$$

and

$$V_{DSp} = V_{out} - V_{DD}$$

Therefore,

$$I_{DSp} = -k_p \left[(V_{in} - V_{DD} - V_{Tp})(V_{out} - V_{DD}) - \frac{(V_{out} - V_{DD})^2}{2} \right] \tag{4.5}$$

where, $k_p = \mu_p C_{ox} W_p/L_p$

$V_{Tp} \rightarrow$ threshold voltage of *p*-device
$\mu_p \rightarrow$ mobility of holes
$W_p \rightarrow$ channel width of *p*-device
$L_p \rightarrow$ channel length of *p*-device

We have, from Eq. (4.1),

$$I_{DSn} = -I_{DSp}$$

Substituting Eqs. (4.4) and (4.5) and simplifying, the above expression becomes

$$V_{\text{out}} = (V_{\text{in}} - V_{Tp}) + \sqrt{\left[(V_{\text{in}} - V_{Tp})^2 - \frac{k_n}{k_p}(V_{\text{in}} - V_{Tp})^2 - 2\left(V_{\text{in}} - \frac{V_{DD}}{2} - V_{Tp}\right)V_{DD}\right]} \tag{4.6}$$

Region C. In this region, both the n- and p-devices are in saturation. This is represented by the schematic in Figure 4.7 which shows two current sources in series.

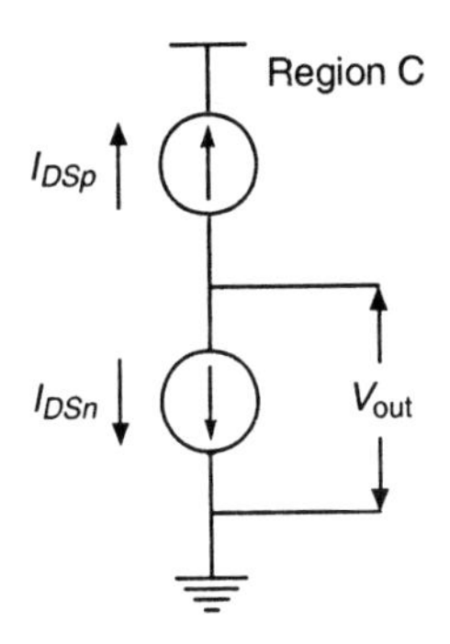

Figure 4.7 Equivalent circuit for region C of a CMOS inverter.

The saturation currents for the two devices are given by substituting V_{GS} values from Eq. (4.1).

$$I_{DSp} = -\frac{k_p}{2}(V_{\text{in}} - V_{DD} - V_{Tp})^2 \tag{4.7}$$

$$I_{DSn} = \frac{k_n}{2}(V_{\text{in}} - V_{Tn})^2 \tag{4.8}$$

Since $I_{DSn} = -I_{DSp}$,

$$\frac{k_n}{2}(V_{\text{in}} - V_{Tn})^2 = \frac{k_p}{2}(V_{\text{in}} - V_{DD} - V_{Tp})^2$$

Substituting, $k_n = k_p$ and $V_{Tn} = -V_{Tp}$;

$$(V_{\text{in}} - V_{Tn})^2 = (V_{\text{in}} - V_{DD} + V_{Tn})^2$$

Taking the square root on both sides,

$$V_{\text{in}} - V_{Tn} = -(V_{\text{in}} - V_{DD} + V_{Tn})$$

$$\Rightarrow \qquad 2V_{\text{in}} = V_{Tn} + V_{DD} - V_{Tn}$$

$$\Rightarrow \qquad V_{\text{in}} = V_{DD}/2 \tag{4.9}$$

This implies that region C exists only for one value of V_{in}. It is seen from Figure 4.5 that the inverter threshold, where $V_{\text{in}} = V_{\text{out}}$, is located in region C.

Region D. The region D is described by $V_{DD}/2 < V_{\text{in}} \leq V_{DD} + V_{Tp}$. Here, the p-device is in saturation while the n-device is operating in its linear region. This condition is represented by the equivalent circuit as in Figure 4.8.

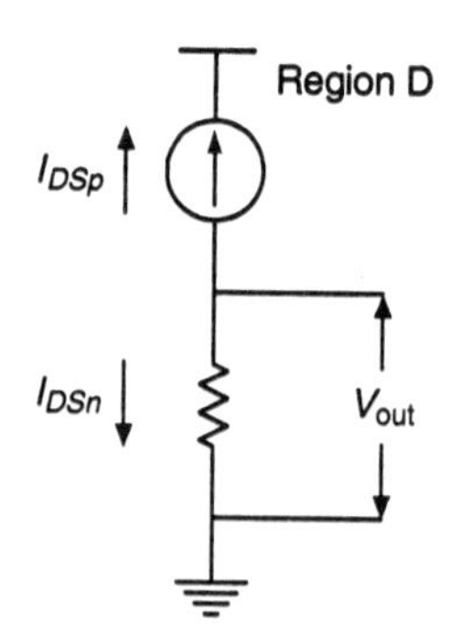

Figure 4.8 Equivalent circuit for region D of a CMOS inverter.

The saturation current for the p-device is obtained as

$$\frac{I_{DSp}}{2} = -k_p (V_{in} - V_{DD} - V_{Tp})^2 \tag{4.10}$$

and the drain current for the linear region of operation of n-device is given by

$$I_{DSn} = k_n \left[(V_{in} - V_{Tn}) V_{out} - \frac{V_{out}^2}{2} \right] \tag{4.11}$$

We have the condition that

$$I_{DSp} = -I_{DSn}$$

By putting Eqs. (4.10) and (4.11) in this condition,

$$\frac{-k_p}{2} (V_{in} - V_{DD} - V_{Tp})^2 = -k_n \left[(V_{in} - V_{Tn}) V_{out} - \frac{V_{out}^2}{2} \right]$$

$$\Rightarrow \qquad \frac{V_{out}^2}{2} - V_{out}(V_{in} - V_{Tn}) + \frac{k_p}{2k_n} (V_{in} - V_{DD} - V_{Tp})^2 = 0$$

Solving for V_{out},

$$V_{out} = (V_{in} - V_{Tn}) - \sqrt{(V_{in} - V_{Tn})^2 - \frac{k_p}{k_n} (V_{in} - V_{DD} - V_{Tp})^2} \tag{4.12}$$

Region E. This region is defined by the input condition $V_{in} \geq V_{DD} - V_{Tp}$ in which the p-device is cut-off ($I_{DSp} = 0$), and the n-device is in the linear mode.
Here, $V_{GSp} \cong V_{in} - V_{DD}$ which is more positive than V_{Tp}

Since $I_{DSp} = 0$, $I_{DSn} = 0$ (as $I_{DSn} = -I_{DSp}$)

$$\Rightarrow \qquad V_{DSn} = 0$$

$$\Rightarrow \qquad V_{out} = 0 \tag{4.13}$$

From Figure 4.5, it may be seen that the transition between the two states is very steep. This characteristic is very desirable because the noise immunity is maximized. Figure 4.5 can be redrawn as below in Figure 4.9 which clearly shows the operating regions of the p- and n-devices.

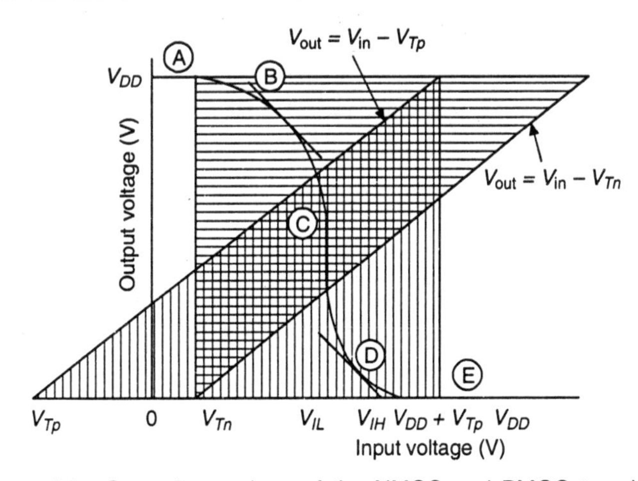

Figure 4.9 Operating regions of the NMOS and PMOS transistor.

We will present an in-depth analysis of the CMOS inverter static characteristics, by calculating the critical voltage points on the VTC. It has already been established that $V_{OH} = V_{DD}$ and $V_{OL} = 0$ V for this inverter, thus, we want to calculate V_{IL}, V_{IH} and the inverter switching threshold, V_{TH}.

Calculation of V_{IL}. By definition, the slope of the Voltage Transfer Characteristic (VTC) is equals (–1), i.e, $(dV_{out}/dV_{in}) = -1$ when the input voltage is $V_{in} = V_{IL}$. In this region, the NMOS transistor operates in the linear region, i.e., Region B of operation is obtained, for the CMOS inverter.

We have,
$$I_{DSn} = -I_{DSp}$$

By substituting the respective drain current expressions for *n*-device and *p*-device in the saturation and linear/resistive regions respectively, we get,

$$\frac{k_n}{2}(V_{GSn} - V_{Tn})^2 = k_p\left[(V_{GSp} - V_{Tp})\,V_{DSp} - \frac{V_{DSp}^2}{2}\right]$$

By making

$$V_{GSn} = V_{in}$$
$$V_{GSp} = V_{in} - V_{DD}$$
$$V_{DSp} = V_{out} - V_{DD}$$

We get,

$$\frac{k_n}{2}(V_{in} - V_{Tn})^2 = \frac{k_p}{2}[2(V_{in} - V_{DD} - V_{Tp})(V_{out} - V_{DD}) - (V_{out} - V_{DD})^2]$$

$$\Rightarrow \quad k_n(V_{in} - V_{Tn})^2 = k_p[2(V_{in} - V_{DD} - V_{Tp})(V_{out} - V_{DD}) - (V_{out} - V_{DD})^2] \tag{4.14}$$

Differentiating on both sides with respect to V_{in},

$$k_n 2(V_{in} - V_{Tn}) = k_p \left[2\left\{ (V_{in} - V_{DD} - V_{Tp}) \frac{dV_{out}}{dV_{in}} + (V_{out} - V_{DD}) \right\} - 2(V_{out} - V_{DD}) \frac{dV_{out}}{dV_{in}} \right]$$

When $V_{in} = V_{IL}$, $\frac{dV_{out}}{dV_{in}} = -1$.

$$\Rightarrow \quad k_n (V_{IL} - V_{Tn}) = k_p [(V_{IL} - V_{DD} - V_{Tp})(-1) + V_{out} - V_{DD} - (V_{out} - V_{DD})(-1)]$$

$$\Rightarrow \quad k_n (V_{IL} - V_{Tn}) = k_p (2V_{out} - V_{DD} + V_{Tp} - V_{IL})$$

$$\Rightarrow \quad \frac{k_n}{k_p} (V_{IL} - V_{Tn}) = 2V_{out} - V_{DD} + V_{Tp} - V_{IL}$$

$$\Rightarrow \quad V_{IL}\left(1 + \frac{k_n}{k_p}\right) = \frac{k_n}{k_p} V_{Tn} + 2V_{out} - V_{DD} + V_{Tp}$$

Let $K_R = k_n/k_p$ = Transconductance ratio

$$\Rightarrow \quad V_{IL}(1 + K_R) = K_R V_{Tn} + 2V_{out} - V_{DD} + V_{Tp}$$

$$\Rightarrow \quad V_{IL} = \frac{2V_{out} - V_{DD} + V_{Tp} + K_R V_{Tn}}{1 + K_R} \tag{4.15}$$

This equation must be solved together with the KCL Eq. (4.14) to obtain the numerical value of V_{IL} and the corresponding output voltage V_{out}.

Calculation of V_{IH}. When the input voltage is equal to V_{IH}, the NMOS transistor operates the linear region and the PMOS transistor operates in saturation, i.e., region D of operation of CMOS inverter is obtained.

When $V_{in} = V_{IH}$ also, $dV_{out}/dV_{in} = -1$

By substituting the drain currents for the *n*-device and *p*-device in the linear and saturation region respectively, the equation $I_{DSn} = -I_{DSp}$ becomes,

$$k_n \left[(V_{GSn} - V_{Tn}) V_{DSn} - \frac{V_{DSn}^2}{2} \right] = \frac{k_p}{2} (V_{GSp} - V_{Tp})^2$$

Making

$$V_{GSn} = V_{in},$$

$$V_{DSn} = V_{out},$$

$$V_{GSp} = V_{in} - V_{DD},$$

We have,

$$k_n \left[(V_{in} - V_{Tn}) V_{out} - \frac{V_{out}^2}{2} \right] = \frac{k_p}{2} (V_{in} - V_{DD} - V_{Tp})^2$$

$$\Rightarrow \quad \frac{k_n}{2} [2(V_{in} - V_{Tn}) V_{out} - V_{out}^2] = \frac{k_p}{2} (V_{in} - V_{DD} - V_{Tp})^2$$

$$\Rightarrow \quad k_n [2(V_{in} - V_{Tn}) V_{out} - V_{out}^2] = k_p (V_{in} - V_{DD} - V_{Tp})^2 \tag{4.16}$$

Differentiating with respect to V_{in}

$$k_n\left[2\left((V_{\text{in}}-V_{Tn})\frac{dV_{\text{out}}}{dV_{\text{in}}}+V_{\text{out}}\right)-2V_{\text{out}}\frac{dV_{\text{out}}}{dV_{\text{in}}}\right]=2k_p(V_{\text{in}}-V_{DD}-V_{Tp})$$

$$\Rightarrow \qquad k_n\left[\left((V_{\text{in}}-V_{Tn})\frac{dV_{\text{out}}}{dV_{\text{in}}}+V_{\text{out}}\right)-V_{\text{out}}\frac{dV_{\text{out}}}{dV_{\text{in}}}\right]=k_p(V_{\text{in}}-V_{DD}-V_{Tp})$$

When $V_{\text{in}} = V_{IH}$, $dV_{\text{out}}/dV_{\text{in}} = -1$.

$$\Rightarrow \qquad k_n[(V_{IH} - V_{Tn})(-1) + V_{\text{out}} - V_{\text{out}}(-1)] = k_p(V_{IH} - V_{DD} - V_{Tp})$$

$$\Rightarrow \qquad k_n(-V_{IH} + V_{Tn} + 2V_{\text{out}}) = k_p(V_{IH} - V_{DD} - V_{Tp})$$

$$\Rightarrow \qquad (V_{IH} - V_{DD} - V_{Tp}) = \frac{k_n}{k_p}(-V_{IH} + V_{Tn} + 2V_{\text{out}})$$

$$\Rightarrow \qquad V_{IH}\left(1+\frac{k_n}{k_p}\right)=V_{DD}+V_{Tp}+\frac{k_n}{k_p}(V_{Tn}+2V_{\text{out}})$$

$$\Rightarrow \qquad V_{IH}=\frac{V_{DD}+V_{Tp}+(k_n/k_p)(V_{Tn}+2V_{\text{out}})}{1+(k_n/k_p)}$$

Taking $k_n/k_p = K_R$ = Transconductance ratio,

$$V_{IH}=\frac{V_{DD}+V_{Tp}+K_R(V_{Tn}+2V_{\text{out}})}{1+K_R} \tag{4.17}$$

Again, this equation must be solved simultaneously with the KCL Eq. (4.16) to obtain the numerical values of V_{IH} and V_{out}.

Calculation of V_{TH}. The inverter threshold voltage is defined as $V_{TH} = V_{\text{in}} = V_{\text{out}}$. Since the CMOS inverter exhibits large noise margins and a very sharp VTC transition, the inverter threshold voltage emerges as an important parameter characterizing the dc performance of the inverter. The inverter threshold voltage is also called the switching point of CMOS inverter or switching threshold.

For $V_{\text{in}} = V_{\text{out}}$, both transistors are in saturation mode, i.e., region C of operation is reached by the inverter.

Now, by substituting the respective drain currents of *p*- and *n*-devices in the equation $I_{DSn} = -I_{DSp}$, we get,

$$\frac{k_n}{2}(V_{GSn}-V_{Tn})^2=\frac{k_p}{2}(V_{GSp}-V_{Tp})^2$$

$$\Rightarrow \qquad k_n(V_{\text{in}} - V_{Tn})^2 = k_p(V_{\text{in}} - V_{DD} - V_{Tp})^2$$

$$\Rightarrow \qquad (V_{\text{in}} - V_{Tn})^2 = \frac{k_p}{k_n}(V_{\text{in}} - V_{DD} - V_{Tp})^2 \tag{4.18}$$

The correct solution for Eq. (4.18) is obtained as below:

$$V_{\text{in}} - V_{Tn} = -\sqrt{\frac{k_p}{k_n}}(V_{\text{in}} - V_{DD} - V_{Tp})$$

$$\Rightarrow \qquad V_{\text{in}}\left(1+\sqrt{\frac{k_p}{k_n}}\right) = V_{Tn} + \sqrt{\frac{k_p}{k_n}}(V_{DD} + V_{Tp})$$

$$\Rightarrow \qquad V_{\text{in}} = \frac{V_{Tn} + \sqrt{(k_p/k_n)}(V_{DD} + V_{Tp})}{1+\sqrt{(k_p/k_n)}}$$

At the switching point, $V_{\text{in}} = V_{\text{out}} = V_{TH}$.

$$V_{TH} = \frac{V_{Tn} + \sqrt{(1/K_R)}(V_{DD} + V_{Tp})}{1+\sqrt{(1/K_R)}} \quad \text{(taking } k_n/k_p = K_R\text{)} \tag{4.19}$$

Figure 4.10 shows the variation of the inversion (switching) threshold V_{TH} as a function of the transconductance ratio, K_R, and for fixed values of V_{DD}, V_{Tn} and V_{Tp}.

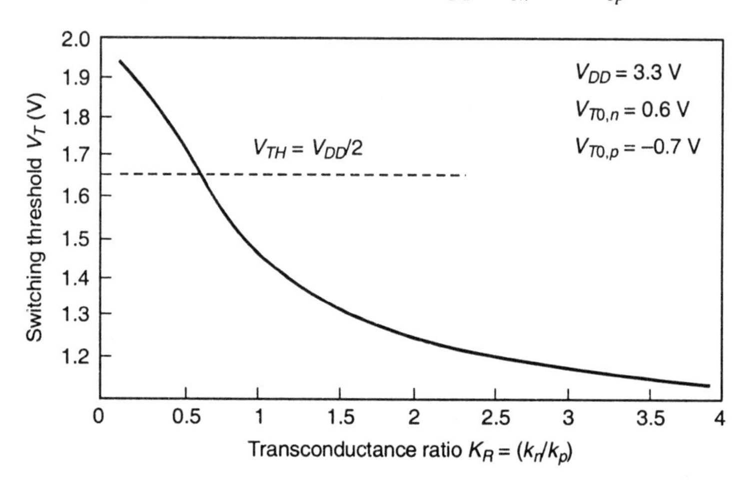

Figure 4.10 Variation of the inversion threshold voltage as a function of K_R.

The inverter threshold voltage is defined as $V_{TH} = V_{\text{in}} = V_{\text{out}}$. However, when the input voltage is equal to V_{TH}, we find that the output voltage can actually attain any value between $(V_{TH} - V_{Tn})$ and $(V_{TH} - V_{Tp})$, without violating the voltage conditions used in this analysis. This is due to the fact that the VTC segment corresponding to Region C in Figure 4.8 becomes completely vertical if the channel-length modulation effect is neglected, i.e., if $\lambda = 0$. In more realistic cases, with $\lambda = 0$, the VTC segment in Region C exhibits a finite, but very large, slope.

It has already been established that the CMOS inverter does not draw any significant current from the power source, except for small leakage and subthreshold currents, when the input voltage is either smaller than V_{Tn} or larger than $(V_{DD} + V_{Tp})$. The NMOS and the PMOS transistors conduct a non-zero current, on the other hand, during low-to-high and high-to-low transitions, i.e., in regions B, C and D. It can be shown that the current being drawn from the power source during transition reaches its peak value when $V_{in} = V_{TH}$. In other words, the maximum current is drawn when both transistors are operating in saturation mode. Figure 4.11 shows the VTC of a typical CMOS inverter circuit and the power supply current, as a function of the input voltage.

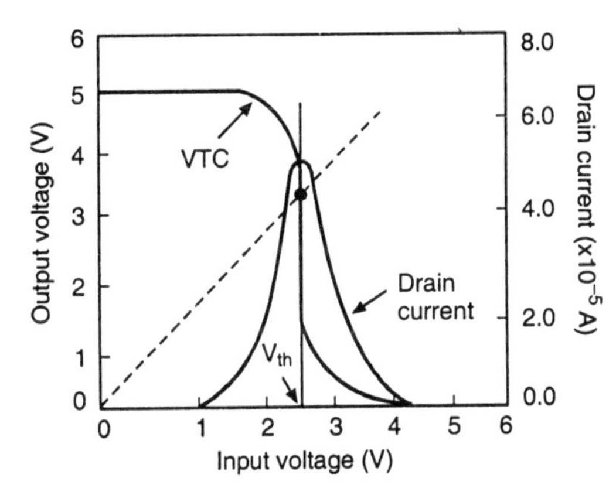

Figure 4.11 Typical VTC and the power supply current of a CMOS inverter circuit.

4.3 DESIGN PARAMETERS OF CMOS INVERTER

The inverter threshold voltage V_{TH} is identified as one of the most important parameters that characterizes the steady-state input-output behaviour of the CMOS inverter circuit. The CMOS inverter can provide a full output voltage swing between 0 and V_{DD}, and, therefore, the noise margins are relatively wide. Thus, the problem of designing a CMOS inverter can be reduced to setting the inverter threshold to a desired voltage value.

Given the power supply voltage, V_{DD}, the NMOS and the PMOS transistor threshold voltages, and the desired inverter threshold voltage, V_{TH}, the corresponding ratio K_R can be found as follows:
From Eq. (4.19), we have,

$$V_{TH} = \frac{V_{Tn} + \sqrt{(1/K_R)}(V_{DD} + V_{Tp})}{1 + \sqrt{(1/K_R)}}$$

Reorganizing,

$$\frac{1}{K_R} = \frac{V_{TH} - V_{Tn}}{V_{DD} + V_{Tp} - V_{TH}} \tag{4.20}$$

Solving for K_R,

$$K_R = \left(\frac{V_{DD} + V_{Tp} - V_{TH}}{V_{TH} - V_{Tn}}\right)^2 \tag{4.21}$$

Recalling that the switching threshold voltage of an ideal inverter is defined as

$$V_{TH,\ \text{ideal}} = \frac{1}{2} V_{DD} \tag{4.22}$$

Substituting Eq. (4.22) in Eq. (4.21),

$$K_{R,\ \text{ideal}} = \frac{k_n}{k_p} = \left(\frac{0.5V_{DD} + V_{Tp}}{0.5V_{DD} - V_{Tn}} \right)^2 \tag{4.23}$$

Since the operations of the NMOS and the PMOS transistors of the CMOS inverter are fully complementary, we can achieve completely symmetric input–output characteristics by setting

$$V_{Tn} = |V_{Tp}|$$

Now, Eq. (4.23) reduces to

$$\left(\frac{k_n}{k_p} \right)_{\text{symmetric inverter}} = 1 \tag{4.24}$$

i.e.,

$$\frac{\mu_n C_{ox} (W/L)_n}{\mu_p C_{ox} (W/L)_p} = 1$$

$\Rightarrow$

$$\frac{\mu_n}{\mu_p} \frac{(W/L)_n}{(W/L)_p} = 1$$

$\Rightarrow$

$$\frac{(W/L)_n}{(W/L)_p} = \frac{\mu_p}{\mu_n} \tag{4.25}$$

The typical values of electron and hole mobilities are given by

$$\mu_p = 580 \text{ cm}^2\text{/V-sec}$$

$$\mu_n = 230 \text{ cm}^2\text{/V-sec}$$

Hence,

$$\left(\frac{W}{L} \right)_p \simeq 2.5 \left(\frac{W}{L} \right)_n \tag{4.26}$$

The VTCs of three CMOS inverter circuits with different K_R ratios are shown in Figure 4.12. It can be clearly seen that the inverter threshold voltage V_{TH}, shifts to lower values with increasing K_R ratio.

4.3.1 Symmetric CMOS Inverter

For a symmetric CMOS inverter with $V_{Tn} = |V_{Tp}|$, $K_R = 1$. Therefore, the input-low level voltage is found from Eq. (4.15) as below:

$$V_{IL} = \frac{1}{8}(3V_{DD} + 2V_{Tn}) \tag{4.27}$$

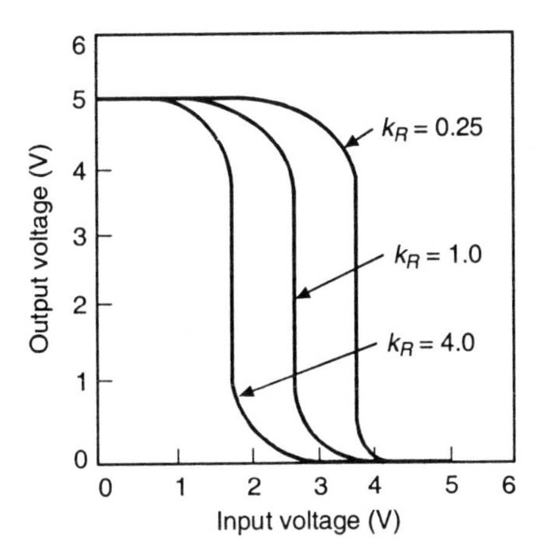

Figure 4.12 VTCs of CMOS inverters with different K_R ratios.

Similarly, the input-high level voltage is obtained from Eq. (4.17) as follows:

$$V_{IH} = \frac{1}{8}(5V_{DD} - 2V_{Tn}) \tag{4.28}$$

The sum of V_{IL} and V_{IH} is always equal to V_{DD} in a symmetric inverter.

i.e.,

$$V_{IL} + V_{IH} = V_{DD} \tag{4.29}$$

4.3.2 Noise Margins of CMOS Inverter

The noise margin represents the amount of noise voltage that can be present in the circuit without affecting the correct normal operation of the circuit. The noise margin is denoted by NM. The noise tolerances or noise immunity of the digital circuits increases with NM. The two noise margins will be defined: the noise margin for low signal levels (NM_L) and the noise margin for high signal levels (NM_H). Figure 4.13 shows the noise margin definitions.

$$NM_L = V_{IL} - V_{OL} \tag{4.30}$$

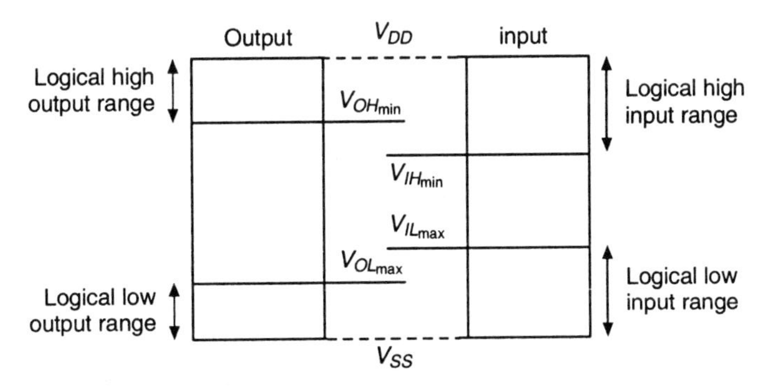

Figure 4.13 Noise margin definitions.

For CMOS inverter,

$$V_{OL} = 0 \text{ V}$$

Hence, Eq. (4.30) becomes,

$$\text{NM}_L = V_{IL} \tag{4.31}$$

Also,

$$\text{NM}_H = V_{OH} - V_{IH} \tag{4.32}$$

For CMOS inverter,

$$V_{OH} = V_{DD} = \text{Supply voltage}$$

Hence, Eq. (4.32) becomes,

$$\text{NM}_H = V_{DD} - V_{IH} \tag{4.33}$$

Thus, from Eqs. (4.31) and (4.33), we see that the low and high noise margins can be calculated by knowing the critical voltages V_{IL} and V_{IH}. Equations (4.31) and (4.33) also represent the design equations for CMOS inverter.

4.3.3 Temperature Dependence of VTC of CMOS Inverter

Temperature also has an effect on the transfer characteristics of an inverter. As the temperature of an MOS device is increased, the effective carrier mobility, μ decreases. This results in decrease in K_n value, which is related to temperature T by

$$K_n \propto T^{-1.5} \tag{4.34}$$

Hence,

$$I_{DS} \propto T^{-1.5} \tag{4.35}$$

Both V_{Tn} and V_{Tp} decrease slightly as temperature increases, and the extent of region A is reduced while the extent of region E is increased. Thus, the overall transfer characteristics shift to the left as temperature increases. It is found that if the temperature rises by 50°C, the thresholds drop by 200 mV each. This would cause a 0.2 V shift in the input threshold of the inverter.

4.3.4 Supply Voltage Scaling in CMOS Inverters

The overall power dissipation of any digital circuit is a strong function of the supply voltage V_{DD}. The reduction (or scaling) of the power supply voltage emerges as one of the most widely practiced measures for low-power design. Here, we examine the effects of supply voltage scaling, i.e., reduction of V_{DD}, upon the static voltage transfer characteristics of CMOS inverters.

Neglecting the second-order effects such as sub-threshold conduction, it can be seen that the CMOS inverter will continue to operate correctly with a supply voltage which is as low as the following limited value:

$$V_{DD}^{\min} = V_{Tn} + |V_{Tp}|$$

This means that the correct inverter operation will be sustained if at least one of the transistors remains in conduction, for any given input voltage. Figure 4.14 shows the VTCs of a CMOS inverter, obtained with different supply voltage levels.

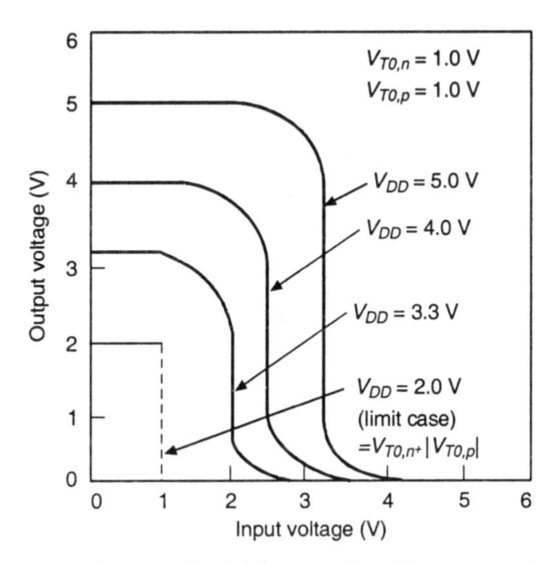

Figure 4.14 VTCs of a CMOS inverter for different supply voltages.

If the supply voltage of the CMOS inverter is reduced below the sum of this two threshold voltages, the VTC will contain a region in which none of the transistors is conducting. The output voltage level within this region is then determined by the previous state of the output, since the previous output level is always preserved as stored charge at the output node. Thus, the VTC exhibits a hysteresis behaviour for very low supply voltage levels (Figure 4.15).

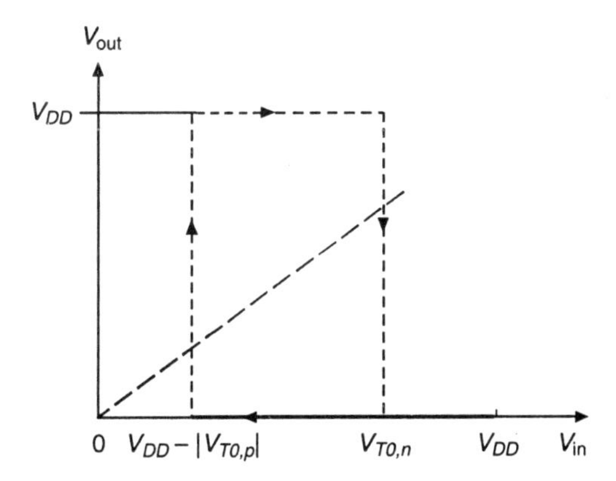

Figure 4.15 Hysteresis behaviour exhibited by VTC of CMOS inverter.

4.3.5 Power and Area Considerations

Since the CMOS inverter does not draw any significant current from the power source in both of its steady-state operating points ($V_{out} = V_{OH}$ and $V_{out} = V_{OL}$), the DC power dissipation of this circuit is almost negligible. This unique property of the CMOS inverter is identified as one of the most important advantages of this configuration. However, it must be noted that the CMOS inverter does conduct a significant amount of current during a switching event, i.e., when the output voltage changes from a low-to-high state, or from a high-to-low state.

Considering the area occupied by the CMOS inverter, it is significantly less compared to other inverter layouts. However, because of the complementary nature of the circuit, CMOS requires more transistors than that for NMOS circuits. Consequently, CMOS logic circuits tend to occupy more area than comparable NMOS logic circuits, which affects the integration density of pure CMOS logic. On the other hand, the actual integration density of NMOS logic is limited by power dissipation and heat generation problems.

4.4 SWITCHING CHARACTERISTICS OF CMOS INVERTER

The switching speed of a CMOS gate is limited by the time taken to charge and discharge the load capacitance C_L. An input transition results in an output transition that either charges C_L toward V_{DD} or discharge C_L toward V_{ss} (Figure 4.16).

Figure 4.16(a) shows the familiar CMOS inverter with a capacitive load, C_L. Figures 4.16(b) shows the output voltage waveform, $V_{\text{out}}(t)$, when the input is driven by a step waveform, $V_{\text{in}}(t)$.

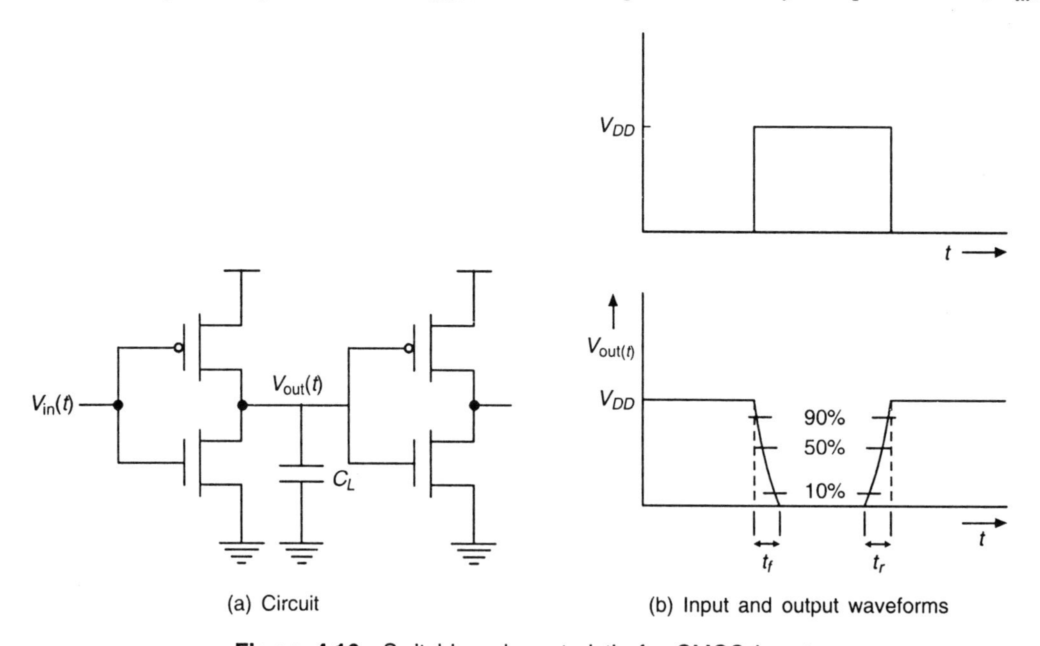

Figure 4.16 Switching characteristic for CMOS inverter.

Definitions

- Rise time, τ_r = time taken for a waveform to rise from 10% to 90% of its steady-state value.
- Fall time, τ_f = time taken for a waveform to fall from 90% to 10% of its steady-state value.
- Delay time, τ_d = time difference between input transition (50%) and the 50% output level. This is the time taken for a logic transition to pass from input to output.

4.4.1 Estimation of CMOS Inverter Delay

A CMOS inverter, in general, either charges or discharges a capacitive load C_L and rise-time τ_r, or fall-time τ_f, can be estimated from the following analysis:

Rise-time estimation. In this analysis, we assume that the p-device stays in saturation for the entire charging period of the load capacitor C_L. The circuit may then be modelled as in Figure 4.17.

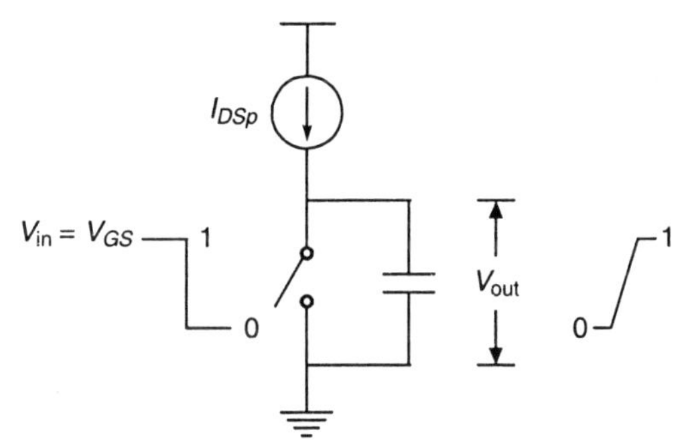

Figure 4.17 Rise-time model of a CMOS inverter.

The saturation current for the p-transistor is given by

$$I_{DSp} = \frac{K_p}{2}(V_{GS} - V_{Tp})^2$$

This current charges C_L and its magnitude is approximately constant.

Now,
$$V_{out} = \frac{I_{DSp}t}{C_L}$$

Substituting for I_{DSp} and rearranging,

We have,
$$t = \frac{2C_L V_{out}}{K_p(V_{GS} - V_{Tp})^2}$$

Assume that $t = \tau_r$ when $V_{out} = +V_{DD}$

Hence,
$$\tau_r = \frac{2V_{DD}C_L}{K_p(V_{GS} - V_{Tp})^2}$$

With $V_{Tp} = 0.2V_{DD}$,

$$\tau_r = \frac{3C_L}{K_p V_{DD}} \tag{4.36}$$

Equation (4.36) gives the rise-time expression for a CMOS inverter.

Fall-time estimation. By applying similar reasoning to the discharge of C_L through the n-transistor, the fall-time may be written as

$$\tau_f = \frac{3C_L}{K_n V_{DD}} \tag{4.37}$$

The circuit model for fall-time during the discharge of C_L is shown in Figure 4.18.

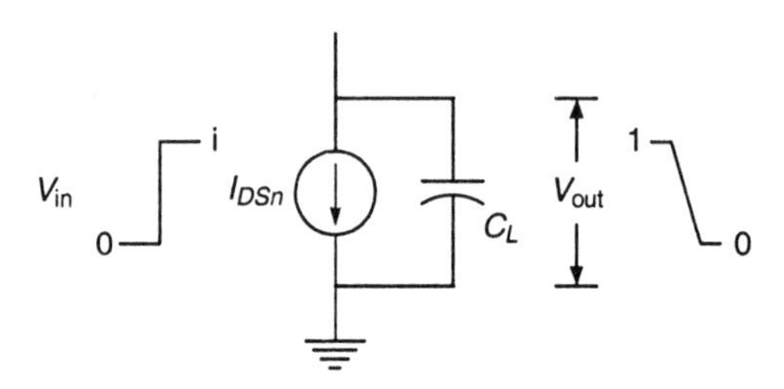

Figure 4.18 Fall-time model of a CMOS inverter.

From Eqs. (4.36) and (4.37), it can be deduced that

$$\frac{\tau_r}{\tau_f} = \frac{k_n}{k_p} \tag{4.38}$$

From Eqs. (4.36) and (4.37), it is observed that the rise time τ_r and the fall time, τ_f depends on the load capacitance C_L, the supply voltage V_{DD}, and the transconductance parameter, k_n (or k_p), i.e.,

- the delay is directly proportional to the load capacitance C_L. Thus to achieve high speed circuits one has to minimize the load capacitance seen by a gate.
- Secondly, it is inversely proportional to supply voltage, i.e., as the supply voltage is raised, the delay time is reduced, thereby increasing the speed of gates in that circuit.
- Finally, the delay is inversely proportional to the k of the driving transistor. The value of k (k_n or k_p) depends on W/L of the transistor.

So, as the transistor width W, is increased or the length L, is decreased, the value of k is increased, the value of k is increased, which in turn, reduces the delay for that transistor.

These three 'knobs' form the major basis by which the CMOS designer can optimize the speed of CMOS logic gates.

Example 4.1 Consider a CMOS inverter circuit with the following parameters:

$$V_{DD} = 3.3 \text{ V}$$
$$V_{Tn} = 0.6 \text{ V}$$
$$V_{Tp} = -0.7 \text{ V}$$
$$k_n = 200 \ \mu\text{A/V}^2$$
$$k_p = 80 \ \mu\text{A/V}^2$$

Calculate the noise margins of the circuit.

The CMOS inverter being considered here has $K_R = 2.5$ and $V_{Tn} \neq |V_{Tp}|$. Hence it is not a symmetric inverter.

To find V_{OL}:

We know that the output low level voltage, $V_{OL} = 0$ V (i)

To find V_{OH}:

Also, the output high level voltage is given by,

$$V_{OH} = V_{DD} = 3.3 \text{ V} \tag{ii}$$

To find V_{IL}:

We have,

$$V_{IL} = \frac{2V_{\text{out}} + V_{Tp} - V_{DD} + K_R V_{Tn}}{1 + K_R}$$

$$= \frac{2V_{\text{out}} - 0.7 - 3.3 + 2.5 \times 0.6}{1 + 2.5}$$

$$= 0.57V_{\text{out}} - 0.71 \tag{iii}$$

Also, we have in the region B of operation of CMOS inverter that

$$\frac{k_n}{2}(V_{GSn} - V_{Tn})^2 = k_p\left[(V_{GSp} - V_{Tp})\left(V_{DSp} - \frac{V_{DSp}^2}{2}\right)\right]$$

Substituting the values of V_{GSn}, V_{GSp} and V_{DSp},
We get,

$$\frac{k_n}{2}(V_{\text{in}} - V_{Tn})^2 = k_p\left[(V_{\text{in}} - V_{DD} - V_{Tp})(V_{\text{out}} - V_{DD}) - \frac{(V_{\text{out}} - V_{DD})^2}{2}\right]$$

When $V_{\text{in}} = V_{IL}$,

$$\frac{k_n}{2}(V_{IL} - V_{Tn})^2 = k_p\left[(V_{IL} - V_{DD} - V_{Tp})(V_{\text{out}} - V_{DD}) - \frac{(V_{\text{out}} - V_{DD})^2}{2}\right]$$

Putting the value of V_{IL} from Eq. (iii) and simplifying, we get,

$$0.66V_{\text{out}}^2 + 0.05V_{\text{out}} - 6.65 = 0$$

Solving for V_{out},

$$V_{\text{out}} = 3.14 \text{ V}$$

(iii) $\Rightarrow$

$$V_{IL} = 0.57 \times 3.14 - 0.71$$

$$= 1.08 \text{ V} \tag{iv}$$

To find V_{IH}:

We have,

$$V_{IH} = \frac{V_{DD} + V_{Tp} + K_R(2V_{out} + V_{Tn})}{1 + K_R}$$

$$= \frac{3.3 - 0.7 + 2.5(2V_{out} + 0.6)}{1 + 2.5}$$

$$= 1.43V_{out} + 1.17 \qquad \text{(v)}$$

In the region D of operation of CMOS inverter, we have the relation

$$\frac{k_n}{2}\left[(V_{GSn} - V_{Tn})V_{DSn} - \frac{V_{DSn}^2}{2}\right] = \frac{k_p}{2}(V_{GSp} - V_{Tp})^2$$

Substituting the values of V_{GSn}, V_{DSn} and V_{GSp} in this equation,

$$k_n\,[(V_{\text{in}} - V_{Tn})\,V_{\text{out}} - V_{\text{out}}^2] = \frac{k_p}{2}(V_{\text{in}} - V_{DD} - V_{Tp})^2$$

When $V_{in} = V_{IH}$,

$$\frac{k_n}{k_p}\left[(V_{IH} - V_{Tn})\,V_{\text{out}} - \frac{V_{\text{out}}^2}{2}\right] = \frac{1}{2}(V_{IH} - V_{DD} - V_{Tp})^2$$

Putting the value of V_{IH} from Eq. (v) and simplifying,

$$2.61V_{\text{out}}^2 + 6.94\ V_{\text{out}} - 204 = 0$$

Solving for V_{out} yields

$$V_{\text{out}} = 0.27\ \text{V}$$

Substituting the value of V_{out} in Eq. (v)
We get,

$$V_{IH} = 1.43 \times 0.27 + 1.17$$

$$= 1.55\ \text{V}$$

The low level and high level noise margins can be calculated using the formulae,

$$N_{ML} = V_{IL} - V_{OL} = 1.08 - 0$$

$$= 1.08\ \text{V}$$

$$N_{MH} = V_{OH} - V_{IH} = V_{DD} - V_{IH} = 3.3 - 1.55$$

$$= 1.75\ \text{V}$$

Exercise

1. Consider a CMOS inverter with the following parameters:

$$\text{NMOS} \Rightarrow V_{Tn} = 0.6 \text{ V}, \mu_n C_{ox} = 60 \ \mu\text{A/V}^2, (W/L)_n = 8$$

$$\text{PMOS} \Rightarrow V_{Tp} = -0.7\text{V}, \mu p C_{ox} = 25 \ \mu\text{A/V}^2, (W/L)_p = 12$$

Calculate the noise margins and the switching threshold (V_{TH}) of this circuit. The power supply voltage is $V_{DD} = 3.3$ V.

4.5 CMOS—GATE TRANSISTOR SIZING

Cascaded complementary inverters. If we want to have approximately the same rise and fall times for an inverter, for current CMOS processes, we must make

$$W_p \simeq (2 \rightarrow 3) \ W_n \tag{4.39}$$

where,

$W_p \rightarrow$ channel width of p-device

$W_n \rightarrow$ channel width of n-device

This, of course, increases layout area and dynamic power dissipation. In some cascaded structures, it is possible to use minimum or equal-sized devices without compromising the switching response.

The delay response for an inverter-pair shown in Figure 4.19(a) with $W_p = 2W_n$ is given by

$$\begin{aligned} t_{\text{inv-pair}} &\propto t_{\text{fall}} + t_{\text{rise}} \\ &\propto R \cdot 3C_{\text{eq}} + 2(R/2) \cdot 3C_{\text{eq}} \\ &\propto 6RC_{\text{eq}} \end{aligned} \tag{4.40}$$

where R = effective 'ON' resistance of a unit-sized n-transistor

$$\begin{aligned} C_{\text{eq}} &= C_g + C_d \\ &= \text{Capacitance of a unit-sized gate and drain region} \end{aligned}$$

Similarly, the inverter pair delay, with $W_p = W_n$ as shown in Figure 4.19(b) is given by

$$\begin{aligned} t_{\text{inv-pair}} &\propto t_{\text{fall}} + t_{\text{rise}} \\ &\propto R \cdot 2C_{\text{eq}} + 2R \cdot 2C_{\text{eq}} \\ &\propto 6RC_{\text{eq}} \end{aligned} \tag{4.41}$$

Thus from Eqs. (4.40) and (4.41), we find that similar responses are obtained for the two different conditions.

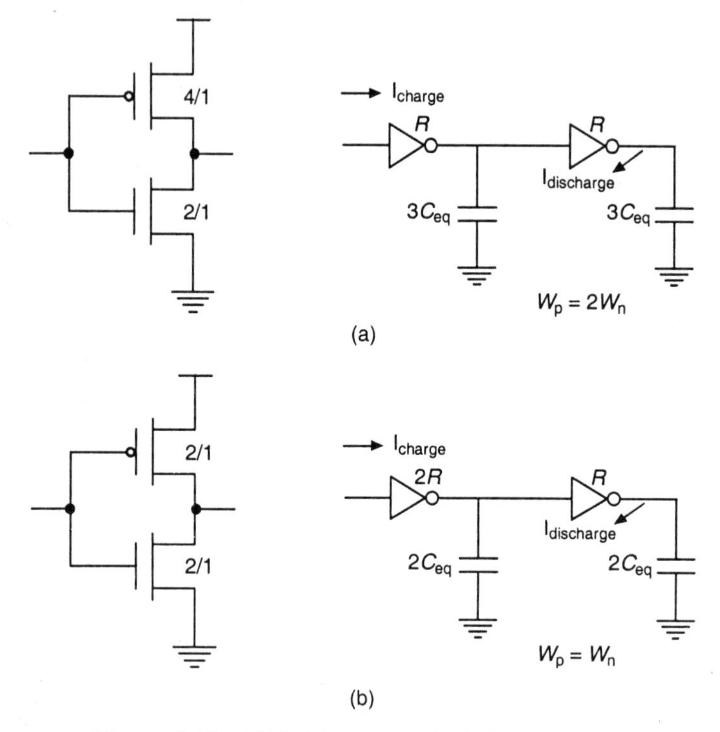

Figure 4.19 CMOS inverter pair timing response.

4.6 STAGE RATIO

Often it is desired to drive large load capacitances such as long buses, I/O buffers, or, ultimately, pads and off-chip capacitive loads. This is achieved by using a chain of inverters where each successive inverter is made larger than the previous one until the last inverter in the chain can drive the large load in the time required. The optimization to be achieved here is to minimize the delay between input and output while minimizing the area and power dissipation. The ratio by which each stage is increased in size is called the stage ratio. The concept of maintaining a good stage ratio is also of importance for a cascaded path through any logic gates where high-speed designs are involved.

4.7 POWER DISSIPATION

There are two components that establish the amount of power dissipated in a CMOS circuit. These are:

(i) Static dissipation due to leakage current or other current drawn continuously from the power supply.

(ii) Dynamic dissipation due to

- Switching transient current
- Charging and discharging of load capacitances.

4.7.1 Static Dissipation

Considering a complementary CMOS gate, if the input is either '0' or '1', one of the transistors is always OFF. Since no current flows into the gate terminal, and there is no dc current path from V_{DD} to V_{SS}, the resultant quiescent (steady-state) current, and hence power P is zero. However, there is some small static dissipation due to reverse bias leakage between diffusion regions conduction can contribute to the static dissipation. The leakage current is described by the diode equation.

$$i_o = i_s(e^{qV/kT} - 1) \tag{4.42}$$

where

i_s = reverse saturation current

V = diode voltage

q = electron charge (1.6×10^{-19} C)

k = Boltzmann's constant (1.38×10^{-23} J/K)

T = temperature

The static power dissipation is the product of the device leakage current and the supply voltage. At room temperature, the leakage current is about 0.1 nA to 0.5 nA per device. Thus, typical static power dissipation due to leakage for an inverter operating at 5 V is between 1 and 2 nW.

4.7.2 Dynamic Dissipation

During transition from either '0' to '1' or from '1', both n- and p-transistors are 'ON' for a short period of time. This results in a short current pulse from V_{DD} to V_{SS}. Current is also required to charge and discharge the output capacitive load. The current pulse from V_{DD} to V_{SS} results in a 'short-circuit' dissipation that is independent of the input rise/fall time, the load capacitance and the gate design. As the capacitive load is increased, the discharge or charge current starts to dominate the current drawn from the power supplies.

The dynamic power dissipation for a repetitive step input is given by

$$P_d = C_L V_{DD}^2 f_p \tag{4.43}$$

where,

$$f_p = 1/t_p = \text{repetition frequency}$$

4.7.3 Short-circuit Dissipation

The short-circuit power dissipation is given by

$$P_{sc} = I_{\text{mean}} V_{DD} \tag{4.44}$$

4.7.4 Total Power Dissipation

The total power dissipation can be obtained from the sum of the three components, so

$$P_{\text{total}} = P_s + P_d + P_{sc} \tag{4.45}$$

4.7.5 Power Economy

Minimizing power in CMOS circuits may be achieved in a number of ways. DC power dissipation may be reduced due to leakage by only using complementary logic gates. The leakage in turn is proportional to the area of diffusion, so the use of minimum-sized devices is advantageous. Dynamic power dissipation may be limited by reducing supply voltage, switched capacitance, and the frequency at which logic is clocked. Supply voltage tends to be a system-design consideration, and low power systems use 1.5 V to 3 V supplies. Minimizing the switched capacitance again tends to favour using minimum-sized devices and optimal allocation of devices such as adders and registers. Manual layout techniques are also of use to minimize routing capacitance. Another big gain can be made by only operating the minimum amount of circuitry at high speeds or having a variable clock depending on how much computation has to be computed.

In summary, for low-power design one should use the lowest supply voltage and operating frequency consistent with achieving the required performance.

SUMMARY

Chapter 4 is concerned with the design of a CMOS inverter. It covers the basic structure of a CMOS inverter and its dc transfer characteristics for various regions of operations (regions A, B, C, D, and E). The various design parameters such as input low level and high level voltages, threshold voltage, aspect ratio, etc., are derived in this chapter. It also looks into the temperature dependence of VTC of the inverter, supply voltage scaling, noise margins and power and area considerations which play a vital role in designing small-sized VLSI circuits. The switching characteristics of CMOS inverter are dealt with the estimation of rise time and fall time values. Finally, the chapter describes the various power dissipations occurring in a CMOS inverter circuit, i.e., static, dynamic, short-circuit dissipation, etc., as low-power design is achieved by the lowest supply voltage.

REVIEW QUESTIONS

1. Explain the dc characteristics of the CMOS inverter.
2. Give short notes on
   (a) Symmetric CMOS inverter
   (b) Noise margins of CMOS inverter
3. Explain the characteristics of CMOS inverter.
4. Explain the concept of power dissipation in detail.

SHORT ANSWER QUESTIONS

1. Guard rings prevent the formation of ____________ and contact cuts.
   Ans. parasitic transistors.
2. What is CMOS inverter?
   Ans. It is a series connection of *p*- and *n*-devices.

3. Give one important advantage of CMOS inverter over the others inverter configurations.
   Ans. The steady state power dissipation of the CMOS inverter circuit is negligible.
4. The slope of the voltage transfer characteristics is equal to ____________.
   Ans. –1.
5. Define inverter threshold voltage.
   Ans. $V_{TH} = V_{in} = V_{out}$.
6. Noise immunity of the digital circuit increases with ____________.
   Ans. noise margin.
7. Define stage ratio.
   Ans. The ratio by which each stage is increased in size is called stage ratio.

5 MOS Circuit Design Processes

5.1 INTRODUCTION

Design processes are aided by simple concepts such as stick and symbolic diagrams. The key element of the design process is a set of design rules. Design rules are the communication link between the designer specifying requirements and the fabricator who materializes them. Design rules are used to produce workable mask layouts from which the various layers in silicon will be formed or patterned.

Design rules specify to the designer certain geometric constraints on the layout artwork so that the patterns on the processed wafer will preserve the topology and geometry of the designs.

5.2 WHY DESIGN RULES

There are several error mechanisms that contribute to deviations in feature shapes from the shapes defined in a designer's artwork. These include mask misalignment, variations in the photoresist edges due to variations in exposure, undercutting of thin oxide under the corners of the photoresist, overetching, spreading of the diffusion and implantation under the gate or near the source–drain end, and tolerance of the field oxide windows. A feature's size may alter during the operation of the circuit.

The purpose of design rules is to guarantee that under the worst cumulative variations of the processes, the circuit does not fail to operate, i.e., separate features do not merge and small features do not split, thereby preserving the original topology of the intended circuit. The rules must also guarantee that the electrical parameters, such as the resistance and capacitance of the wires, which are determined by the physical dimensions of the features, are not altered by the process variations to a point that may cause a serious degradation in performance.

One of the fundamental difficulties in specifying design rules is that the fabrication processes are undergoing rapid evolutionary changes. As a result, the industrial design rules are complex and undergo constant change. The most important parameter that characterizes any process is the feature size. As long as the processing steps are not altered drastically, a set of design rules expressed in terms of feature size will have the best chance of survivability. With improvements in process technology, the feature size will scale down without changing the design rules. Mead and Conway (1980) formulated a set of such design rules for the relatively stable silicon-gate NMOS process.

Advantages of generalized design rules

1. Ease of learning because they are scalable, portable and durable.
2. Longevity of designs that are simple, abstract, and have minimum clutter.
3. Increased designer efficiency due to fewer levels and fewer rules.
4. Design efficiency accomplished by compaction, layout rule checking, electrical rule checking, simulation and verification.
5. Automatic translation to final layout.

5.3 MOS LAYERS

We have seen that MOS circuits are formed on four basic layers—*n*-diffusion, *p*-diffusion, polysilicon, and metal, which are isolated from one another by thick or thin (thinox) silicon dioxide insulating layers. The thinoxide (thinox) mask region includes *n*-diffusion, *p*-diffusion, and transistor channels. Polysilicon and thinox regions interact so that a transistor is formed where they cross one another. In some processes, there may be a second metal layer and also, in some processes, a second polysilicon layer. Layers may be joined together where contacts are formed. The basic MOS transistor properties can be modified by the use of an implant within the thinox region and this is used in NMOS circuits to produce depletion mode transistors.

Bipolar transistors can be included in this design process by adding extra layers to the CMOS process. This is referred to as BiCMOS technology.

5.4 STICK DIAGRAMS

Stick diagrams are used to convey layer information through the use of a colour code for example, in NMOS design, green for *n*-diffusion, red for polysilicon, blue for metal, yellow for implant, and black for contact areas. Here, the designer draws a freehand sketch of a layout, using coloured lines to represent the various process layers such as diffusion, metal, and polysilicon. Where polysilicon crosses the diffusion, transistors are created and where metal wires join diffusion or polysilicon, contacts are formed.

A stick diagram is a cartoon of a chip layout. They are not exact models of layout. The stick diagram represents the rectangles with lines which represent wires and component symbols. In this section, the colour-coding has been complemented by monochrome encoding of the lines so that black and white copies of stick diagrams do not lose the layer information. The encodings chosen are shown and illustrated in colour as in monochrome form (Figures 5.1–5.3). Note that mask layout information, which is also colour coded, may also be hatched for monochrome encoding, also shown in Figures 5.1–5.2.

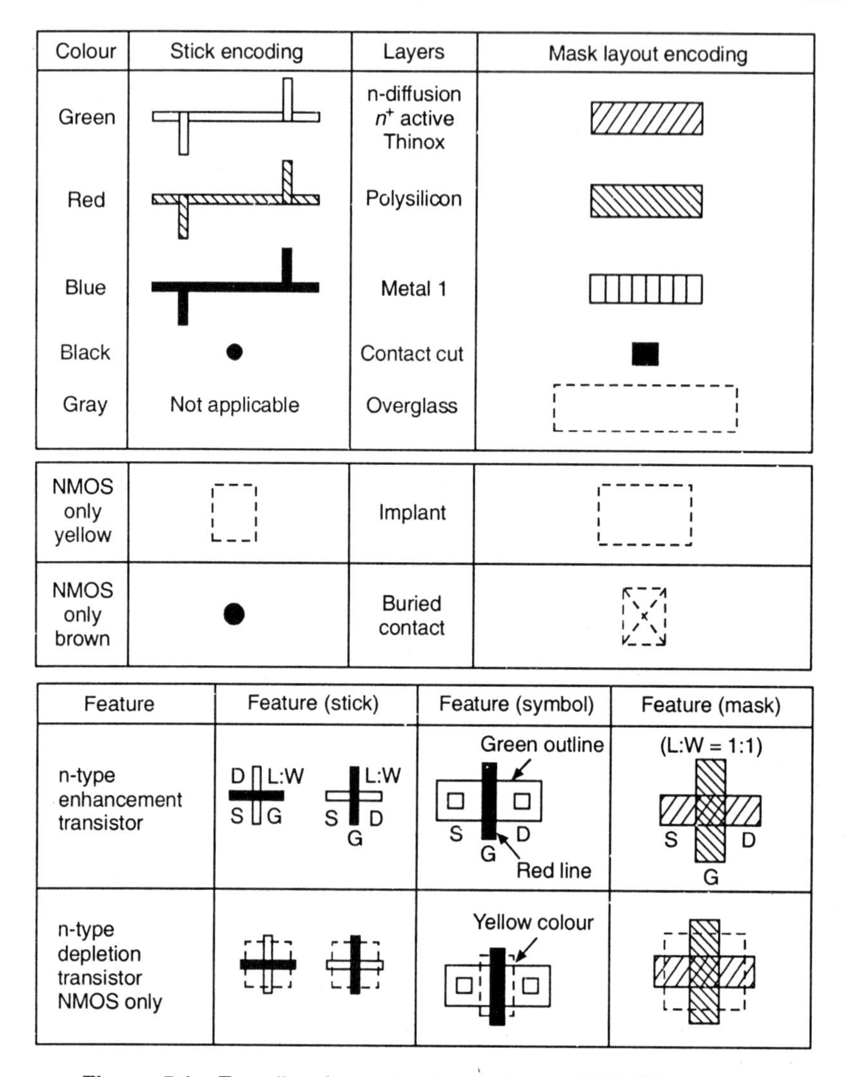

Figure 5.1 Encoding for a simple single metal NMOS process.

The colour and monochrome encoding scheme used has been evolved to cover NMOS and CMOS processes. In order to facilitate the learning and use of the encoding schemes, the simple set required for a single metal NMOS design is set out first as Figure 5.1; for a double-metal CMOS *p*-well process, the required encodings are extended by those given in Figure 5.2. To illustrate stick diagrams, inverter circuits are presented in Figure 5.3, in NMOS, and in *p*-well CMOS technology.

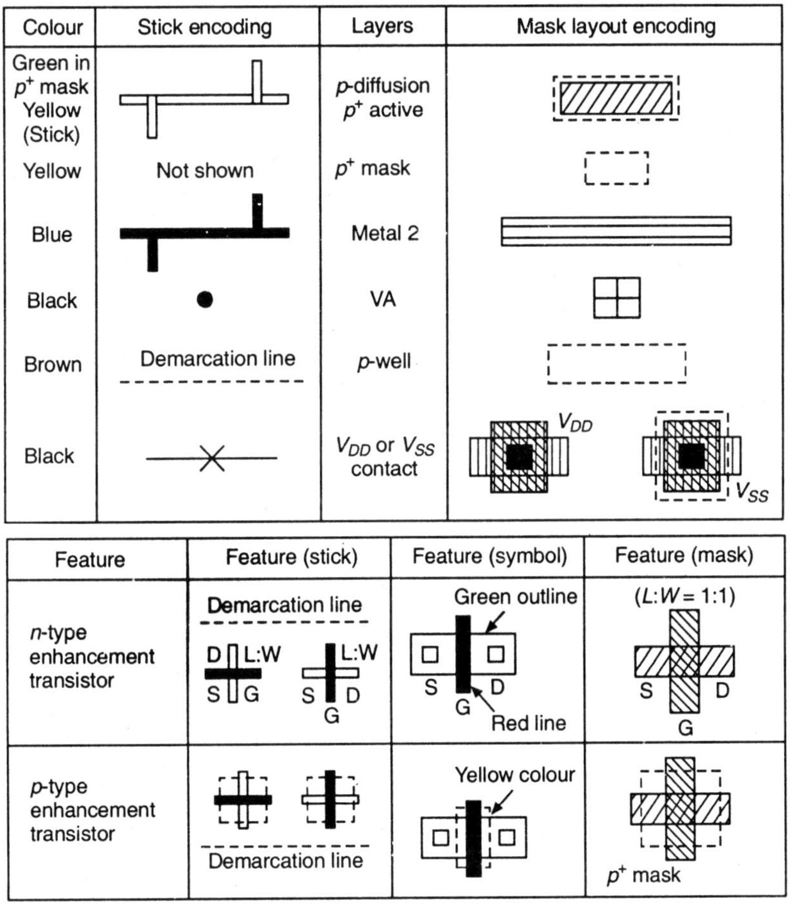

Colour	Stick encoding	Layers	Mask layout encoding
Green in p^+ mask Yellow (Stick)		p-diffusion p^+ active	
Yellow	Not shown	p^+ mask	
Blue		Metal 2	
Black		VA	
Brown	Demarcation line	p-well	
Black		V_{DD} or V_{SS} contact	V_{DD} V_{SS}

Feature	Feature (stick)	Feature (symbol)	Feature (mask)
n-type enhancement transistor	Demarcation line D L:W S G L:W S D G	Green outline S D G Red line	$(L{:}W = 1{:}1)$ S D G
p-type enhancement transistor	Demarcation line	Yellow colour	p^+ mask

Figure 5.2 Encoding for a double metal CMOS p-well process.

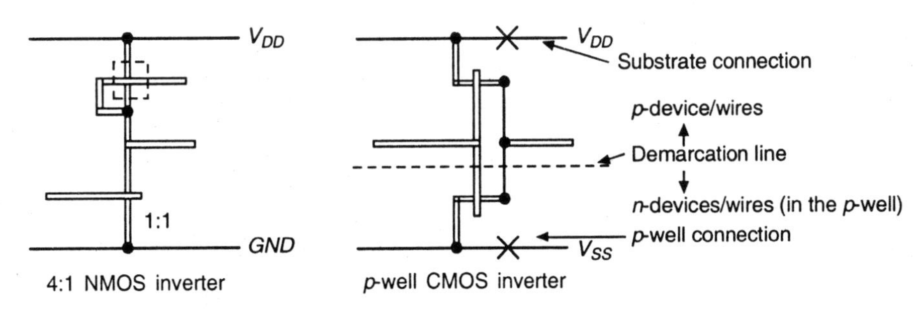

Figure 5.3 Stick diagram and symbolic coding of inverter.

Having conveyed layer information and topology by using stick or symbolic diagrams, these diagrams are relatively easily turned into mask layouts as, for example, the transistor stick diagrams of Figure 5.4. stressing the ready translation into mask layout form. In order that the mask layouts produced during design will be compatible with the fabrication processes, a set of design rules are set out for layouts.

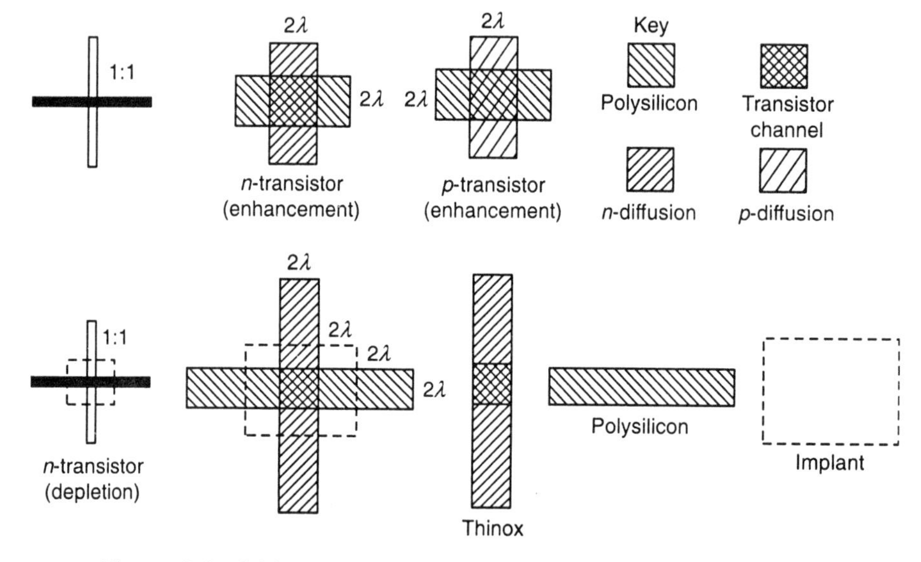

Figure 5.4 Stick diagram and corresponding mask layout examples.

5.4.1 Stick Layout Using NMOS Design

In order to start with a simple process, we consider single metal, single polysilicon NMOS technology. The layout of NMOS involves:

- *n*-diffusion (*n*-diff.) and other thinoxide regions – green
- Polysilicon (poly.) – red
- Metal 1 (metal) – blue
- Implant – yellow
- Contacts – black or brown (buried)

A transistor is formed wherever polysilicon crosses *n*-diffusion (red over green) and all diffusion wires (interconnections) are *n*-type (green). The various steps involved in the design style are:

Step 1: Drawn the metal (blue) V_{DD} and GND rails in parallel allowing enough space between them for the other circuit elements which will be required.

Step 2: Draw the thinox (green) paths between the rails for inverters and inverter based logic [Figure 5.5(a)].

Step 3: Draw the pull-up structure which comprises a depletion mode transistor interconnected between the output point and V_{DD}.

Step 4: Then, draw the pull-down structure comprising an enhancement mode structure interconnected between the output point and GND [Figure 5.5(b)].

Step 5: Signal paths may be switched by pass transistors, and along signal paths often require metal buses (blue). Draw 'leaf-cell' boundaries as shown in Figure 5.5(c).

The inverters and inverter-based logic comprise a pull-up structure and a pull-down structure. While drawing these enhancement or depletion mode transistors, poly. (red) crosses thinox (green). The depletion mode transistors consist of a yellow implant also. It is important to write the length to width (*L*:*W*) ratio for each transistor, particularly in NMOS and NMOS-like circuits.

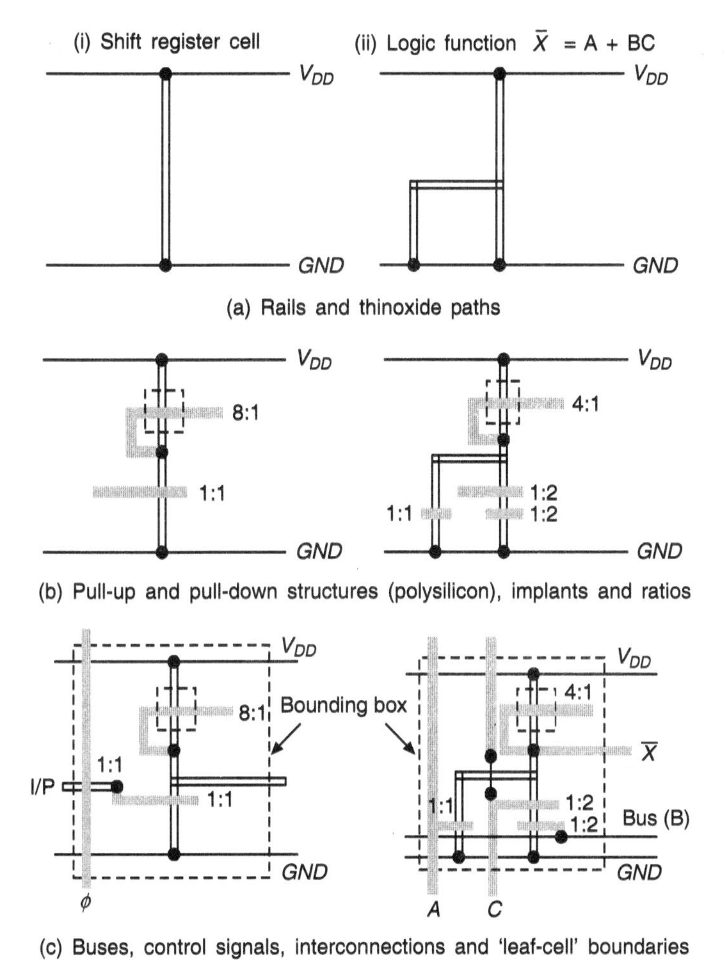

Figure 5.5 Examples of stick diagrams using NMOS design.

5.4.2 Stick Layout Using CMOS Design

The stick and layout representation for CMOS design are a logical extension of the NMOS approach. These are based on the work of Mead and Conway. The two types of transistors used, '*n*' and '*p*' are separated in the stick layout by the demarcation line (representing the *p*-well boundary) above which all *p*-devices are placed (yellow transistor). The *n*-devices (green) are consequently placed below the demarcation line and are thus located in the *p*-well (Figure 5.6).

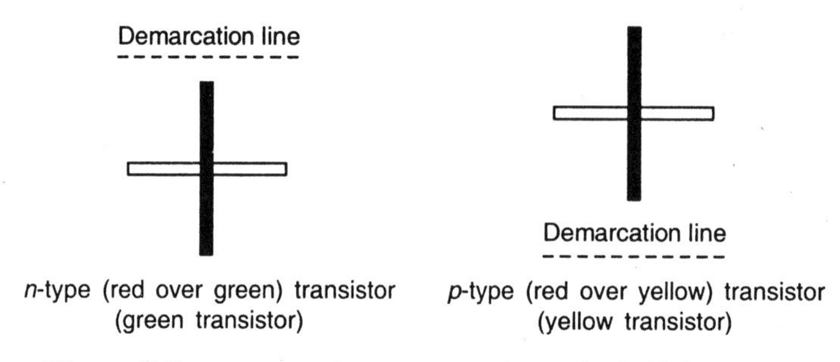

Figure 5.6 *n*-type and *p*-type transistors in CMOS design.

The diffusion paths must not cross the demarcation line and *n*-diffusion and *p*-diffusion wires must not join. The '*n*' and '*p*' features are normally joined by metal where a connection is needed. Apart from the demarcation line, there is no indication of the actual *p*-well topology. The design includes the following steps:

Step 1: Draw the V_{DD} and V_{SS} rails in parallel and in metal and create an (imaginary) demarcation line in between [Figure 5.7(a)].

Step 2: Place the *n*-transistors below the demarcation line close to V_{SS}, and the *p*-transistors above the line and below V_{DD} [Figure 5.7(b)].

Step 3: Interconnect the *n*- with the *p*-transistors as required, using metal and connect to the rails [Figure 5.7(c)]. Only metal and polysilicon can cross the demarcation line, but wires can run to diffusion also.

Step 4: Finally, make the appropriate inter-connections and add the control signals and date inputs [Figure 5.7(d)].

Step 5: Represent the V_{SS} and V_{DD} contact crosses and bounding box for the entire leaf-cell may also be shown if appropriate.

5.5 DESIGN RULES AND LAYOUT

The objective of a set of design rules is to allow a ready translation of circuit design concepts, usually in stick diagram or symbolic form, into actual geometry in silicon. The design rules are the effective interface between the circuit/system designer and the fabrication engineer. Design rules govern the layout of the individual components and the interactions—spacings and electrical connections—between those components. They determine the low-level properties of chip designs: how small individual logic gates can be made; how small the wires connecting gates can be made, and therefore, the parasitic resistance and capacitance which determine delay.

The design rules primarily address two issues:

(i) the geometrical reproduction of features that can be reproduced by the mask-making and lithographical process

(ii) the interactions between different layers

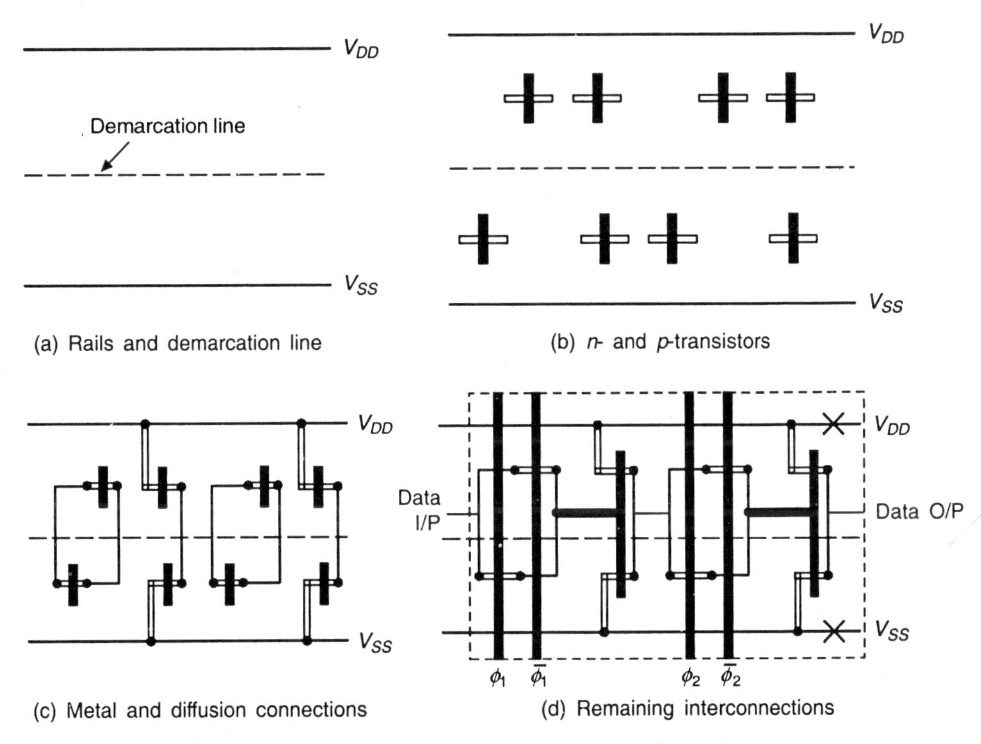

Figure 5.7 Example of stick diagram using CMOS design.

There are several approaches that can be taken in describing the design rules. These include

- Micron design rules
  - stated at some micron resolution
  - usually given as a list of minimum feature sizes and spacings for all masks required in a given process
  - normal style for industry
- Lambda (λ)—based design rules
  - These rules popularized by Mead and Conway are based on a single parameter, λ, which characterizes the linear feature—the resolution of the complete wafer implementation process—and permits first order scaling.
  - They have been widely used, particularly in the educational context and in the design of multiproject chips.

As a rule, they can be expressed on a single page. While these rules have been successfully used for 4–1.2 μm processes, they will probably not suffice for sub-micron processes.

5.5.1 Lambda (λ) Based Design Rules

The lambda, λ design rules are based on Mead and Conway work and in general, design rules and layout methodology are based on the concept of λ which provide a process and feature size—independent way of making mask dimensions to scale. In lambda based design rules, all paths in all layers will be dimensioned in λ units and subsequently λ can be allocated an appropriate value compatible with the feature size of the fabrication process. This concept means that the actual mask layout design takes little account of the value subsequently allocated to the feature size, but the design rules are such that, if correctly obeyed, the mask layouts will produce working circuits for a range of values allocated to λ. For example, λ can be allocated a value of 1.0 μm so that minimum feature size on chip will be 2 μm (2λ).

Design rules, also due to Mead and Conway, specify line widths, separations, and extensions in terms of λ. Design rules can be conveniently set out in diagrammatic form as in Figure 5.8, for the width and separation of conducting paths, and in Figure 5.9 for extensions and separations associated with transistor layouts.

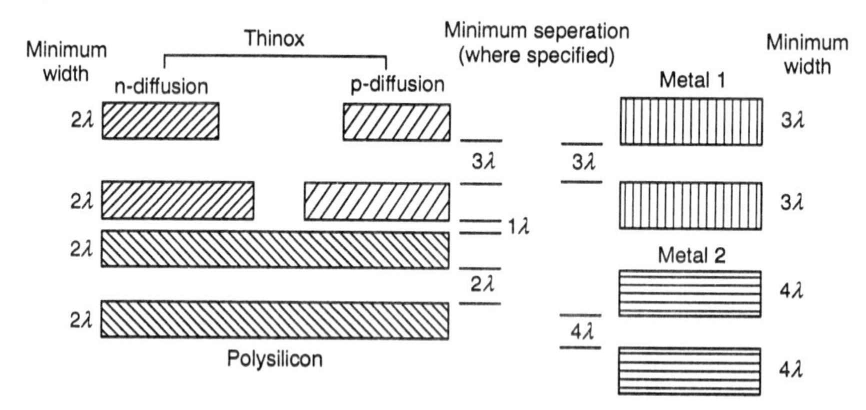

Figure 5.8 Design rules for wires (NMOS and CMOS).

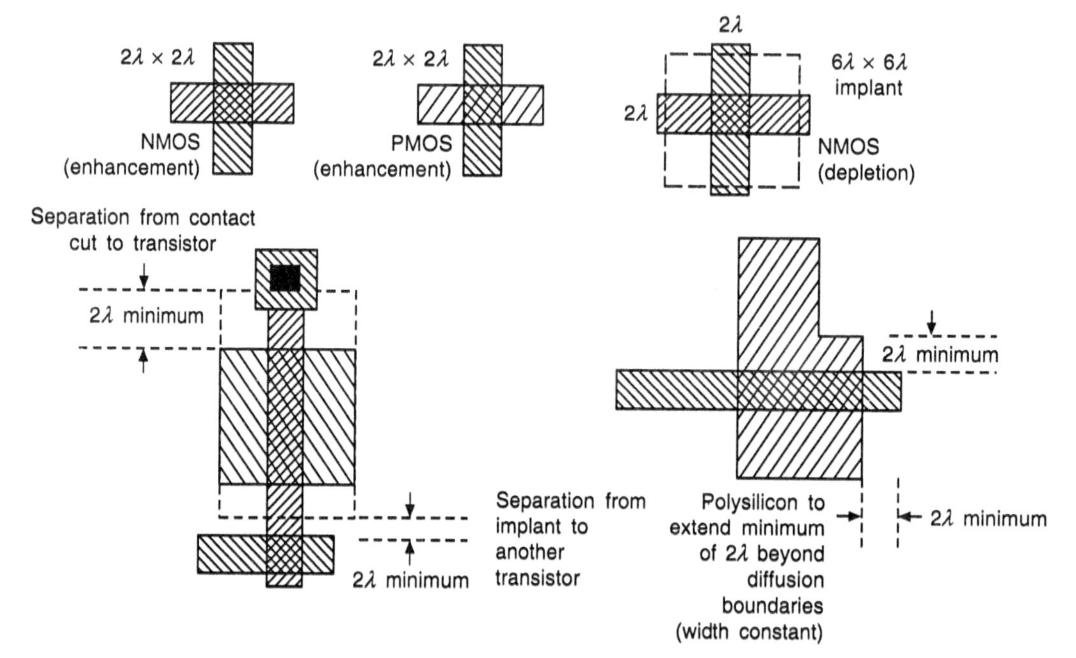

Figure 5.9 Transistor design rules (NMOS, PMOS, and CMOS).

As per design rules, the contacts between layers are set out in Figures 5.10 and 5.11. Here, it will be observed that connection can be made between two or, in the case of NMOS designs, three layers.

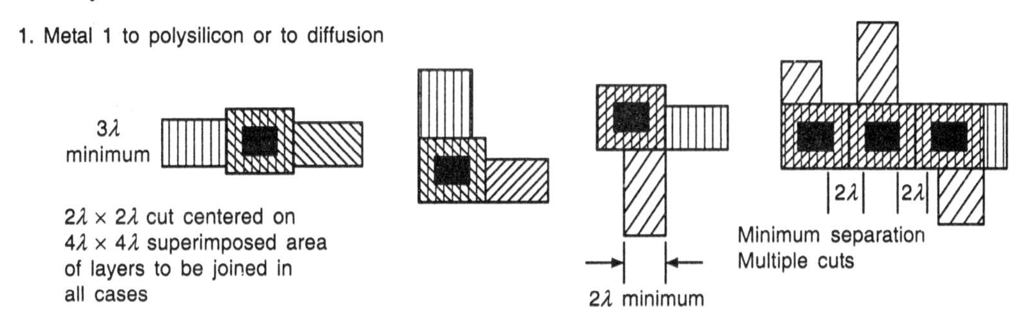

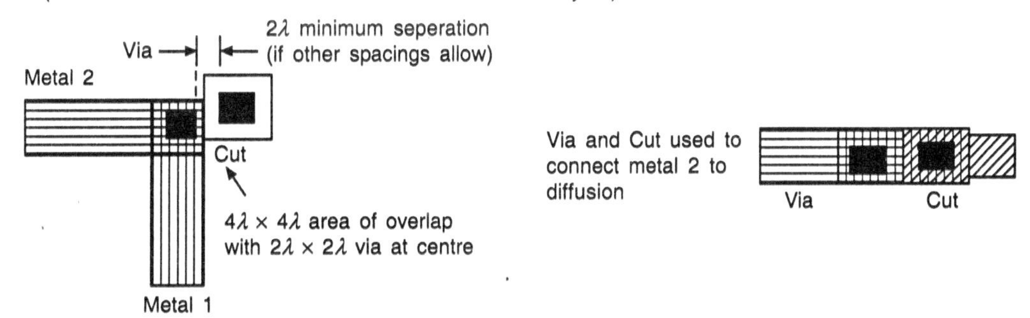

Figure 5.10 Contacts (NMOS and CMOS).

Contact cuts. There are three possible approaches for making contacts between polysilicon and diffusion in NMOS circuits. There are

(i) polysilicon to metal then metal to diffusion

(ii) buried contact polysilicon to diffusion

(iii) butting contact (polysilicon to diffusion using metal)

Among these, the buried contact is the most widely used giving economy in space and a reliable contact. In CMOS designs, poly. to diff. contacts are almost made via metal.

For making connections between metal and either of the other two layers (as in Figure 5.10), the process is quite simple. The $2\lambda \times 2\lambda$ contact cut indicates an area in which the oxide is to be removed down to the underlying polysilicon or diffusion surface. When the deposition of the metal layer takes place, the metal is deposited through the contact cut areas onto the underlying areas so that contact is made between the layers.

For connecting diffusion to polysilicon using the butting contact approach as in Figure 5.11, the process is rather more complex. In effect, a $2\lambda \times 2\lambda$ contact cut is made down to each of the layers to be joined. The layers are butted together in such a way that these two contact cuts become contiguous. Since the polysilicon and diffusion outlines overlap and thinoxide under polysilicon acts as a mask in the diffusion process, the polysilicon and diffusion layers are also

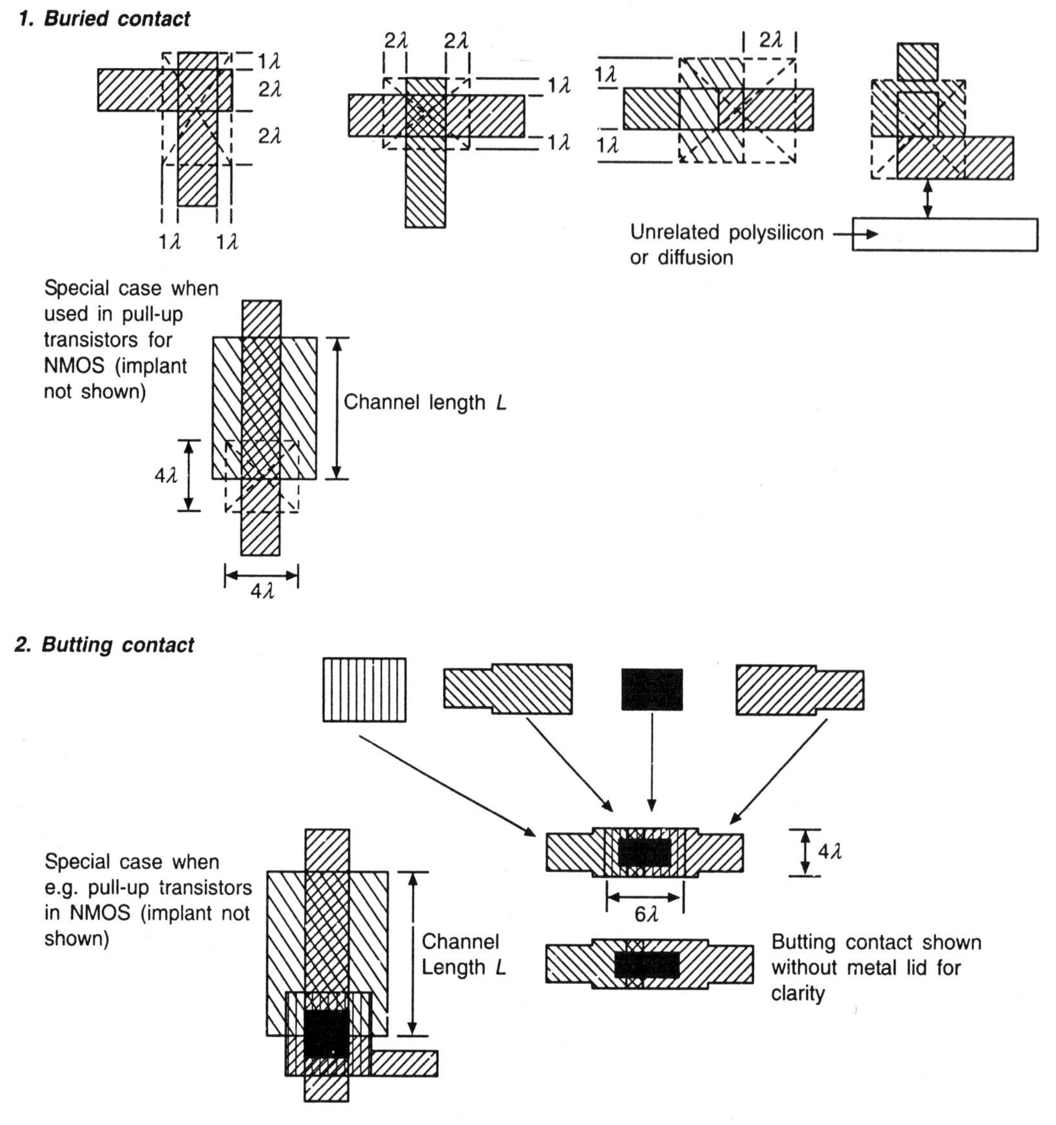

Figure 5.11 Contacts polysilicon to diffusion (NMOS only).

butted together. The contact between the two butting layers is then made by a metal overlay as shown in Figure 5.11. If the butting contact is made as shown in Figure 5.12(b), the cross sectional view of butting contact will make the contact more realistic.

The buried contact approach shown in Figures 5.11 and 5.12(a) is simpler, since the contact cut (broken line) in this case indicating where the thinoxide is to be removed to reveal the surface of the silicon wafer before polysilicon is deposited. Thus, the polysilicon is deposited directly on the underlying crystalline wafer. When diffusion takes place, impurities will diffuse into the polysilicon as well as into the diffusion region within the contact area. Thus, a satisfactory

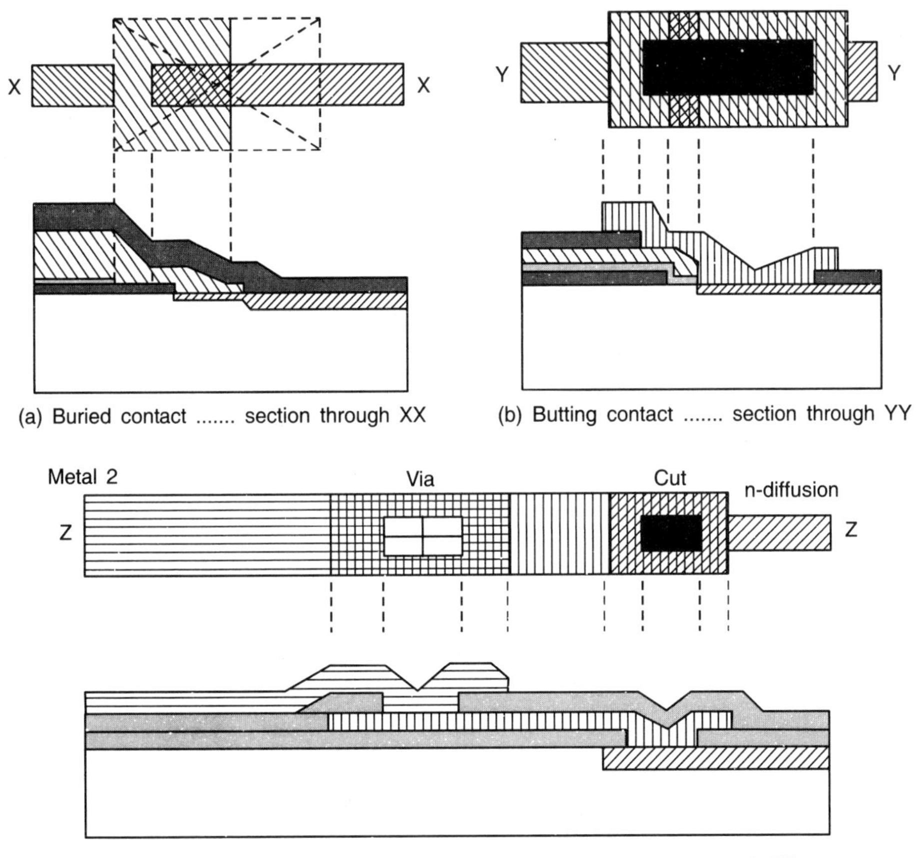

(a) Buried contact .... section through XX

(b) Butting contact .... section through YY

(c) Metal 2-via-metal 1-cut-n-diffusion connection ... section through ZZ.

Figure 5.12 Cross-sections of some contact structures.

connection between polysilicon and diffusion is ensured. Buried contacts can be smaller in area than their butting contact counterpart and, since they use no metal layer, they are subject to fewer design rule restrictions in a layout. Figure 5.12(c) shows the cross-section obtained wing a via contact and a cut.

The design rules in this case ensure that a reasonable contact area is achieved and that there will be no transistor formed unintentionally in series with the contact. The rules are such that they also avoid the formation of unwanted diffusion to polysilicon contacts and protect the gate oxide of any transistor in the vicinity of the buried contact cut area.

5.5.2 Double Metal MOS Process Rules

A powerful extension to the process so far described is provided by a second metal layer. This gives a much greater degree of freedom, for example, in distributing global V_{DD} and V_{SS} (GND) rails in a system. From the overall chip interconnection aspect, the second metal layer in particular is important and, although the use of such a layer is readily envisaged, its disposition relative to (and details of) its connection to other layer using metal 1 to metal 2 contacts, are called vias. Usually, second level metal layers are coarser than the first (conventional) layer and the isolation layer between the layers may also be of relatively greater thickness. To distinguish contacts between first and second metal layers, they are known as vias rather than contact cuts. The second metal layer representation is colour-coded dark blue (purple). For the sake of completeness, the process steps for a two-metal layer process are briefly outlined as follows.

The atmospherical chemical vapour deposition (ATCVD) is used for depositing oxide below the first metal layer and the oxide layer between the metal layers is deposited in a similar fashion. Depending on the process, removal of selected areas of the oxide is accomplished by plasma etching, which is designed to have a high level of vertical ion bombardment to allow high and uniform etch rates. Similarly, the bulk of the process steps for a double polysilicon layer process are similar in nature to those already described, except that a second thinoxide layer is grown after depositing and patterning the first polysilicon layer (poly. 1) to isolate it from the now to be deposited second polysilicon layer (poly. 2). The presence of a second polysilicon layer gives greater flexibility in interconnections and also allows transistors to be formed when second polysilicon crosses *n*- and *p*-diffusion regions.

To revert to the double metal process, it is convenient to consider the layout strategy commonly used with this process. The approach used for double metal process is summarized as follows:

1. Use the second level metal for the global distribution of power buses, i.e., V_{DD} and GND (V_{SS}), and for clock lines.
2. Use the first level metal for local distribution of power and for single lines.
3. Layout the two metal layers so that the conductors are mutually orthogonal wherever possible.

5.5.3 CMOS Lambda-based Design Rules

The CMOS fabrication process is much more complex than NMOS fabrication, which, in turn, has been simplified for ready presentation in this section. The design rules discussed here are complex enough, but in fact they constitute an abstract of the actual processing steps which are used to produce the chip. In a CMOS process, for example, the actual set of industrial design rules may well comprise more than 100 separate rules.

However, extending the Mead and Conway concepts, which have been already set out for NMOS designs, it is possible to add rules peculiar to CMOS (Figure 5.13) to those already setout in Figures 5.8 to 5.12. The additional rules are concerned with those features unique to *p*-well CMOS, such as the *p*-well and p^+ mask and the special 'substrate' contacts.

Although the CMOS rules in total may seem difficult to comprehend for the new designer, once use has been made of the simpler NMOS rules, the transition to CMOS is not hard to achieve. The real key to success in VLSI design is to put it into practice and this text attempts to encourage the students to follow the above said rules.

The design method developed by Carver Mead and Lynn Conway capitalizes on the inherent advantages of NMOS processing by systematizing the NMOS design rules to be pattern—and process independent. The NMOS and CMOS design rules as specified by MOSIS, the MOS

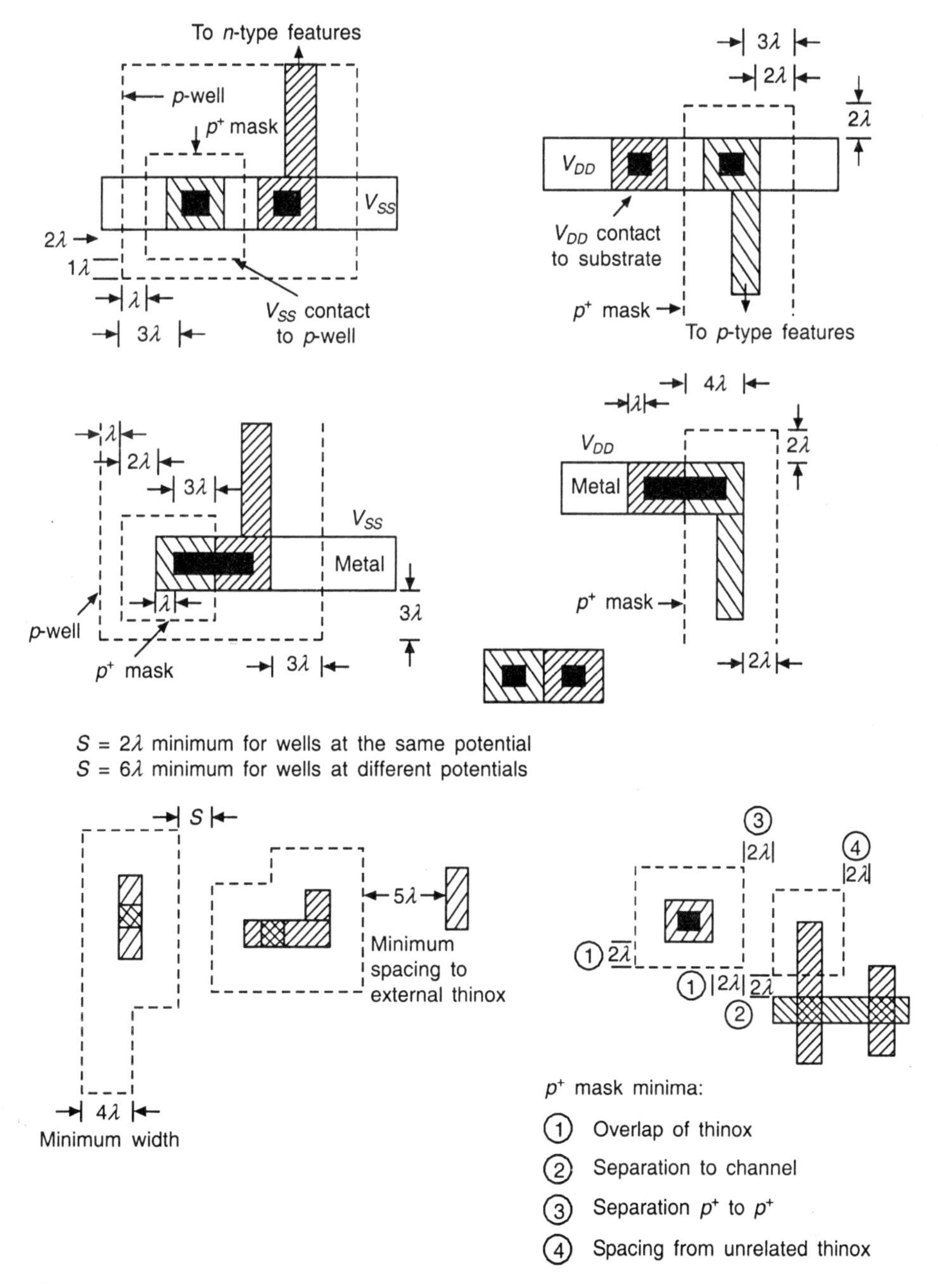

Figure 5.13 Particular rules for p-well CMOS process.

implementation system of the university of Southern California Information Science Institute are given in the following Tables 5.1 and 5.2. Here λ represents a small parameter specifying the feature size of the fabrication process.

Table 5.1 The MOSIS NMOS design rules

N^+ *diffusion mask*
Diffusion width 2λ Diffusion spacing 3λ
Implant mask
Implant gate overlap 1.5λ (MOSIS recommends 2λ) Implant to enhancement gate spacing 1.5λ
Buried contact mask
Buried contact to active device 2λ Overlap in diffusion direction 2λ Overlap in poly or field direction 1λ Buried contact to unrelayed poly. or diffusion spacing 2λ Buried contact poly to unrelated poly spacing 2λ
Poly mask
Poly width 2λ Poly spacing 2λ Poly-diffusion spacing 1λ Poly gate extension beyond diffusion 2λ Diffusion to poly edge 2λ
Contact mask
Contact size $2\lambda \times 2\lambda$ Contact-diffusion overlap 1λ Contact-poly overlap 1λ Contact to contact space 2λ Contact to FET channel 2λ Contact-metal overlap 1λ
Metal mask
Metal width 3λ Metal spacing 3λ Maximum current density 1 mA/μm Bonding pad size 100 μm × 100 μm Probe pad size 75 μm × 75 μm
Overglass mask
Bonding pad overglass cut 90 μm × 90 μm Probe pad overglass cut 65 μm × 65 μm Scribe line width 100 μm Feature distance from scribe line 50 μm Feature distance from bonding pad 40 μm Bonding pad separation 80 μm Bonding pad pitch 180 μm

Table 5.2 The MOSIS portable CMOS design rules

n-well and p-well mask
n-well or *p*-well width 6λ
n-well or *p*-well spacing 6λ
n^+ or p^+ active (diffusion or implant) mask
Active width 3λ
Active to active spacing 3λ
Source/drain active to well-edge 6λ
Substrate/well contact, active to well-edge 3λ
Poly mask
Poly width 2λ
Poly spacing 2λ
Gate overlap of active 2λ
Active overlap of gate 2λ
Field poly to active 1λ
p-select, n-select mask
Select-space (overlap) to (of) channel 3λ
Select-space (overlap) to (of) active 2λ
Select-space (overlap) to (of) contact to well or substrate 1λ
Simpler contact to poly mask
Contact size $2\lambda \times 2\lambda$
Active overlap of contact 2λ
Contact-to-contact spacing 2λ
Contact-to-gate spacing 2λ
Denser contact to poly mask
Contact size $2\lambda \times 2\lambda$
Poly overlap of contact 1λ
Contact spacing on the same poly 2λ
Contact spacing on a different poly 5λ
Contact to non-contact poly 4λ
Space to active short run 2λ
Space to active long run 3λ
Simpler contact to active mask
Contact size $2\lambda \times 2\lambda$
Active overlap of contact 2λ
Contact/contact spacing 2λ
Contact/gate spacing 2λ

(*Contd.*)

Table 5.2 The MOSIS portable CMOS design rules

Denser contact to active mask
Contact size = $2\lambda \times 2\lambda$
Active overlap of contact = 1λ
Contact/contact spacing on same active = 2λ
Contact/contact spacing on different active = 6λ
Contact to different active = 5λ
Contact to gate spacing = 2λ
Contact to field poly, short run = 2λ
Contact to field poly, long run = 2λ
Metal 1 mask
Width = 3λ
Metal 1 to metal 1 = 3λ
Overlap of contact to poly = 1λ
Overlap of contact to active = 1λ
Via mask
Size = $2\lambda \times 2\lambda$
Via-to-Via separation = 2λ
Metal 1/via overlap = 1λ
Space to poly or active edge = 2λ
Via to contact spacing = 2λ
Metal 2 mask
Width = 3λ
Metal 2/Metal 2 spacing = 4λ
Metal overlap of via = 1λ
Overglass mask
Bonding pad = 100 μm × 100 μm
Probe pad = 75 μm × 75 μm
Pad to glass edge = 6 μm

Layout examples

Example 5.1: CMOS inverter.

The CMOS inverter, with its *p*-channel device in an *n*-type well and its *n*-channel device in the *p*-type substrate is shown in Figure 5.14(a). The CMOS inverter always has one transistor OFF, and since the OFF transistor has an extremely high impedance, the output voltage can swing from the positive to the negative voltage rail. The stick diagram and physical layout of CMOS inverter is shown in Figures 5.14(b) and 5.14(c), respectively.

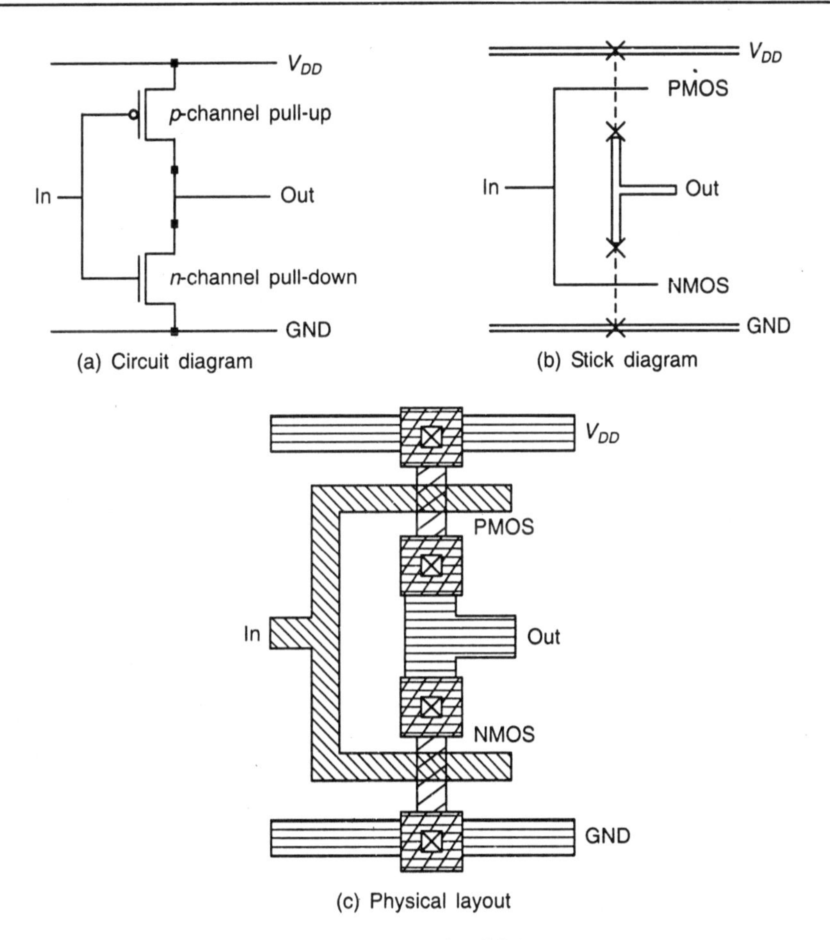

Figure 5.14 The CMOS inverter.

Example 5.2: CMOS NAND gate.

In the CMOS NAND gate, a number of NMOSFETs are connected in series between the output node and the ground. The same number of PMOSFETs are connected in parallel between V_{DD} and the output node. Each input signal is connected to the gates of a pair of n- and PMOSFETs as in the inverter case. In this configuration, the output node is pulled to ground only if all the NMOSFETs are turned ON, i.e., only if all the input voltages are high (V_{DD}). If one of the input signals is low (0 V), the low resistance path between the output node and ground is broken, but one of the PMOSFETs is turned ON, which pulls the output node to V_{DD}. Just like CMOS inverters, there is no static current or standby power dissipation for any combination of inputs in the CMOS NAND circuit. In CMOS-technology NAND gate circuits are much more frequently used than NOR. This is because it is preferable to put the transistors with the higher resistance in parallel and those with the lower resistance in series. By connecting low-resistance NMOSFETs in series and high-resistance PMOSFETs in parallel, a NAND gate achieves better noise immunity as well as a higher overall circuit speed.

The CMOS NAND circuit diagram, its stick layout and the physical layout are shown in Figures 5.15(a), 5.15(b) and 5.15(c) respectively.

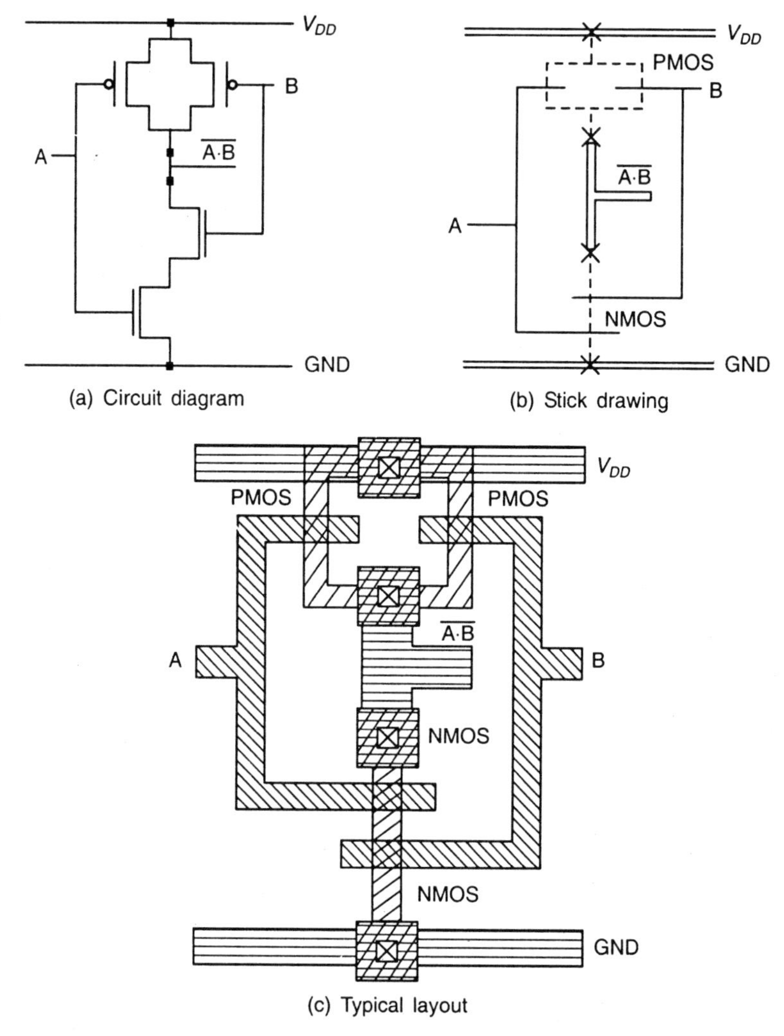

Figure 5.15 The CMOS 2-input NAND gate.

Example 5.3 CMOS NOR gate.

The CMOS NOR gate consists of parallel connected NMOSFETs between the output node and ground, but serially connected PMOSFETs between V_{DD} and the output node. The output voltage is high only if all the input voltages are low, i.e., all the PMOSFETs are ON and all the NMOSFETs are OFF. Otherwise, the output is low. In CMOS NOR circuits, there is no static power dissipation for any combination of inputs. Also, since the high-resistance PMOSFETs are series connected, the overall circuit speed is reduced in CMOS NOR gates.

The circuit diagram, stick representation and physical layout of a 2-input CMOS NOR gate are shown in Figures 5.16(a), 5.16(b) and 5.16(c) respectively.

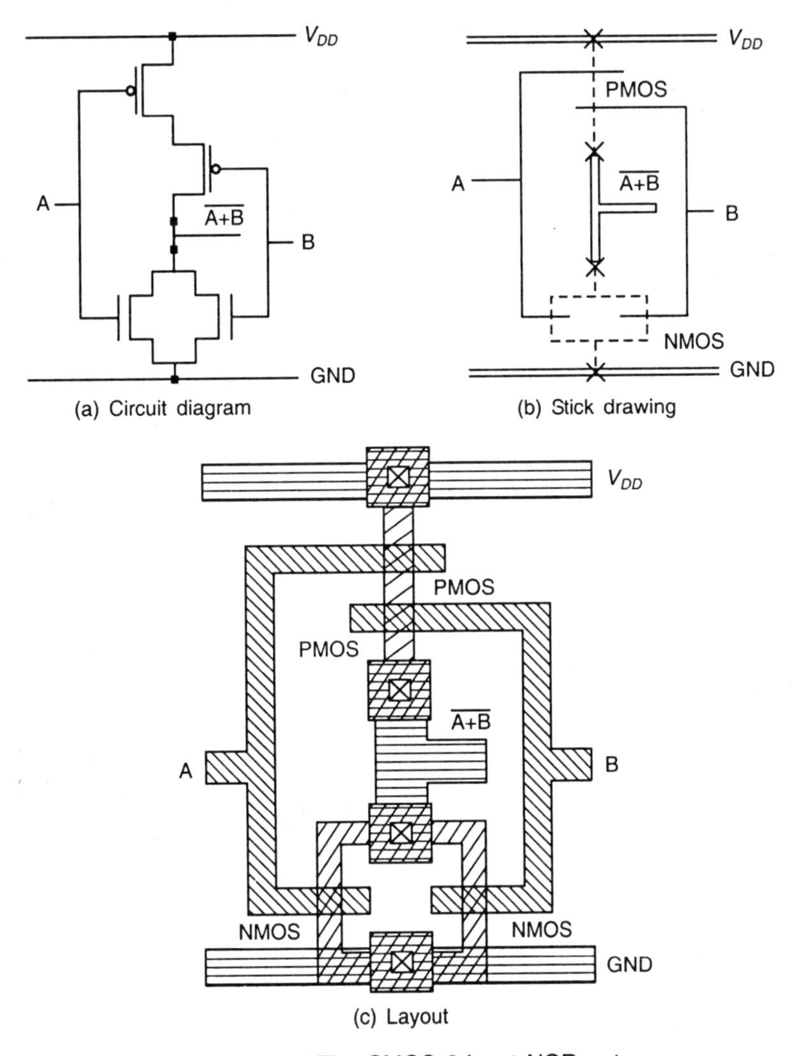

Figure 5.16 The CMOS 2-input NOR gate.

5.6 ELEMENTS OF PHYSICAL DESIGN

Here we study the details of translating logic circuits into silicon, which is called physical design. Details such as the minimum size specifications allowed for a patterned region become critical. However, the important factor in the physical design of VLSI chips revolve around the use of CAD tools and database structures that describe the silicon masks. These give the necessary information for creating the chip, and provide the basis for the hierarchical design of large complex logic networks.

5.6.1 Basic Concepts

Physical design is the actual process of creating circuits on silicon. During this phase, schematic diagrams are carefully translated into sets of geometric patterns that are used to define the on-chip physical structures. Every layer in the CMOS fabrication sequence is defined by a distinct pattern. A patterned layer consists of a group of geometrical objects, called polygons. This naturally includes rectangles and squares, but allows us to include arbitrarily complex n-vertex shapes with specific dimensions. Figure 5.17 shows the examples of polygons in physical design of CMOS processes, where several are superimposed to form the overall layout. When stacked into 3-D structures, the layers are electrically equivalent to the circuit diagram.

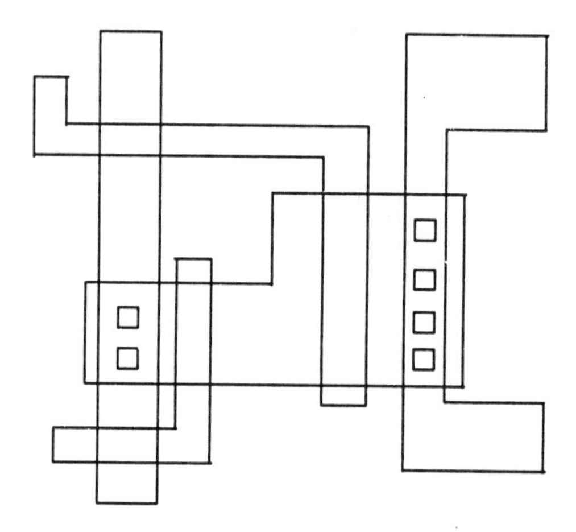

Figure 5.17 Examples of polygons in physical design.

The topology of the transistor network establishes the logic function. In other words, the details of how the FETs are wired together (series, parallel, etc.) are sufficient to determine the binary operations of the circuit. Another aspect of logic is that of switching speed. This is complicated to analyze, but is crucially important to modern chip design. For a given set of processing parameters, the electrical characteristics of a logic gate depend on the aspect ratios of the transistors. This is due to both the current flow levels and the parasitic resistance and capacitance of the devices. Physical design must address both of these areas. The patterns must be created to correctly implement the signal flow network. The complicating factor is that the dimensions of every feature affect the electrical performance of the circuit. In a VLSI chip, the switching speed of some gates will be critical, especially those in long complex logic gates.

The process of physical design is performed using a computer tool called a layout editor. This is a graphics program that allows the designer to specify the shape, dimensions and placement of every polygon on every layer of the chip. Complexity issues are reduced by first designing simple gates and storing their descriptive files in a library subdirectory or folder; the gates constitute cells in the library. Library cells are used as building blocks by creating copies of the basic cells to construct a larger more complex circuit. This process is called instancing of the cells, while a copy of a cell is called an instance.

The Layout must be an accurate representation of the logic network, at the same time achieving the goal of obtaining a fast circuit in the minimum area. Small changes in the shapes or areas of a polygon will affect the resulting electrical characteristics of the circuit. Circuit simulations also help insure that the layout is accurate and provides a network that meets specifications.

5.6.1.1 CAD Toolsets

For the layout design and its verification, Computer Aided Design (CAD) tools may be used. This will not only ease the designing, but also time required for designing will become lesser. A variety of CAD tools are in existence. Some of the important CAD tools are listed below:

(i) Layout editors
(ii) Design Rule Checkers (DRC)
(iii) Circuit extraction, etc.

Layout editors. Layout editor is a graphic program. With the help of this, one can create and delete layout elements. Some layout editors' programs are based on symbolic layouts. On the other hand, most layout editors work on hierarchical layout. In this, the layouts are organized into cells, which include both primitive layout elements and other cells. As an example, a via may be represented as a single rectangle in a symbolic layout, whereas when a final physical layout is requested, the symbolic via is flashed out into all the rectangles required for our process. The symbolic layout is easy to specify, since it is composed of fewer elements. The layout editor also ensures that layouts for symbolic elements are properly constructed. The same layout can be used to generate several variations. Some of these are *n*-tub, *p*-tub, and twin-tub versions of a symbolic design.

Design Rule Checkers (DRCs). The DRC program is used to check the design rule violations in the layout. With DRC program, minimum spacing and sizes are checked. It also ensures that combinations of layers form legal components. Some layouts provide online design rule checking.

Circuit extraction. The circuit extraction can be assumed to be extension of DRC. It also uses the same kind of algorithms as in DRC. The circuit extractor is capable of performing a complete job of component and wire extraction. A netlist is produced in circuit extract, which enlists the transistors in layout. It is also used to generate electrical nets, which connect their terminals.

Other tools are provided to help in large designs. Place and route routines help the layout designer by automatically finding viable wiring routes between two specified points. This is useful when trying to connect two complex units together. Electrical continuity can be seen using an Electrical Rule Checker (ERC) which highlights connecting paths.

5.6.2 Design Hierarchies

VLSI systems are created using the concept of design hierarchies where simple building blocks are used to design more complex units. This nesting continues until the entire chip is complete. The code for a layout editor is structured to provide this type of environment for the chip designer. The key to creating the hierarchy lies in the concept of cells. A cell is defined to be a collection of objects that is treated as a single entity. The characteristics of the objects themselves provide the hierarchical viewpoint.

The simplest cells consist of only polygons. The logic gates such as the NOT and NAND2 examples fall into this category. A cell with this property is said to be flat; this means that every object is independent and not related to any other object. In a flat cell, we can alter any polygon without affecting anything else. To initiate the design process, we create a large number of flat cells, and store them in a library. The most primitive library entries are chosen to be transistors and logic gates that can be used as building blocks in more complex designs. In Figure 5.18, three gate-level designs NOR2, NAND2 and NOT are created at the polygon level. Each is then stored as a separate cell in the library. Each cell is independent of others.

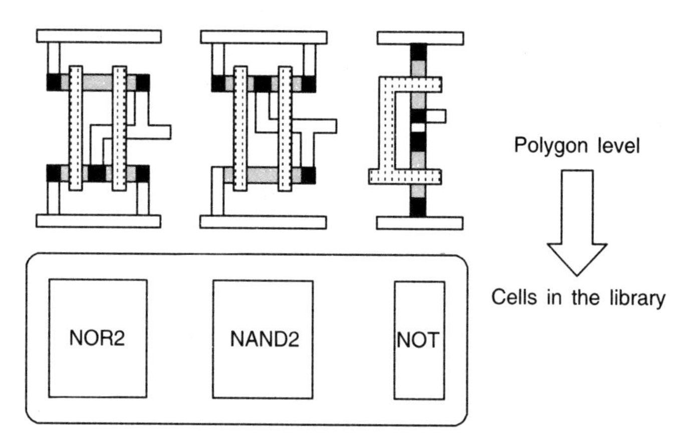

Figure 5.18 Primitive polygon-level library entries.

Once the initial library is established, we can use the cell entries in our design by instancing them into our layout. An instance is a copy of the cell in the library. An instanced object cannot be altered in the new layout. The only way to change the characteristics of an instance is to change the library entry. The new layout will be a more complex object that can itself be stored as a cell in the library. In Figure 5.19, two new cells named cell_1 and cell_2 are designed using instances of the primitive library, plus polygons of their own. We may save the new cells and

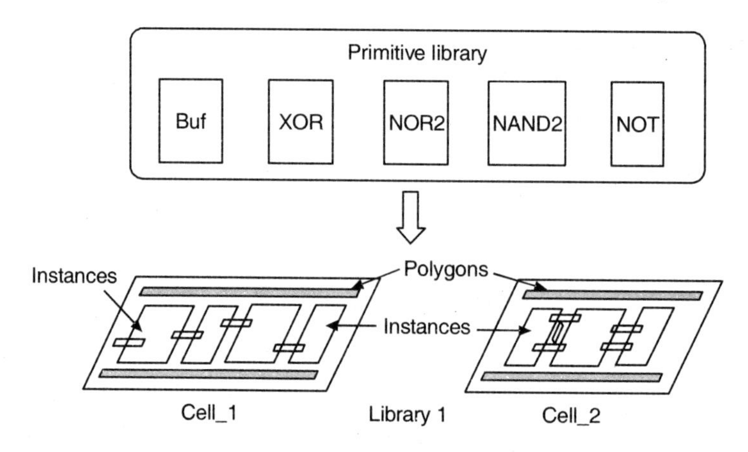

Figure 5.19 Expanding the library with more complex cells.

create a larger library group (Library 1) for use in more complex designs. This process may be repeated as needed. Useful functions are designed into new cells that become a part of the library, and are used to build other cells. The final cell collection chosen for the library should contain the great majority of the cells needed for the design projects.

The concept of cell hierarchy is based on the building of the cell library. Figure 5.20 shows the cell hierarchy structure. At the most primitive level, the cells consist only of polygons representing the material layers (Level_1). The Level_2 cells consist of polygons and instances of Level_1 cells. The next group is designated as Level_3 cells. These consist of polygons and may contain instances from Level_1 and Level_2 entries. The last groups shown in the drawing are the Level_4 cells. They are made up of polygons and instances of any cells from Level_1 to Level_3.

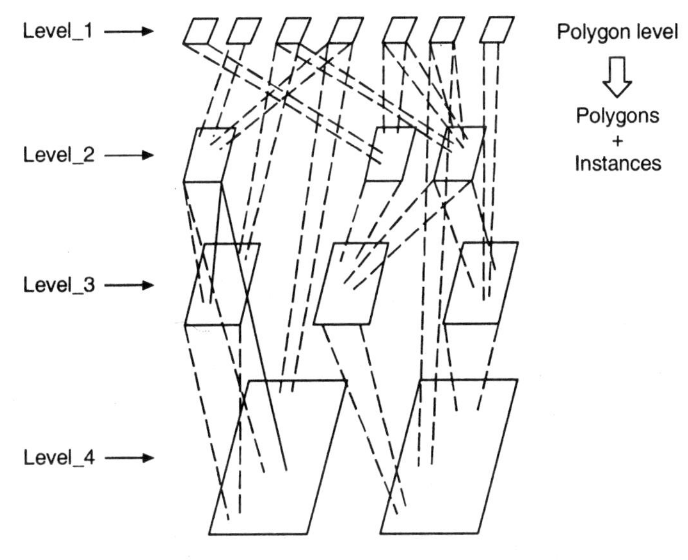

Figure 5.20 Cell hierarchy.

An instance is only a copy of a simpler entity and its internal structure cannot be altered at a higher level. For example, if a Level_2 cell is instanced into a Level_4 cell, the Level_4 design treats it as being invariant. To alter the Level_2 cell, one must return to the original Level_2 design. Any changes in the cell will propagate to all higher levels, where the cell was instanced. In practice, library is used by a large number of designers, but most users do not have access privileges to the central group of cells. This prevents someone from changing a characteristic that may be critical in another's design.

Although the contents of an instance to be changed, it is possible to decompose it into polygons by the flatten command. After a cell is flattened, all references to the original cell are lost and individual features of the circuit can be modified. Figure 5.21 shows the effect of the flatten operation. A flattened cell cannot be restored to its original instanced form.

The concept of design hierarchies allows us to build up complex networks by starting at primitive levels and adding cells as deemed useful. In this manner, various libraries can be built and maintained for use in many different projects. Complex systems are broken down into manageable sections, and the concept of building chips with millions of transistors becomes a reality.

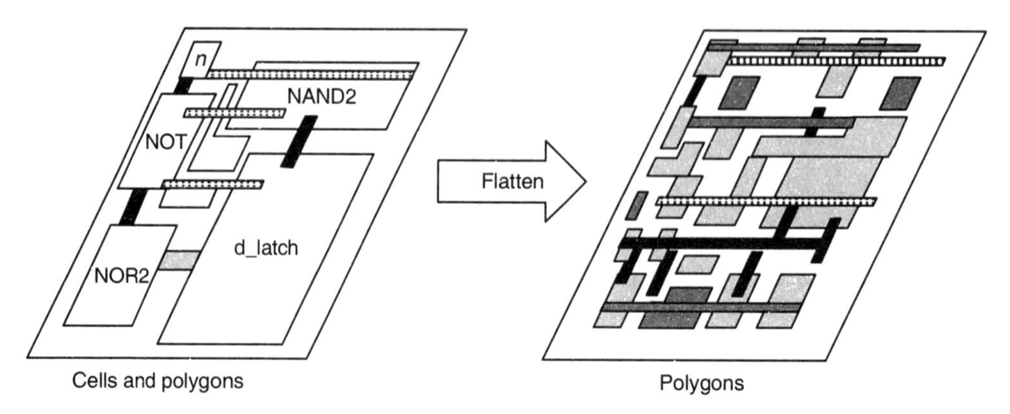

Figure 5.21 Effect of the flatten operation.

SUMMARY

In Chapter 5, design processes of MOS circuit are described in detail. The different MOS layers and their stick diagrams in NMOS and CMOS design are listed first and then various steps involved in drawing the stick representations are explained. The stick diagram is the cartoon of the stick layout. It actually conveys layer information through the use of colour code. The design rules are the translation of circuit design concepts in stick diagram or symbolic form into actual geometry in silicon. They determine the low level properties of chip designs: how small individual logic gates can be made; how small the wires connecting gates can be made, etc. The different layout approaches, micron-based and lambda-based design rules are described in this chapter. The layout examples of logic gates such as CMOS inverter, CMOS NAND gate and CMOS NOR gate are given. The chapter also deals about the elements of physical design which give the necessary information for creating the chip, and provides for hierarchical design.

REVIEW QUESTIONS

1. Explain stick diagram and draw stick layout using NMOS design.
2. Explain Lambda (λ) based design rules.
3. Explain the CAD toolsets.

SHORT ANSWER QUESTIONS

1. What is the purpose of design rule?

   Ans. Design rule is to guarantee that under the worst cumulative variations of the processes, the circuit does not fail to operate.

2. What is the use of stick diagram?

   Ans. It is used to convey layer information through the use of a colour code.

3. What are the different approaches for describing the design rule?

   Ans. (a) Micron design rule.

   (b) Lambda (λ) based design rule.

4. List some of the important CAD toolsets

   Ans. (i) Layout editors.

   (ii) Design Rule Checkers (DRC).

   (iii) Circuit extraction.

5. Give three possible approaches for making contacts between polysilicon and diffusion in NMOS circuit.

   Ans. (i) Polysilicon to metal then metal to diffusion.

   (ii) Buried contact polysilicon to diffusion.

   (iii) Butting contacts.

6 Special Circuit Layouts

6.1 INTRODUCTION

While highly complex logic is often best implemented with symmetric, repeatable, and expandable structures such as PLAs and ROMs, many simple logic problems can be solved with special circuit approaches. There are a number of circuits for which unique, clever layouts have been devised, some of which will be investigated in this chapter.

The tally circuit is a special application of pass-transistor logic, and will be studied first. Two-level logic will be examined next: NAND–NAND, NOR–NOR, and AND–OR–INVERT (AOI) will be compared. The AND–OR–INVERT circuit will perform two-level logic at almost the speed to one-level logic.

The Exclusive-OR (XOR) gate is functionally equivalent to an AOI structure, as is the Exclusive-NOR (XNOR), and all three are topologically identical. The exclusive-OR circuit is obtained from the Exclusive-NOR circuit by interchanging one pair of inputs, and both are special cases of AND–OR–INVERT logic.

NMOS and CMOS multiplexing (MUX) circuits will also be examined. The CMOS MUX with clocked precharge is structurally, equivalent to an NMOS MUX, except for the *n*-well required by the pull-up circuitry. The general-purpose function block discussed in Sec. 6.8 is also structurally related to the MUX. The Mead–Conway barrel shifter will also be studied.

Layout and routing problems and techniques will be studied next. The problem of interconnecting a large number of nodes on a chip is increasingly taking more design time and chip area and many designs are limited by the metallization routing. A large effort is being invested in solutions to routing problems at present. Two-metal and multimetal designs are being investigated in attempts to case routing limitations.

The goal of wire routing is to maximize metal while minimizing the total wire length, the number of crossovers, and the number of interconnects. The basic tradeoff is between chip area and routability. If sufficient wiring space is not available, the chip cannot be wired as is, and the layout must be redone. If too much area is allocated for wiring, the chip cost is too high.

The success of any algorithm is measured by its completion rate and computation time. Acceptable solutions must guarantee completion. Most algorithms do well for small layouts, but many fail to complete complex routing layouts. When the computer is unable to finish the routing, a good designer may be better off to ignore the routing attempt and start over.

Power and clocking should be routed first. When laying out an array of alternating V_{DD} and GND grids, channels must be laid out in such a way as to allow routing nets through them. Module placement on the chip is related to routing success, and some routing algorithms allow the designer to adjust placement interactively in order to assist the router.

6.2 TALLY CIRCUITS

A tally circuit counts the number of inputs that are high and outputs the answer. If there are N inputs there are $N + 1$ possible outputs, corresponding to 0, 1, 2, ..., N inputs that are high. Output S_i, is the sum of the number of inputs that are high. A 4-input, 5-output tally circuit is shown in Figure 6.1.

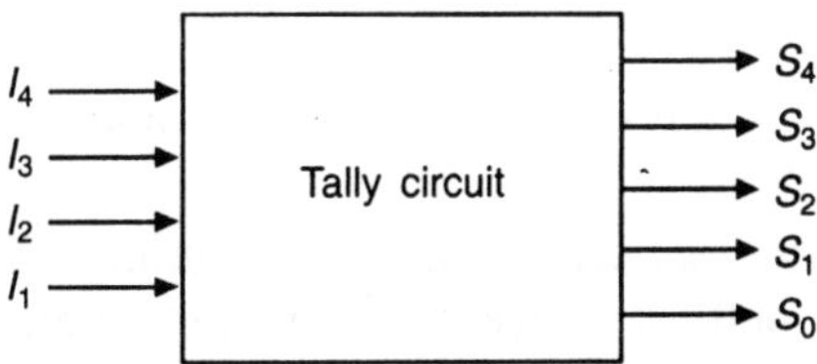

Figure 6.1 4-input, 5-output tally circuit.

There are several conceptual approaches to laying out the circuit. The first method that comes to mind might be to write a minimum sum-of-products solution for each of the $N + 1$ output functions. For the 4-input tally circuit shown in Figure 6.1, the output sum-of-products solutions are:

S_0 is true if no inputs are high: 1 term

S_1 is true if only one input is high: 4 terms

S_2 is true if any two inputs are high: 6 terms

S_3 is true if all the inputs but one are high: 4 terms

S_4 is true if all the inputs are high: 1 term

The sum-of-products solutions consist of the binomial expansions of N items taken M at a time, as M ranges from 0 to N. In the above example, there are a total of sixteen product terms, requiring sixteen 4-input AND gates. One 6-input and two 4-input OR gates are also required to obtain the desired outputs.

This approach yields a total of 19 gates and 78 inputs, with the 5 outputs in a canonical sum-of-products form. This is a worst-case, or upper-bound, solution to the problem.

Another convenient approach would be to use a ROM look-up table. For a 4-input tally circuit, this requires 4 inputs, 16 words of 5 bits each, and 5 outputs. This is also quite uneconomical. A PLA design will be smaller than an equivalent ROM layout, but it is still not an ideal solution.

A third approach might consist of an iterative network such as a chain of simple adders. Parity checkers are often designed this way.

A fourth approach would be a hierarchy of small tally circuits. For 4 inputs, this could be done with two 2-input tally circuits, with their outputs combined in a third 2-input tally circuit, and run through a decoder to yield the correct output.

A fifth attempt might be a tree network, which would have the layout shown in Figure 6.2. Each node requires a yes/no circuit to provide the signal to an output branch according to whether the number of high inputs is greater than a given amount or not. In Figure 6.2 if the input is 4 highs (logic 1s), the signal flow is from node a to d to S_3. An input of 2 sends the signal from node a to b to S_2, etc.

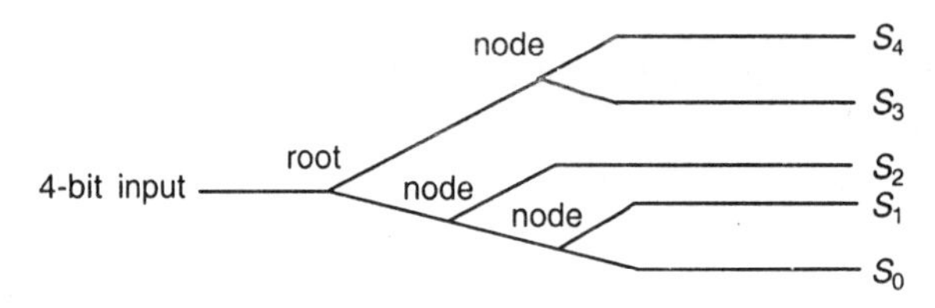

Figure 6.2 A tree network realization of a 4-input tally circuit.

What is desired is a circuit that can be easily laid out in a regular and expandable manner, with a minimum number of pull-up devices and a minimum area requirement.

Charge-steering (pass-transistor) logic is ideally suited to this type network, and the basic layout of a 3-input, 4-output tree network is shown in Figure 6.3. Each tree branch requires two choices, depending upon the inputs. Two pass-transistors can be gated to accomplish each branching operation. The circuit must be readily expandable to any reasonable size.

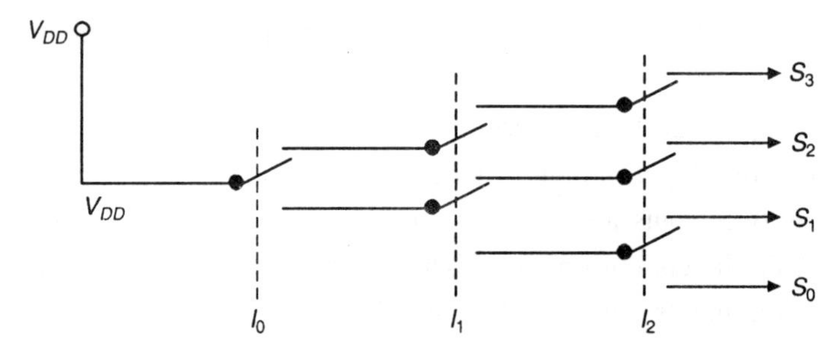

Figure 6.3 A conceptual layout of a 3-input tally circuit.

A stick-diagram of a 3-input tally circuit designed with pass transistors and only one pull-up load device is shown in Figure 6.4. This circuit is highly repetitive, easily expandable to any number of inputs, and requires no additional pull-ups. Since all the pull-down devices are pass transistors, they can minimize geometry also.

The circuit allows a shift right, or a shift up and right. When I_0 is false, $\overline{I_0}$ shifts the signal to the right, and when I_0 is true, it shifts the data up one row and to the right. Each input that is true causes the input to shift up and to the right, each input that is complemented causes the input to shift right only.

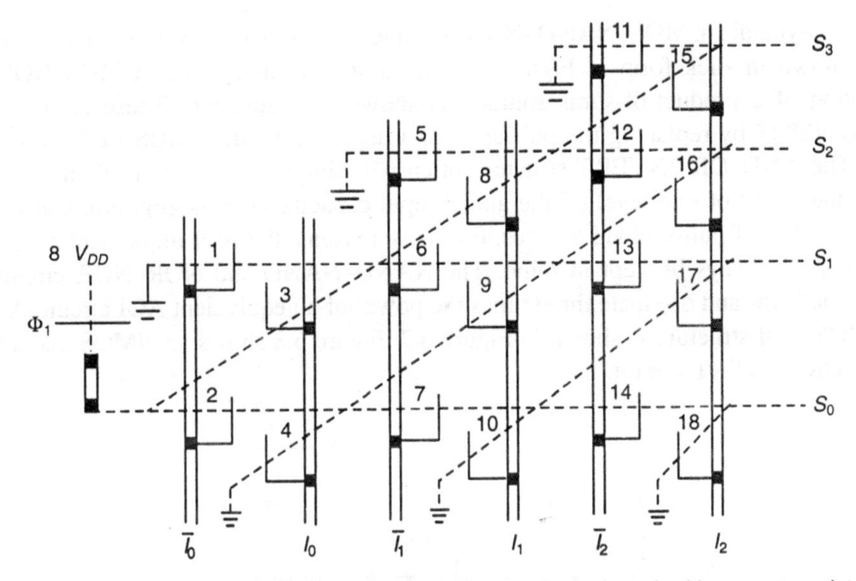

Figure 6.4 A stick-diagram of a 3-input tally circuit designed with pass transistors.

Label the inputs I_0, I_1, I_2, plus their complements, as shown in Figure 6.4. When the inputs are 0, 0, 0, the only path from the pull-up to an output is through transistors 2, 7, and 14. The pull-up is now connected to S_0, which goes high, for a tally of 0. When the input is 1, 0, 0, the only path from pull-up to an output is through transistors 3, 6, and 13. The output is S_1 high for a tally of 1.

All eight combinations of inputs and the paths they activate are shown in Table 6.1, and the length of each pass-transistor string is equal to the number of inputs. The tree structure is ideally suited to small or slow circuits where propagation delays are not important. However, the circuit can be expanded to larger size by allowing for the insertion of restoring inverters at suitable intervals.

Table 6.1 Eight combinations of inputs and the paths they activate

Input I_0	I_1	I_2	*Enabled path PU to output*	*High output*
0	0	0	2, 7, 14	S_0
0	0	1	2, 7, 17	S_1
0	1	0	2, 9, 13	S_1
0	1	1	2, 9, 16	S_2
1	0	0	3, 6, 13	S_1
1	0	1	3, 6, 16	S_2
1	1	0	3, 8, 12	S_2
1	1	1	3, 8, 15	S_3

6.3 NAND–NAND, NOR–NOR, AND AOI LOGIC

All Boolean functions can be represented in a canonical sum-of-products form (AND–OR), as well as in a canonical product-of-sums form (OR–AND). AND–OR logic converts directly to NAND–NAND logic and OR–AND logic converts directly to NOR–NOR logic.

A typical dynamic CMOS NAND–NAND implementation of a simple sum-of-products function is shown in stick form in Figure 6.5, and a stick-form dynamic CMOS NOR–NOR implementation of a product-of-sums solution is shown in Figure 6.6. These circuits can be converted to NMOS by replacing the pull-up with a depletion-mode NMOS FET, and deleting the *n*-well. The AND–OR–INVERT structure often will simplify the circuit, giving two levels of logic for the speed of one level (if the added input capacitance is negligible), and requiring a minimum number of pull-up load devices. For these reasons, the AOI implementation of two-level logic should always be kept in mind. The NAND–NAND and NOR–NOR circuits both require three pull-ups and dissipate three times the power of an equivalent AOI circuit. A typical clocked CMOS AOI structure is shown in Figure 6.7. Figure 6.8 shows an NMOS and a CMOS static AND–OR–INVERT circuit.

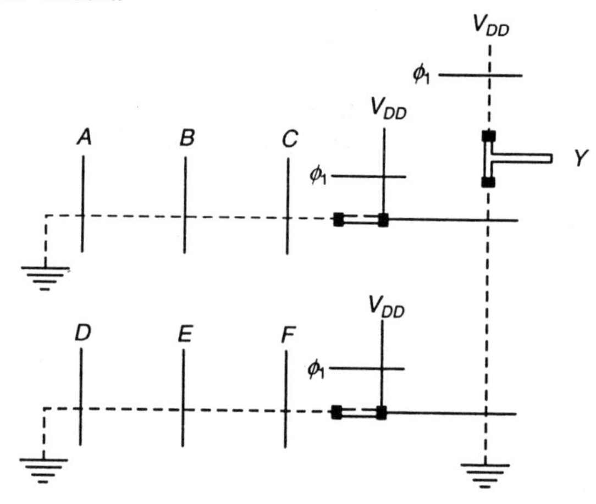

Figure 6.5 NAND–NAND implementation of $Y = ABC + DEF$.

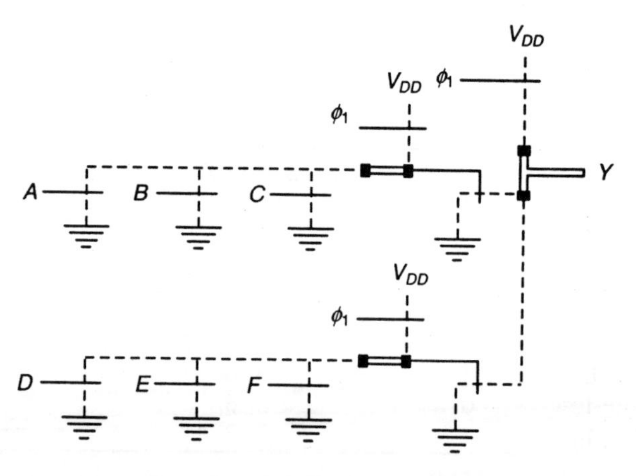

Figure 6.6 NOR–NOR implementation of $Y = (A + B + C)(D + E + F)$.

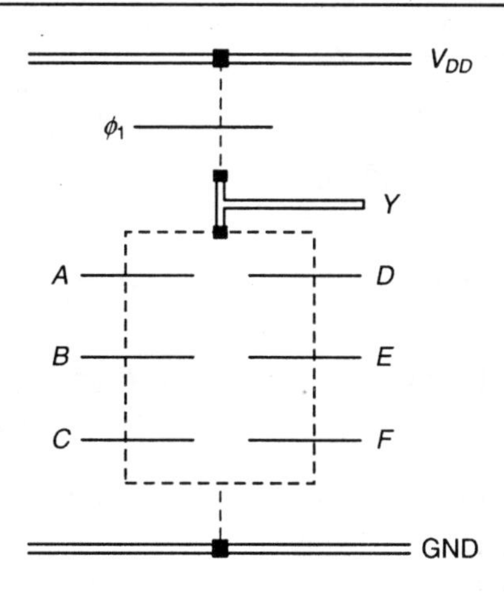

Figure 6.7 AOI implementation of $Y = (\overline{ABC + DEF})$

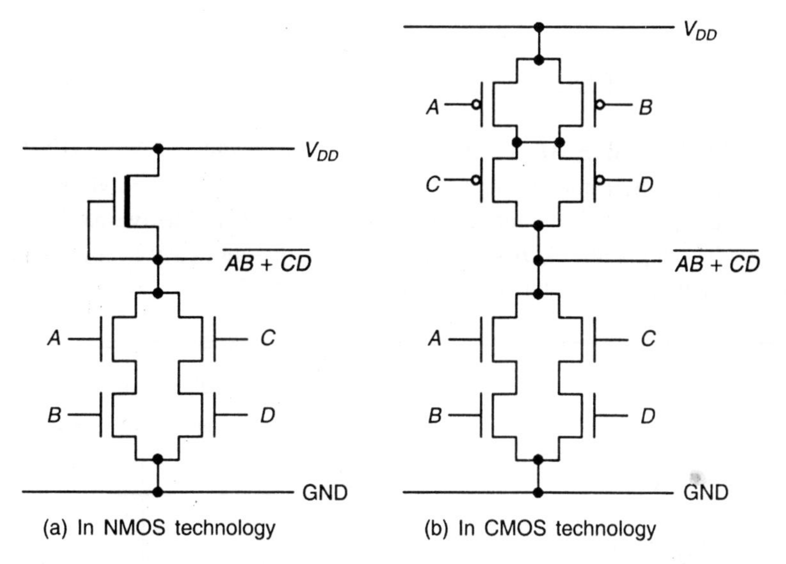

Figure 6.8 Static AOI gates to realize $Y = (\overline{AB + CD})$.

Dynamic logic dominates VLSI design due to its much lower power dissipation. Dynamic NMOS circuitry suffers from both circuit complications and loss of packing density with respect to static NMOS. These are the main reasons for choosing NMOS in the first place. Dynamic precharge-high evaluate-low NMOS and CMOS AOI circuits are shown in Figure 6.9.

Memory design is still largely done in NMOS because it is faster than CMOS, and because the savings in real estate due to the utilization of dynamic logic more than offset the increased clocking circuitry needed. This allows the designer time that can be applied to the more complicated clocking circuitry and substrate pump design. Often a satisfactory compromise is to be CMOS peripheral circuitry with standard NMOS memory matrices (the mostly NMOS approach), in order to maintain the low standby power dissipation of CMOS, while approaching the speed of NMOS.

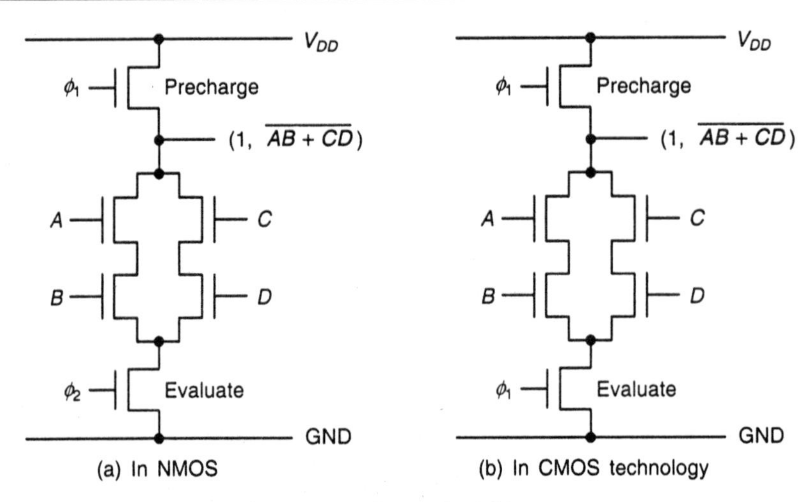

Figure 6.9 Dynamic AOI gates.

6.4 EXCLUSIVE-OR STRUCTURES

Since $\overline{A \oplus B} = \overline{A\bar{B} + \bar{A}B} = \bar{A}\bar{B} + AB = \bar{A} \oplus B = A \oplus \bar{B}$, the Exclusive-NOR (Ex-NOR) circuit can be implemented with the same circuit as the Exclusive-OR (Ex-OR) by simply interchanging two inputs. They are topologically identical. The Exclusive-OR (Exclusive-NOR) circuit, shown in Figure 6.10, is identical to the AOI structure and is simple and easy to implement. Modified Ex-OR/Ex-NOR structure are shown in Figures 6.11 and 6.12.

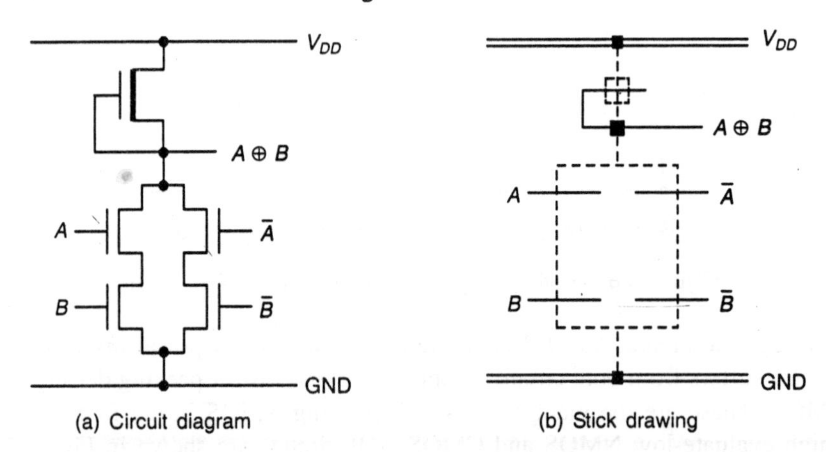

Figure 6.10 The NMOS Ex-OR/Ex-NOR structure with inputs.

The circuits shown in Figures 6.10(a), (b) and 6.11 are NMOS. They are structurally similar to precharged CMOS circuits, and can be converted to CMOS by replacing the pull-up by as clocked PMOS device and adding an n-well. A CMOS Ex-OR equivalent of the NMOS circuit of Figures 6.12(a) and (b) is shown in Figure 6.13. It consists of one conventional inverter, one transmission gate, and one 'floating' inverter whose pull-up is input A and whose pull-down is the output of the other inverter.

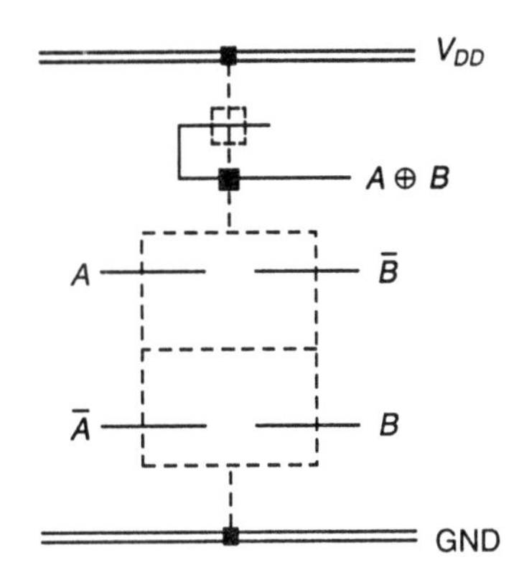

Figure 6.11 A modified NMOS Ex-OR/Ex-NOR Circuit.

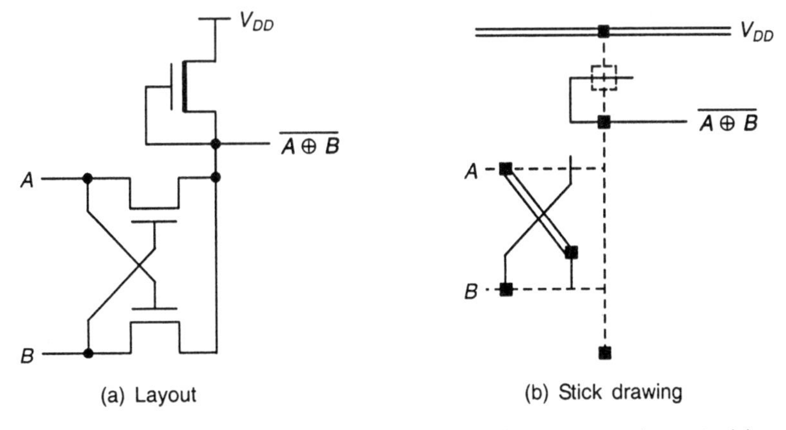

Figure 6.12 An Ex-OR/Ex-NOR gate which requires no complemented inputs.

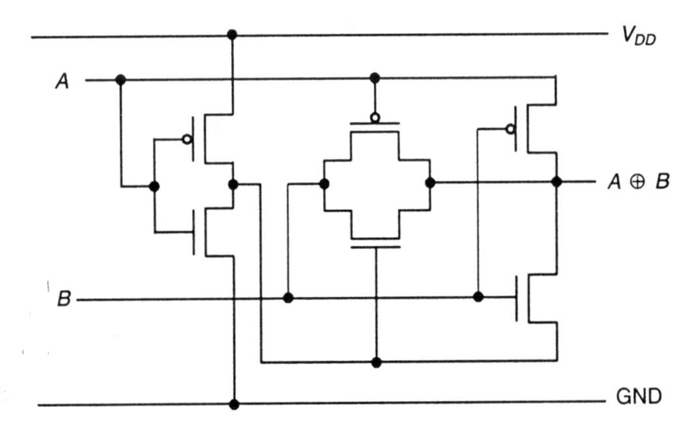

Figure 6.13 A static CMOS Ex-OR circuit.

Multiplexer structures. The basic symbol of a 4-to-1 multiplexer (MUX) and an equivalent circuit for it are shown in Figures 6.14(a) and (b). The goal is to implement a simple, cost-effective multiplexer circuit. The MUX can be seen to consist of switched paths, as did the tally circuit. It is thus ideally suited to pass-transistor logic, and can be easily implemented in NAND form or NOR form.

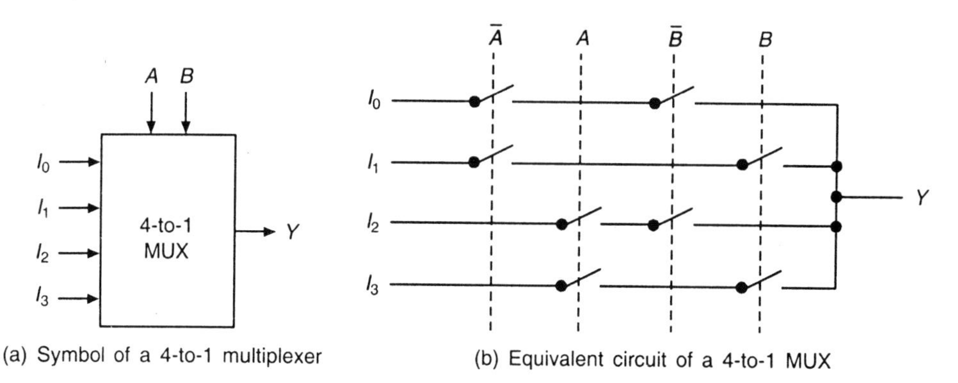

(a) Symbol of a 4-to-1 multiplexer

(b) Equivalent circuit of a 4-to-1 MUX

Figure 6.14

NMOS multiplexers. A NOR implementation requires eight enhancement-mode transistors and eight contacts, not counting any driving inverters which are external to the circuit. The circuit is shown in Figure 6.15. The NOR approach is preferred for large multiplexers.

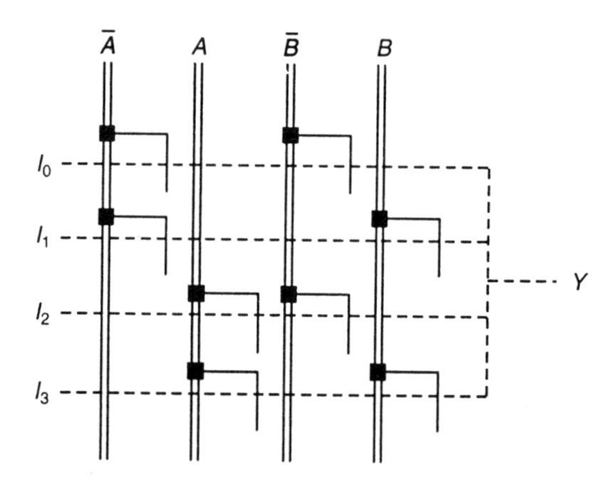

Figure 6.15 The NOR implementation of an NMOS 4-to-1 MUX.

An alternative approach to multiplexers design consists of a NAND structure, using depletion-mode devices to short circuit the undesired connections, while using enhancement-mode devices to form the switches that exert control. The MUX is defined by the locations of the ion implants required to form the depletion-mode transistors, as shown in Figure 6.16. If a large NAND multiplexer is required, restoring inverters must be incorporated in the layout.

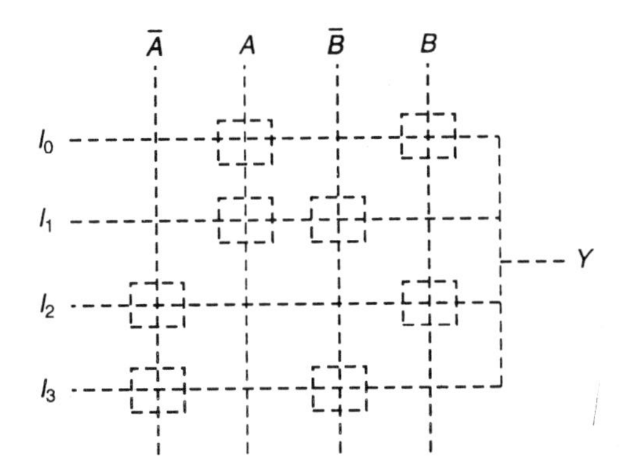

Figure 6.16 The NAND implementation of an NMOS 4-to-1 MUX.

CMOS multiplexers. A CMOS multiplexer is shown in Figure 6.17. The structure is the same as for the NMOS NOR multiplexer, except that the NMOS pass transistors have been replaced by CMOS transmission gates. The NMOS 4-to-1 multiplexer has a faster fall time, and the CMOS MUX has the faster rise time.

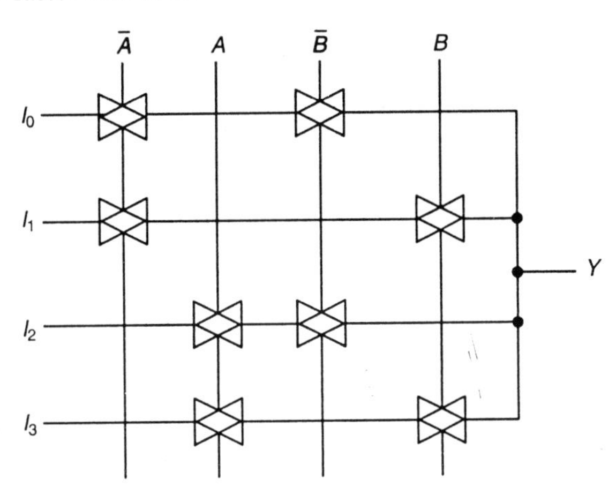

Figure 6.17 CMOS implementation of a 4-to-1 MUX.

To layout a CMOS MUX, the first step is to redraw the circuit with all the NMOS pull-down devices in the bottom half of the layout, and all the PMOS pull-up transistors in the top half of the layout, as shown in Figure 6.18. This allows one to put all the PMOS devices in a single isolation well and all the NMOS devices in another single isolation region.

While simple to layout, conventional CMOS multiplexers take lots of real estate. The pull-down part of the circuit is an NMOS multiplexer, and pass-transistor logic must supply both pull-up and pull-down paths. A static pull-up can be obtained by running the output, *Y*, through an inverter and feeding the inverted signal, *Y*-BAR, to the gate of a PMOS pull-up transistor. Whenever *Y* is low, the pull-up device will be OFF, and whenever *Y* is high, the pull-up transistor is ON. This circuit is shown in Figure 6.19, and has negligible dc power dissipation.

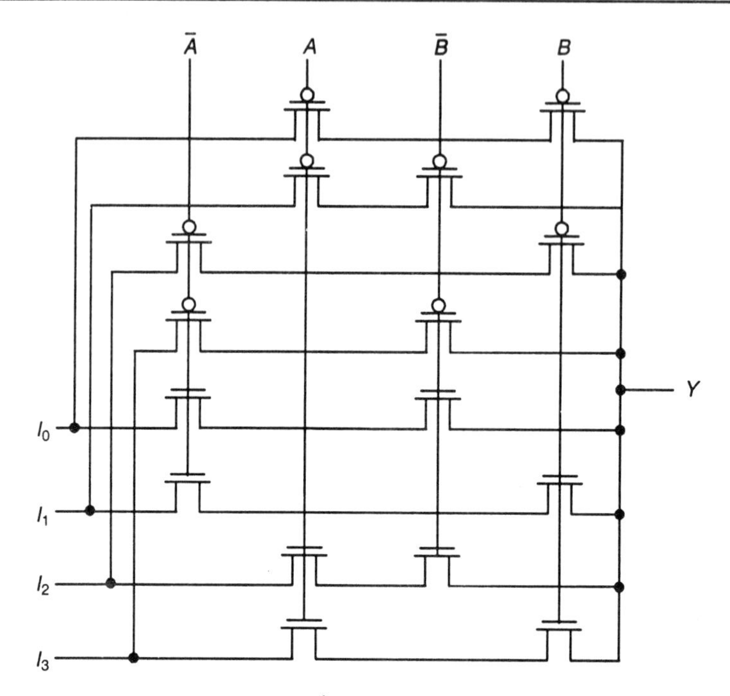

Figure 6.18 CMOS 4-to-1 MUX with full pull-up and pull down circuitry.

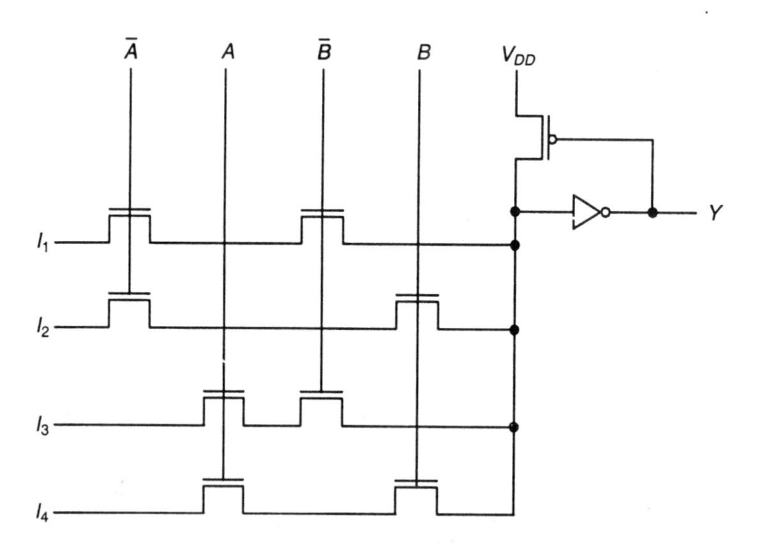

Figure 6.19 CMOS 4-to-1 MUX with a static pull-up driven by Y-BAR.

A dynamic CMOS pull-up circuit can be realized by precharging the pull-up on phase 1 of the clock as shown in the circuit of Figure 6.20. The dynamic circuit is comparable in speed to the all-NMOS multiplexer.

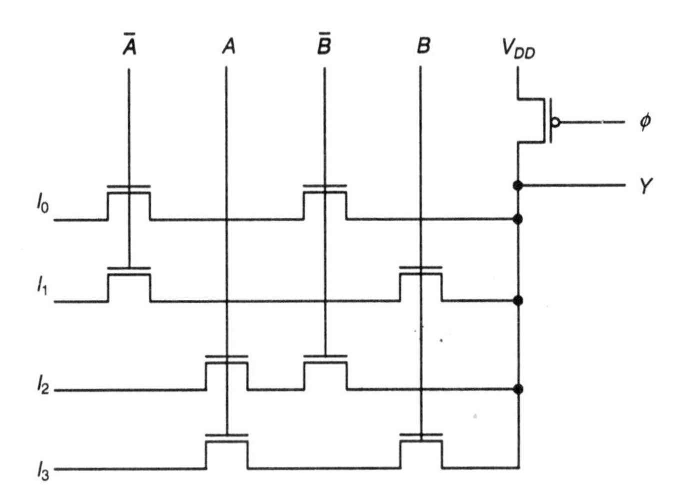

Figure 6.20 Dynamic CMOS 4-to-1 MUX, using a clocked precharge pull-up.

6.5 BARREL SHIFTER

A barrel shifter is a wrap around or end around shifter that is a very useful switch array. It uses only combinational logic, and can be easily implemented in silicon. The basic layout is shown in Figure 6.21. The inputs are labelled I_i, the shift controls S_i, and the outputs Y_i. If a shift of one or two is desired, then the outputs will be as indicated in Figures 6.22(a) and (b).

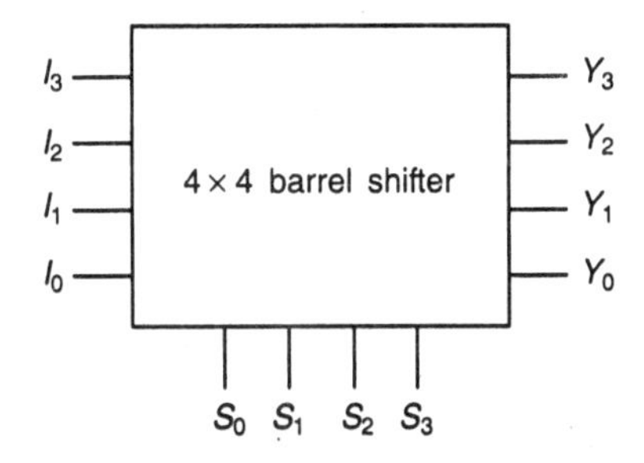

Figure 6.21 Block diagram of a 4-bit barrel shifter.

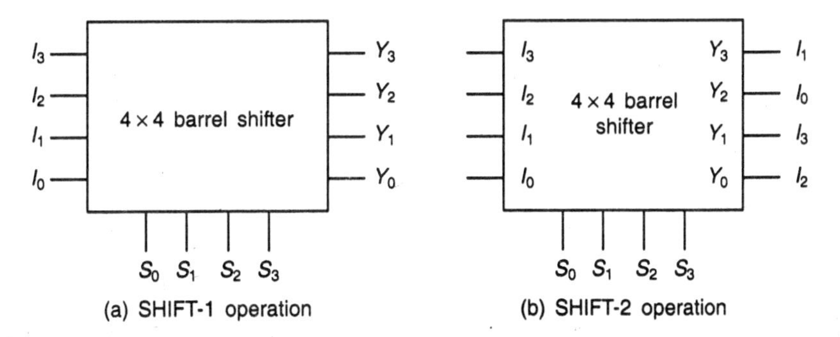

Figure 6.22 Conceptual picture.

The barrel shifter will be laid out with two 4-bit buses running horizontally through it, and the data paths running vertically through it. A way of connecting any bus with any output bit is needed, and the 4×4 crossbar switch is a simple circuit from which to start.

The switching is done with transistors labelled S_{ij}, where switch ij connects bus i to output j. All sorts of shifting and interchanging of data can be done with this structure, but it requires N^2 control lines which limits the design to reasonably small values of N.

To convert the crossbar shifter to a barrel shifter, one must add a third horizontal bus line to handle the shift signals. Next, add all the FET switches to connect bus i to output i as shown in Figure 6.23, and connect the gate of all these transistors together with a line labelled SHIFT-0.

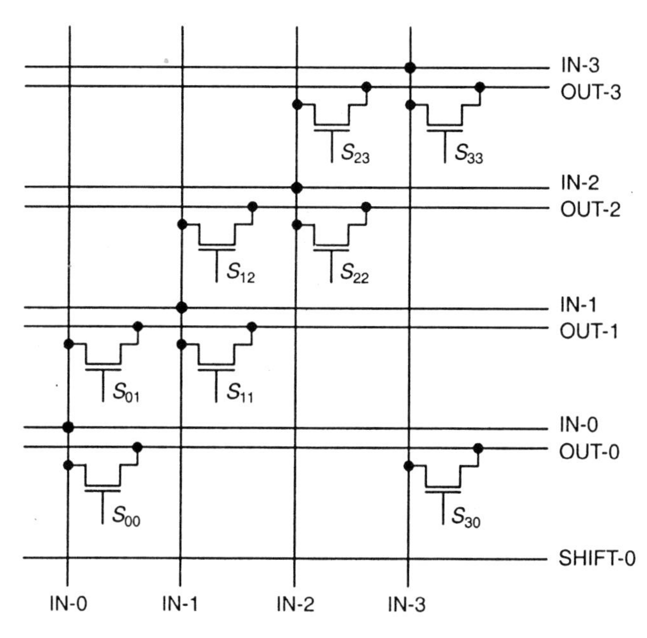

Figure 6.23 Starting layout for a 4×4 crossbar shifter with SHIFT-0 and SHIFT-1 transistors in place.

Add the FET switches to connect bus i to output $i + 1$, and connect the gates of all these transistors together with a line labelled SHIFT-1. The 4×4 barrel shifter at this point is shown in Figure 6.24. Continue in this manner until finished. Only one of the shift lines may be high at any one time. The finished circuit is shown in Figure 6.25.

The barrel shifter is now ready to be laid out. The buses can run horizontally through the shifter in polysilicon, the output lines can run through horizontally in diffusion, and the control lines for shifting can run vertically in metal. The lines connecting the shift signal to the gates of the switching transistors can then run horizontally in polysilicon also.

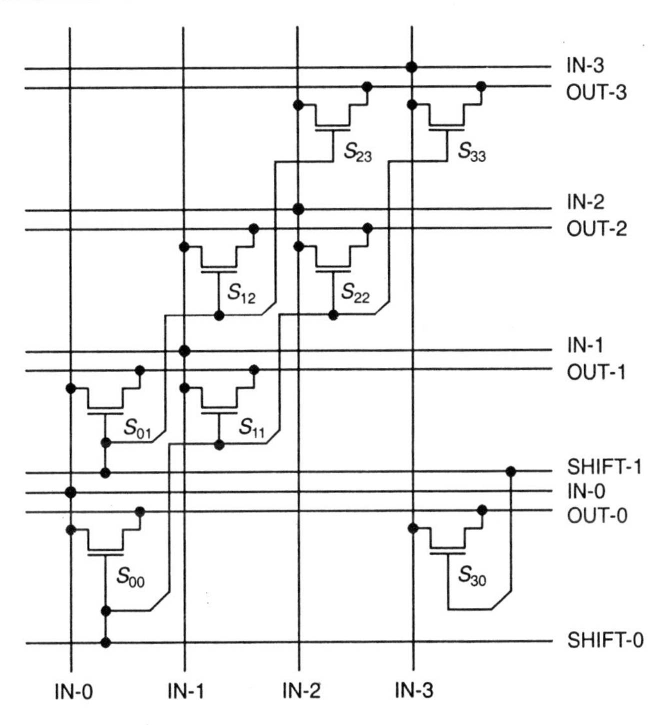

Figure 6.24 The crossbar shifter starting with SHIFT-0 and SHIFT-1 control lines added.

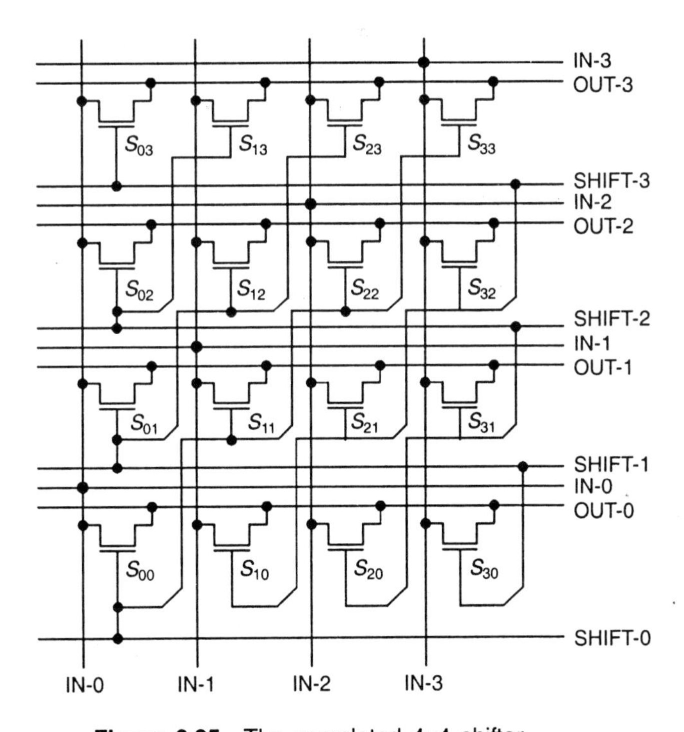

Figure 6.25 The completed 4×4 shifter.

6.6 TRANSMISSION GATES

The VLSI designer must minimize power, delay and chip area. Good pass transistor design can often give faster logic than conventional MOS design, with very low power loss and much less chip area. An NMOS pass transistor can pull down to the negative rail (Figure 6.26(b)), but it can only pull up to a threshold voltage below the positive rail (Figure 6.26(a)). It can output a

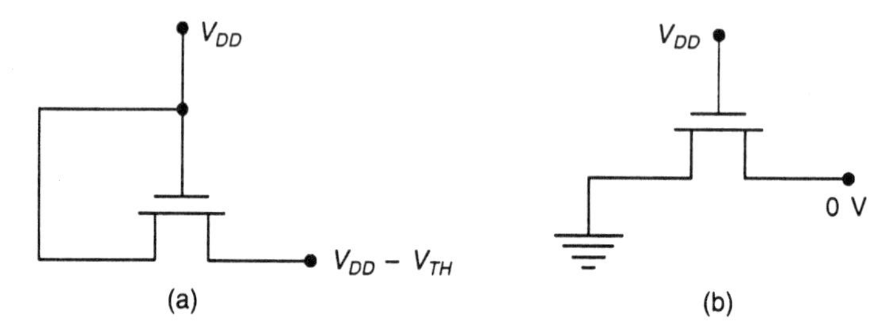

Figure 6.26 The NMOS pass transistor passes.

strong zero, but only a weak one. By contrast, a PMOS pass transistor can pull up to the positive rail (Figure 6.27(a)), but can only pull down to a threshold voltage above the negative rail. It can output a strong one, but a weak zero (Figure 6.27(b)).

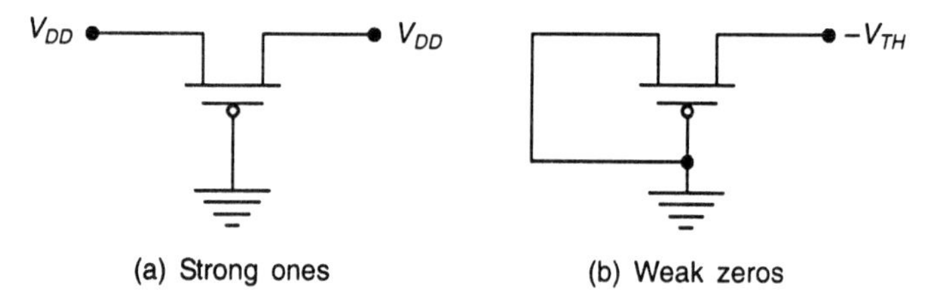

Figure 6.27 The PMOS pass transistor passes.

The NMOS and PMOS depletion—mode devices can both pass strong ones and zeros, but they never turn off and can only be used to remove control from a gate signal.

Transmission gates (TGs) are logic—controlled switches that can be used to construct a wide variety of logic networks. The symbol and switch models of transmission gate are shown in Figure 6.28(a), (b) and (c).

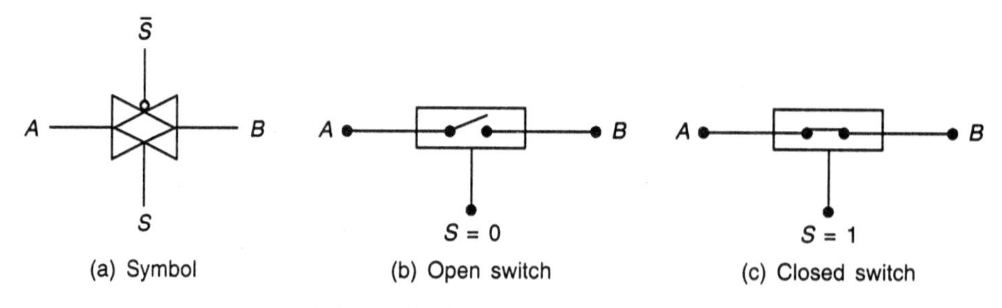

Figure 6.28 Transmission gate.

The value of the control bit S determines whether the path between the left and right sides (A and B) is open or closed.

When $S = 0$, no path exists, hence no relationship between A and B

When $S = 1$, TG provides a closed path and $B = A$

A CMOS transmission gate consists of a PMOS and an NMOS pass transistors connected in parallel and gate terminals of the transistors are supplied with complementary control signals. The circuit diagram, stick layout and physical layout of the CMOS transmission gate are shown in Figures 6.29(a), (b) and (c) respectively.

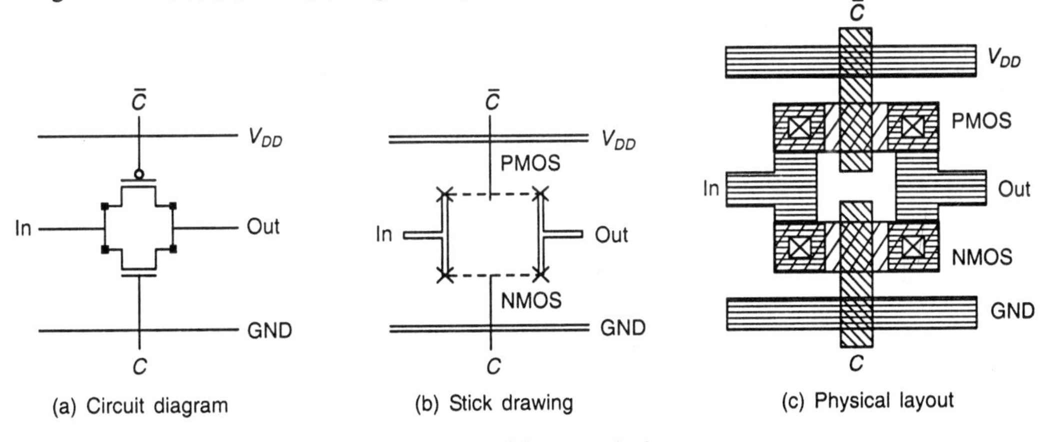

(a) Circuit diagram (b) Stick drawing (c) Physical layout

Figure 6.29 CMOS transmission gate.

The CMOS transmission gates are superior to NMOS pass-transistors in two ways:

1. They output both strong 1s and strong 0s.
2. The TGs consists of two transistors in parallel, this reduces the resistance value to only half the value of a single pass transistor.

But the disadvantages of CMOS transmission gates are that

(i) they require more area than NMOS pass circuitry.
(ii) they require complemented control signals.

6.7 LATCHES AND FLIP-FLOPs

A generic memory element has an internal memory and some circuitry to control access top the internal memory. In CMOS circuits, the memory is formed by some kind of capacitance or by positive feedback of energy from the power supply. Access to the internal memory is controlled by the clock input—the memory element reads its data input value when instructed by the clock and stores that value in its memory.

Latches are transparent; while the internal memory is being set from the data input the (possibly changing) input value is transmitted to the output.

Flip-flops are not transparent—reading the input value and changing the flip-flop's output are two separate events.

The two most commonly quoted parameters of a memory element are its set-up time and hold time (Figure 6.30). The set-up time is the minimum time the data input must be stable before the check signal changes, while the hold time is the minimum time the data must remain stable after the clock changes. The set-up and hold times, along with the delay times through the combinational logic, determine how fast the system can run.

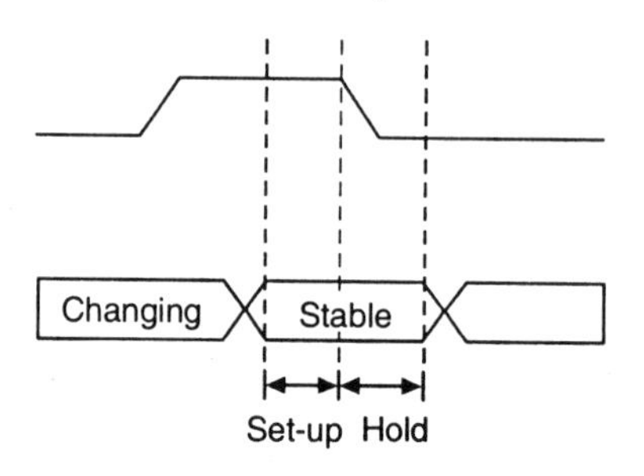

Figure 6.30 Set-up and hold times.

6.7.1 CMOS Static Latches

Figures 6.31(a) and (b) show various forms of CMOS static latches, and the symbolic layouts.

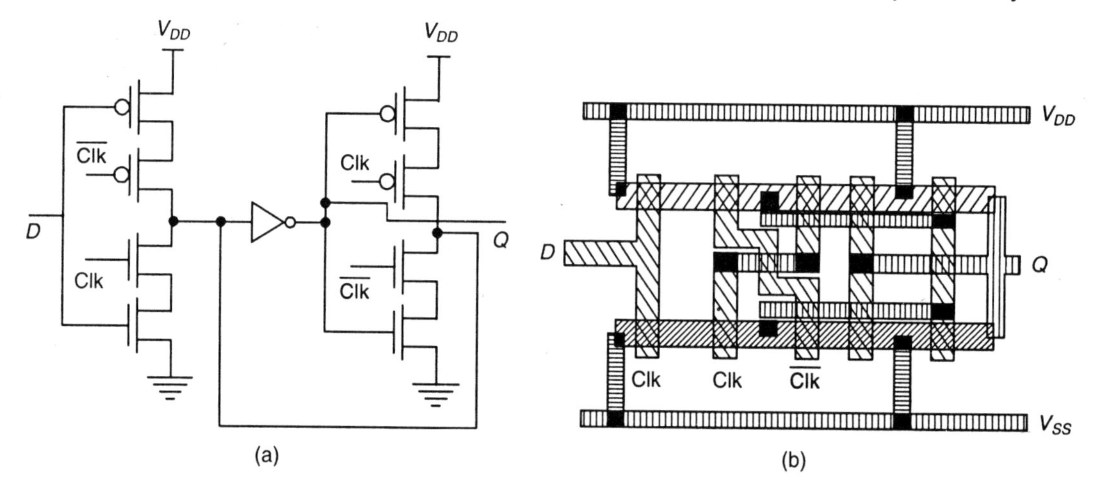

Figure 6.31 Typical CMOS static latch symbolic layouts.

6.7.2 CMOS Dynamic Latches

The simplest memory element in MOS technology is the dynamic latch. It consists of a transmission gate and a capacitance. It is called dynamic because the memory value is not refreshed by the power supply and a latch because its output follows its input under some conditions (Figure 6.32).

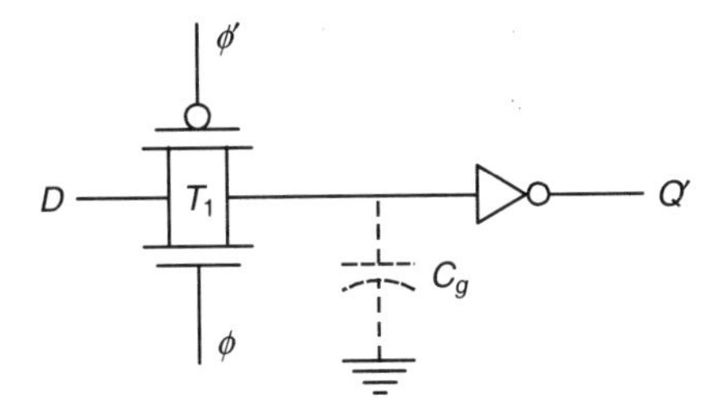

Figure 6.32 Dynamic latch.

The latch capacitance is guarded by a fully complementary transmission gate. Dynamic latches generally require *p-n* pair switches because they transmit logic 0 and 1 equally well and provide better storage on the capacitor. The transmission gate is controlled by two clock signals, ϕ and ϕ'. Figure 6.33 shows one possible stick diagram for a dynamic latch and Figure 6.34 shows a layout of the latch.

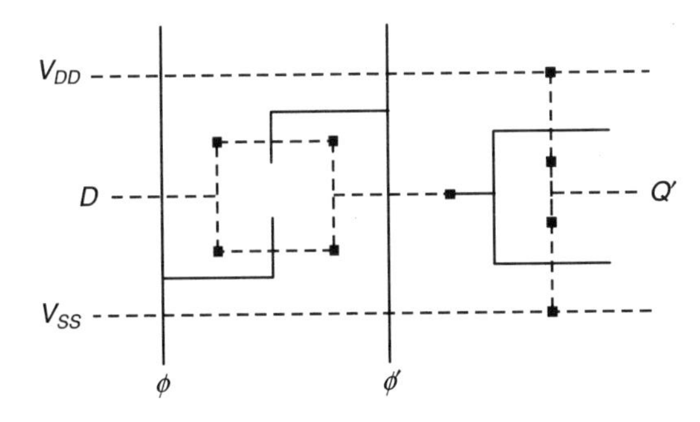

Figure 6.33 Stick diagram of a dynamic latch.

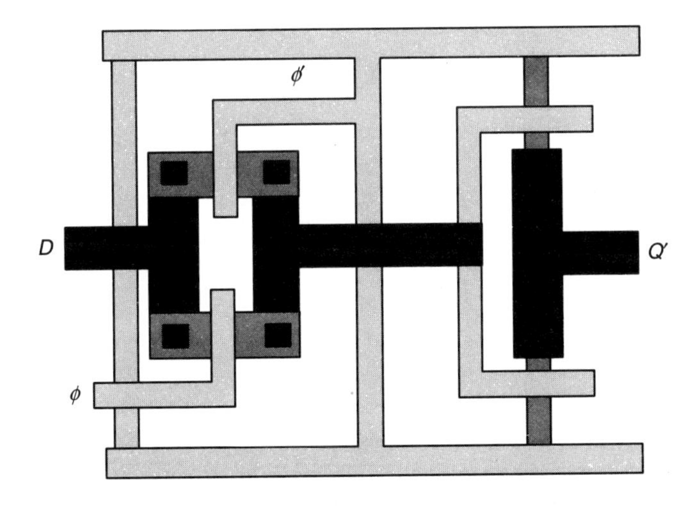

Figure 6.34 A layout of a dynamic latch.

6.8 FAN-IN AND FAN-OUT OF CMOS LOGIC DESIGN

The fan-in of a logic gate is the number of inputs the gate has in the logic path being exercised. The fan-out of a logic gate is the total number of gate inputs that are driven by a gate output.

The fan-in of a gate affects the speed of the gate. If two identical transistors are connected in series, the rise (or fall) time will be approximately double that for a single transistor with the same capacitive load. When gates with large number of inputs have to be implemented, the high speed performance may be obtained by using gates where the number of series inputs ranges between 2 and 5.

To illustrate this point, a very simple analysis will be presented. We will consider t_{dr} the worst-case rise delay time for an m-input NAND gate to be

$$t_{dr} = \frac{R_P}{n}(mn\, C_d + C_r + kC_g) \tag{6.1}$$

ignoring the body effect
where,

R_P = the effective resistance of p-device in a minimum-sized inverter
n = width multiplier for p-devices in this gate
k = fan-out
m = fan-in of the gate
C_g = gate capacitance of a minimum-sized inverter
C_d = Source–drain capacitance of a minimum-sized inverter
C_r = routing capacitance

Equation (6.1) can be reformulated as

$$t_{dr} = \frac{R_p}{n}(mnrC_g + q(k)\, C_g + kC_g)$$

$$= \frac{R_pC_g}{n}(mnr + q(k) + k) \tag{6.2}$$

where,

$r = C_d/C_g$ = ratio of the intrinsic drain capacitance of an inverter to the gate capacitance.
$q(k)$ = a function of the fan-out representing the routing capacitance as a multiplier times the gate capacitance.

$$t_{dr} = t_{(\text{internal} - r)} + k \times t_{(\text{output} - r)} \tag{6.3}$$

with

$$t_{\text{internal} - r} = R_pC_g\, mr$$

$$t_{\text{output} - r} = \frac{R_pC_g}{n}\left(1 + \frac{q(k)}{k}\right)$$

Similarly, the fall delay time t_{df} is approximated by

$$t_{df} = \frac{mR_n}{n}(mnrC_g + q(k)\,C_g + kC_g)$$

$$= R_nC_gm^2r + \frac{mkR_nC_g}{n}\left(1+\frac{q(k)}{k}\right)$$

$$= t_{(\text{internal}-f)} + k \times t_{(\text{output}-f)} \tag{6.4}$$

where, R_n = the effective resistance of an n-device in a minimum-sized inverter.

The previous equations assume an "equal-sized" gate strategy often used in standard cells and gate arrays where the p- and n- transistors in gates are fixed in size with relation to each other. This condition is usually enforced to automate the layout in a straightforward manner. Another equally valid strategy would be to use an 'equal-delay' method where the rise and fall times are equalized. This may allow somewhat smaller gate.

Hence with

$$t_{dr} = t_{df}$$

$$\frac{R_P}{n}(mnrC_g + q(k)\,C_g + kC_g) = \frac{mR_n}{n}(mnrC_g + q(k)C_g + kC_g)$$

$$\Rightarrow \qquad R_P = mR_n$$

Thus,

$$k_pW_p = \frac{k_nW_n}{m}$$

The equations for an m-input NOR gate for the equal-sized options are given as

$$t_{dr} = \frac{mR_p}{n}(mnrC_g + q(k)C_g + kC_g) \tag{6.5}$$

$$t_{df} = \frac{R_n}{n}(mnrC_g + q(k)\,C_g + kC_g) \tag{6.6}$$

Thus, from the discussion, it is seen that when designing CMOS complementary logic with speed as a concern, some basic guidelines may be listed as below:

- Use NAND structures where possible.
- Place inverters at high fan-out modes, if possible
- Avoid the use of NOR structure in high-speed circuits especially with a fan-in greater than 4 and where the fan-out is large.
- Use a fan-out below 5–10.
- Use minimum-sized gates on high fan-out nodes to minimize the load presented to the driving gate.
- Keep rising and falling edges sharp

When designing with power or area as a constraint, large fan-in complementary gates will always work given enough time.

SUMMARY

Special NMOS and CMOS circuits have been investigated. Clever designs can greatly facilitate VLSI layouts. The Mead–Conway tally circuit and barrel shifter are examined with this in mind. NAND–NAND and NOR–NOR logic are also examined. The NMOS and CMOS AND–OR–INVERT logic and Ex-OR/Ex-NOR logic circuits are shown to be the same technology, and to be superior in performance to NAND–NAND and NOR–NOR logic. Next, NMOS and CMOS Multiplexer circuits are examined. The general purpose block is structurally related to this Multiplexer. The chapter also describes the basic circuits and layouts of transmission gates and static and dynamic latches.

REVIEW QUESTIONS

1. Explain tally circuit and draw the tree network realization and stick diagram of the same.
2. Explain barrel shifter.
3. Explain different CMOS latches.
4. Give brief notes on Fan-in and Fan-out of CMOS logic design.

SHORT ANSWER QUESTIONS

1. ____________ allow the designer to adjust placement interactively.
   Ans. Routing algorithms.
2. What is the use of tally circuit?
   Ans. It is used to count the number of inputs that are high and outputs the answer.
3. What is barrel shifter?
   Ans. A barrel shifter is a wrap-around or end-around shifter which is a very useful switch array.
4. The simplest memory element in MOS technology is____________ .
   Ans. dynamic latch.
5. What are the two most commonly quoted parameters of a memory element?
   Ans. (a) Setup time
   (b) Hold time.

7 Super Buffers, BiCMOS and Steering Logic

7.1 INTRODUCTION

When driving a very large capacitive load, serious delays can occur unless current buffers are used to source and sink large amounts of charge in short times. Devices that can do this are referred to as super buffers, and can be either inverting or non-inverting, NMOS or CMOS. Often a small-signal gate must drive either a large pad or a long line, usually of polysilicon or metal, either presents a large capacitive load to the driver. The line acts as a transmission line, with a delay proportional to the square of its length.

Bond pads are required on every chip to interface it with the rest of the world, and probe pads are often required for testing. Bonding pads are larger than probe pads, but both present a very heavy capacitive load to the pad-drivers. One of the main applications of super buffers is in interfacing between small-signal gates and large pad drivers. Since the stage that actually drives the bonding pad is very large, it should be capable of being turned off, to avoid dissipating standby power.

BiCMOS high current drivers act as alternatives to super buffers. Pass-transistor logic is an alternative to conventional logic. Pass-transistors logic offers no direct path from the power supply to ground, and thus uses no standby power. Pass-transistors can be designed to be minimum size, with channel dimensions of $2\lambda \times 2\lambda$. Pass-transistor circuits must be designed to allow both a pull-up and a pull-down path in order to guarantee that any node driven by them can be discharged as well as charged.

This chapter discusses super buffers and methods of scaling them properly will be investigated. Pre-charging is studied as an alternative method of improving performance. Precharging circuitry requires atleast two-phase clocking, as well as specially designed precharge drivers. The chapter also describes BiCMOS high current drivers and pass-transistor design. Depletion-mode, enhancement-mode, and CMOS function blocks will be discussed as general forms of pass-transistor logic.

The NMOS pass transistors output a logic high that is one threshold voltage below the supply rail, and PMOS pass transistors output a logic low that is one threshold voltage above ground. This leads to strong and weak 1s and 0s, as well as restrictions on driving NMOS and PMOS pass transistors.

7.2 RC DELAY LINES

A long metal line is a low-loss distributed capacitance. A long polysilicon line can be treated as a lumped RC transmission line or delay line as shown in Figure 7.1.

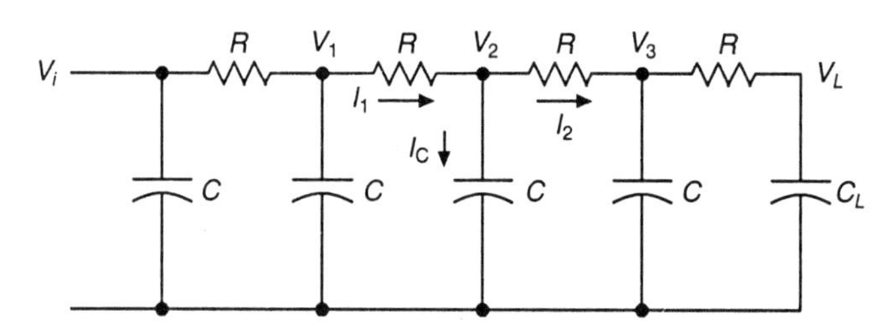

Figure 7.1 Lumped equivalent circuit representation of a long polysilicon line.

The current charging the capacitance is $C(dV_2/dt)$ which can be approximated as $C(\Delta V_2/dt)$. The current entering node 2 is

$$I_1 = \frac{V_1 - V_2}{R} \tag{7.1}$$

The current leaving node 2 is

$$I_2 = \frac{V_2 - V_3}{R} \tag{7.2}$$

Applying Kirchoff's current law at the node,

$$I_1 = I_2 + I_C$$

$\Rightarrow$

$$I_C = I_1 - I_2$$

$$C\frac{\Delta V_2}{\Delta t} = \frac{(V_1 - V_2) - (V_2 - V_3)}{R}$$

Let

$$V_1 - V_2 = \frac{\Delta V_{\text{left}}}{\Delta x}$$

$$V_1 - V_3 = \frac{\Delta V_{\text{right}}}{\Delta x}$$

$\therefore$

$$RC\frac{\Delta V_2}{\Delta t} = \frac{\Delta V_{\text{left}} - \Delta V_{\text{right}}}{\Delta x} = \frac{\Delta^2 V_2}{\Delta x^2} \tag{7.3}$$

Equation (7.3) represents a finite-difference equation. As the number of sections becomes large, the finite-difference equation can be approximated by the differential equation of a distributed model. Then, Eq. (7.3) reduces to

$$RC\frac{dV}{dt} = \frac{d^2V}{dx^2} \tag{7.4}$$

where, R and C are the resistance and capacitance per lumped element respectively. The propagation delay is proportional to the total resistance, R_t, and the total capacitance, C_t, of the line, both of which are proportional to the length of the line. Hence, the propagation time is proportional to the square of the length of the line.

From a discrete analysis of the circuit, the signal delay of N sections with a matched load $C_L = C$ is

$$t_d = 0.7\frac{N(N+1)}{2}\frac{R_t}{N}\frac{C_t}{N} \tag{7.5}$$

Let r and c be the resistance and capacitance per length of line respectively. Then,

$$R_t = rL$$

$$C_t = cL$$

Now, the propagation time over a wire of length L is

$$t_d = 0.7\frac{N(N+1)}{2}\frac{L^2rc}{N^2} \tag{7.6}$$

As the number of sections of line (N) approaches infinity,

$$t_d = 0.7\frac{rcL^2}{2} \tag{7.7}$$

As L becomes very large, the delay becomes unacceptably large. To improve the performance of a very long line, one can insert buffers along the line to restore the signal. The buffer delay depends upon the resistance of the segment driving it, and the capacitance of the segment it drives.

The line resistance can be minimized by using silicide or metal, but the capacitance remains. It may even get larger if metal is run over poly. To drive these large capacitances, a buffer capable of supplying and sinking large currents is required.

7.3 SUPER BUFFERS

A super buffer is a symmetric inverting or noninverting gate that can supply or remove large currents and switch large capacitive loads faster than a standard inverter. Basically, a super buffer consists of a totem-pole or push–pull output driven by an inverting or noninverting input that supplies the signal and its complement to the totem-pole. The super buffer speeds switching both ways, and by proper choice of geometry can be made almost independent of the inverter ratio.

The ratio-type logic suffers from asymmetry in driving capacitive loads because the pull-up drive has less drive capability than the pull-down device, and is quite slow when charging a load. The pull-down device has limited current sinking ability also, and must fight the pull-up device when the output falls. The pull-down is faster than the pull-up by approximately the factor R_{inv}.

An NMOS super buffer is rendered ratioless by designing for an inverter ratio of 4-to-1 and driving the totem-pole pull-up with twice the gate bias of a standard depletion-mode pull-up. This gives the inverter equal pull-up and pull-down current capability and identical rise and fall times. The CMOS super buffer is rendered ratioless by designing the pull-up aspect ratio to be twice the pull-down aspect ratio, so that it can supply or remove the same amount of current and $t_{rise} = t_{fall}$.

Schematically, the two configurations of an **NMOS super buffer** can be represented as shown in Figure 7.2, with the inverted input driving either the pull-up device or the pull-down device. Stick drawings of an inverting and a noninverting super buffer are shown in Figure 7.3(a) and (b). Transistors PU_1 and PD_1 form the driving inverter, and transistors PU_2 and PD_2 are the output totem-pole.

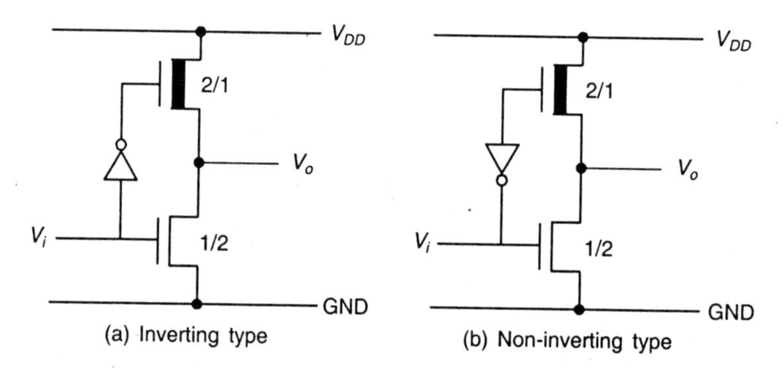

Figure 7.2 Mixed-mode diagrams of NMOS super buffers.

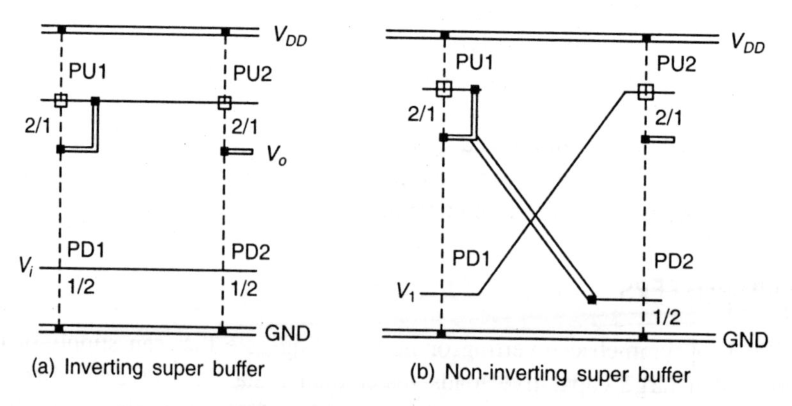

Figure 7.3 Stick drawings.

To analyze the behaviour of the inverting super buffer circuit, consider the following. For the inverting super buffer of Figure 7.3(a), when the input voltage is low, the gates of both pull-down FETs are low, the gates of both pull-up FETs are high, and the output is at the positive rail. When the input goes high, the gates of both pull-down FETs switch high, and the output switches from high to low in approximately t_{fall} seconds.

When the gate of pull-down FET PD_1 goes low, the gate of pull-up FET PU_1 goes to the positive rail very fast, since the load is the gate capacitance of pull-up FET PU_2 only. This drives the push–pull stage very quickly, and the output switches from low to high in approximately t_{fall} seconds also. The super buffer thus has a propagation delay of t_{fall}, which improves the switching speed.

To examine the switching speed of the noninverting super buffer, notice that when the input voltage is high PU_2, is ON, PD_2 is OFF, and the output is at the positive rail. When the input switches from high to low, PU_2 is turned OFF immediately. With input V_i low, the gate of PU_1, which is driving the capacitive load of PD_2 only, rises rapidly to V_{DD}, turning PD_2 ON.

In either the inverting or the non-inverting case, the depletion-mode FET, PU_2, is turned ON with about twice the drive of a FET with its gate connected to its source. In the saturation region, the drain current is almost proportional to the square of the gate-to-source voltage, and doubling the voltage drive of the depletion-mode FET allows it to supply about four times the current of a standard inverter.

The increase in pull-up current drive of the super buffer allows it to be almost ratioless when it is designed to have a nominal inverter ratio of 4-to-1, and the effective delay time, t_p, is almost the same whether the output switches from high to low or low to high.

7.3.1 NMOS Super Super Buffer

An NMOS super super buffer can be formed as shown in Figure 7.4, by combining an inverting super buffer composed of transistors Q_{1A} through Q_{4A}, with a non-inverting super buffer, composed of Q_{1B} through Q_{4B}, both driving the totem-pole output stage Q_5 and Q_6.

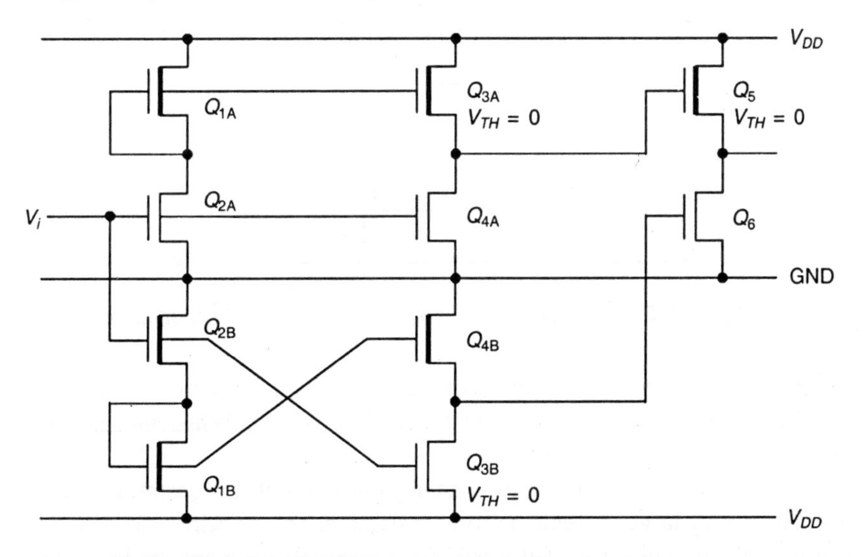

Figure 7.4 An NMOS Super super buffer circuit layout.

Transistors Q_{3A}, Q_{3B} and Q_5 are zero-threshold devices that have replaced the normal depletion-mode devices. The zero-threshold device has a threshold voltage that is nominally zero volts. When the zero-threshold device is used as a source–follower, it pulls up to V_{DD}. When the gate bias is low, the zero-threshold device is at or quite near cut-off, and the power dissipation of Q_5 approximates that of a PMOS pull-up, rather than a depletion-mode pull-up. This buffer can drive large capacitive loads faster than either the inverting or non-inverting super buffer and it exhibits low power consumption under no-load conditions. An extra masking step is needed to create the zero-threshold transistors.

7.3.2 NMOS Tristate Super buffers and Pad-Drivers

Tristate drivers are desirable to multiplex a bus, and to drive large capacitive loads such as pads. A tristate pad-driver circuit consists of a suitable number of tristate buffer stages followed by a pad-driver stage. An NMOS tristate super buffer stage is shown in Figure 7.5. A disable signal, i.e., $\overline{\text{Enable}}$, forces both V_o and $\overline{V_o}$ to be low, but it does not cause all current to cease, since the pull-up devices are depletion-type, and their sources go to ground when the gate is disabled.

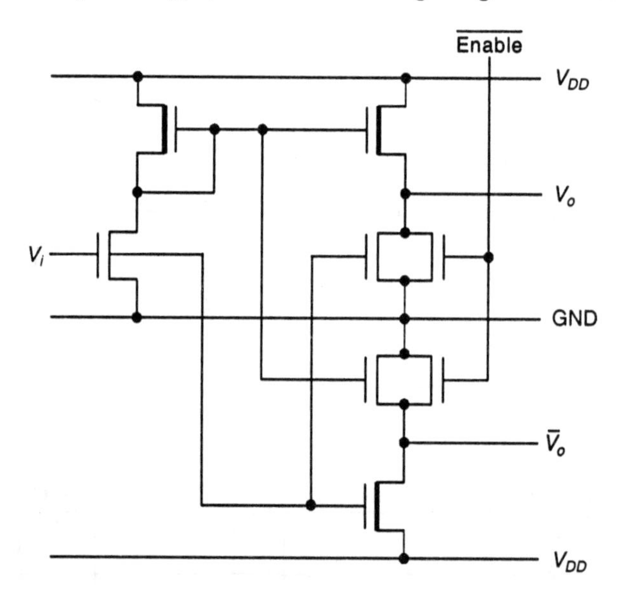

Figure 7.5 An NMOS non-inverting, translatable super buffer.

The outputs of this stage feed a pad-driver load such as the circuit shown in Figure 7.6, and will tristate the pad driver when the super buffer is disabled. When the gate is enabled, V_o follows V_i.

The NMOS tristate super buffer draws standby current, but it is desirable that the pad driver which must handle very large currents be truly tristatable. The pad driver is not an NMOS super buffer because both the pull-up and the pull-down are enhancement mode devices. However, when both transistor gates are low, the output pad is truly tristated and draw negligible standby current.

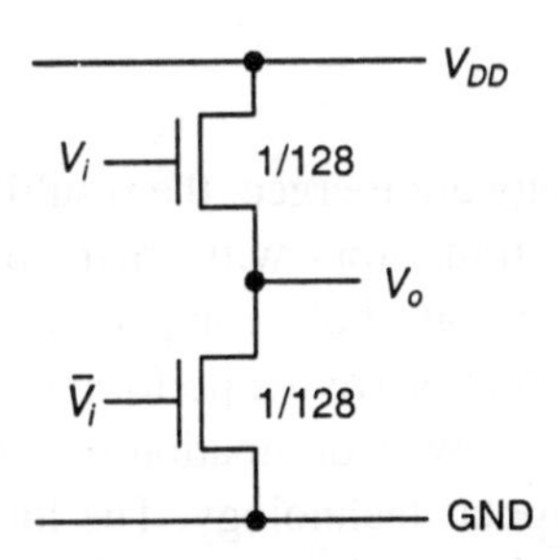

Figure 7.6 A tristatable pad-driver stage.

As in the output stage of a super buffer, the pull-up and pull-down of the pad driver form a totem-pole. The pad driver devices are normally wrapped around the pads they are driving in order to accommodate the extremely large width of both the pull-up and the pull-down gates.

7.3.3 CMOS Super Buffers

By doubling the width of the PMOS channel, the CMOS inverter can be made ratioless with respect to performance. A CMOS super buffer is simply a wide-channel CMOS inverter, or pair of inverters, that can supply or remove large currents. The CMOS super buffer should have the capability of tristating the pad-driver also.

A CMOS tristatable super buffer is shown in Figure 7.7. When the enable signal, EN is true, $\overline{EN}$ is false, Q_{1B} and Q_{3B} are OFF, Q_{2A} and Q_{2B} are ON, V_{o1} is connected V_{o2}, and the single output signal is the complement of the input signal. When EN is false, the inverter is disabled, Q_{2A} and Q_{2B} are both OFF, while Q_{1B} and Q_{3B} are both ON, and V_{o1} is pulled to the positive rail, while V_{o2} is pulled to the negative rail. This forces both the PMOS pull-up and the NMOS pull-down transistors of the CMOS pad-driver totem-pole to be OFF. The CMOS super buffer is truly tristated.

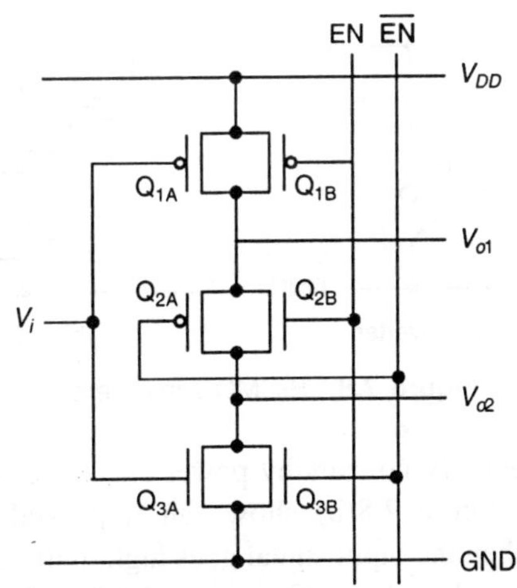

Figure 7.7 A CMOS tristatable inverting super buffer.

7.3.4 BiCMOS Gates

When bipolar and MOS technology are merged, the resulting circuits are referred to as BiCMOS circuits. High gain vertical *npn* transistors with their collectors tied to the positive rail, and medium-gain lateral *npn* transistors are both compatible with conventional CMOS processing.

BiCMOS gates can be used to improve the performance of line drivers and sense amplifiers. They combine the traditional low power dissipation of CMOS with the output drive capability and low propagation delay of bipolar technology. The load degradation is practically the same for all circuit functions because the bipolar push–pull devices isolate the CMOS circuits from the loading. Bipolar devices also make it much easier to interface with the current-mode or emitter-coupled logic (ECL), which is required in high speed systems.

7.3.4.1 BiCMOS inverters

BiCMOS inverters are shown in Figure 7.8, and use a pair of push–pull bipolar devices to provide driving capability. The MOS devices provide a high input impedance, while the BJT devices provide current drive and low output impedence. When the input signal is high, the pull-down FET short-circuits the base of Q_2 to its collector, converting it to a diode. The pull-up FET is OFF and Q_1 has no base drive. The output sees a very high resistance to the positive rail and a low resistance to the ground rail, resulting in a low output value. Q_2 can sink large amounts of current. The reverse happens when the input signal goes low, causing Q_1 to source current and pull the output high, while Q_2 is cut off (Figure 7.8(a)).

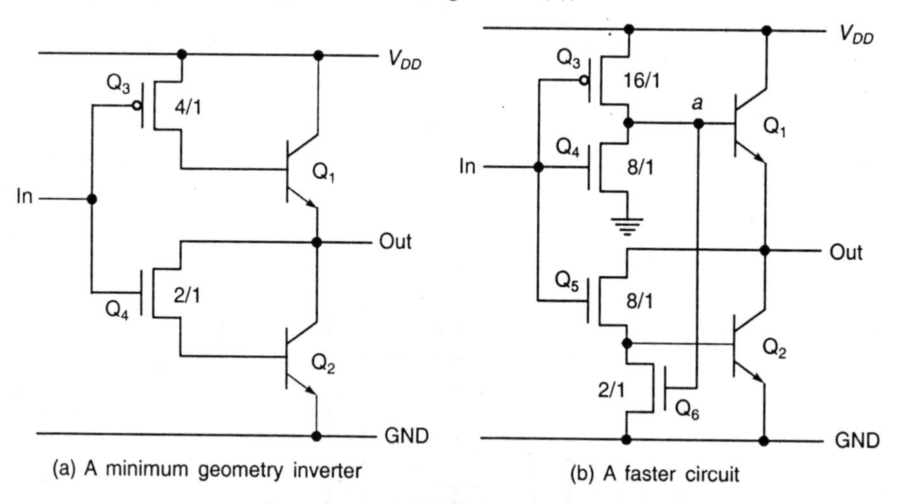

Figure 7.8 BiCMOS inverters.

The circuit dissipates essentially no standby power and the low power consumption of the CMOS circuit is preserved. Figure 7.8(b) shows an improved BiCMOS Inverter. In this improved type, Q_4 turns ON when the input signal goes high, pulling node *a* down and discharge Q_1 quickly. As the potential at node *a* drops, Q_6 is turned OFF, allowing Q_5 to drive Q_2 on hard and pull the output low. When the input goes low, Q_3 turns ON while Q_4 turns OFF. This allows Q_1 to turn ON fast. Q_5 is OFF and Q_6 turns ON when the voltage at node a rises, allowing Q_2 to discharge quickly.

The output high level voltage is given by $V_{OH} = V_{DD} - V_{BE}$, where V_{BE} is the voltage drop across the base-emitter junction of Q_1, while the output low level voltage is given by $V_{OL} = V_{BE}$ of Q_2 when the output is short circuited to the base of Q_2 through Q_5 which conducts when the input is high. When a load capacitance connected to the inverter output is fully charged (discharged) transistor Q_1 (Q_2) is drawing a low collector current, and its V_{BE} is small. The noise margins are then a few tenths of a volt less than half of the V_{DD} supply voltage, or almost as good as those of a CMOS gate.

When the BiCMOS gate is switching, all FET devices are conducting and there is a path from V_{DD} to ground. During transition, the two BJT devices are also ON, providing another low-impedence path between the supply rails, which is in parallel with the MOS path. This causes the switching current spikes to be larger than those of a conventional CMOS gate. Because of the low output impedence of a BiCMOS gate, the output switches much faster than the output of a CMOS gate. This causes a load gate driven by a bipolar output to switch faster than it would normally switch when driven by a CMOS gate, and the switching current spikes to be narrower for the BiCMOS. The overall result is a lowered power dissipation per gate per frequency.

7.3.4.2 BiCMOS NAND and NOR gates

The BiCMOS implementation of a 2-input NAND gate and a 2-input NOR gate are shown in Figures 7.9 and 7.10 respectively. Typical FET width-to-length conductance ratios are shown in Figures 7.8, 7.9, and 7.10.

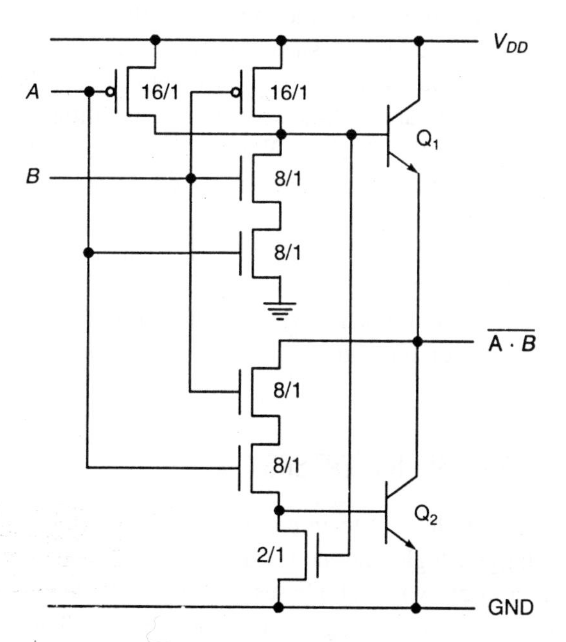

Figure 7.9 A BiCMOS 2-input NAND gate.

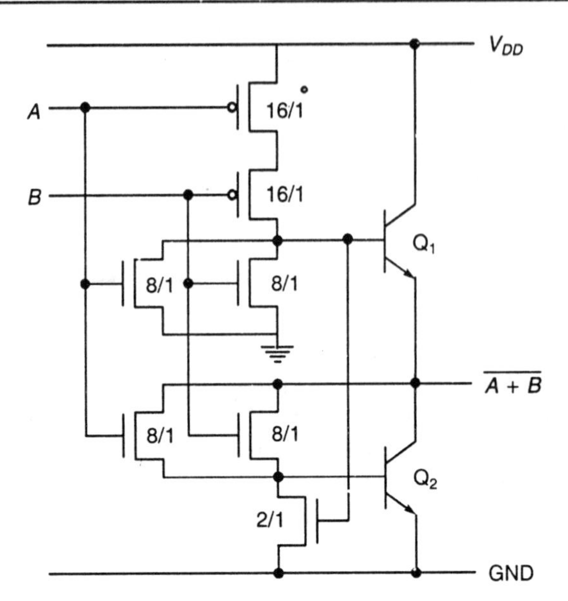

Figure 7.10 A BiCMOS 2-input NOR gate.

Under no-load conditions, CMOS is faster than BiCMOS, but as the circuit loading increases, the BiCMOS easily outperforms the standard CMOS gate due to its superior current capability. The BiCMOS circuits consume only slightly more power then their CMOS counterparts while providing much better switching response.

The BiCMOS circuits can be used wherever super buffers can be used. One advantage of BiCMOS gates is that the pull-up and pull-down current capabilities are determined only by the current gains (betas) of the bipolar devices. When driving heavy loads, BiCMOS has a big advantage over super buffers which do best when scaled up by a factor of no more than four each, and requires many buffers in series for heavy loads.

The BiCMOS process merges bipolar transistors that are built into a CMOS process structure without an epitaxial layer. The basic technology is an *n*-well, 2-micron, double-metallization CMOS process, the bipolar devices built with polysilicon emitters. The complete process requires the addition of three mask steps to the ten-mask CMOS *n*-well fabrication process.

7.4 DYNAMIC RATIOLESS INVERTERS

Precharging is another approach used to improve switching performance. For NMOS circuits, the output can be precharged high (the slow transition) and selectively discharged low. This method requires a minimum of two clock phases, so that all the loads can be precharged on one phase, and selective loads can be discharged on the other phase. This is a dynamic approach to avoiding ratio logic.

A typical example is shown in Figure 7.11, where a busline is precharged high. The pull-down devices must be multiplexed since they are in parallel, and only one pull-down at a time can be allowed to selectively discharge the load. Both the pull-up and pull-down should have minimum geometry, which saves space and increases speed.

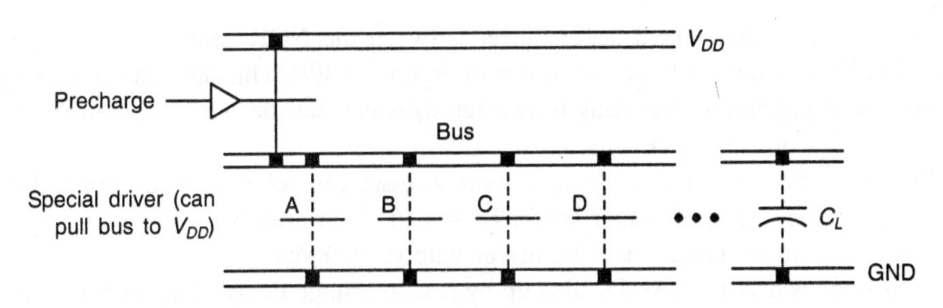

Figure 7.11 A mixed-mode drawing of a precharge driver and bus line.

A basic problem with NMOS precharging circuitry is that the driver that precharges the line must have an output that is at least a threshold voltage above plus supply, so that the output line is pulled up to plus supply. A signal on any input line discharges the bus fast if its input is high, while it has no effect if its input is low.

7.5 LARGE CAPACITIVE LOADS

During large capacitive loads such as output pads present a severe mismatch problem. For example, a two-input NOR gate driving a capacitive load, C_L, of one hundred times the NOR gate's capacitance will have an effective propagation delay of $100t_p$ as shown in Figure 7.12. This is very slow.

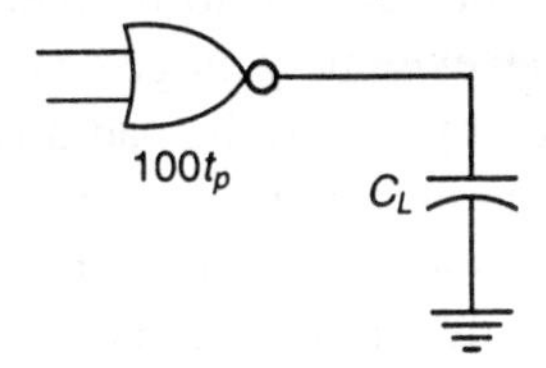

Figure 7.12 A NOR gate driving an effective fan-out of 100.

An inverting or non-inverting super buffer can supply much more current than a standard MOS gate, and can be designed to charge the load capacitance with a delay of t_{pLH}, or discharge the load with a delay of $t_{pHL} = t_{pLH} = t_p$, as shown in Figure 7.13(a). However, the super buffer presents a load equivalent to a fan-out of 100 to the NOR gate, and the effective propagation delay of the super buffer plus load is now at least $101t_p$. This is worse than the original situation.

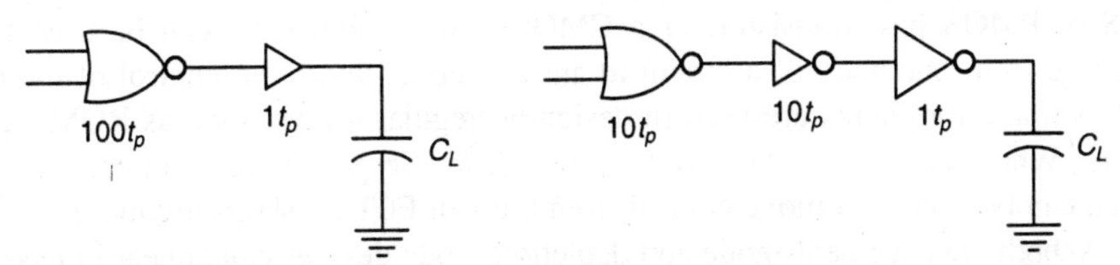

(a) A NOR gate driving a super buffer which drives a fan-out of 1

(b) A NOR gate driving a 10X larger super buffer, which drives a 100X larger super buffer

Figure 7.13 Driving capability of NOR gate for one or more super buffers.

Inserting an intermediate-sized super buffer between the NOR gate and the output super buffer gives an effective delay of $21t_p$, as shown in Figure 7.13(b). This indicates that using more than one buffer and gradually increasing the buffer sizes is faster because it gradually increases the load at all nodes of the circuit.

Since super buffers are not ratio devices, their propagation delays low-to-high and high-to-low are both the same. Let each super buffer be scaled up in area by a factor of f, so that its gate capacitance is f times larger than the driver gate capacitance.

Then each stage is driving a load scaled up by f, with a stage delay of ft_p. For N such stages, the delay is Nft_p.

Let γ be the ratio of the load capacitance to one gate capacitance

$$\gamma = \frac{C_L}{C_g} \tag{7.8}$$

Now, the minimum total delay can be given as

$$Nft_p = et_p \ln\left(\frac{C_L}{C_g}\right) \tag{7.9}$$

where e is the base of natural logarithms.

Often the real estate saved by using a ratio of 4-to-1 when scaling up, more than justifies the 6% penalty in delay time. In fact, when driving an output pad or other very large capacitive load, it might be worthwhile to scale the super buffers by 5-to-1 in order to use fewer buffers and save space. The speed penalty for this is seen to be 14%.

The capacitance of a pad can be extremely large, can require very large pad-driving circuitry. Since buffers should be scaled by 4-to-1, the pad driver should also be scaled to drive an effective fan-out of 4. When power dissipation is more important than speed, the NMOS pad-driver can be scaled up by more than 4-to-1 with respect to the super buffer driving it. The CMOS super buffer of Figure 7.7 is truly tristatable, and should be scaled up by 4-to-1 with respect to its driver.

A NOR gate driving a fan-out of 100 with three super buffers, each scaled up by a factor of 4-to-1, can charge the load capacitance in a time of $4t_p$ of the NOR gate, $4t_p$ delay for each of the first two super buffers, and $100/64 = 1.56t_p$ for the third super buffer, giving a total time of $13.6t_p$, which is much better than the $21t_p$ of Figure 7.13(b).

7.6 PASS-TRANSISTOR LOGIC

An NMOS or PMOS pass-transistor, or a CMOS transmission gate, can be used to steer or transfer charge from one node of a circuit to another node, under the control of the FETs gate voltage. Pass transistor chains are used in designing regular arrays, such as ROMs, PLAs, and multiplexers. When used in regular arrays, depletion-mode pass transistors created by an ion-implant step can be used to remove control from a given FET by short-circuiting its drain to its source. Thus, both enhancement-mode and depletion-mode devices can appear in pass-transistor chains.

Some problems associated with pass-transistor logic will be discussed in this section, followed by design aspects in Sec. 7.7. Pass transistors have several advantages over inverters. Two major advantages of pass transistors over standard NMOS gate logic are:

1. They are not 'ratioed' devices and can be minimum geometry.
2. They do not have a path from plus supply to ground, and do not dissipate stand by power.

If the gate and drain of a pass transistor are both high, the source will rise to the lower of the two potentials V_{DD} and $V_{GS} - V_{TH}$. If the gate and drain are both at V_{DD}, the source can only rise to one threshold voltage below the gate. If the source tries to rise higher, the device cuts off. If the gate is at least a threshold voltage higher than the drain, the source will rise to within a few millivolts of the drain potential. This is shown in Figure 7.14.

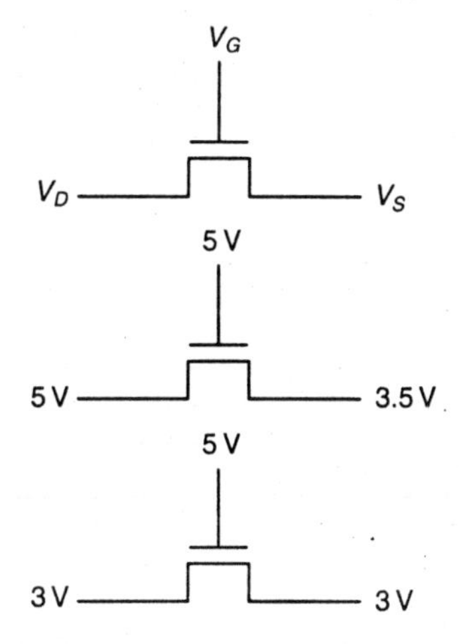

Figure 7.14 The source voltage is always the lower of voltages V_D and $V_G - V_{TH}$.

Three pass transistors driving an inverter are shown in Figure 7.15. Let the threshold voltage of enhancement-mode pass transistors be 1 V with no back-gate bias, and 1.5 V with a back-gate bias of 3.5 V. Then the node voltages are as shown in Figure 7.15. With the gate and drain of the first pass transistor at 5 V, its source rises to 3.5 V and the device is at the onset of pinching off. With the gate bias of the second FET at 5 V and its drain at 3.5 V, its source rises to the drain potential 3.5 V, and the device passes the input voltage to its output. Any depletion-mode devices are short-circuits and can be omitted from the discussion.

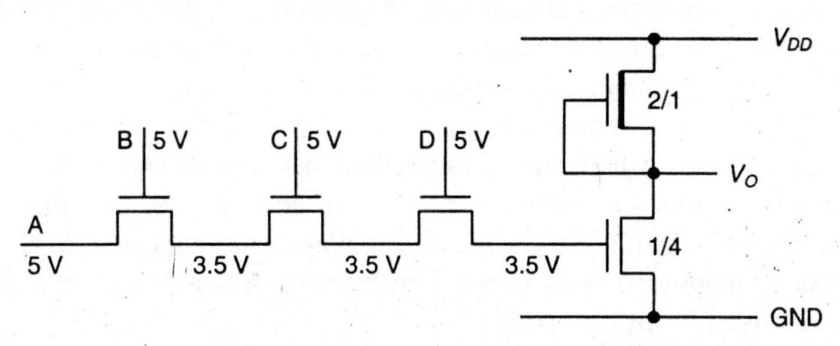

Figure 7.15 NMOS transmission gates driving an inverter.

Each additional input requires only a minimum-geometry FET, and adds no dc power dissipation to the circuit. However, there are disadvantages to charge steering. For the circuit shown in Figure 7.15, and inputs A, B, C, and D all at 5 V, the voltage presented to the inverter input is only 3.5 V. This must be sufficient to drive the inverter output low.

In fact, if a pass transistor is driven by a voltage less than 5 V, the source of that FET will be at one threshold voltage less than the corresponding gate voltage, and the voltage presented to the inverter input will be lower than the lowest gate voltage of the pass-transistor chain by one threshold voltage drop. Figure 7.16 shows the situation when input C is 3.5 V, which would be the case if it were drawn by a pass-transistor. Even though input D is 5 V, the voltage to the inverter is only 2 V. This is below the inverter threshold voltage, and will most likely be treated by the inverter as a logic 0 when it should have been a logic 1.

A sufficiently high inverter ratio will solve this problem, but at the expense of a long pull-up. If NMOS pass-transistor logic circuits are designed such that one pass-transistor never drives the gate of another pass-transistor, the output high from a pass-transistor logic string will be about 3.5 V. An inverter with a ratio of 8-to-l, as shown in Figures 7.15 and 7.16, will suffice to restore the signal to a logic high of 5 V. This leads to the first major rule in designing NMOS pass transistor logic.

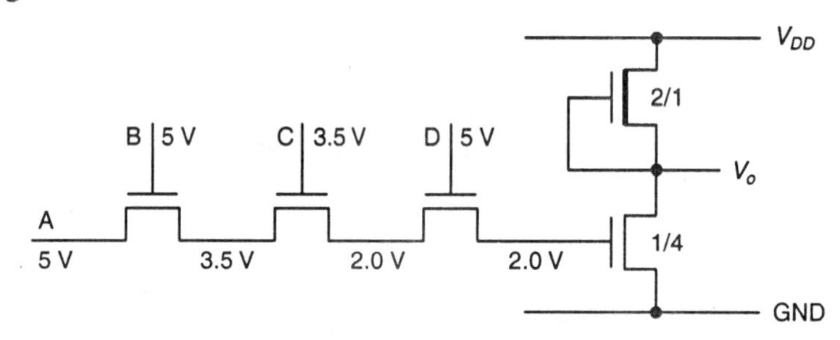

Figure 7.16 The circuit of Figure 7.15 with input C at 3.5 V.

In designing NMOS pass-transistor logic, one must never drive a pass-transistor with the output of another pass-transistor. This is not a problem when using CMOS transmission gates.

There is another problem associated with pass-transistor logic that the designer must be aware of. If control *D* in Figure 7.15 goes from logic 1 to logic 0, the gate-to-drain capacitance of the pass transistor driven by signal *D* couples a negative voltage step to the inverter input. When the voltage at *D* drops from 5 to 0 V, the input voltage to the inverter drops from 3.5 to −1.5 V. The pass transistor is turned off by the logic 0 at *D* and there is no discharge path for the pull-down device of inverter. Thus the second major rule in designing NMOS pass-transistor chains: in designing pass-transistor logic, care must be taken to ensure the existence of both charging and discharging paths to the inputs of all inverters.

Charge sharing is a serious problem which occurs when two or more capacitors at different potentials are tied together. A node of a network must never be driven simultaneously by signals of opposed polarity, as this can leave the node in an erroneous or undefined state. One must beware of 'sneak paths' which allow charge to leak. Pass transistors are bilateral, and charge can flow from output to input also. This is not a problem if all the inputs are designed to connect to the output via mutually disjoint paths.

A sneak path is created when two pass transistors are both on at the same time and one is connected to V_{DD} while the other is connected to GND, as shown in Figure 7.17. Pass transistors are usually designed to be of minimum size, $2\lambda \times 2\lambda$. If further, the two devices have the same gate-to-source bias, their on resistances will be approximately equal and the output voltage will be about $V_{DD}/2$.

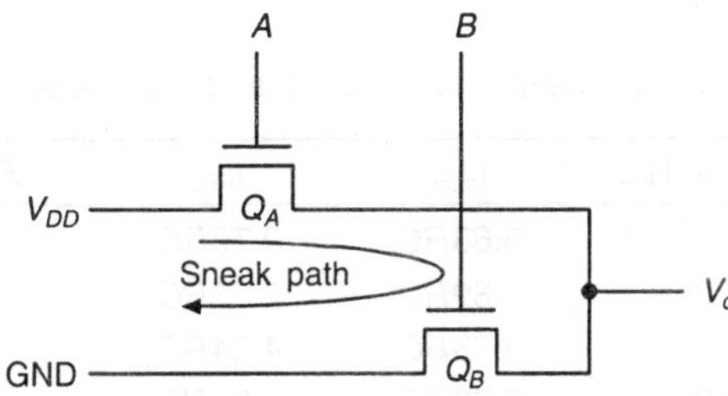

Figure 7.17 A sneak path exists when *A* and *B* are high simultaneously.

There is also timing problem associated with the circuit shown in Figure 7.15. If all four signals are high, the inverter input is high. If *B*, *C*, or *D* goes low while *A* is high, the inverter input will remain high when A goes low, and the output will be low. However, if A goes low first, a low is presented to the inverter, and when any of the pass transistors turn off, a low will be latched to the inverter input, keeping its output high. This problem is avoided by either tying A to the positive rail and not using it as a data input, or by providing A with both charging and discharging paths to the inverter. One must always provide both a charging and a discharging path for all input variables.

In designing pass-transistor circuits, both true and false input variables must be passed to the output in order to provide pull-up and pull-down paths and avoid undefined states. Inverters are needed to drive control inputs because they output strong ones, and super buffers are required to drive long chains.

A pass-transistor chain driving a capacitive load is shown in Figure 7.18. As with the long transmission line of Sec. 7.2, the pass-transistor chain can be approximated as a lumped RC equivalent circuit, as shown in Figure 7.1. In this approximation, *C* is the gate-to-channel capacitance plus any parasitic capacitance of the gate node; and *R* is the series resistance of the channel with the FET operating in the linear region, plus any parasitic resistance between nodes. The distributed model is as discussed in Sec. 7.2. The rise time is proportional to the total resistance, R_t, and the total capacitance, C_t, of the line, both of which double when the number of pass transistors doubles. Hence the propagation time is proportional to the square of the

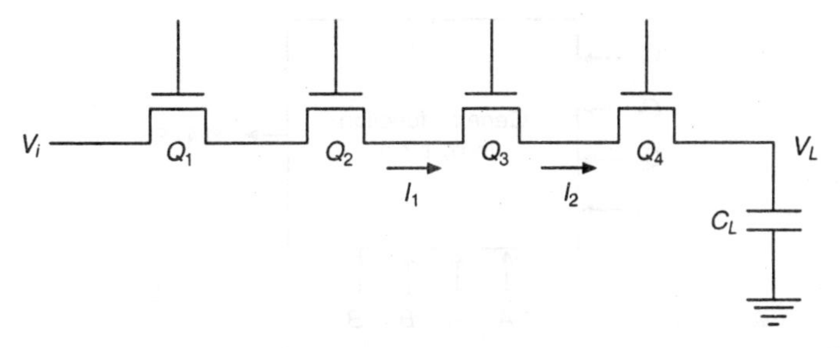

Figure 7.18 A pass-transistor chain driving a capacitive load C_L.

number of pass transistors in the chain. As with the finite transmission line consisting of N elements, a string of N identical pass transistors driving a matched load, $C_L = C_g$, exhibits a delay of approximately

$$t_p = 0.7\frac{N(N+1)}{2}RC_L \tag{7.10}$$

Table 7.1 Rising and falling propagation delay times and pair delay for short pass transistors

N	$N(N + 1)/2$	t_{pLH}	t_{PHL}	$P = t_{pLH} + t_{pHL}$
1	1	$0.63RC_L$	$0.79RC_L$	$1.42RC_L$
2	3	$1.89RC_L$	$2.37RC_L$	$4.26RC_L$
3	6	$3.78RC_L$	$4.74RC_L$	$8.52RC_L$
4	10	$6.30RC_L$	$7.90RC_L$	$14.20RC_L$
5	15	$9.45RC_L$	$11.85RC_L$	$21.30RC_L$

Again, more accurate estimates for the propagation delay times low-to-high, t_{pLH}, and high-to-low, t_{pLH} are

$$t_{pLH} = 0.63\frac{N(N+1)}{2}RC_L \quad \text{and} \quad t_{pHL} = 0.79\frac{N(N+1)}{2}RC_L \tag{7.11}$$

In Table 7.1 rising and falling propagation delay times and pair delay are calculated for short pass-transistor chains. Periodic restoration of the full signal voltage swing will improve the speed of the circuit. A cascade of more than four steering gates produces a very slow circuit, and the signal should be restored by an inverter after every three or four steering gates. This can be a severe problem in designing large PLAs.

7.7 GENERAL FUNCTION BLOCKS

One application of pass-transistor logic is the *universal logic module*, or general *function block*, which is a close relative of the PLA. It is often less costly in area, time, and power to implement a general function block than it is to implement a specific function. The function block can be either NMOS or CMOS and, if implemented, the details of its operation can be left unbounded until later. This provides a cleaner interface to the next level of design. The general two-variable function block that implements all 16 logic functions of two input variables A and B, and is controlled by inputs C_0 through C_3, is shown in Figure 7.19.

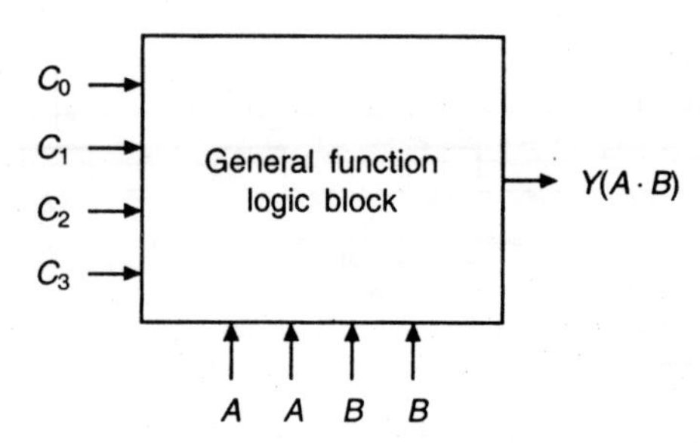

Figure 7.19 Symbol for a 2-variable function block for variables A and B, controlled by C_0 through C_3.

7.7.1 NMOS Function Blocks

Examination of 2-input NAND, NOR, AND, OR, Ex-OR, and Ex-NOR pass networks shows that they all are of the same topological structure, with only the inputs, controls, and outputs changing. These six functions are special cases of the sixteen possible logic functions realizable with a 2-input function block, or logic block, as shown in Figure 7.20.

(a) Circuit

General Function Block

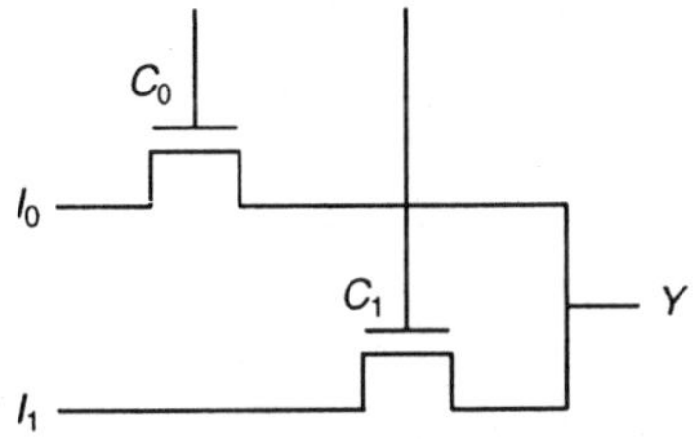

Output y	*Inputs* I_0	I_1	*Controls* C_0	C_1	*Logic Function*
AB	0	B	$\bar{A}$	A	AND
$\overline{AB}$	1	B	$\bar{A}$	A	NAND
$A + B$	B	1	$\bar{A}$	A	OR
$\overline{A + B}$	B	0	$\bar{A}$	A	NOR
$A \oplus B$	B	B	$\bar{A}$	A	EOR
$A \otimes B$	B	B	$\bar{A}$	A	ENOR

(b) Control and input variables for six output functions

Figure 7.20 The general two-variable function block.

There are two possible realizations of a two-input function block, one consisting of enhancement-mode transistors only and one made with both depletion-mode and enhancement-mode transistors. The first approach is the conventional steering logic, the second realization is an alternative approach. The two possible realizations are shown in Figures 7.21 and 7.22.

The mask sets for the circuits in Figures 7.21 and 7.22 are complementary in the sense that an enhancement-mode device loses control if it is missing, while an ion-implant removes control from a depletion-mode device. Thus, implants in Figure 7.22 are placed where transistors were omitted in Figure 7.21.

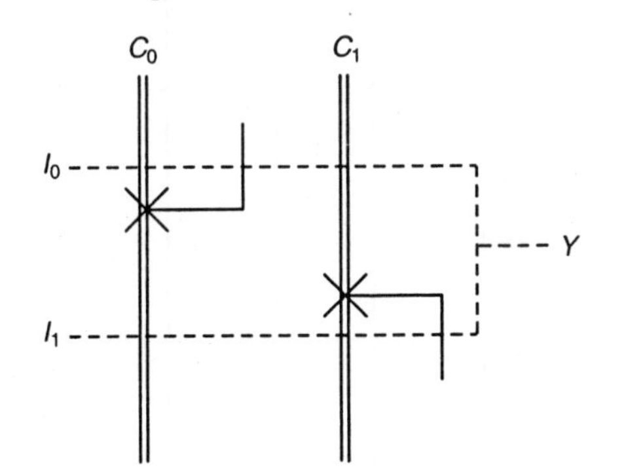

Figure 7.21 The two-input function block: stick drawing.

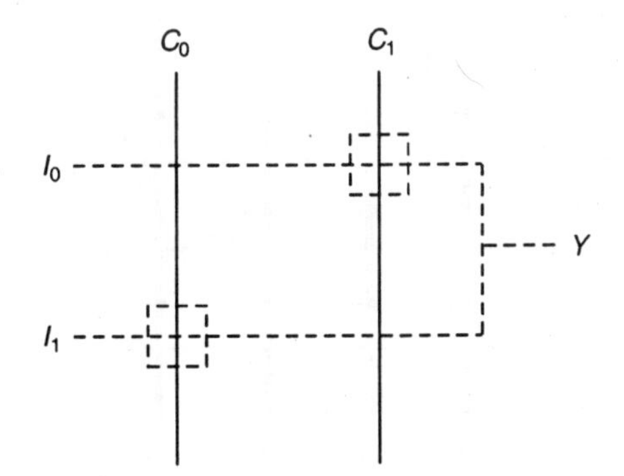

Figure 7.22 An alternative two-input function block: stick drawing.

Timing problems can occur when variables are used as inputs to pass-transistor chains as well as control signals. The timing problems can be eliminated by using only V_{DD} and GND as inputs to the pass-transistor chains, but this doubles the size of the gate. For example, to design an exclusive OR gate with 0 and 1 inputs requires four controls: A, $\bar{A}$, B and $\bar{B}$. This requires a 4-to-1 multiplexer in either NAND or NOR form. Two realisations of an Ex-OR gate with fixed inputs are shown in Figure 7.23(a) and (b) and are seen to require a 4 × 4 matrix of transistors.

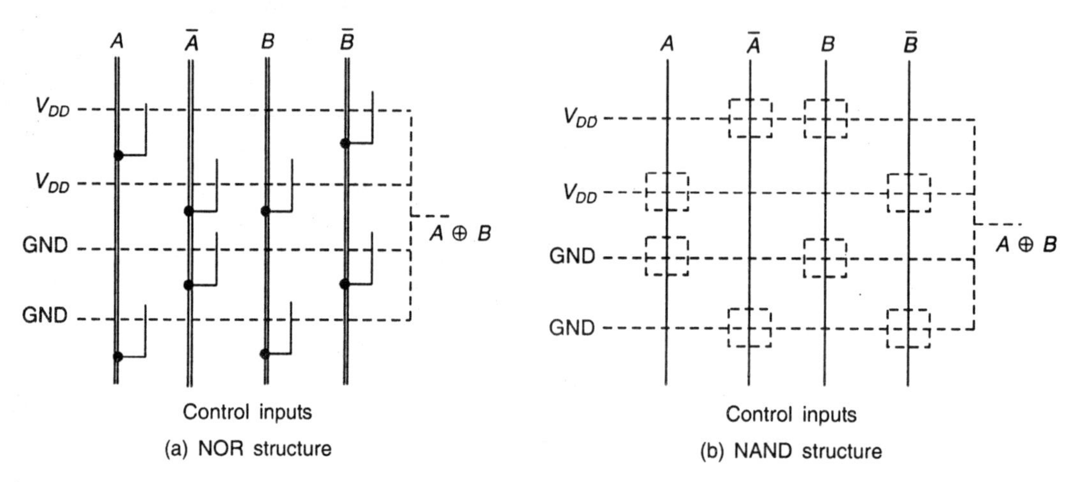

Figure 7.23 NMOS Ex-OR gates realized with fixed inputs.

To complete the discussion, stick drawings of NMOS pass-transistor 2-variable function blocks are shown in Figure 7.24. Figure 7.24(a) shows the function block realized with NOR logic consisting of enhancement-mode devices only, and Figure 7.24(b) shows the function block realized with NAND logic consisting of enhancement-mode and depletion-mode transistors.

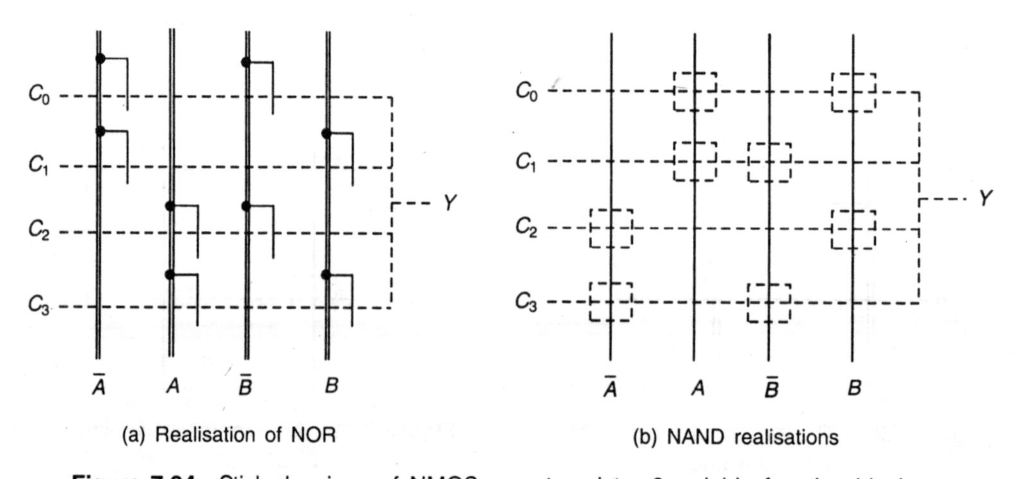

Figure 7.24 Stick drawings of NMOS pass-transistor 2-variable function block.

7.7.2 CMOS Function Blocks

A CMOS function block that implements all 16 logic functions of two input variables, A and B, is controlled by inputs C_0 through C_3, would be structurally similar to the NMOS block in Figure 7.24(a). A complementary CMOS design would consist of eight transmission gates replacing the eight pass transistors of the NMOS design. When the CMOS function block is laid out, the transmission gates would be split. The PMOS transistors would be grouped in an n-well and the NMOS transistors would be grouped in a p-well. This gives rise to a structure twice the size of an NMOS function block, containing twice as many transistors. Depletion mode devices are not used in CMOS design, and there is no CMOS equivalent of the NMOS NAND structure of Figure 7.24(b).

The CMOS function block can be reduced in size by using precharge logic. It can be precharged either high or low. A precharged-high mostly-NMOS function block and a precharged-low mostly-PMOS function block are shown in Figure 7.25. The output of the function block is precharged on phase 1, and is controlled by the four signals labelled C_0, C_1, C_2, and C_3. During precharge, the controls are kept at the same logic level as the output, thus eliminating the need to disconnect the evaluation block during precharge. The four controls are kept low in the case of mostly-NMOS circuit of Figure 7.25(a), and high in the case of the mostly-PMOS circuit of Figure 7.25(b).

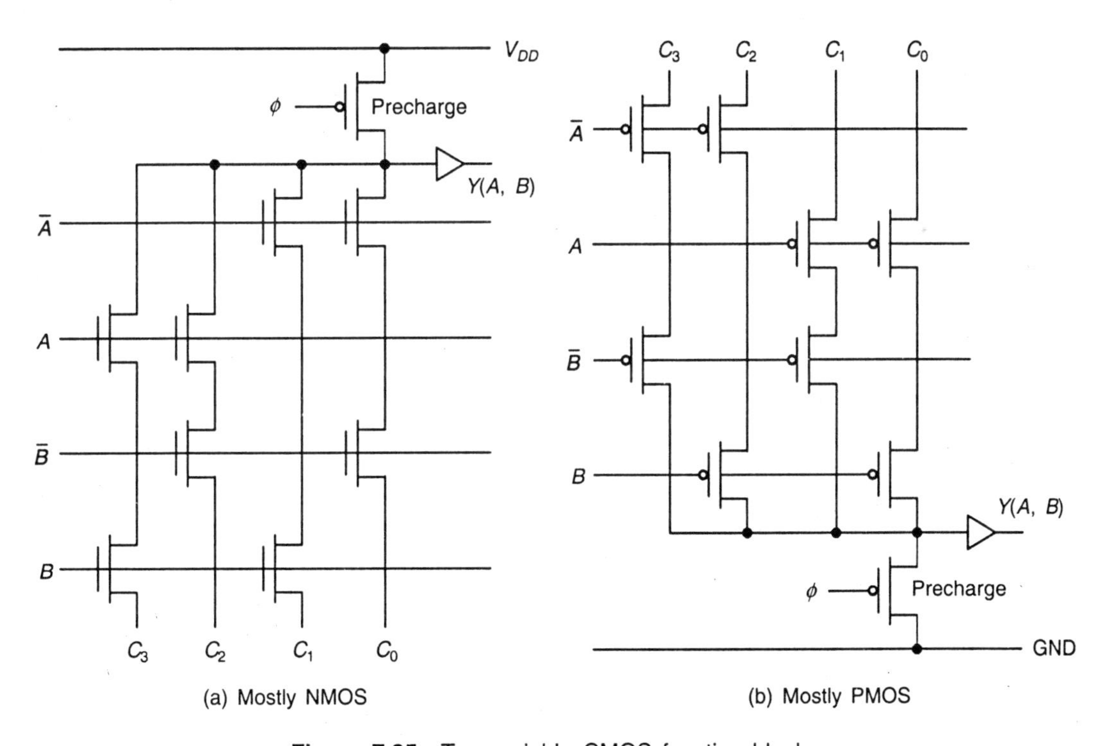

Figure 7.25 Two-variable CMOS function blocks.

SUMMARY

MOS transistors often must drive large capacitive loads. Super buffers are inverting and non-inverting gates built of larger transistors which can source and sink more current than standard-sized devices. If an exponentially increasing driving circuitry is scaled such that each FET drives an effective load of $e = 2.72$, the propagation delay is a minimum. For most applications scaling up by 4 is satisfactory and saves space.

BiCMOS draws negligible standby power and uses bipolar transistors with high current gain to enhance switching speed without causing power-dissipation problems. This approach does increase input loading, but interconnect capacitance often dominates layouts that require BiCMOS drivers.

Increasing the transistor width, as in super buffer design, can make the circuit run faster without significantly increasing the overall power consumption, but BiCMOS can source and sink large currents while taking up much less space than a series of super buffers. BiCMOS does add three masking steps to the process.

Pass transistors can be of minimum size and they dissipate no standby power. A pass-transistor is faster than an inverter, but a long string of steering logic begins to act as transmission line and delays become unacceptably large. Inserting inverters to restore the signal after about four pass-transistors gives shorter propagation delay.

The NMOS steering gates output a weak 1 of about 3.5 V, and PMOS steering gates output a weak 0 of about 1.5 V. NMOS pass transistors must never be driven by weak 1s, nor for that matter, can PMOS pass transistors be driven by weak 0s. Inverters driven by pass transistors must have at least an 8-to-1 inverter ratio, and one must ensure that both charging and discharging paths are available at the input of each inverter.

Steering logic can be mapped into general function blocks which can realize all the possible combinations of the input variables. They can be designed with enhancement-mode devices only, or with enhancement-mode and depletion-mode devices. To avoid timing problems, function blocks should have inputs connected to the appropriate supply rails, and use data signals only for controls. For two variables, this approach gives a 4×4 matrix. Larger matrices will require restoring logic after every four pass-transistors.

REVIEW QUESTIONS

1. Explain RC delay lines with a suitable equivalent circuit.
2. Draw stick diagrams for super buffers and explain them.
3. Give a brief notes on BiCMOS gates and inverters.
4. Explain pass-transistor logic with suitable example.

SHORT ANSWER QUESTIONS

1. Define zero threshold voltage device.

   Ans. A device which has a threshold voltage that is normally zero.

2. What is the use of CMOS super buffers?

   Ans. CMOS super buffers can supply or remove large currents.

3. ____________ mode and ____________ mode devices can appear in pass-transistor chains.

   Ans. Enhancement, depletion.

4. What is charge sharing?

   Ans. When two or more capacitors at different potentials are tied together, charge sharing will take palce.

5. ____________ allows the charge to leak.

   Ans. Sneak paths.

6. When a sneak path is created?

   Ans. This is created when two pass transistors, one is connected to V_{DD} and the other is connected to GND.

8 CMOS Combinational Logic Circuits

8.1 INTRODUCTION

Logic design deals with the creation of a digital network that performs a particular task. In terms of Boolean logic, this is equivalent to implementing a particular function. Combinational logic deals with networks that use logic gates to combine the input variables as needed to produce logic functions. In a combinational circuit, the value of output is determined by the current value of the input. If any of the inputs are changed, then the value of the output may change as specified by the function. There is no intentional connection from outputs back to inputs (Figure 8.1).

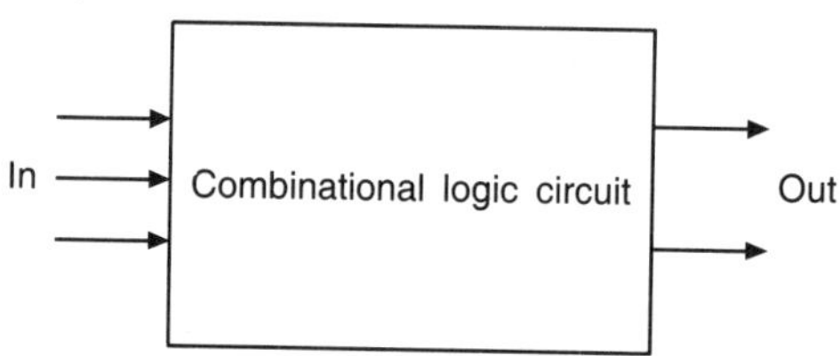

Figure 8.1 Combination logic circuit.

The common design metrics by which a gate is evaluated are area, speed, energy and power. For example, the switching speed of digital circuits is the primary metric in a high-performance processor, while in a battery operated circuit, it is energy dissipation. Recently, power dissipation also has become an important concern and considerable emphasis is placed on understanding the sources of power and approaches to dealing with power. In addition to these metrics, robustness to noise and reliability are also very important considerations. We will see that certain logic styles can significantly improve performance, but they usually are more sensitive to noise.

8.2 STATIC CMOS DESIGN

The most widely used logic is static complementary CMOS. The primary advantage of the CMOS structure is robustness (i.e., low sensitivity to noise), good performance, and low power consumption with no static power dissipation. Most of those properties are carried over to large fan-in logic gates implemented using a similar circuit topology.

In this section, we address the design of various static circuits including complementary CMOS, ratioed logic (pseudo-NMOS and DCVSL), and pass-transistor logic.

8.2.1 Complementary CMOS

A static CMOS gate is a combination of two networks—the pull-up network (PUN) and the pull-down network (PDN) as in Figure 8.2. The function of the PUN is to provide a connection between the output and V_{DD} anytime the output of the logic gate is meant to be 1 (based on the inputs). Similarly, the function of the PDN is to connect the output to V_{SS} when the output of the logic gate is meant to be 0. The PUN and PDN networks are constructed in a mutually exclusive fashion such that one and only one of the networks is conducting in steady-state.

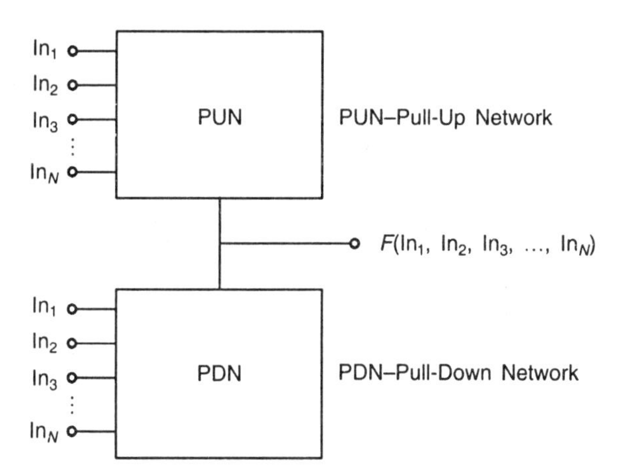

Figure 8.2 Complementary logic gate as a combination of pull-down and pull-up networks.

In constructing the PDN and PUN networks, the following points should be noted:

- A transistor can be thought of as a switch controlled by its gate signal. An NMOS switch assumes positive logic and PMOS switch assumes negative logic for their operation.
- The PDN is constructed using NMOS devices while PMOS transistors are used in the PUN. This is due to the fact that NMOS transistors produce 'strong 0s' and PMOS devices generate 'strong 1s'.

Consider the examples shown in Figure 8.3. In Figure 8.3(a), the output capacitance is initially charged to V_{DD}. An NMOS device pulls the output down to GND all the way, while a PMOS lowers the output no further than $|V_{Tp}|$—the PMOS turns off at that point and stops contributing discharge current. The NMOS transistors are thus preferred in the PDN. In

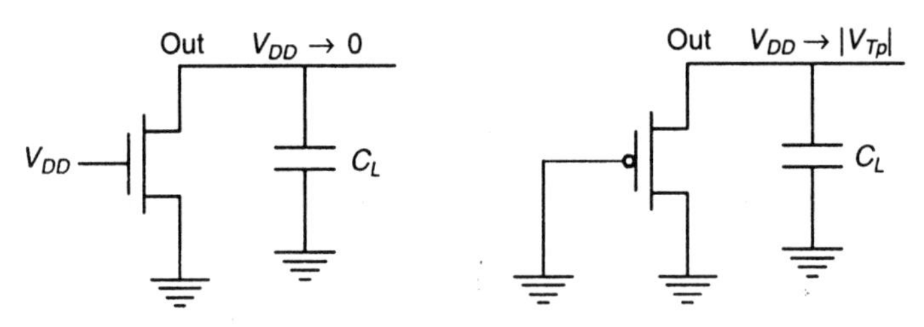

(a) Pulling down a node by using NMOS and PMOS switches

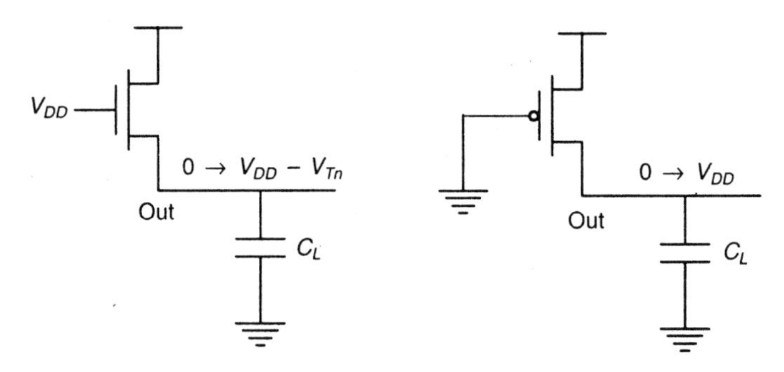

(b) Pulling up a node by using NMOS and PMOS switches

Figure 8.3 Illustrations of NMOS switches in PDN and a PMOS switch in PUN.

Figure 8.3(b), the output is initially at GND. A PMOS switch succeeds in charging the output all the way to V_{DD}, while the NMOS device fails to raise the output above $V_{DD} - V_{Tn}$. Thus, PMOS transistors are preferably used in a PUN.

- A set of rules can be derived to construct logic function. NMOS devices connected in series correspond to an AND function, while NMOS transistors connected in parallel represent an OR function.

  Similarly, a series connection of PMOS represents a NOR function whereas PMOS transistors in parallel implement a NAND function.
- The pull-up and pull-down networks of a complementary CMOS structure are dual networks. This means that a parallel connection of transistors in the pull-up network corresponds to a series connection of the corresponding devices in the pull-down network, and vice versa.
- The complementary gate is naturally inverting, implementing only functions such as NAND, NOR and XNOR. The realization of a non-inverting Boolean function (such as AND, OR or XOR) in a single stage is not possible, and requires the addition of an extra inverter stage.
- The number of transistors required to implement an N-input logic gate is $2N$.

Static properties of complementary CMOS gates

- They exhibit rail-to-rail swing with $V_{OH} = V_{DD}$ and V_{OL} = GND.
- The circuits have no static power dissipation, since the circuits are designed such that the pull-down and pull-up networks are mutually exclusive.
- The analysis of the dc voltage transfer characteristics and the noise margins is more complicated than for the inverter, as these parameters depend upon the data input patterns applied to the gate.

Propagation delay of complementary CMOS gates. For the delay analysis, each transistor is modelled as a resistor in series with an ideal switch. The value of the resistance is dependent on the power supply voltage and an equivalent large signal resistance, scaled by the ratio of device width over length, must be used. The logic is transformed into an equivalent RC network that includes the effect of internal node capacitances. Figures 8.4(a) and (b) show the two-input NAND gate and its equivalent RC switch level model.

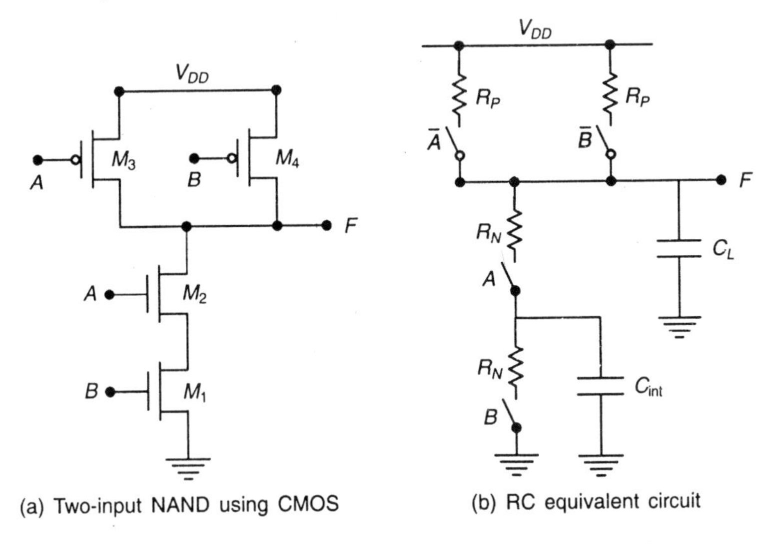

(a) Two-input NAND using CMOS (b) RC equivalent circuit

Figure 8.4 Equivalent RC model for a two-input NAND gate.

The internal node capacitance C_{int}—attributable to the source/drain regions and the gate overlap capacitance of M_2 and M_1—is included. The propagation delay depends on the input patterns.

If both inputs A and B are driven low, the two PMOS devices are ON. The delay in this case is 0.69 $(R_P/2)C_L$, since the two resistors are in parallel. The worst-case low-to-high transition occurs when only one device turns ON, and is given by $0.69R_PC_L$. For the pull-down path, the output is discharged only if both A and B are switched high, and the delay is given by $0.69(R_P/2)C_L$. In other words, adding devices in series slows down the circuit, and devices must be made wider to avoid a performance penalty.

Consider a four-input NAND gate as in Figure 8.5, which shows the equivalent RC model of the gate, including the internal node capacitances.

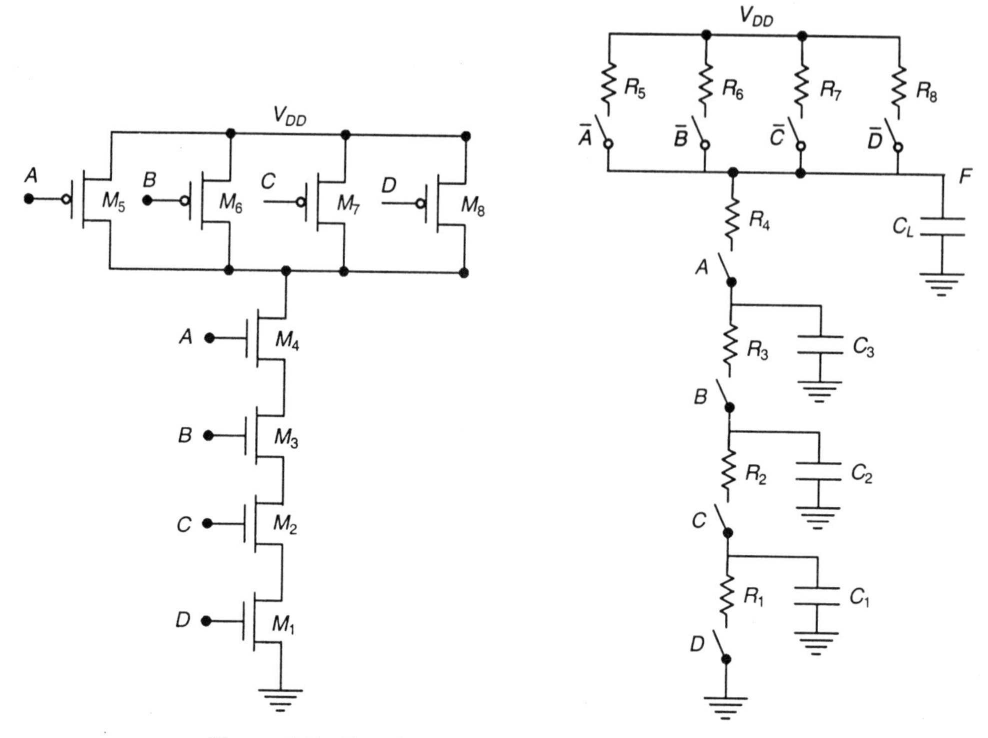

Figure 8.5 Four-input NAND gate and its RC model.

The propagation delay can be computed by using the Elmore delay model,

$$t_{pHL} = 0.69\,[R_1C_1 + (R_1 + R_2)C_2 + (R_1 + R_2 + R_3)C_3 + (R_1 + R_2 + R_3 + R_4)C_L] \quad (8.1)$$

Assuming that all NMOS devices have an equal size, Eq. (8.1) simplifies to

$$t_{pHL} = 0.690R_N(C_1 + 2C_2 + 3C_3 + 4C_L) \quad (8.2)$$

The main advantage of complementary CMOS is that it is a very robust and simple approach for implementing logic gates. Two major problems associated with using this style as the complexity of the gate (i.e., fan-in) increases are:

1. The number of transistors required to implement an N fan-in gate is $2N$. This can result in a significantly large implementation area.
2. The propagation delay of a complementary CMOS gate deteriorates rapidly as a function of the fan-in.

Design techniques for large fan-in. The designer has a number of techniques to reduce the delay of large fan-in circuits.

- **Transistor sizing:** The most obvious solution is to increase the transistor sizes. This lowers the resistance of devices in series and lowers the time constants.
- **Progressive Transistor Sizing:** An alternate approach to uniform sizing is to use progressive transistor sizing. This approach reduces the dominant resistance, while keeping the increase in capacitance within bounds.

- **Input Reordering:** Some signals in complex combinational logic blocks might be more critical than others. An input signal to a gate is called critical if it is the last signal of all inputs to assume a stable value. The path through the logic which determines the ultimate speed of the structure is called the critical path. Putting the critical path transistors closer to the output of the gate can result in a speed up.
- **Logic Restructuring:** Manipulating the logic equations can reduce the fan-in requirements and thus reduce the gate delay.

Power consumption in CMOS Logic gates. The power dissipation is a strong function of transistor sizing (which affects physical capacitance), input and output rise–fall times (which determine the short-circuit power), device thresholds and temperature (which impact leakage power) and switching activity. The dynamic power dissipation is given by $C_L V_{DD}^2 f$. Making a gate more complex mostly affects the switching activity, which consists of two components: a static component that is only a function of the topology of the logic network, and a dynamic one that results from the timing behaviour of the circuit. The latter factor is also called *glitching*.

The CMOS logic described is highly robust and scalable with technology, but requires $2N$ transistors to implement an N-input logic gate. Also, the load capacitance is significant, since each gate drives two devices (a PMOS and an NMOS) per fan-out. This has opened the door for alternative logic families that are either simpler or faster.

8.2.2 Ratioed Logic

Ratioed logic attempts to reduce the number of transistors required to implement a given logic function, often at the cost of reduced robustness and extra power dissipation. In ratioed logic, a gate consists of an NMOS pull-down network that realizes the logic function and a simple load device, which replaces the entire pull-up network (Figure 8.6(a)). A ratioed logic which uses a grounded PMOS load is referred to as a pseudo-NMOS gate (Figure 8.6(b)).

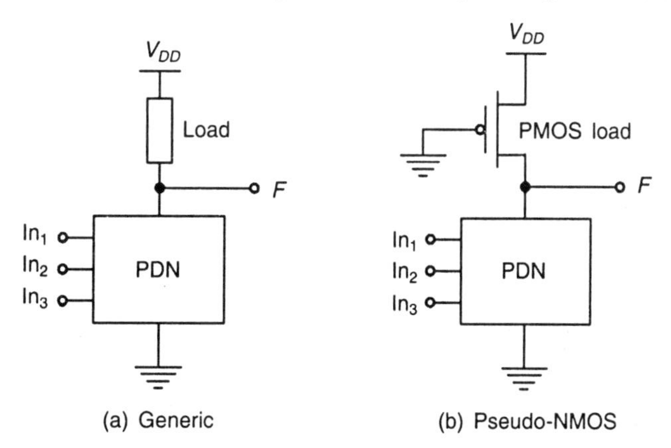

Figure 8.6 Ratioed logic gate.

The advantage of pseudo-NMOS gate is the reduced number of transistors ($N + 1$) versus $2N$ for complementary CMOS. The nominal high output voltage (V_{OH}) for this gate is V_{DD}, but the nominal low output voltage (V_{OL}) is not 0 V. This results in reduced noise margins and more static power dissipation. Since the voltage swing on the output and the overall functionality of the gate depend on the NMOS and PMOS sizes, the circuit is called ratioed. This is in contrast to the ratioless logic such as complementary CMOS, where the low and high levels do not depend on transistor sizes.

The static power dissipation of pseudo-NMOS limits its use. When area is most important, however, its reduced transistor count compared with complementary CMOS is quite attractive. Pseudo-NMOS thus still finds occasional use in large fan-in circuits.

Differential Cascade Voltage Switch Logic (DCVSL). This logic family combines two concepts differential logic and positive feedback. A differential gate requires that each input is provided in complementary format, and it produces complementary outputs in turn. The feedback mechanism ensures that the logic device is turned off when not needed. This logic completely eliminates static currents and provides rail-to-rail swing (Figure 8.7).

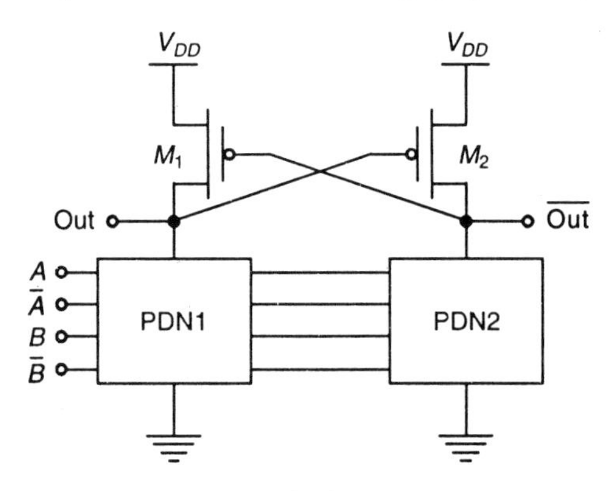

Figure 8.7 DCVSL logic gate.

The pull-down networks PDN1 and PDN2 use NMOS devices and are mutually exclusive, i.e., when PDN1 conducts, PDN2 is off, and when PDN1 is off, PDN2 conducts—such that the required logic function and its inverse are simultaneously implemented. The resulting circuit exhibits a rail-to-rail swing, and the static power dissipation is eliminated. The circuit is still ratioed since the sizing of the PMOS devices relative to the pull-down devices is critical to functionality, not just performance. In addition to the problem of increased design complexity, this circuit style has a power-dissipation problem due to crossover currents. During the transition, there is a period of time when PMOS and PDN are turned ON simultaneously, producing a short-circuit path.

8.2.3 Pass-Transistor Logic

A popular and widely used alternative to complementary CMOS is pass-transistor logic, which attempts to reduce the number of transistors required to implement logic by allowing the primary inputs to drive gate terminals as well as source–drain terminals. Figure 8.8 shows an implementation of the AND function constructed that way, using only NMOS transistors. In this gate, if the B input is high, the top transistor is turned ON and copies the input A to the output F. When B is low, the bottom pass-transistor is turned ON and passes a 0. The promise of this approach is that fewer transistors are required to implement a given function. The reduced number of devices has the additional advantage of lower capacitance.

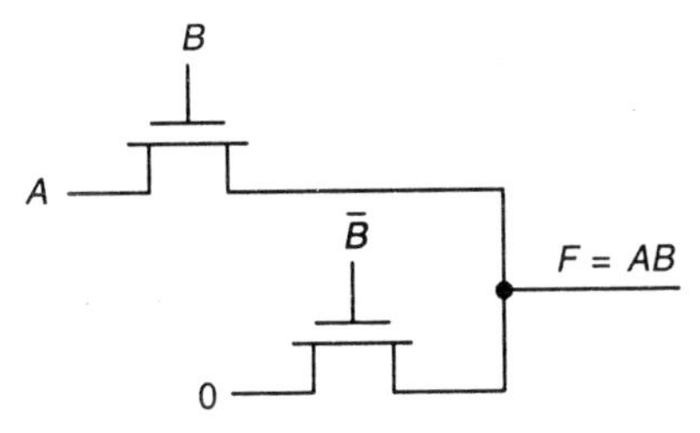

Figure 8.8 Pass-transistor implementation of an AND gate.

The pass-transistor gates cannot be cascaded by connecting the output of a pass gate to the gate input of another pass-transistor. Also, it is observed that a pure pass-transistor is not regenerative. A gradual signal degradation will be observed after passing through a number of subsequent stages.

While the circuit exhibits lower switching power, it may also consume static power when the output is high—the reduced voltage level may be insufficient to turn OFF the PMOS transistor of the subsequent CMOS inverter.

Differential pass-transistor logic. For high performance design, a differential pass-transistor logic family, called complementary pass-transistor logic (CPL) or DPL is commonly used. It accepts true and complementary inputs and produces true and complementary outputs. Several CPL gates (AND/NAND, OR/NOR, and XOR/XNOR) are shown in Figure 8.9.

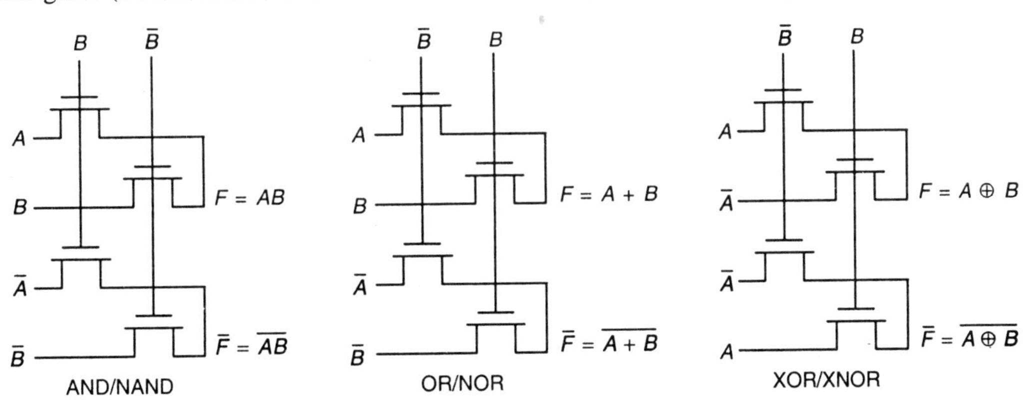

Figure 8.9 Complementary pass-transistor logic (CPL) examples.

The CPL gates possess some interesting properties:

(i) Since the circuits are differential, complementary data inputs and outputs are always available. Thus, some complex gates such as XORs and adders can be realized efficiently with smaller number of transistors. Furthermore, the availability of both polarities of every signal eliminates the need for extra inverters.

(ii) CPL belongs to the class of static gates, because the output-defining nodes are always connected to either V_{DD} or GND through a low-resistance path.

(iii) The design is very simple. More complex gates can be built by cascading the standard pass-transistor modules.

Unfortunately, differential pass-transistor logic suffers from static power dissipation and reduced noise margins, since the high input to the signal-restoring inverter only charges up to $V_{DD} - V_{Tn}$. This problem can be dealt with several solutions by the use of level restorer, multiple-threshold transistors and/or by the use of transmission gates.

8.3 DYNAMIC CMOS DESIGN

In the previous section, it was noted that static CMOS logic with a fan-in of N requires $2N$ devices. A variety of approaches including pseudo-NMOS, pass-transistor logic, etc., were presented to reduce the number of transistors required to implement a given logic function. The pseudo-NMOS logic style requires only $N + 1$ transistors to implement an N input logic gate, but unfortunately it has static power dissipation. In this section, an alternative logic style called dynamic logic is presented that obtains a similar result, while avoiding static power consumption. With the addition of a clock input, it uses a sequence of *precharge* and conditional *evaluation* phases.

8.3.1 Dynamic Logic: Basic Principles

The basic construction of an (n-type) dynamic logic gate is shown in Figure 8.10. The PDN is constructed exactly as in complementary CMOS. The operation of this circuit is divided into two phases—precharge and evaluation—with the mode of operation determined by the *clock* signal, Clk.

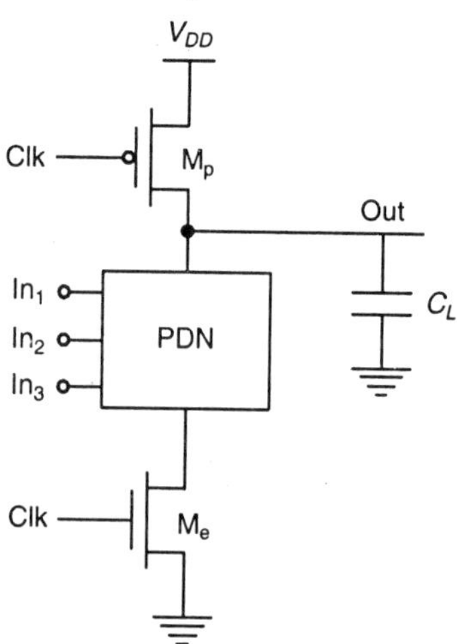

Figure 8.10 Basic dynamic logic circuit.

Precharge. When Clk = 0, the output node *Out* is precharged to V_{DD} by the PMOS transistor, M_p. During that time, the evaluate NMOS transistor, M_e is OFF, so that the pull-down path is disabled. The evaluation FET eliminates any static power that would be consumed during the precharge period.

Evaluation. For Clk = 1, the precharge transistor M_p is OFF, and the evaluation transistor M_e, is turned ON. The output is conditionally discharged based on the input values and the pull-down topology. If the inputs are such that the PDN conducts, then a low resistance path exists between *Out* and GND, and the output is discharged to GND. If the PDN is turned OFF, the precharged value remains stored on the output capacitance, C_L. During the evaluation phase, the only possible path between the output node and a supply rail is to GND. Consequently, once *Out* is discharged, it cannot be charged again until the next precharge operation. The inputs to the gate can thus make at most one transition during evaluation. The output can be in the high-impedance state during the evaluation period if the pull-down network is turned OFF.

Properties of dynamic logic gates

- The logic function is implemented by the NMOS pull-down network.
- The number of transistors (for complex gates) is lower than in the static case: $N + 2$ versus $2N$.
- It is non-ratioed.
- It only consumes dynamic power. Ideally, no static current path ever exists between V_{DD} and GND. The overall power dissipation, however, can be significantly higher compared with a static logic gate.
- The logic gates have faster switching speeds.

8.3.2 Speed and Power Dissipation of Dynamic Logic

The main advantages of dynamic logic are increased speed and reduced implementation area. Fewer devices to implement a given logic function implies that the overall load capacitance is much smaller. After the *precharge* phase, the output is high. For a low input signal, no additional switching occurs. As a result, $t_{pLH} = 0$. On the other hand, the high-to-low transition requires the discharging of the output capacitance through the pull-down network. Therefore, t_{pHL} is proportional to C_L and the current-sinking capabilities of the pull-down network. The presence of evaluation transistor slows the gate somewhat, as it presents an extra series resistance. Omitting this transistor, while functionally not forbidden, may result in static power dissipation and potentially a performance loss.

The dynamic logic presents a significant advantage from a power perspective, due to the following reasons:

- First, the physical capacitance is lower since dynamic logic uses fewer transistor to implement a given function. Also, the load seen for each fan-out is one transistor instead of two.
- Second, by construction, the dynamic logic gates can have at most one transistor per clock cycle. Glitching (or dynamic hazards) does not occur in dynamic logic.
- Finally, dynamic gates do not exhibit short-circuit power since the pull-up path is not turned ON when the gate is evaluating.

8.3.3 Signal Integrity Issues in Dynamic Design

Dynamic logic clearly can result in high-performance solutions compared to static. However, there are several important considerations that must be taken into account for dynamic circuits to function properly. These include charge leakage, charge sharing, capacitive coupling, and clock feedthrough.

Charge leakage. The operation of a dynamic gate depends on the dynamic storage of the output value on a capacitor. If the pull-down network is OFF, ideally, the output should remain at the precharged state of V_{DD}, during the *evaluation* phase. However, this charge gradually leaks away due to leakage currents finally resulting in a malfunctioning of the gate. Figure 8.11(a) shows the sources of leakage for the basic dynamic inverter circuit.

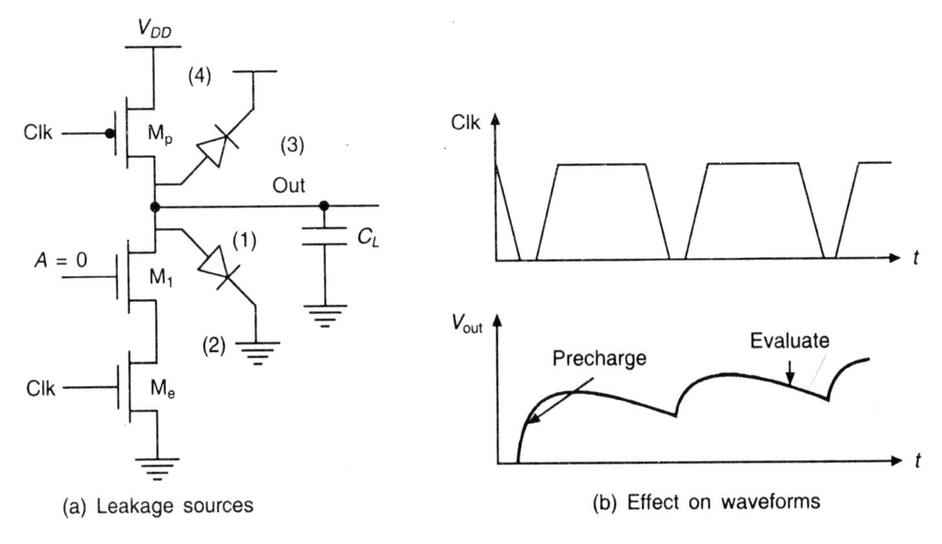

(a) Leakage sources

(b) Effect on waveforms

Figure 8.11 Leakage issues in dynamic circuits.

Sources 1 and 2 are the reverse-biased diode and subthreshold leakage of the NMOS pull-down device M_1, respectively. The charge stored on C_L will slowly leak away through these leakage channels, causing a degradation in the high level [Figure 8.11(b)]. Therefore, dynamic circuits require a minimal clock rate, which is typically of the order of a few kHz. This makes the usage of dynamic techniques unattractive for low-performance products such as watches or processors that use conditional clocks. The PMOS precharge device also contributes some leakage current due to the reverse bias diode (source 3) and the subthreshold conduction (source 4). To some extent, the leakage current of the PMOS counteracts the leakage of the pull-down path. As a result, the output voltage is going to be set by the resistive divider composed of the pull-down and pull-up path.

Leakage is caused by the high-impedance state of the output node during the *evaluate* mode, when the pull-down path is turned OFF. The leakage problem may be counteracted by reducing the output impedance on the output node during evaluation. This often is done by adding a bleeder transistor as shown in Figure 8.12(a). The function of the bleeder is to compensate for

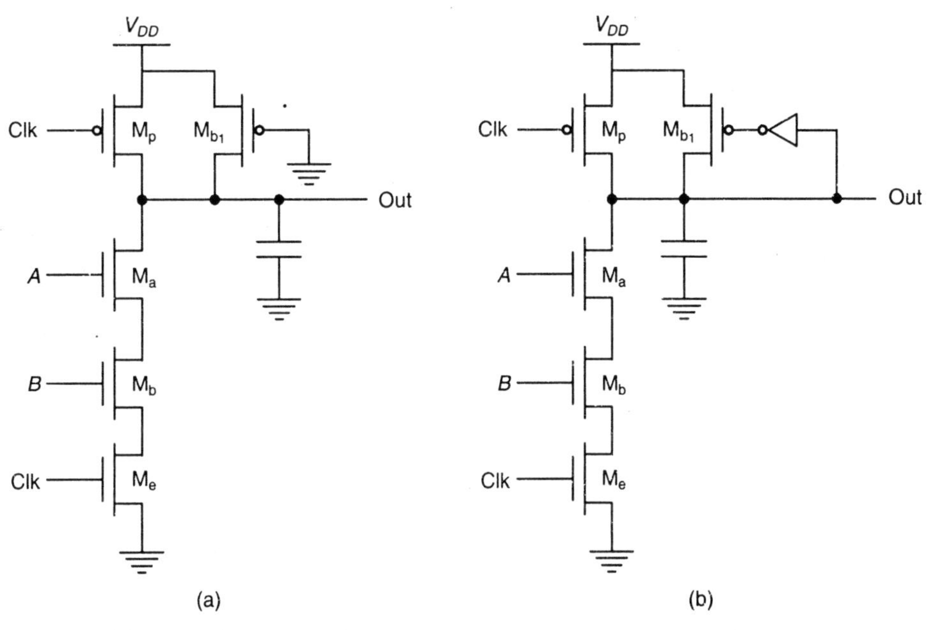

Figure 8.12 Static bleeders compensate for the charge leakage.

the charge lost due to the pull-down leakage paths. To avoid the ratio problems associated with this style of circuit and the associated static power consumption, the bleeder resistance is made high. This allows the (strong) pull-down devices to lower the *Out* node substantially below the switching threshold of the next gate. Often, the bleeder is implemented in a feedback configuration to eliminate the static power dissipation altogether [Figure 8.12(b)].

Charge sharing. Another important concern in dynamic logic is the impact of charge sharing. Consider the circuit in Figure 8.13. During the *precharge* phase, the output node is precharged to V_{DD}. Assume that all inputs are set to 0 during precharge, and that the capacitance C_a is

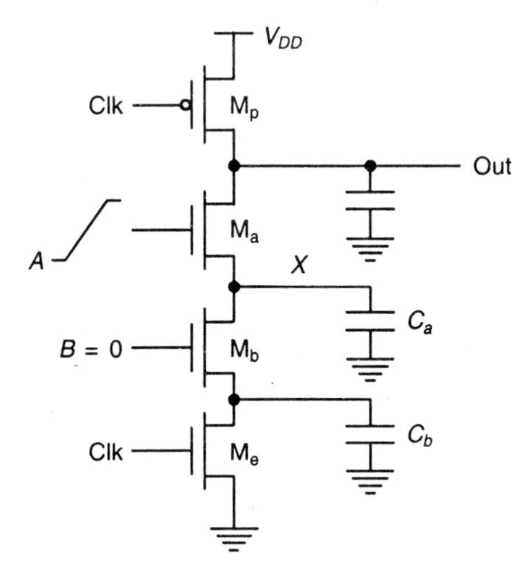

Figure 8.13 Charge sharing on dynamic networks.

discharged. Assume further that input B remains at 0 during evaluation, while input A makes a 0 to 1 transition, turning transistor M_a ON. The charge stored originally on capacitor C_L is redistributed over C_L and C_a. This causes a drop in the output voltage, which cannot be recovered due to the dynamic nature of the circuit.

The most common and effective approach to deal with the charge redistribution is to also precharge critical internal nodes, as shown in Figure 8.14. Since the internal nodes are charged to V_{DD} during precharge, charge sharing does not occur. This solution comes at the cost of increased area and capacitance.

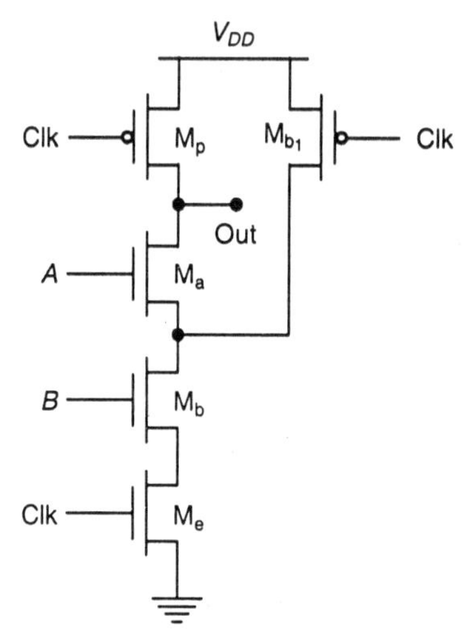

Figure 8.14 Dealing with charge sharing by precharging internal nodes.

An NMOS precharge transistor may also be used, but this requires an inverted clock.

Capacitive coupling. The relatively high impedance of the output node makes the circuit very sensitive to crosstalk effects. A wire routed over or next to a dynamic node may couple capacitively and destroy the state of the floating node. Another form of capacitive coupling is backgate (or output-to-input) coupling. When designing and laying out dynamic circuits, special care is needed to minimize capacitive coupling.

Clock feedthrough. A special case of capacitive coupling is clock feedthrough, an effect caused by the capacitive coupling between the clock input of the precharge device and the dynamic output node. This coupling capacitance consists of the gate-to-drain capacitance of the precharge device, and includes both the overlap and channel capacitances. This capacitive coupling causes the output of the dynamic node to rise above V_{DD} on the low-to-high transition of the clock, assuming that the pull-down network is turned OFF. Subsequently, the fast rising and falling edges of the clock couple onto the signal node.

The danger of clock feedthrough is that it may cause the normally reverse biased junction diodes of the precharge transistor to become forward biased. This causes electron injection into the substrate, which can be collected by a nearby high-impedance node in the 1 state, finally resulting in faulty operation. CMOS latch up may be another result of this injection. For all purposes, high-speed dynamic circuits should be carefully simulated to ensure that clock feedthrough effects stay within bounds.

8.3.4 Cascading Dynamic Gates

There is one major problem that complicates the design of dynamic circuits during straight forward cascading of dynamic gates to create multilevel logic structures. The problem is best illustrated with two cascaded n-type dynamic inverters as in Figure 8.15(a).

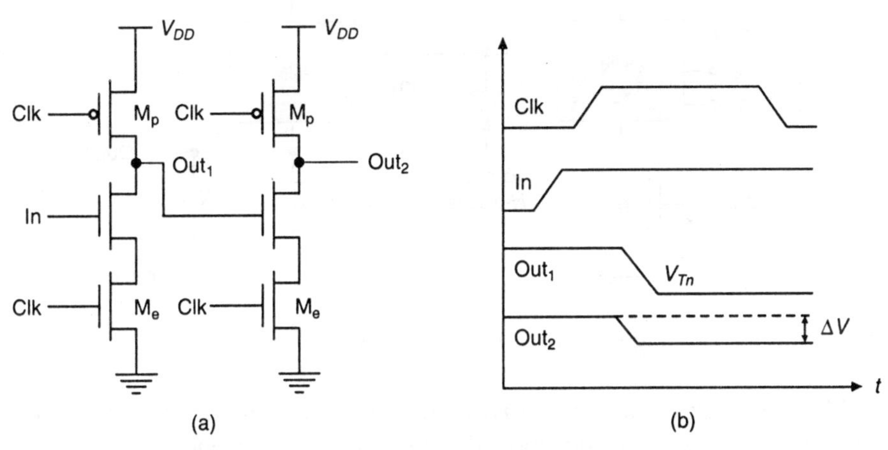

Figure 8.15 Cascade of dynamic n-type network.

During the precharge phase (Clk = 0), the outputs of both inverters are precharged to V_{DD}. Assume that the primary input. In makes a 0 to 1 transition [Figure 8.14(b)]. On the rising edge of the clock, output Out_1 starts to discharge. The second output should remain in the precharged state of V_{DD} as its expected value is 1 (Out_1 transitions to 0 during evaluation). However, there is a finite propagation delay for the input to discharge Out_1 to GND. Therefore, the second output also starts to discharge. The charge loss leads to reduced noise margins and potential malfunctioning.

The correct operation is guaranteed as long as the inputs can only make a single $0 \rightarrow 1$ transition during the evaluation period.

Domino logic. A domino logic module consists of an n-type dynamic logic block followed by a static inverter. Domino CMOS has the following properties:

- Since each dynamic gate has a static inverter, only non-inverting logic can be implemented.
- Very high speeds can be achieved.

Several optimizations can be performed on domino logic gates. The most obvious performance optimization involves the sizing of the transistors in the static inverter. A designer should consider reduced noise margin and performance impart simultaneously during the device sizing.

NP-CMOS. An alternative approach to cascading dynamic logic is provided by NP-CMOS, which uses two flavours (n-tree and p-tree) of domino logic, and avoids an extra static inverter in the critical path that comes with domino logic. In a p-tree logic gate, PMOS devices are used to build a pull-up logic network, including a PMOS evaluation transistor (Figure 8.16).

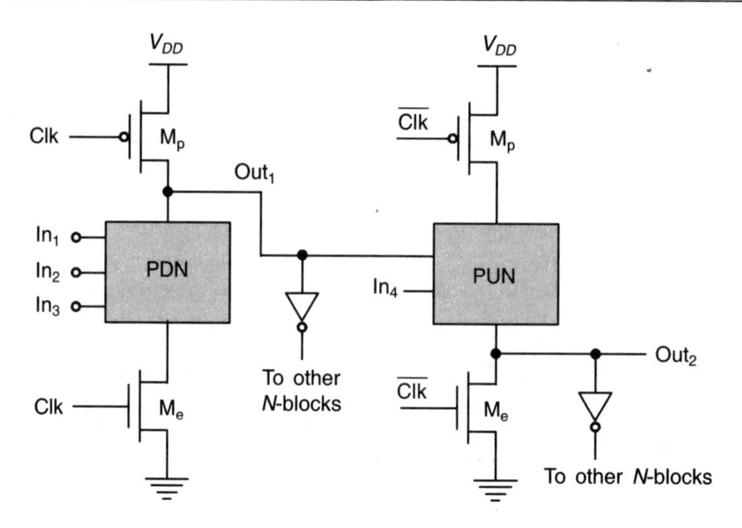

Figure 8.16 The NP-CMOS logic circuit.

The NMOS precharge transistor drives the output low during precharge. The output conditionally makes a $0 \rightarrow 1$ transition during evaluation depending on its inputs. The NP-CMOS logic exploits the duality between n-tree and p-tree logic gates to eliminate the cascading problem. If the n-tree gates are controlled by Clk, and p-tree gates are controlled using $\overline{\text{Clk}}$, n-tree gates can directly drive p-tree gates, and vice versa.

8.4 COMPLEX LOGIC GATES IN CMOS

A complex logic gate is one that implements a function that can provide the basic NOT, AND and OR operation but integrates them into a single circuit. CMOS is ideally suited for creating gates that have logic equations exhibiting the following:

(i) AND–OR–INVERT – AOI form
(ii) OR–AND–INVERT – OAI form

An AOI logic equation is equivalent to a complemented SOP form, while an OAI equation is equivalent to a complemented POS structure.

Example 8.1

AOI Example: Implement the following complex logic function in CMOS,

$$G(A, B, C) = \overline{A \cdot B + C}$$

CMOS implementation procedure

Step 1: Connect two NFETs in series and apply inputs A and B to form the AND function $A \cdot B$

Step 2: Connect another NFET in parallel to this combination from step 1 and apply the input C to obtain the logic function $(A \cdot B) + C$

Step 3: Now, to draw the PMOS logic, since the term $(A \cdot B)$ is ORed (in parallel) with C in NMOS logic, in the corresponding PMOS network, both of them must be connected in series.

Step 4: Since AB in NMOS is two transistors in series, the corresponding p-transistors must be connected in parallel.

Step 5: Apply inputs A, B and C to the respective PMOS transistors.

Step 6: Apply V_{DD} to the top of PMOS logic and GND to the bottom of NMOS logic.

Step 7: Obtain the output expression from the point at which PMOS and NMOS logics are connected together (Figure 8.17).

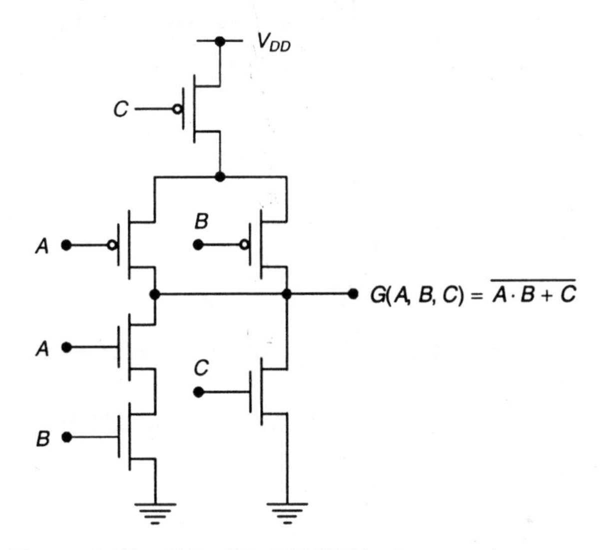

Figure 8.17 AND–OR–INVERT logic example.

Example 8.2

OAI Example: Implement the following logic function in CMOS.

$$F = \overline{(A+B).C}$$

CMOS implementation procedure

Step 1: Connect two NFETS in parallel and apply inputs A and B to form the OR function, $A + B$

Step 2: Connect another NFET in series with this combination from step 1 and apply the input C to obtain the function $(A + B) \cdot C$

Step 3: To draw the PMOS logic, since the term $A + B$ is ANDed (in series) with the input C in the NMOS logic, both of them must be connected in parallel in the corresponding PMOS logic.

Step 4: Since $A + B$ in NMOS is two NFETS in parallel, the corresponding PFETS must be connected in series.

Step 5: Apply inputs A, B and C to the respective PMOS transistors.

Step 6: Connect V_{DD} to the top of PMOS logic and GND to the bottom of NMOS logic.

Step 7: Obtain the output from the junction at which PMOS and NMOS logic join together (Figure 8.18).

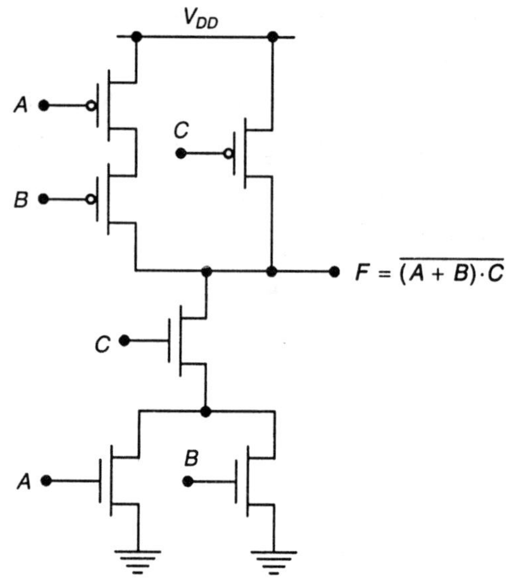

Figure 8.18 CMOS OR–AND–INVERT gate.

Example 8.3 Implement the following logic function in CMOS.

$$h = \overline{AB + CD}$$

Using the procedure described above, CMOS implementation of function h is shown in Figure 8.19.

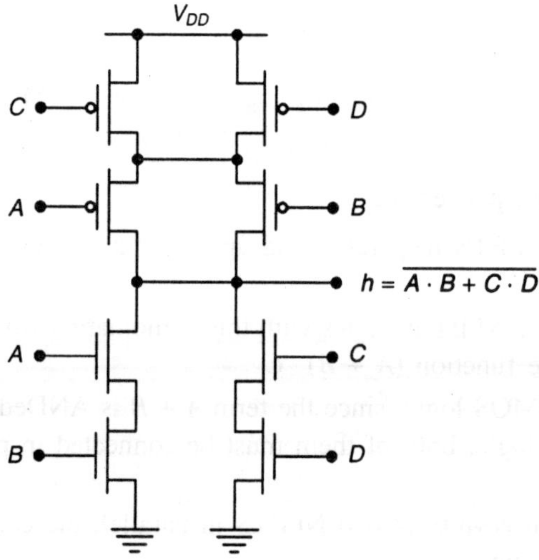

Figure 8.19 CMOS implementation of function *h*.

Example 8.4: Implement the circuits of XOR and XNOR gates in CMOS.

CMOS Implementation of XOR gate

$$\begin{aligned} h_1 &= A \oplus B \\ &= \bar{A}\cdot B + A\cdot \bar{B} \\ &= \bar{A}\cdot A + \bar{A}\cdot B + A\cdot \bar{B} + \bar{B}\cdot B \\ &= (\bar{A} + \bar{B})\cdot (A + B) \\ &= \overline{A\cdot B}\cdot \overline{\bar{A}\cdot \bar{B}} \\ &= \overline{A\cdot B + \bar{A}\cdot \bar{B}} \end{aligned} \tag{8.3}$$

Equation (8.3) can be realized using CMOS as in the following Figure 8.20.

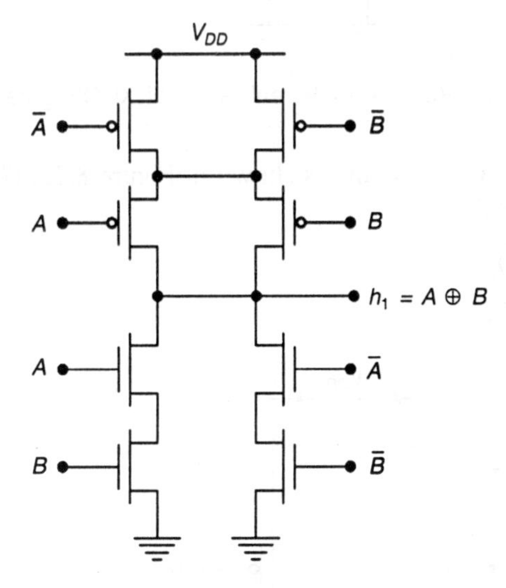

Figure 8.20 CMOS realization of XOR gate.

CMOS Implementation of XNOR gate

$$\begin{aligned} h_2 &= \overline{A \oplus C} \\ &= \overline{\bar{A}\cdot C + A\cdot \bar{C}} \end{aligned} \tag{8.4}$$

Equation (8.4) can be realized in CMOS as in Figure 8.21 given below.

Example 8.5: Implement the following logic functions in CMOS:

(i) $F = \overline{(A + B)\cdot (C + D)}$

(ii) $G = \overline{\overline{(A + B)}C\cdot Y + X}$

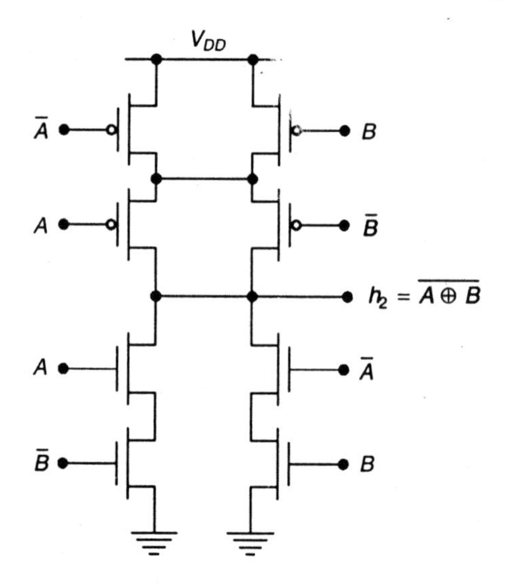

Figure 8.21 CMOS realization of XNOR gate.

F represents a 4-input OAI logic gate and is shown in Figure 8.22. *G* is an example of a logic cascade in CMOS (Figure 8.23).

(i) $F = \overline{(A+B)\cdot(C+D)}$

(ii) $G = \overline{\overline{(A+B)}C\cdot Y + X}$

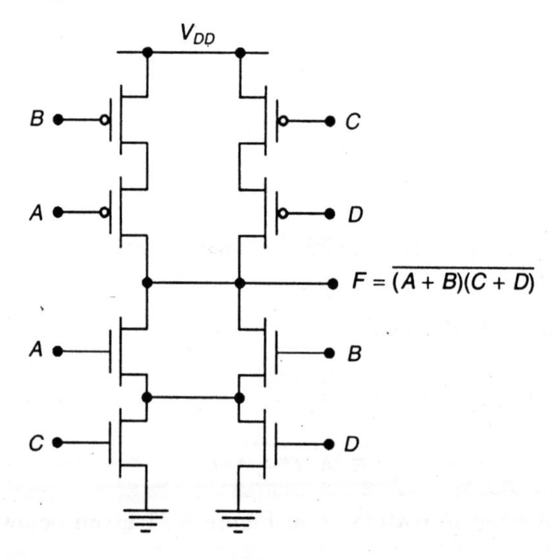

Figure 8.22 A 4-input OAI logic gate circuit.

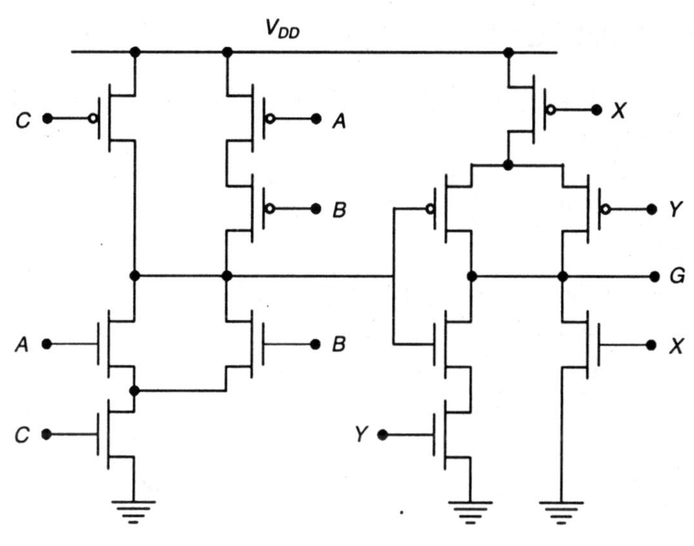

Figure 8.23 Example of logic cascade in CMOS.

SUMMARY

In this chapter, the behaviour and performance of CMOS combinational digital circuits are analyzed extensively with regard to area, speed and power. Static complementary CMOS combines dual pull-down and pull-up networks, only one of which is enabled at any time. The performance of a CMOS gate is a strong function of fan-in. Techniques to deal with fan-in include transistor-sizing, input reordering and partitioning. The speed is a linear function of the fan-out. The ratioed logic style consists of an active pull-down (up) network connected to a load device. This results in a substantial reduction in gate complexity at the expense of static power consumption and an asymmetrical response. Careful transistor sizing is necessary to maintain sufficient noise margins. The most popular approaches in this class are the pseudo-NMOS techniques and the differential DCVSL, which require complementary signals. Pass-transistor logic implements a logic gate as a simple switch network. NMOS only pass transistor logic produces even simpler structures, but might suffer from static power consumption and reduced noise margins. The operation of a dynamic logic is based on the storage of charge on a capacitive node and the conditional discharging of that node as a function of the inputs. This calls for a two-phase scheme, consisting of a precharge followed by an evaluation step. Dynamic logic trades off noise margin for performance. The power consumption of a logic network is strongly related to the switching activity of the network. By careful circuit design and transistor sizing, sources of power consumption such as glitches and short-circuit currents can be minimized.

REVIEW QUESTIONS

1. Explain pull-up network using NMOS and PMOS switches.
2. What is dynamic logic and explain speed and power dissipation of dynamic logic.
3. Give brief notes on leakage issues in dynamic circuits.
4. What are the advantagous of cascading dynamic gates and explain it by cascading a dynamic *n*-type network.
5. With suitable example explain complex logic gates in CMOS.

SHORT ANSWER QUESTIONS

1. The primary advantage of the CMOS structure is ____________ .

   Ans. robustness.

2. PUN (pull-up network) is to provide a connection between the ____________ and ____________, anytime the output of logic gate is meant to be 1.

   Ans. output, V_{DD}.

3. The propagation delay can be computed by ____________ .

   Ans. Elmore delay model.

4. Define ratioed logic.

   Ans. It is to reduce the number of transistors required to implement a given logic function.

5. How many transistors are required to implement N-input logic gates in the preudo NMOS logic.

   Ans. $N + 1$ transistor.

6. The function of bleeder transistor is to ____________ .

   Ans. compensate for the charge lost due to the pull-down leakage paths.

7. The ____________ drives the output low during precharge.

   Ans. NMOS precharge transistor.

9 CMOS Sequential Logic Circuits

9.1 INTRODUCTION

A sequential logic circuit, also called a regenerative circuit, is one in which the output at any point in time is a function of the current input data as well as of previous values of the input signals. This can be accomplished by connecting one or more outputs intentionally back to some inputs. Consequently, the circuit 'remembers' past events and has a sense of history. A sequential circuit includes a combinational logic portion and a module that holds the state. Example circuits are registers, counters, oscillators, and memory. Figure 9.1 shows the block diagram of a sequential logic circuit. A generic finite-state machine (FSM) consisting of combinational logic and registers, which hold the system state, is shown in Figure 9.2.

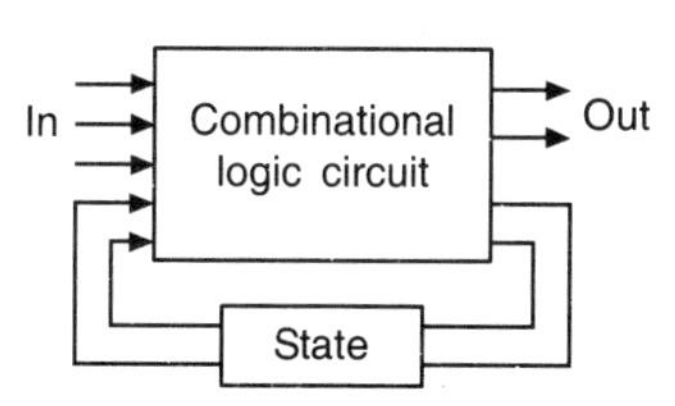

Figure 9.1 Sequential logic circuit—block schematic.

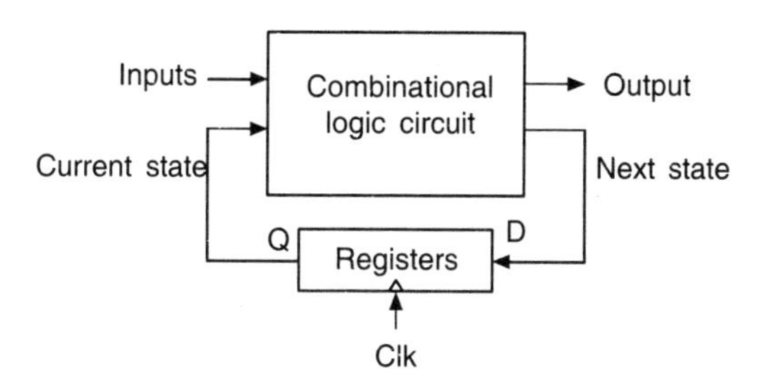

Figure 9.2 Block diagram of a finite-state machine, using positive edge-triggered registers.

The system, depicted in Figure 9.2, belongs to the class of synchronous sequential circuits, in which all registers are under control of a single global clock. The output of the FSM is a function of the current inputs and the current state. The next state is determined based on the current state and the current inputs and is fed to the inputs of registers. On the rising edge of the clock, the next state bits are copied to the outputs of the registers (after some propagation delay), and a new cycle begins. The register then ignores changes in the input signals until the next rising edge. In general, registers can be positive edge triggered (where the input data is copied on the rising edge of the clock) or negative edge triggered (where the input data is copied on the falling edge, as indicated by a small circle at the clock input).

A pipelined system is shown in Figure 9.3. Pipelined systems use storage devices to capture the output of each processing stage at the end of each clock period, and in general have no feedback. The majority of VLSI systems are a combination of pipelined and finite-state machines.

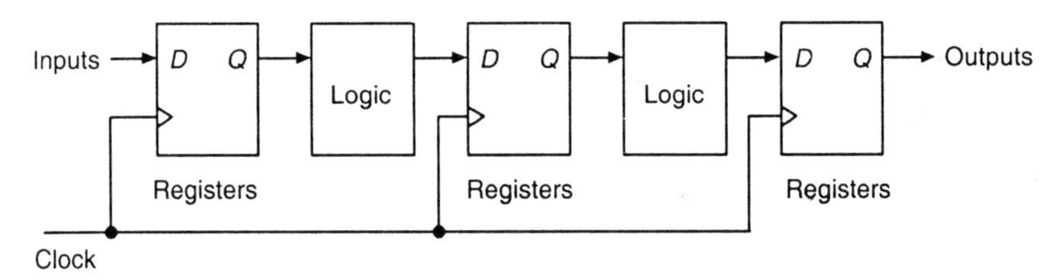

Figure 9.3 A pipelined system.

This chapter discusses the CMOS implementation of the most important sequential building blocks. A variety of choices in sequential primitives and clocking methodologies exist; making the correct selection is getting increasingly important in modern digital circuits, and can have a great impact on performance, power, and/or design complexity.

9.2 TIMING METRICS FOR SEQUENTIAL CIRCUITS

Consider a register circuit shown in Figure 9.4. There are three important timing parameters associated with a register. The setup time (t_{su}) is the time that the data inputs (D) must be valid before the clock transition (i.e., the $0 \rightarrow 1$ transition for a positive edge-triggered register). The hold time (t_{hold}) is the time the data input must remain valid after the clock edge. Assuming that the setup and hold times are met, the data at the D input is copied to the Q output after a worst case propagation delay (with reference to the clock edge) denoted by $t_{c\text{-}q}$.

The clock period T, at which the sequential circuit operates, must thus accommodate the longest delay of any stage in the network. Assume that t_{plogic} denotes the worst-case propagation delay of the logic, t_{cd}, the contamination delay. Then, the minimum clock period T required for proper operation of the sequential circuit is given by

$$T \geq t_{c\text{-}q} + t_{plogic} + t_{su} \tag{9.1}$$

The hold time of the register imposes an extra constraint for proper operation, namely

$$t_{cd\ register} + t_{cd\ logic} \geq t_{hold} \tag{9.2}$$

where $t_{cd\ register} \rightarrow$ minimum propagation delay (or contamination delay) of the register.

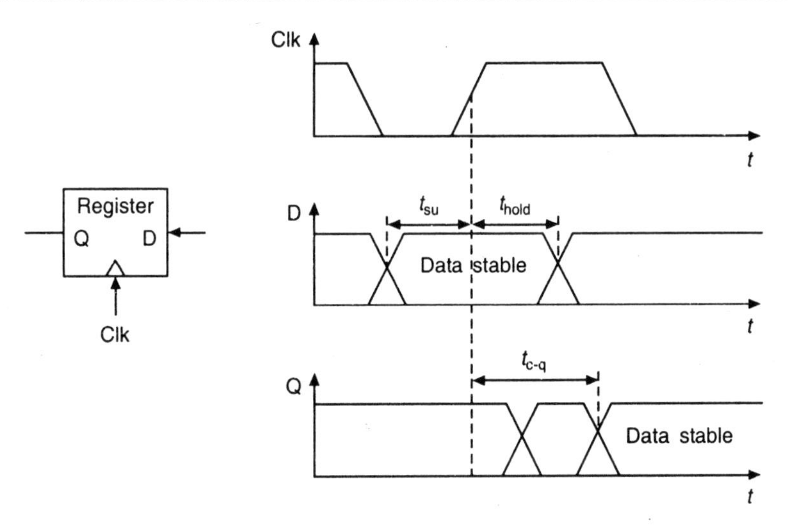

Figure 9.4 Definition of setup time, hold time, and propagation delay of a synchronous register.

This constraint ensures that the input data of the sequential elements is held long enough after the clock edge and is not modified too soon by the new wave of data coming in.

As seen from Eq. (9.1), it is important to minimize the values of the timing parameters associated with the register, as these directly affect the rate at which a sequential circuit can be clocked. In fact, modern high-performance systems are characterized by a very low logic depth and the register propagation delay and setup times account for a significant portion of the clock period. In general, the requirement of Eq. (9.2) is not difficult to meet, although it becomes an issue when there is little or no logic between registers.

9.3 CLASSIFICATION OF MEMORY ELEMENTS

Foreground and background memory. At a high level, memory is classified into background and foreground memory. Memory that is embedded into logic is foreground memory and is most often organized as individual registers or register banks. Large amounts of centralized memory core are referred to as background memory. Background memory achieves higher area densities through efficient use of array structures and by trading off performance and robustness for size.

Static vs. dynamic memory. Memories can be either static or dynamic. Static memories preserve the state as long as the power is turned ON. They are built by using positive feedback or regeneration, where the circuit topology consists of intentional connections between the output and the input of a combinational circuit. Static memories are most useful when the register will not be updated for extended periods of time. Configuration data, loaded at power-up time, is a good example of static data. This condition also holds for most processors that use conditional clocking (i.e., gated clocks) where the clock is turned OFF for unused modules. In that case, there are no guarantees on how frequently the registers will be clocked, and static memories are needed to preserve the state information. Memory based on positive feedback falls under the class of elements called multivibrator circuits. The bistable element is its most popular representative, but other elements such as monostable and astable circuits are also frequently used.

Dynamic memories store data for a short period of time, perhaps milliseconds. They are based on the principle of temporary charge storage on parasitic capacitors associated with MOS devices. The capacitors have to be refreshed periodically to compensate for charge leakage. Dynamic memories tend to be simpler, resulting in significantly higher performance and lower power dissipation. They are more useful in datapath circuits that require high performance levels and are periodically clocked. It is possible to use dynamic circuitry even when circuits are conditionally clocked, if the state can be discarded when a module goes into idle mode.

Latches vs. registers. A latch is a level-sensitive circuit that passes the D input to the Q output when the clock signal is high. This latch is said to be in transparent mode. When the clock is low, the input data sampled on the falling edge of the clock is held stable at the output for the entire phase, and the latch is in hold mode. The inputs must be stable for a short period around the falling edge of the clock to meet setup and hold requirements. A latch operating under these conditions is a positive latch. Similarly, a negative latch passes the D input to the Q output when the clock signal is low. Positive and negative latches are called transparent high or transparent low, respectively. The signal waveforms for a positive and negative latch are shown in Figure 9.5. A wide variety of static and dynamic implementations exists for the realization of latches.

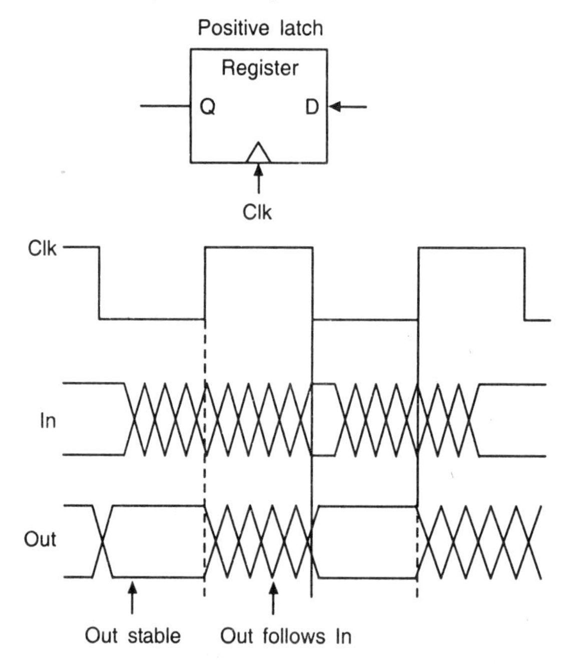

Figure 9.5 Timing of positive and negative latches.

A latch is an essential component in the construction of an edge-triggered register. Contrary to level-sensitive latches, edge-triggered registers only sample the input on a clock transition—that is, $0 \rightarrow 1$ for a positive edge-triggered register, and $1 \rightarrow 0$ for a negative edge-triggered register. An often-recurring configuration is the master–slave structure, which cascades a positive and negative latch. Registers also can be constructed by using one-shot generators of the clock signal ('glitch' registers), or by using other specialized structures. The definitions for the different types of storage elements (i.e., register, flip-flop, and latch) are given below.

- An edge-triggered storage element is called a register.
- A latch is a level-sensitive device.
- Any bistable component formed by the cross-coupling of gates is called a flip-flop.

9.4 STATIC LATCHES AND REGISTERS

9.4.1 Bistability Principle

Static memories use positive feedback to create a bistable circuit—a circuit having two stable states that represent 0 and 1. Figure 9.6 shows two inverters connected in cascade, i.e., cross coupling of two inverters results in a bistable circuit. The circuit serves as a memory, storing either a 1 or a 0. In the absence of any triggering, the circuit remains in a single state (assuming that the power supply remains applied to the circuit) and thus remembers a value. A bistable circuit is also called flip-flop. A flip-flop is useful only if there also exists a means to bring it from one state to the other one. In general, two different approaches may be used to accomplish the following:

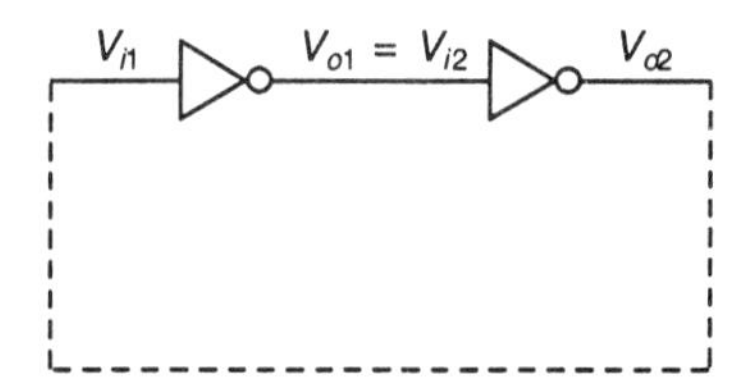

Figure 9.6 Two cascaded inverters.

- **Cutting the feedback**
  Once the feedback loop is open, a new value can easily be written into Out (or Q). Such a latch is called multiplexer based, as it realizes that the logic expression for a synchronous latch is identical to the multiplexer equation:

$$Q = \overline{\text{CLK}} \cdot Q + \text{CLK} \cdot \overline{Q} \tag{9.3}$$

  This approach is the most popular in today's latches.
- **Over powering the feedback loop**
  By applying a trigger signal at the input of the flip-flop, a new value is forced into the cell by overpowering the stored value. This needs a careful signing of the transistors in the feedback loop and the input circuitry.

9.4.2 Multiplexer-Based Latches

The most robust and common technique to build a latch involves the use of transmission-gate multiplexers. Figure 9.7 shows an implementation of positive and negative static latches based on multiplexers. For a negative latch, input 0 of the multiplexer is selected when the clock is low, and the D input is passed to the output. When the clock signal is high, input 1 of the

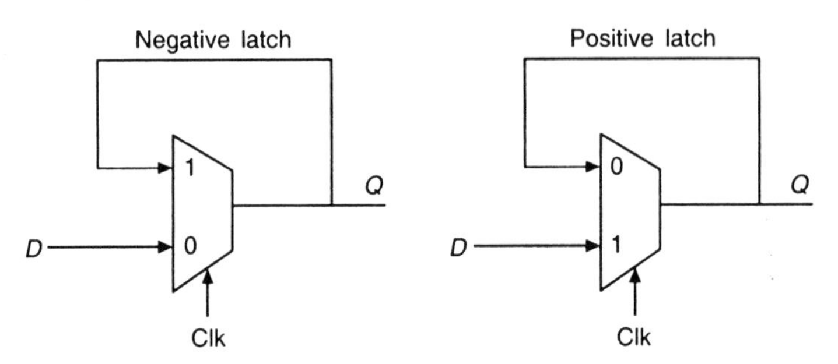

Figure 9.7 Negative and positive latches based on multiplexers.

multiplexer, which connects to the output of the latch, is selected. The feedback ensures a stable output as long as the clock is high. Similarly in the positive latch, the *D* input is selected when the clock signal is high, and the output is held (using feedback) when the clock signal is low.

A transistor-level implementation of a positive latch based on multiplexers is shown in Figure 9.8. When Clk is high, the bottom transmission gate is ON and the latch is transparent—i.e, the *D* input is copied to the output. During this phase, the feedback loop is open, since the top transmission gate is OFF. This particular latch implementation is not very efficient from power perspective. It presents a load of four transistors to the Clk signal.

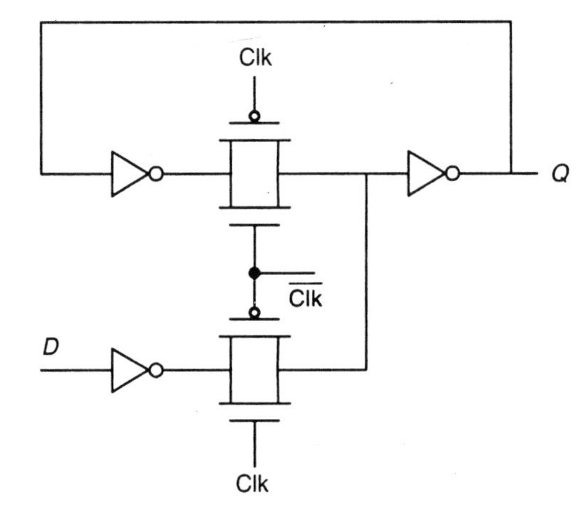

Figure 9.8 Transistor-level implementation of a positive latch built by using transmission gates.

The clock load can be reduced to two transistors by implementing multiplexers that use NMOS-only pass transistors, as shown in Figure 9.9.

When CLK is high, the latch samples the *D* input, while a low clock signal enables the feedback loop, and puts the latch in the hold mode. Even though the circuit is simple, the use of NMOS-only pass transistors results in the passing of a degraded high voltage of $V_{DD} - V_{Tn}$ to the input of the first inverter. This impacts both noise margin and the switching performance, especially in the case of low value of V_{DD} and high values of V_{Tn}. It also causes static power dissipation in the first inverter, because the maximum input voltage to the inverter equals $V_{DD} - V_{Tn}$, and the PMOS device of the inverter is never fully turned OFF.

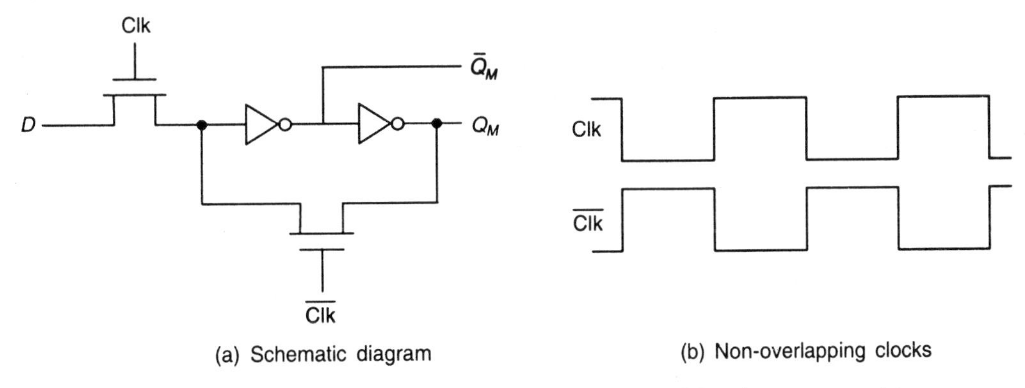

(a) Schematic diagram (b) Non-overlapping clocks

Figure 9.9 Multiplexer-based NMOS latch by using NMOS-only pass transistors.

9.4.3 Master–Slave Edge-Triggered Register

The most common approach for constructing an edge-triggered register is to use a master–slave configuration, as in Figure 9.10. The register consists of cascading a negative latch (master stage) with a positive latch (slave stage). A multiplexer-based latch is used in this particular implementation. On the low phase of the clock, the master stage is transparent, and the D input is passed to the master stage output Q_M. During this period, the slave stage is in the hold mode,

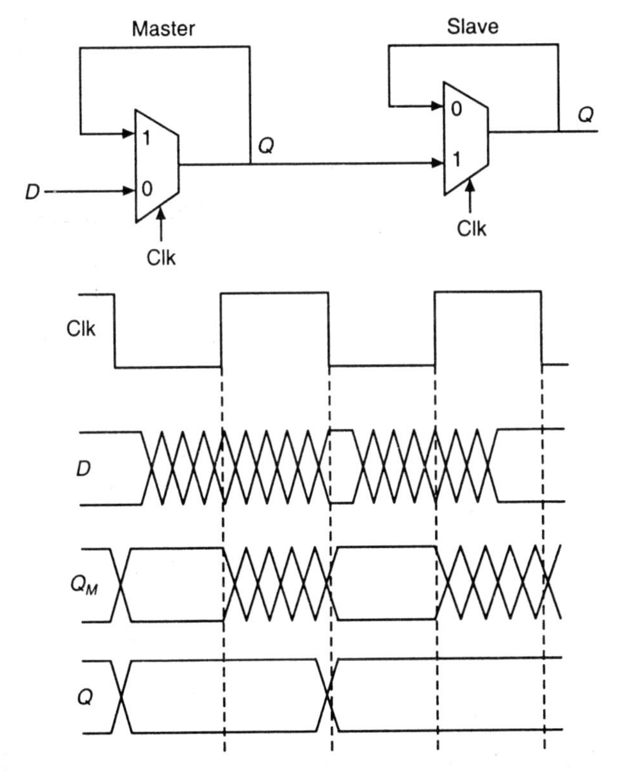

Figure 9.10 Positive edge-triggered register based on a master–slave configuration.

keeping its previous value by using feedback. On the rising edge of the clock, the master stage stops sampling the input, and the slave stage starts sampling. During the high phase of the clock, the slave stage samples the output of the master stage (Q_M), while the master stage remains in a hold mode. Since Q_M is constant during the high phase of the clock, the output Q makes only one transition per cycle. The value of Q is the value of D input right before the rising edge of the clock, achieving the positive-edge triggered effect. A negative edge-triggered register can be constructed by using the same principle by simply switching the order of the positive and negative latches (i.e., placing the positive latch first).

The master–slave positive edge-triggered register can also be implemented with transmission gate logic. The drawback of the transmission-gate register is the high capacitive load presented to the clock signal. The clock load per register is important, since it directly impacts the power dissipation of the clock network. One approach to reduce the clock load at the cost of robustness is to make the circuit ratioed, but with increased design complexity. Using minimum or close-to-minimum size devices in the transmission gate is desirable to reduce the power dissipation in the latches and the clock distribution network.

Another problem with this scheme is reverse conduction—the second stage can affect the state of the first latch.

Non-ideal clock signals. So far, it is assumed that $\overline{\text{CLK}}$ is a perfect inversion of CLK, or in other words, that the delay of the generating inverter is zero. This is not true always, variations can exist in the wires used to route the two clock signals, or the load capacitances can vary based on data stored in the connecting latches. This effect is known as the clock skew and is a major problem, causing the two clock signals to overlap as in Figure 9.11. Clock overlap can cause two types of failures:

1. When the clock goes high, the slave stage should stop sampling the master stage output and go into a hold mode. However, since $\overline{\text{CLK}}$ and CLK are both high for a short period of time (the overlap period), and there is a direct path from the D input to the Q output. As a result, the data at the output can change on the rising edge of the clock, which is undesired for a negative edge-triggered register. This is known as a race condition.
2. In a multiplexer-based register, clock overlap between $\overline{\text{CLK}}$ and CLK results in an undefined state between the input and the output.

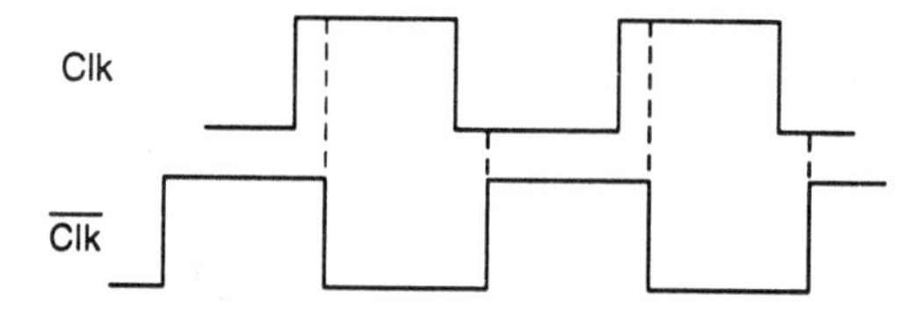

Figure 9.11 Overlapping clock phases.

The above mentioned problems can be avoided by using two non-overlapping clocks (Figure 9.12) and by keeping the non-overlap time between the clock larger enough so that no overlap occurs even in the presence of clock-routing delays. During the non-overlap time, the

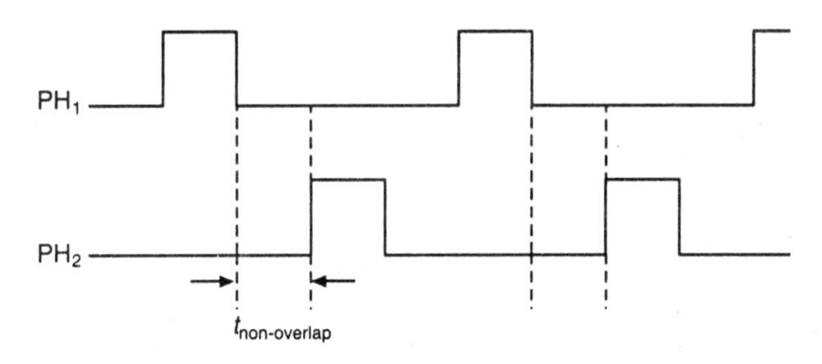

Figure 9.12 Two-phase non-overlapping clocks.

flip-flop is in the high-impedance state. Leakage will destroy the state if this condition holds for too long. The register employs a combination of static and dynamic storage approaches, depending upon the state of the lock.

Generating non-overlapping clocks. Figure 9.13 shows one possible implementation of the clock generation circuitry for generating a two-phase non-overlapping clock.

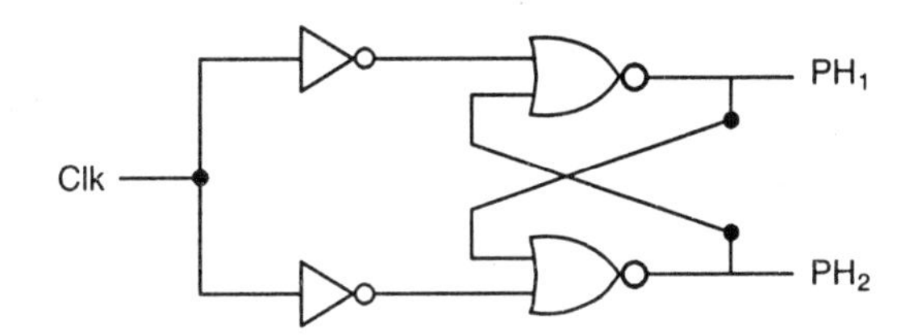

Figure 9.13 Circuitry for generating a two-phase non-overlapping clock.

9.4.4 Low Voltage Static Latches

The scaling of supply voltages is critical for low-power operation. Unfortunately, certain latch structures do not function at reduced supply voltages. At very low power supply voltages, the input to the inverter cannot be raised above the switching threshold, resulting in incorrect evaluation. Even with the use of transmission gates, performance degrades significantly at reduced supply voltages.

Scaling to low supply voltages thus requires the use of reduced threshold voltages. However, this has the negative effect of exponentially increasing the subthreshold leakage power. One solution to this problem involves the use of multiple threshold devices, where the low-threshold inverters are gated by using high-threshold devices to eliminate leakages.

9.5 DYNAMIC LATCHES AND REGISTERS

In a static sequential circuit, the stored value remains valid as long as the supply voltage is applied to the circuit, hence the name static. The major limitation is its complexity. When registers are used for computational structures that are constantly clocked (such as a pipelined datapath), the requirement that the memory should hold state for extended periods of time can be significantly relaxed.

This results in a class of circuits based on temporary storage of charge on parasitic capacitors. The principle is that the charge stored on a capacitor can be used to represent a logic signal. The absence of charge denotes a 0, while its presence stands for a stored 1. Unfortunately, no capacitor is ideal, and some charge leakage is always present. A stored value can, thus, be kept for a limited amount of time, typically in the range of milliseconds. A periodic refresh of the value is necessary for signal integrity; hence the name dynamic storage. Reading the value of the stored signal from a capacitor without disrupting the charge requires the availability of a device with high input impedance.

9.5.1 Dynamic Transmission-Gate Edge-Triggered Registers

A fully dynamic positive edge-triggered register based on the master–slave concept is shown in Figure 9.14. When CLK = 0, the input data is sampled on storage node 1, which has an equivalent capacitance of C_1, consisting of the gate capacitance of I_1, the junction capacitance of T_1 and the overlap gate capacitance of T_2. During this period, the slave stage is in the hold mode, with node 2 in a high-impedance (floating) state. On the rising edge of the clock, the

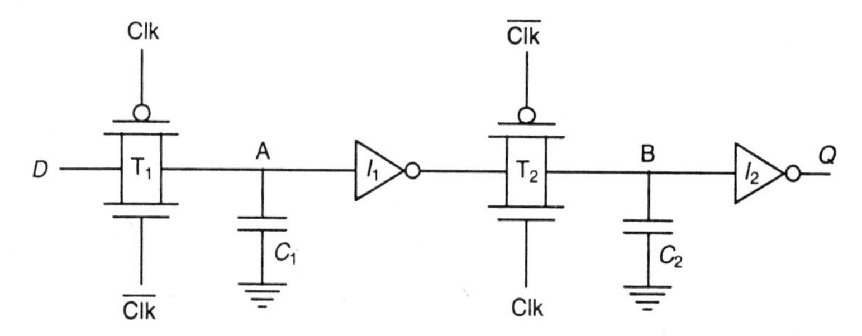

Figure 9.14 Dynamic edge-triggered register.

transmission gate T_2 turns ON, and the value sampled on Node 1 right before the rising edge propagates to the output Q. Node 2 now stores the inverted version of Node 1. This implementation of an edge-triggered register is very efficient because it requires only eight transistors. The sampling switches can be implemented using NMOS-only pass transistors, resulting in an even simple six-transistor implementation. The reduced transistor count is attractive for high-performance and low-power systems.

The setup time of this circuit is simply the delay of the transmission gate, and it corresponds to the time it takes node 1 to sample the D input. The hold time is approximately zero, since the transmission gate is turned OFF on the clock edge and further inputs changes are ignored. The propagation delay is equal to the two-inverter delays plus the delay of the transmission gate T_2.

One important consideration for such a dynamic register is that the storage nodes (i.e., the state) have to be refreshed at periodic intervals to prevent losses due to charge leakage, diode leakage, or sub-threshold currents. In datapath circuits, the refresh rate is not an issue, since the registers are periodically clocked, and the storage nodes are constantly updated.

Clock overlap is an important concern for this register. This can be addressed by making sure that there is enough delay between the D input and node B, ensuring that new data sampled by the master stage does not propagate through to the slave stage.

The dynamic circuits shown in this section are very appealing from the perspective of complexity, performance, and power. Unfortunately, robustness considerations limit their use. In a fully dynamic circuit like that shown in Figure 9.14, a signal net that is capacitively coupled to the internal storage node can inject significant noise and destroy the state. This is especially important in ASIC flows, where there is little control over coupling between signal nets and internal dynamic nodes. Leakage currents cause another problem: most modern processors require that the clock can be slowed down or completely halted, to conserve power in low activity periods. Finally, the internal dynamic nodes do not track variations in power supply voltage. This results in reduced noise margins.

Most of these problems can be adequately addressed by adding a weak feedback inverter and making the circuit pseudo static (Figure 9.15). While this slight results in delay, it improves the noise immunity significantly.

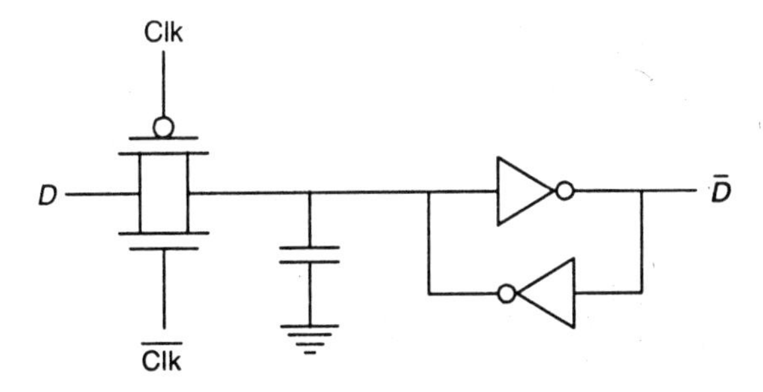

Figure 9.15 Making dynamic latch pseudo static.

9.5.2 C²MOS—A Clock Skew Insensitive Approach

The C²MOS register. Figure 9.16 shows a positive edge-triggered register that is based on a master–slave concept insensitive to clock overlap. This circuit is called the C²MOS (Clocked CMOS) register, and operates in two phases:

1. CLK = 0 ($\overline{\text{CLK}}$ = 1): The first tri-state driver is turned ON, and the master stage acts as an inverter sampling the inverted version of D on the internal node X. The master stage is in the evaluation mode. Meanwhile, the slave section is a high impedance mode, or in a hold mode. Both transistors M_7 and M_8 are OFF, decoupling the output from the input. The output Q retains its previous value stored on the output capacitor C_{L2}.
2. The roles are reversed when CLK = 1: The master stage is in hold mode (M_3-M_4 OFF) while the second section evaluates (M_7-M_8 ON). The value stored on C_{L1} propagates to the output node through the slave stage, which acts as an inverter.

The advantage of this circuit is that the C²MOS register with $\overline{\text{CLK}}$-CLK clocking is insensitive to overlap, as long as the rise and fall times of the clock edges are sufficiently small.

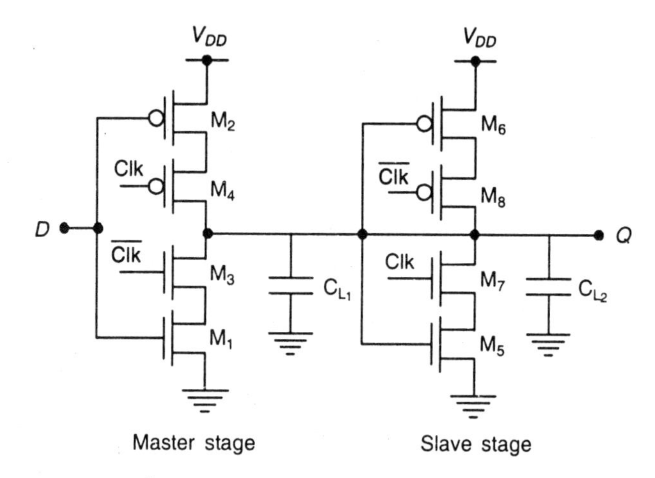

Figure 9.16 C²MOS master–slave positive edge-triggered register.

Dual-edge registers. In this, the sequential circuits are designed to sample the input on both edges (rising or falling) of the clock. The advantage of this scheme is that a lower clock frequency half the original rate is distributed for the same functional throughput, resulting in power savings in the clock distribution network.

9.5.3 True Single-Phase Clocked Register (TSPCR)

The true single-phase clocked register (TSPCR) uses a single clock, CLK. The basic single-phase positive and negative latches are shown in [Figures 9.17(a) and (b)]. For the positive latch, when CLK is high, the latch is in the transparent mode and corresponds to two cascaded inverters; the latch is non-inverting, and propagates the input to the output [Figure 9.17(a)]. On the other hand, when CLK = 0, both inverters are disabled, and the latch is in the hold mode. Only the pull-up networks are still active, while the pull-down circuits are deactivated. As a result of the dual-stage approach, no signal can ever propagate from the input of the latch to the output in this mode. A register can be constructed by cascading positive and negative latches.

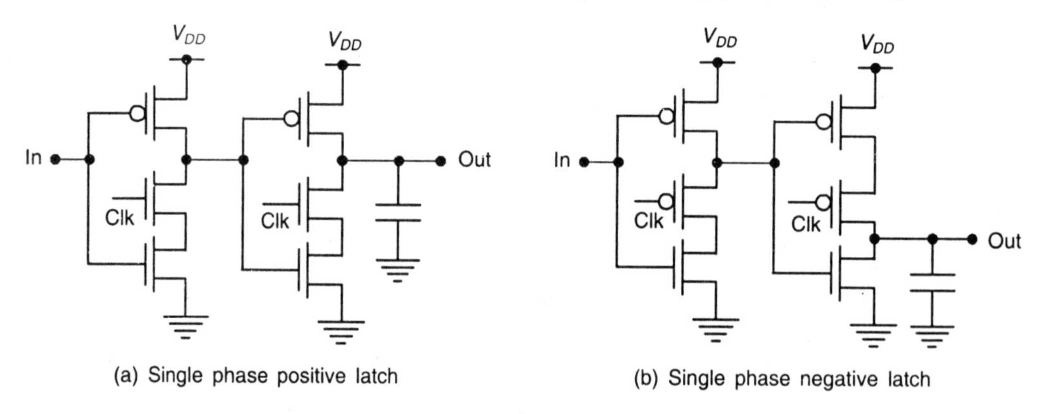

Figure 9.17 True single phase clocked latches.

The advantage of this circuit is the use of single clock phase. The TSPC offers an additional advantage of the possibility of embedding logic functionality into the latches. This reduces the delay overhead associated with the latches. Figure 9.18 shows an example of positive latch that implements the AND of In_1 and In_2 in addition to performing the latching function. This approach of embedding logic into latches has been used extensively in the design of many high-performance processors.

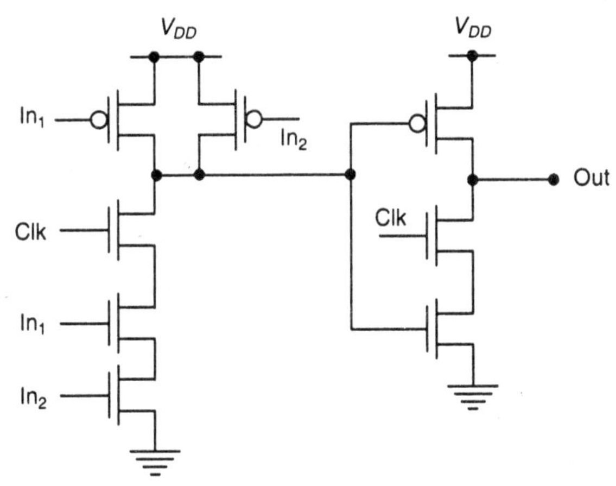

Figure 9.18 Adding logic to the TSPC approach (AND latch).

The disadvantage of TSPC is the slight increase in the number of transistors. Also, when the clock is low (for the positive latch), the output node may be floating, and it is exposed to coupling from other signals. Charge sharing can also occur if the output node drives transmission gates. Dynamic nodes should be isolated with the aid of static inverters, or made pseudo-static for improved noise immunity.

The TSPC circuit can be reduced in complexity as shown in Figure 9.19, where the first inverter is controlled by the clock. These circuits have the additional advantage that clock load is reduced by half. On the other hand, not all the node voltages in the latch experience the full logic swing.

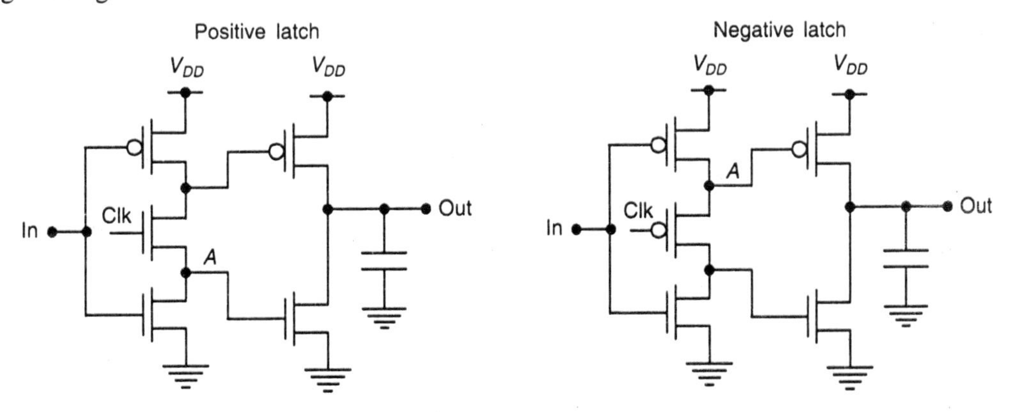

Figure 9.19 Simplified TSPC latch.

9.6 ALTERNATIVE REGISTER STYLES

9.6.1 Pulse Registers

A fundamentally different approach for constructing a register uses pulse signals [Figure 9.20(a)]. The idea is to construct a short pulse around the rising (or falling) edge of the clock. This pulse acts as the clock input to a latch, sampling the input only in a short window [Figure 9.20(c)]. Thus race conditions are avoided by keeping the opening time of the latch very short. The combination of glitch generation circuitry [Figure 9.20(c)] and the latch results in a positive edge-triggered register (Figure 9.20).

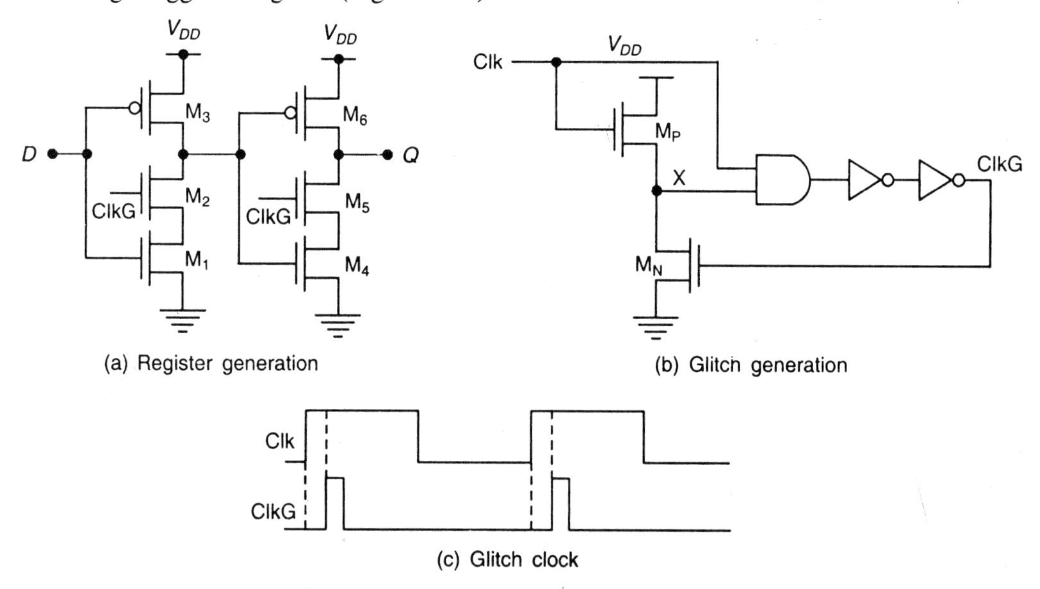

Figure 9.20 TSPC-based glitch latch—timing generation and register.

The advantage of the approach is the reduced clock load and the small number of transistors. The disadvantage is a substantial increase in verification complexity.

9.6.2 Sense Amplifier-Based Registers

An edge-triggered register can also be implemented based on sense amplifiers (Figure 9.21). Sense amplifier circuits accept small input signals and amplify them to generate rail-to-rail swings. They are used extensively in memory cores and in low-swing bus drivers to either improve performance or reduce power dissipation. There are many techniques to construct these amplifiers. A common approach is to use feedback—for instance, through a set of cross-coupled amplifiers.

The circuit shown in Figure 9.21 uses a precharged front-end amplifier that samples the differential input signal on the rising edge of the clock signal. The outputs of the front end are fed into a NAND cross-coupled SR flip-flop that holds the data and guarantees that the differential outputs switch only once per clock cycle. The differential inputs in this implementation don't have to have rail-to-tail swing.

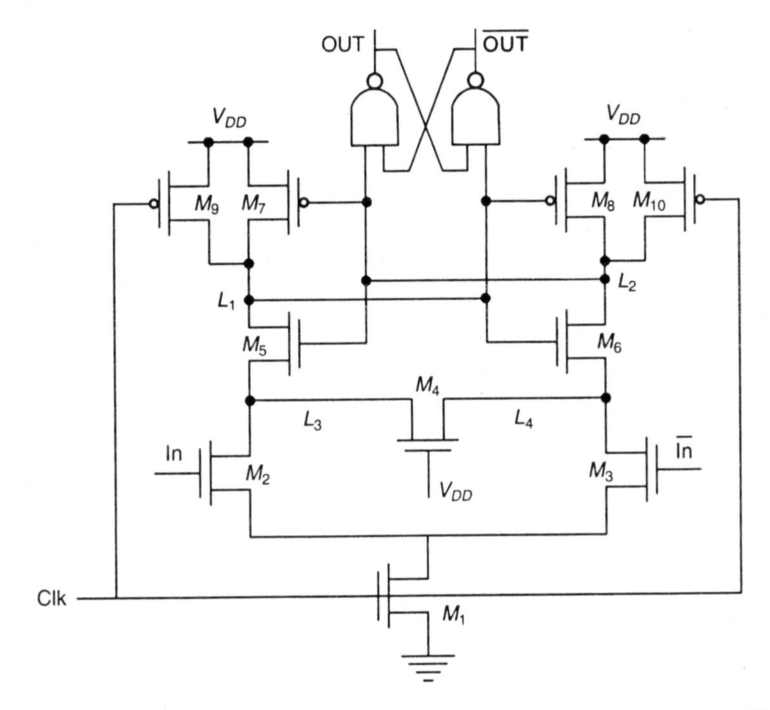

Figure 9.21 Positive edge-triggered register based on sense amplifier.

9.7 NON-BISTABLE SEQUENTIAL CIRCUITS

Besides the bistable element, other regenerative circuits can be catalogued as astable and monostable circuits. The astable circuits act as oscillators and can be used for on-chip clock generation. The monostable circuit serves as pulse generators, also called one-shot circuits. Another interesting regenerative circuit is the Schmitt trigger. This component has the useful property of showing hysteresis in its dc characteristics—its switching threshold is variable and depends upon the direction of transition (low-to-high or high-to-low). This peculiar feature can come handy in noisy environments.

9.7.1 The Schmitt Trigger

A Schmitt trigger is a device with two important properties:

1. It responds to a slowly changing input waveform with a fast transition time at the output.
2. The VTC of the device displays different switching thresholds for positive- and negative-going input signals [Figure 9.22(a)]. The switching thresholds for low-to-high and high-to-low transitions are called V_{M+} and V_{M}- respectively. The hysteresis is defined as the difference between the two values. The schematic symbol for the non-converting schmitt trigger is shown in Figure 9.22(b).

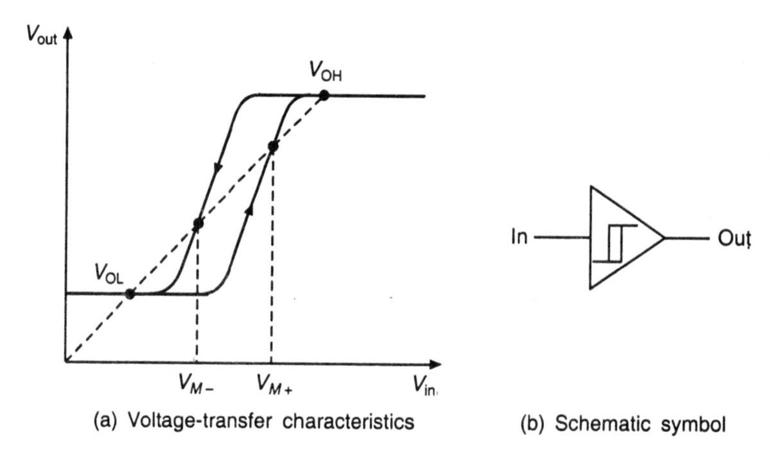

(a) Voltage-transfer characteristics (b) Schematic symbol

Figure 9.22 Non-inverting Schmitt trigger.

One of the main uses of Schmitt trigger is to turn a noisy or slowly varying input signal into a clean digital output signal. This is illustrated in Figure 9.23. Notice how the hysteresis suppresses the ringing on the signal. At the same time, the fast low-to-high (and high-to-low) transitions of the output signal should be observed. In general, steep signal slopes are advantageous in reducing power consumption by suppressing direct-path currents. The Schmitt trigger concept uses positive feedback.

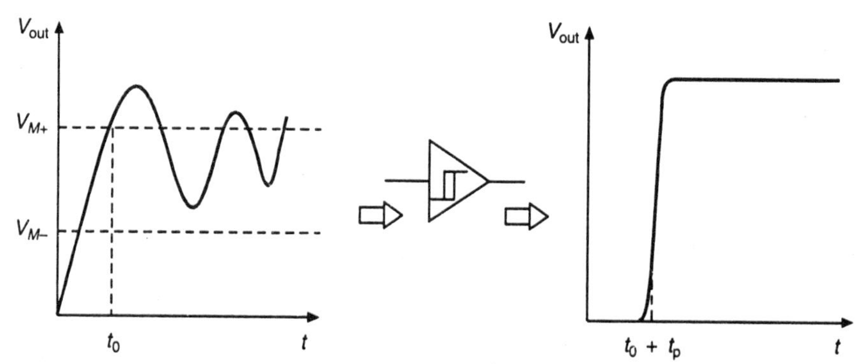

Figure 9.23 Noise suppression using a Schmitt trigger.

CMOS implementation of Schmitt trigger. Figure 9.24 shows the CMOS implementation of a Schmitt trigger. The idea is that the switching threshold of a CMOS inverter is determined by the (k_n/k_p) ratio between the PMOS and NMOS transistors. Increasing this ratio raises the switching threshold, and while decreasing, it lowers V_M.

9.7.2 Monostable Sequential Circuits

A monostable circuit is a circuit that generates a pulse of predetermined width every time the quiescent circuit is triggered by a pulse or transition event. It is called monostable because it has only one stable state. A trigger event, which is either a signal transition or a pulse, causes the

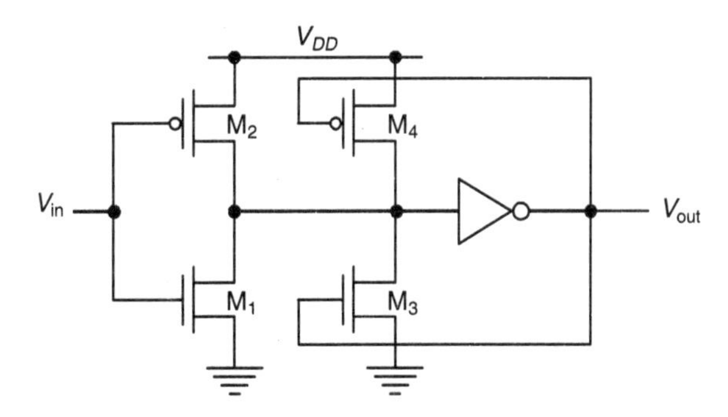

Figure 9.24 CMOS Schmitt trigger.

circuit to go temporarily into another quasi-stable state. This means that it eventually returns to its original state after a time period determined by the circuit parameters. This circuit, also called a one-shot, is useful in generating pulses of a known length.

The most common approach to the implementation of one-shots is the use of a simple delay element to control the duration of the pulse. This concept is illustrated in Figure 9.25.

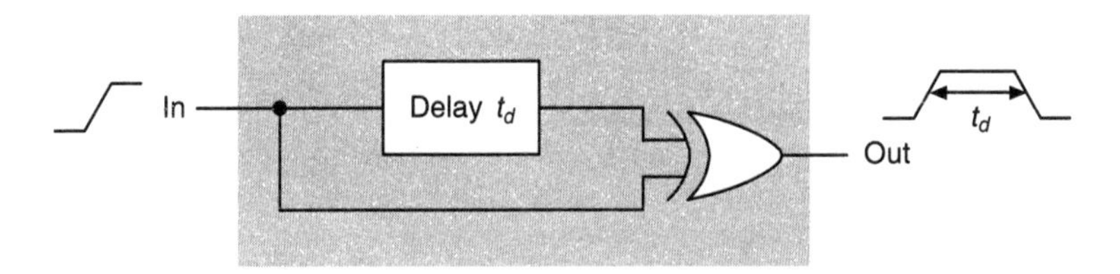

Figure 9.25 Transition-triggered one shot.

9.7.3 Astable Circuits

An astable circuit has no stable states. The output oscillates back and forth between two quasi-stable states, with a period determined by the circuit topology and parameters. One of the main applications of oscillators is the on-chip generation of clock signals. The ring oscillator is a simple example of an astable circuit.

The ring oscillator composed of cascaded inverters produces a waveform with a fixed oscillating frequency determined by the delay of the inverter in the CMOS process. In many applications, it is necessary to control the frequency of the oscillator. An example of such a circuit is the Voltage-Controlled Oscillator (VCO), whose oscillation frequency is a function of a control voltage (Figure 9.26). The standard ring oscillator can be modified into a VCO by replacing the standard inverter with a current starved inverter like the one shown in Figure 9.26. The mechanism for controlling the delay of each inverter is to limit the current available to discharge the load capacitance of the gate.

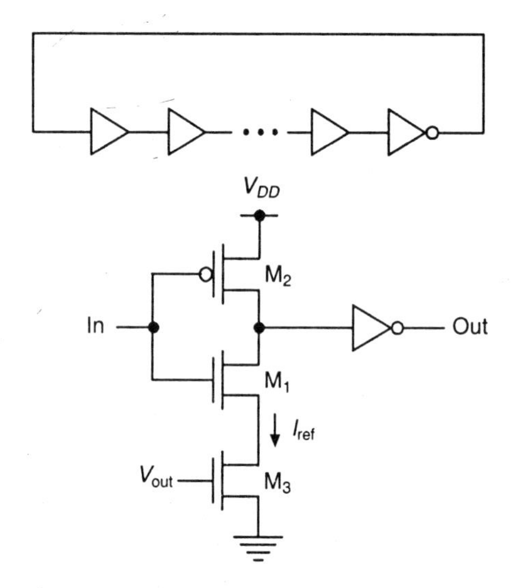

Figure 9.26 Voltage-controlled oscillator based on current-starved inverters.

SUMMARY

This chapter has explored the subject of sequential digital circuits. The cross-coupling of two inverters forms a flip-flop. From a transient perspective, a flip-flop is characterized by three parameters: the setup time, the hold time, and the propagation delay. The classification of memory elements are dealt in detail in this chapter. It also describes the latches and registers for both static and dynamic cases. The master–slave and edge triggered flip-flops are designed to avoid race conditions. The static flip-flop use positive feedback as the storage mechanism. The resulting structure is robust and retains its charge as long as the supply voltage is applied. Static flip-flops are slow and require a large area, however. The dynamic flip-flop approach relies on charge storage. It has the advantage of yielding simpler and faster flip-flops, but it is unfortunately more susceptible to error due to leakage, dynamic artifacts, and synchonization discrepancies. The problem of race conditions can be addressed either by using multiple clock phases, or by employing special register structures. The C^2MOS and the TSPC latches are the examples of the latter. The combination of dynamic logic with dynamic latches can produce extremely fast computational structures. Monostable structures have only one stable state. They are useful as pulse generators. Astable multivibrators possess no stable state. Schmitt triggers display hysteresis in their dc characteristics and fast transitions in their transient response. They are used to suppress noise.

REVIEW QUESTIONS

1. Explain different types of latches.
2. Explain master–slave edge triggered register.
3. With the help of a diagram, explain the operation of Schmitt trigger.

SHORT ANSWER QUESTIONS

1. What is setup time (SUT)?

   Ans. The time that the data input must be valid before clock transition.

2. Memories can be eithter ____________ or ____________ .

   Ans. static, dynamic.

3. An edge triggered storage element is called ____________ .

   Ans. register.

4. A latch is a ____________ .

   Ans. level sensitive device.

5. Any bistable component formed by the cross coupling of gates is called ____________ .

   Ans. flip flop.

6. Give the multiplexer equation

   Ans. $Q = \overline{\text{Clk}} \cdot Q + \text{Clk} \cdot \overline{Q}$

7. What is race condition?

   Ans. The change of data at the output on the rising edge of clock.

8. A circuit of positive edge-triggered register that is based on master–slave concept is called ____________ .

   Ans. C^2MOS (Clocked CMOS) register.

9. One of the main use of Schmitt trigger circuit is ____________ .

   Ans. changing a noisy or slowly varying input signal into a clean digital output signal.

10. Monostable vibrator is otherwise also called ____________ .

    Ans. One shot vibrator.

11. How to convert standard ring oscillator into VCO.

    Ans. By replacing the standard inverter with a current starved inverter.

10 Design of Arithmetic Building Blocks

10.1 INTRODUCTION

Most chips are built from a collection of subsystems: adders, register files, state machines, etc. Of course, to do a good job of designing a chip, we must be able to properly design each of the major components. Studying individual sub-systems is also a useful prelude to the study of complete chips because a single component is a focused design problem. When designing a complete chip, we often have to perform several different types of computation, each with different cost constraints. On the other hand, a single component performs a single task; as a result, the design choices are much more clear.

As always, the cost of a design is measured in area, delay and power. For most components, we have a family of designs where all of them perform the same basic function, but with different area/delay trade-offs. Having access to a variety of ways to implement a function gives us architectural freedom during chip design. Area and delay costs can be reduced by optimization at each level of abstraction.

- **Layout** We can make microscopic changes to the layout to reduce parasitics: Moving wires to pack the layout or to reduce source–drain capacitance, adding vias to reduce resistance, etc. We can also make macroscopic changes by changing the placement of gates, which may reduce wire parasitics, reduce routing congestion, or both.
- **Circuit** Transistor sizing in the first line of defence against circuits that inherently require long wires. Advanced logic circuits, such as precharged gates, may help reduce the delay within logic gates.
- **Logic** Redesigning the logic to reduce the gate depth from input to output can greatly reduce delay, though usually at the cost of area.
- **Register-transfer and above** Proper placement of memory elements makes maximum use of the available clock period. Proper encoding of signals allows clever logic designs that minimize logic delay. Pipelining provides trade-offs between the clock period and latency.

Logic and circuit design are at the core of subsystem design. Many important components, foremost being the adder, have been so extensively studied in this chapter that specific optimizations and trade-offs are well understood. Before analyzing the design of the arithmetic modules, a short discussion of the role of the datapath in the digital-processor picture is appropriate. This not only highlights the specific design requirements for the datapath, but also puts the rest of this book in perspective.

10.2 DATAPATHS

A datapath is a logical and a physical structure. It is built from components which perform typical data operations such as addition, multiplication, comparison, and shift, AND, OR and XOR, etc. Datapaths typically include several types of components: registers (memory elements) store data; adders and ALUs perform arithmetic operations; shifters perform bit operations; counters may be used for program counters. A datapath may include point-to-point connections between pairs of components, but the typical datapath has too many connections for every component to be connected to every other component. Data is often passed between components on one or more buses. The number of buses determines the maximum number of data transfers on a clock cycle and is a primary design parameter of datapaths.

Datapaths are often arranged in a bit-sliced organization as in Figure 10.1. A bit-slice is a one-bit version of the complete datapath, and the *n*-bit data path is constructed by replicating the bit-slice. Typically, data flows horizontally through the bit-slice along point-to-point connections or buses, while control signals flow vertically. Bit-slice layout design requires careful, simultaneous design of the cells that comprise the datapath. Since the bit-slice must be stacked

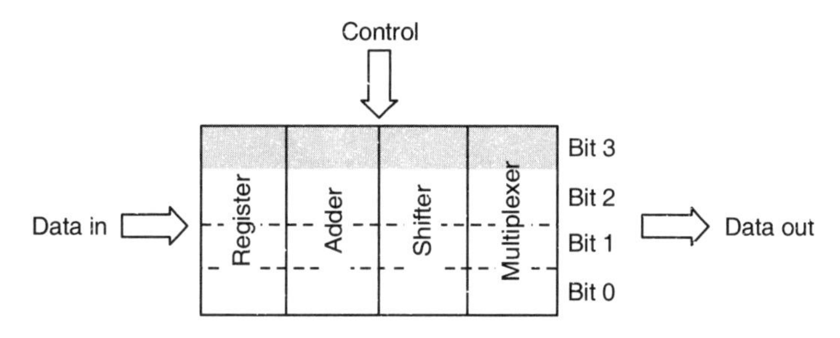

Figure 10.1 Bit-sliced datapath organization.

vertically, signals that pass through the cells must be aligned at top and bottom. Horizontal constraints are often harder to satisfy. The V_{DD} and V_{SS} lines must run horizontally through the cells, as must in buses. Signals between adjacent cells must also be aligned. While the vertical wires often distribute signals, the horizontal wires are often interrupted by logic gates.

Instead of operating on single-bit digital signals, the data in a processor are arranged in a word-based fashion. Typical microprocessor datapaths are 32 or 64 bits wide, while the dedicated signal processing datapaths, such as those in DSL modems, magnetic disk drives, or compact-disc players are of arbitrary width typically 5 to 24 bits. For instance, a 32-bit processor operates on data words that are 32-bits wide. This is reflected in the organization of datapath. Since the operation frequently has to be performed on each bit of the data word, the datapath consists of 32-bit slices, each operating on a single bit—hence the name bit sliced. Bit slices are either identical or resemble a similar structure for all bits. The datapath designer can concentrate on the design of a single slice that is repeated 32 times.

10.3 THE ADDER

Addition forms the basis for many processing operations from counting to multiplication to filtering. As a result, adder circuits that add binary numbers are of great interest to digital system designers. A wide variety of adder implementations are available to serve different speed/density requirements. Careful optimization of adder is of utmost importance. This optimization can proceed either at the logic or circuit level. Typical logic-level optimizations try to rearrange the Boolean equations so that a faster or smaller circuit is obtained. Example circuit is carry lookahead adder discussed later in the chapter. Circuit optimization, on the other hand, manipulates transistor sizes and circuit topology to optimize the speed. Before considering both optimization processes, we provide a short summary of the basic definitions of an adder circuit.

10.3.1 The Binary Adder: Definitions

Table 10.1 shows the truth table of a binary full-adder. A and B are the adder inputs. C_i is the carry input. S is the sum output and C_o is the carry output. The Boolean expressions for S and C_o are given in Eq. (10.1).

Table 10.1 Truth table for full-adder

A	B	C_i	S	C_o	*Carry status*
0	0	0	0	0	delete
0	0	1	1	0	delete
0	1	0	1	0	propagate
0	1	1	0	1	propagate
1	0	0	1	0	propagate
1	0	1	0	1	propagate
1	1	0	0	1	generate/propagate
1	1	1	1	1	generate/propagate

$$\begin{aligned} S &= A \oplus B \oplus C \\ &= A\overline{B}\overline{C}_i + \overline{A}B\overline{C}_i + \overline{A}\overline{B}C_i + ABC_i \\ C_o &= AB + BC_i + AC_i \end{aligned} \tag{10.1}$$

It is often useful from an implementation perspective to define S and C_o as functions of some intermediate signals G (generate), D (delete), and P (propagate). $G = 1 (D = 1)$ ensures that a carry bit will be generated (deleted) at C_o independent of C_i, while $P = 1$ guarantees that an incoming carry will propagate to C_o. Expressions for these signals can be derived from inspection of the truth table:

$$\begin{aligned} G &= AB \\ D &= \overline{A}\overline{B} \\ P &= A \oplus B \end{aligned} \tag{10.2}$$

We can rewrite S and C_o as functions of P and G (or D):

$$C_o(G, P) = G + PC_i$$

$$S(G, P) = P \oplus C_i \tag{10.3}$$

Notice that G and P are only functions of A and B and are not dependent upon C_i. In a similar way, we can also derive expressions for $S(D, P)$ and $C_o(D, P)$.

An N-bit adder can be constructed by cascading N full-adder (FA) circuits in series, connecting $C_{o,\,k-1}$ to $C_{i,\,k}$ for $k = 1$ to $N - 1$, and the first carry in $C_{i,\,0}$ to 0 (Figure 10.2). This configuration is called a ripple-carry adder, since the carry bit 'ripples' from one stage to the other. The delay through the circuit depends upon the number of logic stages that must be traversed and is a function of the applied input signals. For some input signals, no rippling effect occurs at all, while for others, the carry has to ripple all the way from the least significant bit (LSB) to the most significant bit (MSB). The propagation delay of such a structure (also called the critical path) is defined as the worst-case delay over all possible input patterns.

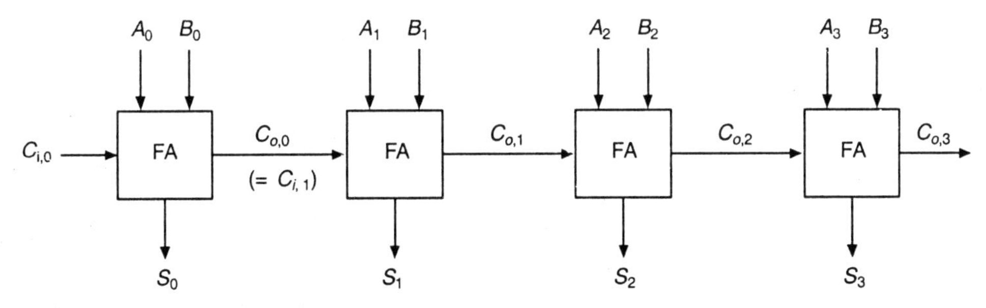

Figure 10.2 Four-bit ripple-carry adder: topology.

In the case of the ripple-carry adder, the worst-case delay happens when a carry generated at the least significant bit position propagates all the way to the most significant bit position. This carry is finally consumed in the last stage to produce the sum. The delay is proportional to the number of bits in the input words N and is approximated by

$$t_{\text{adder}} \approx (N - 1)\, t_{\text{carry}} + t_{\text{sum}} \tag{10.4}$$

where t_{carry} and t_{sum} equal the propagation delays from C_i to C_o and S, respectively.

Two important conclusions can be drawn from Eq. (10.4).

- The propagation delay of the ripple-carry adder is linearly proportional to N. This property becomes increasingly important when designing adders for the wide datapaths ($N = 16, \ldots, 128$) that are desirable in the present and future computers.
- When designing the full-adder cell for a fast ripple-carry adder, it is far more important to optimize t_{carry} than t_{sum} since the latter has only a minor influence on the total value of t_{adder}.

Before starting an indepth discussion on the circuit design of full-adder cells, the following additional logic property of the full-adder is worth mentioning.

Inverting all inputs to a full-adder results in inverted values for all outputs. This property, also called the inverting property is expressed as a pair of equations:

$$\overline{S}(A, B, C_i) = S(\overline{A}, \overline{B}, \overline{C}_i)$$

$$\overline{C}_o(A, B, C_i) = C_o(\overline{A}, \overline{B}, \overline{C}_i) \tag{10.5}$$

and will be extremely useful when optimizing the speed of the ripple-carry adder. It states that the circuits of Figure 10.3 are identical.

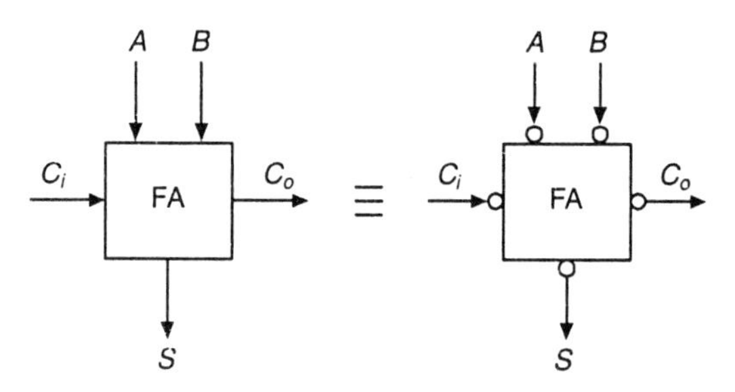

Figure 10.3 Inverting property of the full-adder (the circles indicate inverters).

10.3.2 The Full-Adder: Circuit Design Considerations

Static adder circuit. One way to implement the full-adder circuit is to take the logic equations of Eq. (10.1) and translate them directly into complementary CMOS circuitry. Some logic manipulations can help to reduce the transistor count. For instance, it is advantageous to share some logic between the sum- and carry-generation sub-circuits, as long as this does not slow down the carry generation, which is the most critical part as stated preciously. The following is an example of such a reorganized equation set:

$$C_o = AB + BC_i + AC_i$$

and

$$S = ABC_i + \overline{C}_o(A + B + C_i) \tag{10.6}$$

The equivalence with the original equation is easily verified. The corresponding adder design, using complementary static CMOS is shown in Figure 10.4 and requires 28 transistors. In addition to consuming a large area, this circuit is slow:

- Tall PMOS transistor stacks are present in both carry and sum generation circuits.
- The intrinsic load capacitance of the C_o signal is large and consists of two diffusion and six gate capacitances plus the wiring capacitance.
- The signal propagates through two inverting stages in the carry generation circuit. As mentioned earlier, minimizing the carry-path delay is the prime goal of the designer of high-speed adder circuits. Given the small load (fan-out) at the output of the carry chain, having two logic stages is too high a number and leads to extra delay.
- The sum generation requires one extra logic stage, but that is not that important since a factor appears only once in the propagation delay of the ripple-carry adder of Eq. (10.4).

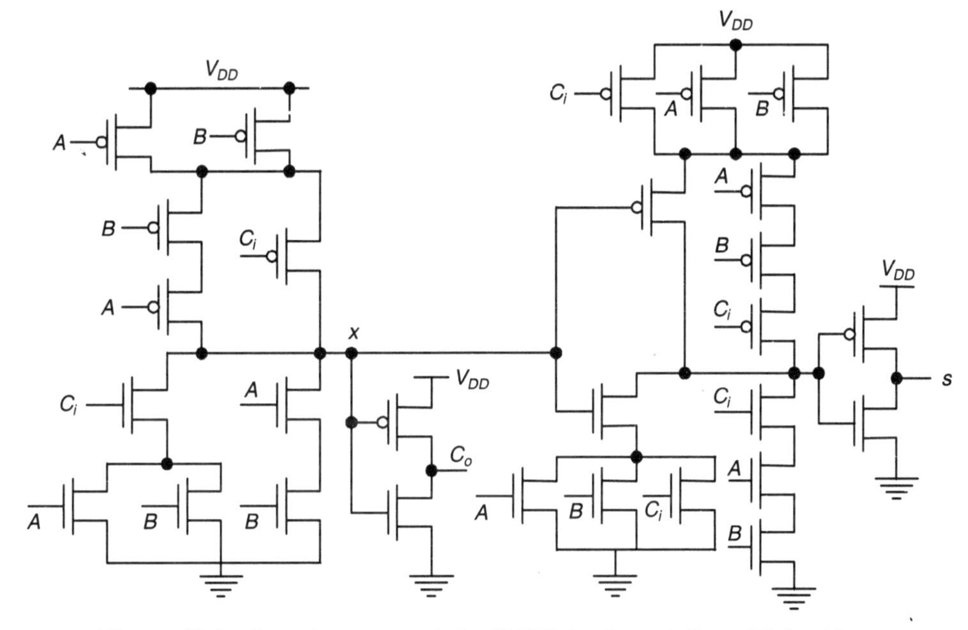

Figure 10.4 Complementary static CMOS implementation of full-adder.

Although slow, the circuit includes some smart design tricks. Notice that the first gate of the carry-generation circuit is designed with the C_i signal on the smaller PMOS stack, lowering its logical effort to 2. Also, the NMOS and PMOS transistors connected to C_i are placed as close as possible to the output of the gate. This is a direct application of a circuit-optimization technique—transistors on the critical path should be placed as close as possible to the output of the gate. For instance, in stage k of the adder, signals A_k and B_k are available and stable long before $C_{i,k}$ (=$C_{0,k-1}$) arrives after rippling through the previous stages. In this way, the capacitances of the internal nodes in the transistor chain are precharged or discharged in advance. On arrival of $C_{i,k}$ only the capacitance of node X has to be (dis)charged. Putting the $C_{i,k}$ transistors closer to V_{DD} and GND would require not only the (dis)charging of the capacitance of node X, but also of the internal capacitances.

The speed of this circuit can now be improved gradually by using some of the adder properties discussed in the previous section. First, the number of inverting stages in the carry path can be reduced by exploiting the inverting property—inverting all the inputs of a full-adder cell also inverts all the outputs. This rule allows us to eliminate an inverter in a carry chain as demonstrated in Figure 10.5.

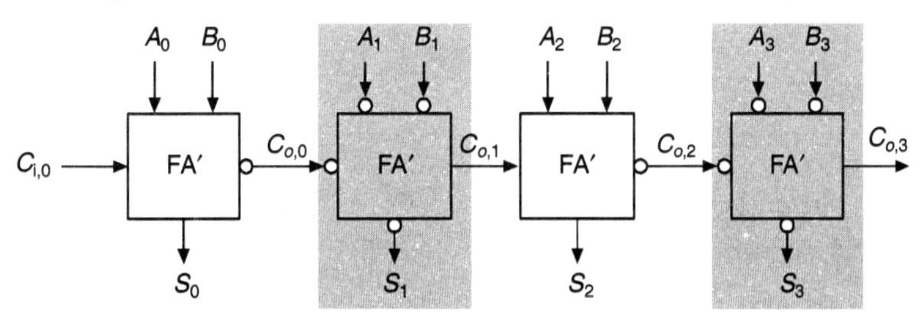

Figure 10.5 Inverter elimination in carry path (FA′ stands for a full-adder without the inverter in the carry path).

Mirror adder design. An improved adder circuit, also called the mirror adder is shown in Figure 10.6. Its operation is based on Eq. (10.3). The carry-generation circuitry is worth analyzing. First, the carry-inverting gate is eliminated. Secondly, the PDN and PUN networks of the gate are not dual. Instead, they form a clever implementation of the propagate/generate/

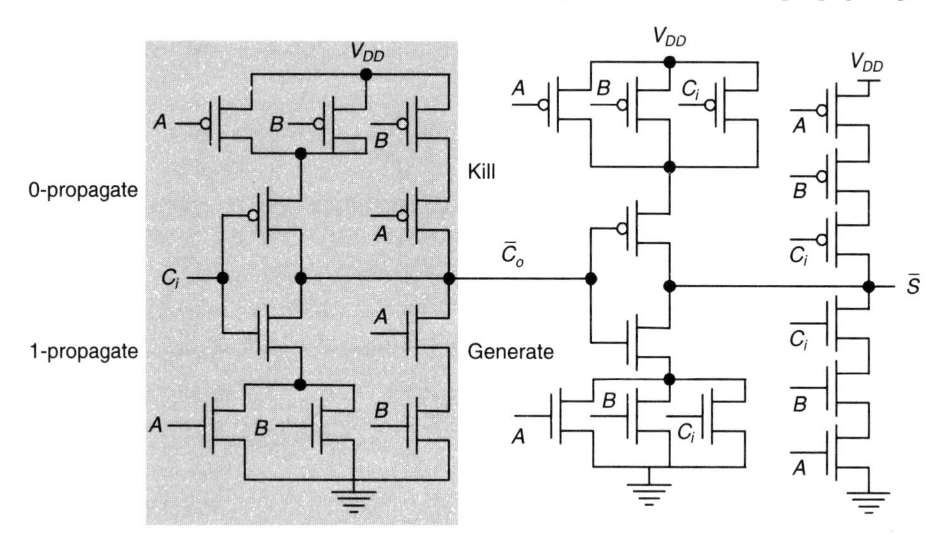

Figure 10.6 Mirror adder—circuit schematics.

delete function—when either D or G is high, $\bar{C}_o$ is set to V_{DD} or GND, respectively. When the conditions for a propagate are valid (or P is 1), the incoming carry is propagated (in inverted format) to $\bar{C}_o$. This results in a considerable reduction in both area and delay. The sum circuitry can also be analyzed in the same manner. The following observations are worth considering:

- This full-adder cell requires only 24 transistors.
- The NMOS and PMOS chains are completely symmetrical, which still yields correct operation due to self-duality of both the sum and carry functions. As a result, a maximum of two series transistors can be found in the carry-generation circuitry.
- The transistors connected to C_i are placed closest to the output of the gate.
- Only the transistors in the carry stage have to be optimized for speed. All transistors in the sum stage can be of minimum size. When laying out the cell, the most critical issue is the minimization of the capacitance at node $\bar{C}_o$. Shared diffusions reduce the stack node capacitances.
- In the adder cell of Figure 10.4, the inverter can be sized independently to drive the C_i input of the adder stage that follows. If the carry circuit in Figure 10.6 is symmetrically sized, each of its inputs has a logical effort of 2. This means that the optimal fan-out sized for minimum delay should be (4/2) = 2. However, the output of this stage drives two internal gate capacitances and six gate capacitances in the connecting adder cell. A clever solution to keep the transistor sizes the same in each stage is to increase the size

of the carry stage to about three to four times the size of the sum stage. This maintains the optimal fan out of 2. The resulting transistor sizes are annotated on Figure 10.6 where a PMOS/NMOS ratio of 2 is assumed.

Transmission-gate based adder. A full-adder can be designed to use multiplexers and XORs. While this is impractical in a complementary CMOS implementation, it becomes attractive when the multiplexers and XORs are implemented as transmission gates. A full-adder implementation based on this approach is shown in Figure 10.7 and uses 24 transistors. It is based on the *propagate–generate* model, introduced in Eq. (10.3). The propagate signal, which is the XOR of inputs A and B, is used to select the true or complementary value of the input carry as the new sum output. Based on the propagate signal, the output carry is either set to the input carry, or either one of inputs A or B. One of increasing features of such an adder is that it has similar delays for both sum and carry outputs.

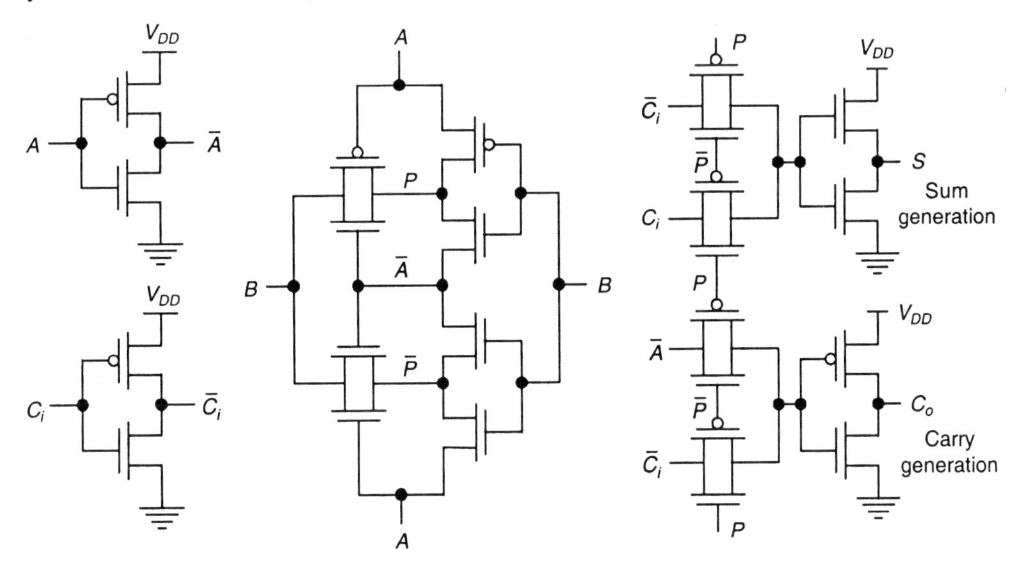

Figure 10.7 Transmission-gate based full adder cell.

Manchester carry-chain adder. The carry propagation circuitry in Figure 10.7 can be simplified by adding generate and delete signals, as shown in Figure 10.8(a). The propagate path is unchanged, and it passes C_i to the C_o output if the propagate signal ($A_i \oplus B_i$) is true. If the propagate condition is not satisfied, the output is either pulled low by the D_i signal or pulled up by $\bar{G}_i$. The dynamic implementation [Figure 10.8(b)], makes even further simplification possible. Since the transitions in a dynamic circuit are monotonic, the transmission gates can be replaced by NMOS-only pass transistors. Precharging the output eliminates the need for the kill signal (for the case in which the carry chain propagates the complementary values of the carry signals).

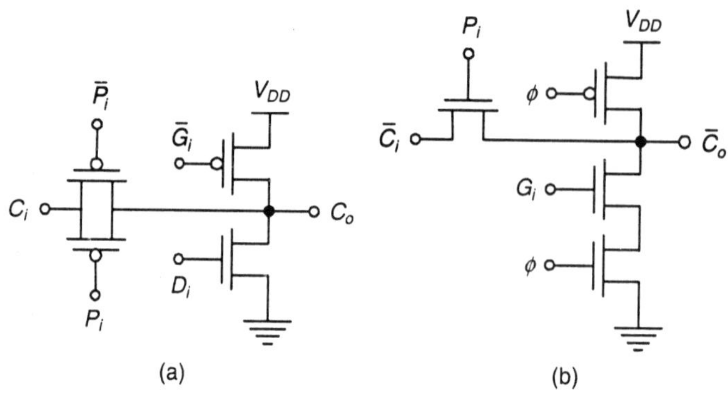

Figure 10.8 Manchester carry gates. (a) Static, using propagate, generate, and kill (delete) signals, (b) Dynamic implementation, using only propagate and generate signals.

A *Manchester carry-chain adder* uses a cascade of pass transistors to implement the carry chain. An example, based on the dynamic circuit version introduced in Figure 10.8, is shown in Figure 10.9. During the precharge phase ($\varphi = 0$), all intermediate nodes of the pass transistor carry chain are precharged to V_{DD}. During evaluation, the A_k node is discharged when there is an incoming carry and the propagate signal P_k is high, or when the generate signal for stage $k(G_k)$ is high.

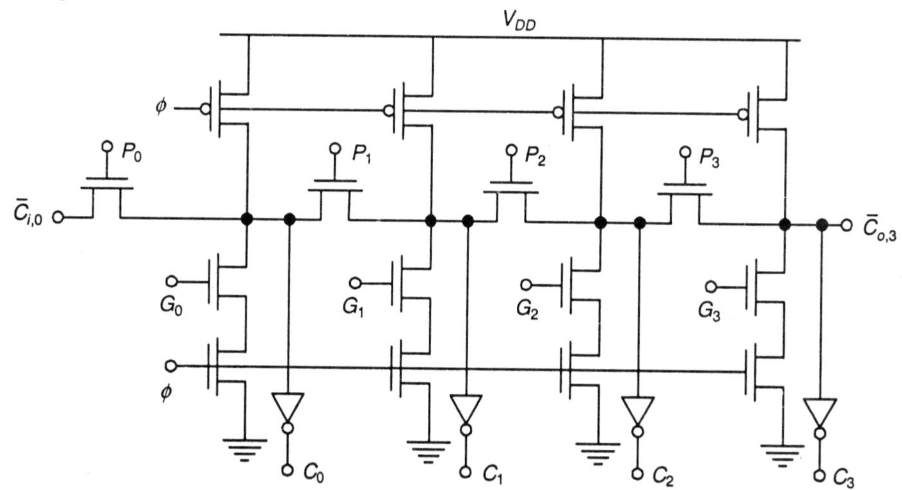

Figure 10.9 Manchester carry-chain adder in dynamic logic.

The worst-case delay of the carry chain of the adder in Figure 10.9 is modelled by the linearized RC network of Figure 10.10. The propagation delay of such a network equals

$$t_p = 0.69 \sum_{i=1}^{N} C_i \left(\sum_{j=1} R_j \right) = 0.69 \frac{N(N+1)}{2} RC \tag{10.7}$$

when all $C_i = C$ and $R_j = R$

Increasing the transistor width reduces this time constant, but it also loads the gates in the previous stage. Therefore, the transistor size is limited by the input loading capacitance.

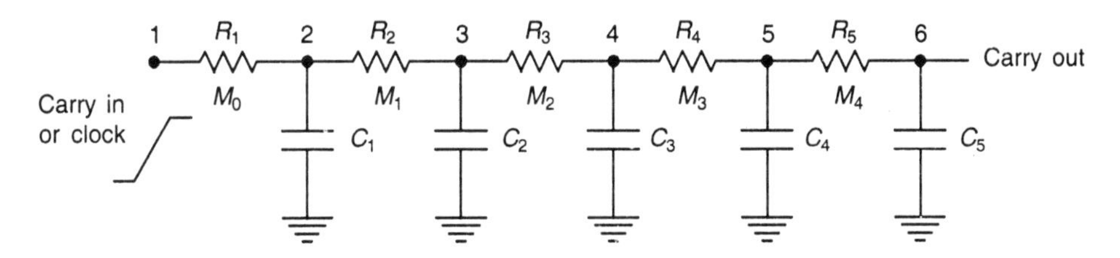

Figure 10.10 Equivalent network to determine propagation delay of a carry chain.

Unfortunately, the distributed RC-nature of the carry chain results in a propagation delay that is quadratic in the number of bits N. To avoid this, it is necessary to insert signal-buffering inverters. The optimum number of stages per buffer depends on the equivalent resistance of the inverter and the resistance and capacitance of the pass transistors. In most practical cases, this number is between 3 and 4. Adding the inverter makes the overall propagation delay a linear function of N, as is the case with ripple carry adders.

10.3.3 The Binary Adder: Logic Design Considerations

The ripple-carry adder is only practical for the implementation of additions with a relatively small word length. Most desktop computers use word lengths of 32 bits, while servers require 64: very fast computers, such as mainframes, supercomputers, or multimedia processors (e.g. the Sony Play Station 2) [Suzuoki99], require word lengths of up to 128 bits. The linear dependence of the adder speed on the number of bits makes the usage of ripple adders rather impractical. Logic optimizations are therefore necessary, resulting in adders with $t_p < O(N)$.

The carry-bypass adder. Consider the four-bit adder block of Figure 10.11(a). Suppose that the values A_k and B_k ($k = 0 \ldots 3$) are such that all propagate signals P_k ($k = 0 \ldots 3$) are high. An incoming carry $C_{i,0} = 1$ propagates under those conditions through the complete adder chain and causes an outgoing carry $C_{0,3} = 1$. In other words,

If ($P_0P_1P_2P_3 = 1$), then $C_{0,3} = C_{i,0}$
Else either DELETE or GENERATE occurred (10.8)

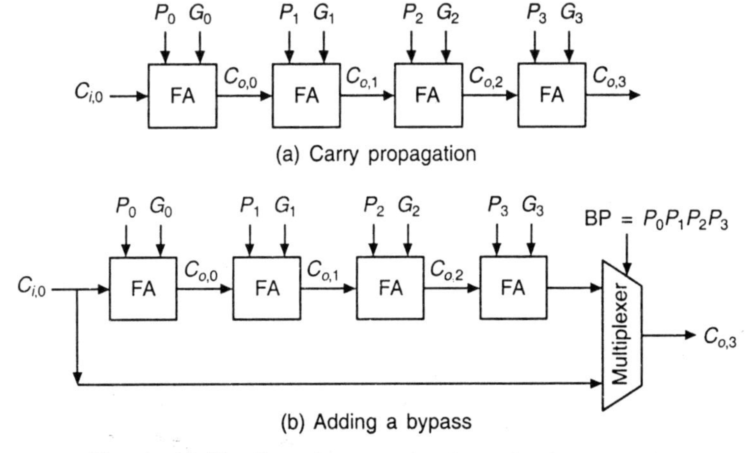

Figure 10.11 Carry-bypass structure—basic concept.

This information can be used to speed up the operation of the adder, as shown in Figure 10.11(b). When BP = $P_0P_1P_2P_3$ = 1, the incoming carry is forwarded immediately to the next block through the bypass transistor M_b—hence the name carry bypass adder or carry-skip adder. If this is not the case, the carry is obtained by way of the normal route.

Carry-bypass in Manchester carry-chain adder. Figure 10.12 shows the possible carry-propagation paths when the full-adder circuit is implemented in Manchester-carry style. This picture demonstrates how the bypass speeds up the addition. The carry propagates either through the bypass path, or a carry is generated somewhere in the chain. In both cases, the delay is smaller than the normal ripple configuration. The area overhead incurred by adding the bypass path; is small and typically ranges between 10 and 20%. However, adding the bypass path breaks the regular bit-slice structure.

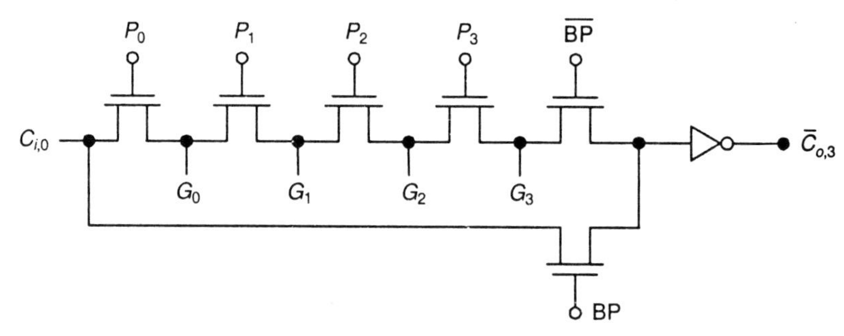

Figure 10.12 Manchester carry-chain implementation of bypass adder.

Let us now compute the delay of an N-bit adder. At first, we assume that the total adder is divided into (N/M) equal-length bypass stages, each of which contains M bits. An approximating expression for the total propagation time can be derived from Figure 10.13 and is given by Eq. (10.9), namely,

$$t_p = t_{\text{setup}} + Mt_{\text{carry}} + [(N/M) - 1]\, t_{\text{bypass}} + [M - 1]\, t_{\text{carry}} + t_{\text{sum}} \tag{10.9}$$

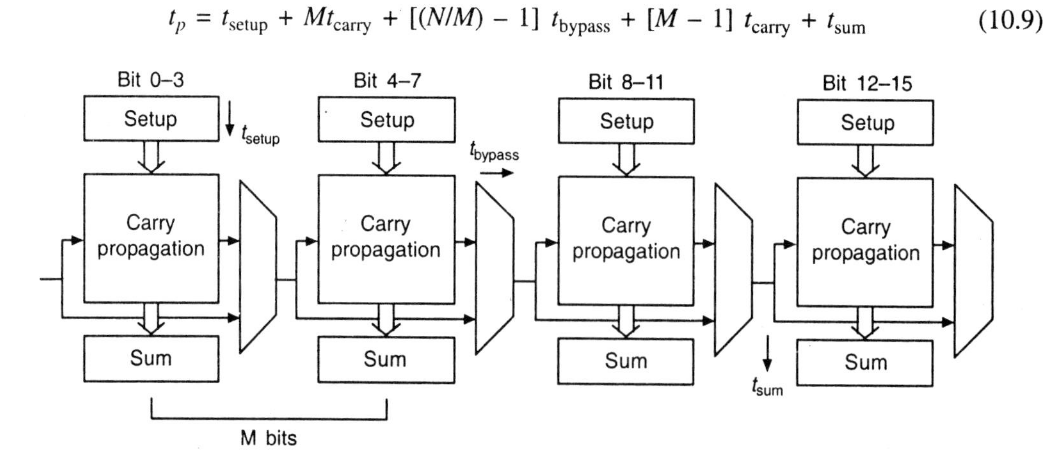

Figure 10.13 (N = 16) carry-bypass adder composition.

where

- t_{setup}: the fixed overhead time to create the generate and propagate signals.
- t_{carry}: the propagation delay through a single bit. The worst-case carry-propagation delay through a single stage of M bits is approximately M times larger.
- t_{bypass}: the propagation delay through the bypass multiplexer of a single stage.
- t_{sum}: the time to generate the sum of the final stage.

From Eq. (10.9), it follows that t_p is still linear in the number of bits N, since in the worst case, the carry is generated at the first bit position, ripples through the first block, skips around $[(N/M) - 2]$ bypass stages, and is consumed at the last bit position without generating an output carry. The optimal number of bits per skip block is determined by technological parameters such as the extra delay of the bypass selecting multiplexer, the buffering requirements in the carry chain, and the ratio of the delay through the ripple and the bypass paths.

Although still linear, the slope of the delay function increases in a more gradual fashion than for the ripple-carry adder, as pictured in Figure 10.14. This difference is substantial for large adders. The ripple adder is actually faster for small values of N, for which the overhead of the extra bypass multiplexer makes the bypass structure not interesting. The crossover point depends upon technology considerations and is normally situated between four and eight bits.

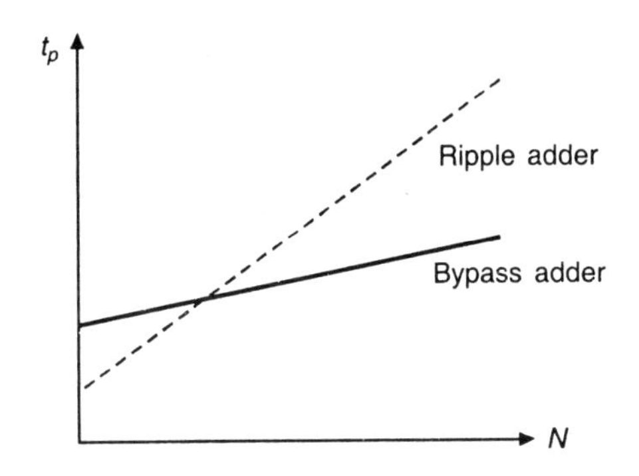

Figure 10.14 Propagation delay of ripple-carry vs. carry-bypass adder.

The linear carry-select adder. In a ripple carry adder, every full-adder cell has to wait for the incoming carry before an outgoing carry can be generated. One way to get around this linear dependency is to anticipate both possible values of the carry input and evaluate the result for both possibilities in advance. Once the real value of the incoming carry is known, the correct result is easily selected with a simple multiplexer stage. An implementation of this idea, appropriately called the carry-select adder, is demonstrated in Figure 10.15: Consider the block of adders, which is adding bits k to $k + 3$. Instead of waiting for the arrival of the output carry of bit $k - 1$, both the 0 and 1 possibilities are analyzed. From a circuit point of view, this means that two carry paths are implemented. When $C_{0,k-1}$ finally settles, either the result of the 0 or the 1 path is selected by the multiplexer, which can be performed with a minimal delay. As is evident from Figure 10.15, the hardware overhead of the carry-select adder is restricted to an additional carry path and a multiplexer, and equals about 30% with respect to a ripple-carry structure.

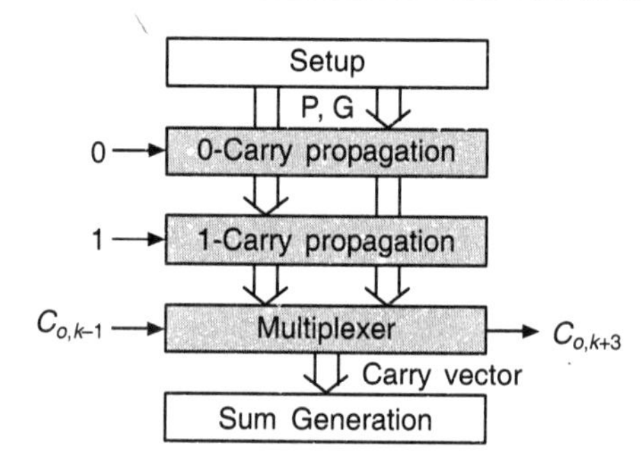

Figure 10.15 Four-bit carry-select module-topology.

A full carry-select adder is now constructed by chaining a number of equal-length adder stages, as in the carry-bypass approach (see Figure 10.16). The critical path is shaded in gray. From inspection of the circuit, we can derive a first-order model of the worst-case propagation delay of the module, written as

$$t_{add} = t_{setup} + Mt_{carry} + \left(\frac{N}{M}\right) t_{mux} + t_{sum} \tag{10.10}$$

where t_{setup}, t_{sum} and t_{mux} are fixed delays and N and M represent the total number of bits, and the number of bits per stage, respectively. t_{carry} is the delay of the carry through a single full-adder cell. The carry delay through a single block is proportional to the length of that stage or equals Mt_{carry}.

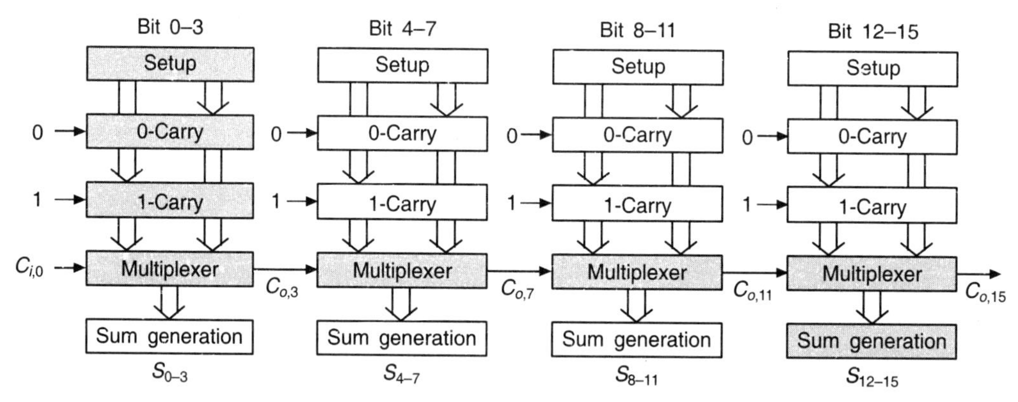

Figure 10.16 Sixteen-bit, linear carry select adder (the critical path is shaded in gray).

The propagation delay of the adder is, again, linearly proportional to N [Eq. (10.10)]. The reason for this linear behaviour is that the block-select signal that selects between the 0 and 1 solutions still has to ripple through all stages in the worst-case.

Square-root carry-select adder. The next structure illustrates how an alert designer can make a major impact. To optimize a design, it is essential to locate the critical timing path first. Consider the case of a 16-bit linear carry-select adder. To simplify the discussion, assume that the full-adder and multiplexer cells have identical propagation delays equal to a normalized value of 1. The worst-case arrival times of the signals at the different network nodes with respect to the time the input is applied are marked and annotated on Figure 10.17(a). This analysis demonstrates that the critical path of the adder ripples through the multiplexer networks of the subsequent stages.

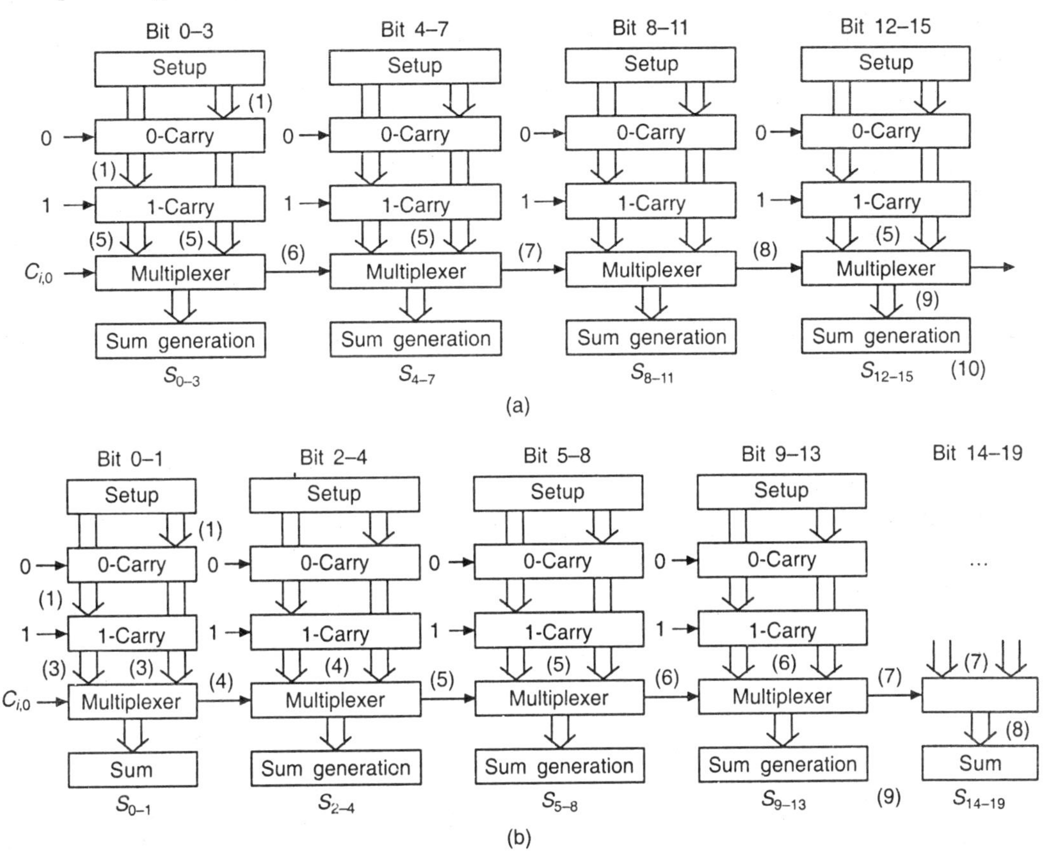

Figure 10.17 Worst case signal arrival times in carry-select adders (the signal arrival times are marked in parentheses).

One striking opportunity is readily apparent. Consider the multiplexer gate in the last adder stage. The inputs to this multiplexer are the two carry chains of the block and the block-multiplexer signal from the previous stage. A major mismatch between the arrival times of the signals can be observed. The results of the carry chains are stable, long before the multiplexer signal can be observed. It makes sense to equalize the delay through both paths. This can be achieved by progressively adding more bits to the subsequent stages in the adder, requiring more time for the generation of the carry signals. For example, the first stage can add 2 bits, the second contains 3, the third has 4, and so forth as demonstrated in Figure 10.17(b). The annotated arrival times show that this adder topology is faster than the linear organization, even though an extra stage is needed. In fact, the same propagation delay is also valid for a 20-bit adder. Observe that the discrepancy in arrival times at the multiplexer nodes has been eliminated.

In effect, the simple trick of making the adder stages progressively longer results in an adder structure with sublinear delay characteristics. This is illustrated by the following analysis:

Assume that an N-bit adder contains P stages, and the first stage adds M bits. An additional bit is added to each subsequent stage. The following relation then holds:

$$N = M + (M + 1) + (M + 2) + (M + 3) + \cdots + (M + (P - 1))$$

$$= MP + \frac{P(P-1)}{2}$$

$$= \frac{P^2}{2} + P\left(M - \frac{1}{2}\right) \tag{10.11}$$

If $M << N$ (e.g., $M = 2$ and $N = 64$), the first term dominates, and Eq. (10.11) can be simplified to

$$N = \frac{P^2}{2} \tag{10.12}$$

or

$$P = \sqrt{2N} \tag{10.13}$$

Equation (10.13) can be used to express t_{add} as a function of N by rewriting Eq. (10.10):

$$t_{add} = t_{setup} + Mt_{carry} + (\sqrt{2N})\, t_{mux} + t_{sum} \tag{10.14}$$

The delay is proportional to $\sqrt{N}$ for large adders ($N >> M$), or $t_{add} = O(\sqrt{N})$. This square-root relation has a major impact, which is illustrated in Figure 10.18, where the delays of both the linear and square root select adders are plotted as a function of N. It can be observed that for large values of N, t_{add} becomes almost a constant.

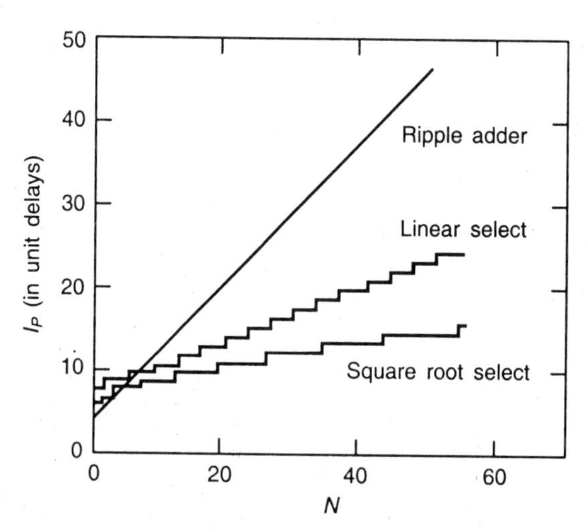

Figure 10.18 Propagation delay of square-root carry-select adder versus linear ripple and select adders.

Carry-lookahead adder

(i) Monolithic Lookahead Adder

When designing even faster adders, it is essential to get around the rippling effect of the carry that is still present in one form or the other in both the carry-bypass and carry-select adders. The carry lookahead principle offers a possible way to do so. The following relation holds for each bit position in an N-bit adder:

$$C_{0,k} = f(A_k, B_k, C_{0,k-1}) = G_k + P_k C_{0,k-1} \tag{10.15}$$

The dependency between $C_{0,k}$ and $C_{0,k-1}$ can be eliminated by expanding $C_{0,k-1}$:

$$C_{0,k} = G_k + P_k(G_{k-1} + P_{k-1}C_{0,k-2}) \tag{10.16}$$

In a fully expanded form,

$$C_{0,k} = G_k + P_k[G_{k-1} + P_{k-1} (\ldots + P_1(G_0 + P_0 C_{i,0}))] \tag{10.17}$$

with $C_{i,0}$ typically equal to 0.

This expanded relationship can be used to implement an N-bit adder. For every bit, the carry and sum outputs are independent of the previous bits. The ripple effect has thus been effectively eliminated, and the addition time should be independent of the number of bits. A block diagram of the overall composition of a carry-lookahead adder is shown in Figure 10.19.

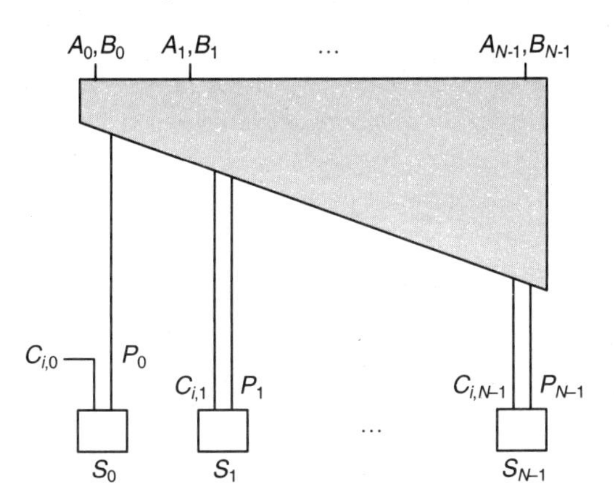

Figure 10.19 Conceptual diagram of a carry-lookahead adder.

Such a high-level model contains some hidden dependencies. When we study the detailed schematics of the adder, it becomes obvious that the constant addition time is wishful thinking and that the real delay is at least increasing linearly with the number of bits. This is illustrated in Figure 10.20, where a possible circuit implementation of Eq. (10.17) is shown for $N = 4$. Note that the circuit exploits the self-duality and the recursivity of the carry-lookahead equation to build a mirror structure, similar in style to the single-bit full-adder of Figure 10.6. The large fan-in of the circuit makes it prohibitively slow for larger values of N. Implementing it with simpler gates requires multiple logic levels. In both cases, the propagation delay increases. Furthermore, the fan-out on some of the signals tends to grow excessively, slowing down the adder even more. For instance, the signals G_0 and P_0 appear in the expression for every one of

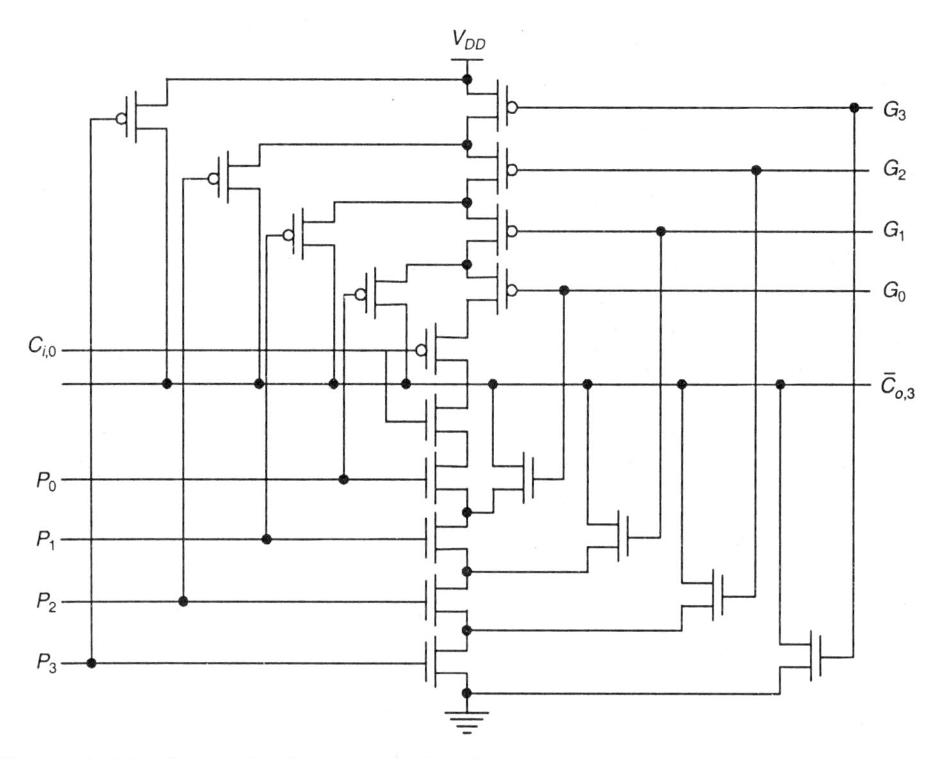

Figure 10.20 Schematic diagram of mirror implementation of four-bit lookahead adder.

the subsequent bits. Hence, the capacitance on these lines is substantial. Finally, the area of the implementation grows progressively with N. Therefore, the lookahead structure suggested by Eq. (10.16) is only useful for small values of $N(\leq 4)$.

(ii) Logarithmic Lookahead Adder

For a carry-lookahead group of N bits, the transistor implementation has $N + 1$ parallel branches with up to $N + 1$ transistors in the stack. Since wide gates and large stacks display poor performance, the carry lookahead computation has to be limited to up to two or four bits in practice. In order to build very fast adders, it is necessary to organize carry propagation and generation into recursive trees. A more effective implementation is obtained by hierarchically decomposing the carry propagation into subgroups of N bits:

$$\begin{aligned}
C_{o,0} &= G_0 + P_0C_{i,0} \\
C_{o,1} &= G_1 + P_1G_0 + P_1P_0C_{i,0} = (G_1 + P_1G_0) + (P_1P_0)C_{i,0} = G_{1,0} + P_{1,0}C_{i,0} \\
C_{o,2} &= G_2 + P_2G_1 + P_2P_1G_0 + P_2P_1P_0C_{i,0} = G_2 + P_2C_{0,1} \\
C_{o,3} &= G_3 + P_3G_2 + P_3P_2G_1 + P_3P_2P_1G_0 + P_3P_2P_1P_0C_{i,0} \\
&= (G_3 + P_3G_2) + (P_3P_2)\ C_{o,1} = G_{3,2} + P_{3,2}C_{o,1}
\end{aligned} \tag{10.18}$$

In Eq. (10.18), the carry-propagation process is decomposed into subgroups of two bits. $G_{i,j}$ and $P_{i,j}$ denote the generate and propagate functions, respectively, for a group of bits (from bit positions i to j). Therefore, we call them block generate and propagate signals. $G_{i,j}$ equals 1 if the group generates a carry, independent of the incoming carry. The block propagate

$P_{i,j}$ is true if an incoming carry propagates through the complete group. This condition is equivalent to the carry bypass, discussed earlier. For example, $G_{3,2}$ is equal to 1 when a carry either is generated at bit position 3 or is generated at position 2 and propagated through position 3, or $G_{3,2} = G_3 + P_3G_2$. $P_{3,2}$ is true when an incoming carry propagates through both bit positions, or $P_{3,2} = P_3P_2$.

Note that the format of the new expression for the carry is equivalent to the original one, except that the generate and propagate signals are replaced with block generate and propagate signals. The notation $G_{i,j}$ and $P_{i,j}$ generalizes the original carry equations, since $G_i = G_{i,i}$ and $P_i = P_{i,i}$. Another generalization is possible by treating the generate and propagate functions as a pair $(G_{i,j}, P_{i,j})$, rather than considering them as separate functions. A new Boolean operator, called the dot operator $(\cdot)$, can be introduced. This operator on the pairs and allows for the combination and manipulation of blocks of bits:

$$(G, P) \cdot (G', P') = (G + PG', PP') \tag{10.19}$$

Using this operator we can now decompose $(G_{3,2}, P_{3,2}) = (G_3, P_3) \cdot (G_2, P_2)$. The dot operator obeys the associative property, but it is not commutative.

By exploiting the associative property of the dot operator, a tree can be constructed that effectively computes the carries at all $2^i - 1$ positions (that is, 1, 3, 7, 15, etc.) for $i = 1 \ldots \log_2(N)$. The crucial advantage is that the computation of the carry at position $2^i - 1$ takes only $\log_2(N)$ steps. In other words, output carry of an N-bit adder can be computed in $\log_2(N)$ time. This is a major improvement over the previously described adders. For example, for an adder of 64 bits, the propagation delay of a linear adder is proportional to 64. For a square-root select adder, it is reduced to 8, while, for a logarithmic adder, the proportionality constant is 6. This is illustrated in Figure 10.21, which shows the block diagram of a 16-bit logarithmic adder. The carry at position 15 is computed by combining the results of blocks (0, 7) and (8, 15). Each of these, in turn, is composed hierarchically. For instance, (0, 7) is the composition of (0, 3) and (4, 7), while (0, 3) consists of (0, 1) and (2, 3), etc.

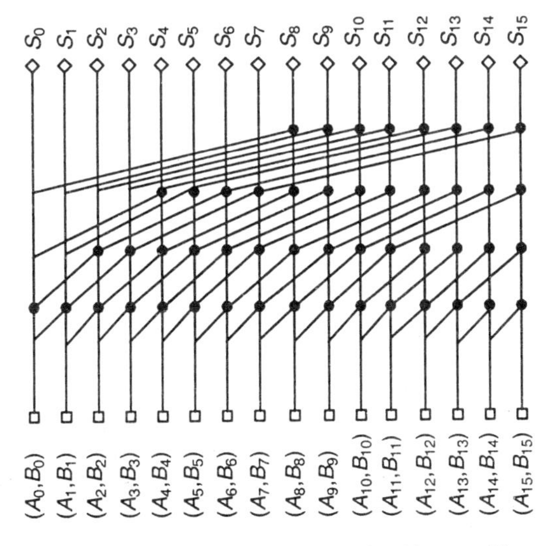

Figure 10.21 Schematic diagram for Kogge–Stone 16-bit lookahead logarithmic adder.

Computing the carries at just the $2^i - 1$ positions is obviously not sufficient. It is necessary to derive the carry signals at the intermediate positions as well. One way to accomplish this is by replicating the tree at every bit position, as illustrated in Figure 10.21 for $N = 16$. For instance, the carry at position 6 is computed by combining the results of blocks (6, 3) and (2, 0). This complete structure, which frequently is referred to as a Kogge-Stone tree, is a member of the radix-2 class of trees. Radix-2 means that the tree is binary: it combines two carry words at a time at each level of hierarchy. The total adder requires 49 complex logic gates each to implement the dot operator. In addition, 16 logic modules are needed for the generation of the propagate and generate signals at the first level (P_i and G_i), as well as 16 sum-generation gates.

Logarithmic lookahead adder—alternatives. Designers of fast adders sometimes revert to other styles of tree structures as they trade off for area, power or performance. We briefly discuss the Brent–Kung adder and the radix-4 adder, two of the common alternative structures.

The Kogge–Stone tree of Figure 10.21 has some interesting properties. First, its interconnect structure is regular, which makes implementation quite easy. Furthermore, the fan-out thoughout the tree is fairly constant, especially on the critical paths. The task of sizing the transistors for optimal performance is therefore simplified. At the same time, however, the replication of the carry trees to generate the intermediate carries comes at a large cost in terms of both area and power. Designers sometimes trade off some delay for area and power by choosing less complex trees. A simpler tree structure computes only the carries to the powers of 2-bit positions, as illustrated in Figure 10.22 for $N = 16$.

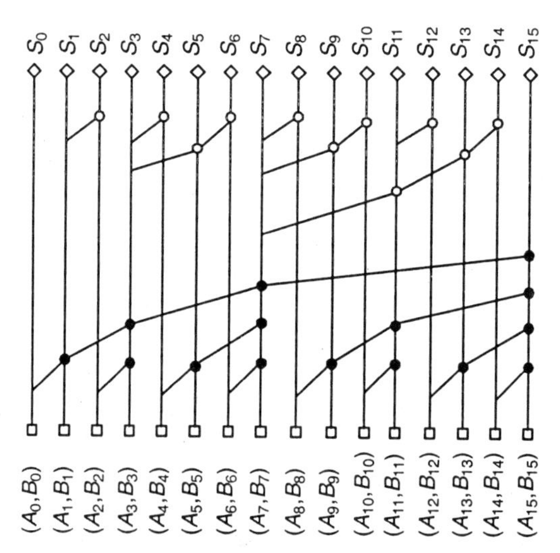

Figure 10.22 A 16-bit Brent–Kung tree.

The forward binary tree realizes the carry signals only at positions $2^N - 1$:

$$(C_{o,0},\ 0) = (G_0,\ P_0) \cdot (C_{i,0},\ 0)$$

$$(C_{o,1},\ 0) = [(G_1,\ P_1) \cdot (G_0,\ P_0)] \cdot (C_{i,0},\ 0) = (G_{1,0},\ P_{1,0}) \cdot (C_{i,0},\ 0)$$

$$(C_{0,3},\ 0) = [(G_{3:2},\ P_{3:2}) \cdot (G_{1:0},\ P_{1:0})] \cdot (C_{i,0}, 0) = (G_{3:0},\ P_{3:0}) \cdot (C_{i,0},\ 0)$$

$$(C_{o,7},\ 0) = [(G_{7:4},\ P_{7:4}) \cdot (G_{3:0},\ P_{3:0})] \cdot (C_{i,0},\ 0) = (G_{7:0},\ P_{7:0}) \cdot (C_{i,0},\ 0) \qquad (10.20)$$

The forward binary tree structure is not sufficient to generate the complete set of carry bits. An *inverse binary tree* is needed to realize the other carry bits (shown in Figure 10.22). This structure combines intermediate results to produce the remaining carry bits.

The resulting structure, commonly called the Brent–Kung adder, uses 27 dot gates, or almost half of the 49 needed for a full radix-2 tree, and it needs fewer wires as well. The wiring structure is less regular, however, and fan-out varies from gate-to-gate, making performance optimization more difficult. Especially the fan-out of the middle node ($C_{0,7}$), which equals one sum and five dot operations for this example, is of major concern. This observation makes the Brent–Kung adder rather unsuited for very large adders (>32 bits).

An option to reduce the depth of the tree is to combine four signals at a time at each level of the hierarchy. The resulting tree is now of class radix-4, because it uses building blocks of order 4 as shown in Figure 10.23. A 16-bit addition needs only two stages of carry logic.

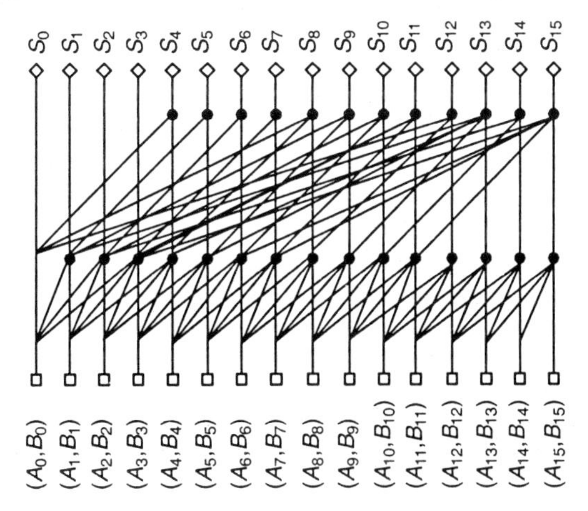

Figure 10.23 Radix-4 Kogge–Stone tree for 16-bit operands.

On average, a lookahead adder is several times larger than a ripple adder, but has dramatic speed advantages for larger operands. The logarithmic behaviour makes it preferable over bypass or select adders for larger values of N. The exact value of the cross point depends heavily on technology and circuit design factors.

The discussion of adders is by no means complete. Due to its impact on the performance of computational structures, the design of fast circuits has been the subject of many publications. It is even possible to construct adder structures with a propagation delay that is independent of the number of bits. Examples of those are the carry save structures and the redundant binary arithmetic structures. These adders require number-encoding and decoding steps, whose delay is a function of N. Therefore, they are only interesting when embedded in larger structures such as multipliers or high-speed signal processors.

10.4 THE MULTIPLIER

Multiplications are expensive and slow operations. The performance of many computational problems often is dominated by the speed at which a multiplication operation can be executed. This observation has, for instance, prompted the integration of complete multiplication units in state-of-the-art digital signal processors and microprocessors.

Multipliers are, in effect, complex adder arrays. The analysis of the multiplier gives us some further insight into how to optimize the performance (or the area) of complex circuit topologies. After a short discussion of the multiply operation, we discuss the basic array multiplier. We also discuss different approaches to partial product generation, accumulation and their final summation.

10.4.1 The Multiplier: Definitions

Consider two unsigned binary numbers X and Y that are M and N bits wide, respectively. To introduce the multiplication operation, it is useful to express X and Y in the binary representation:

$$X = \sum_{i=0}^{M-1} X_i 2^i \qquad Y = \sum_{j=0}^{N-1} Y_j 2^j \tag{10.21}$$

with X_i, $Y_j \in \{0, 1\}$. The multiplication operation is then defined as follows:

$$Z = X \times Y = \sum_{k=0}^{M+N-1} Z_k 2^k$$

$$= \left(\sum_{i=0}^{M-1} X_i 2^i\right)\left(\sum_{j=0}^{N-1} Y_j 2^j\right) = \sum_{i=0}^{M-1}\left(\sum_{j=0}^{N-1} X_i Y_j 2^{i+j}\right) \tag{10.22}$$

The simplest way to perform a multiplication is to use a single two-input adder. For inputs that are M and N bits wide, the multiplication takes M cycles, using an N-bit adder. This shift-and-add algorithm for multiplication adds together M partial products. Each partial product is generated by multiplying the multiplicand with a bit of the multiplier—which, essentially, is an AND operation—and by shifting the result on the basis of the multiplier bit's position.

A faster way to implement multiplication is to resort to an approach similar to manually computing a multiplication. All the partial products are generated at the same time and organized in an array. A multi-operand addition is applied to compute the final product. The approach is illustrated in Figure 10.24. This set of operations can be mapped directly into hardware. The resulting structure is called an array multiplier and combines the following three functions: *partial product generation, partial product accumulation, and final addition.*

```
            1 0 1 0 1 0    Multiplicand
          ×     1 0 1 1    Multiplier
      -------------------
            1 0 1 0 1 0
          1 0 1 0 1 0      Partial products
        0 0 0 0 0 0
      1 0 1 0 1 0
      -------------------
      1 1 1 0 0 1 1 1 0    Result
```

Figure 10.24 Binary multiplication—an example.

10.4.2 Partial-Product Generation

Partial products result from the logical AND of multiplicand X with a multiplier bit Y_i (see Figure 10.25). Each row in the partial-product array is either a copy of the multiplicand or a row of zeros. Careful optimization of the partial product generation can lead to some substantial delay and area reductions. Note that in most cases the partial product array has many zero rows that have no impact on the result and thus represent a waste of effort when added. In the case of a multiplier consisting of all 1s, all the partial products exist, while in the case of all 0s, there is none. This observation allows us to reduce the number of generated partial products by half.

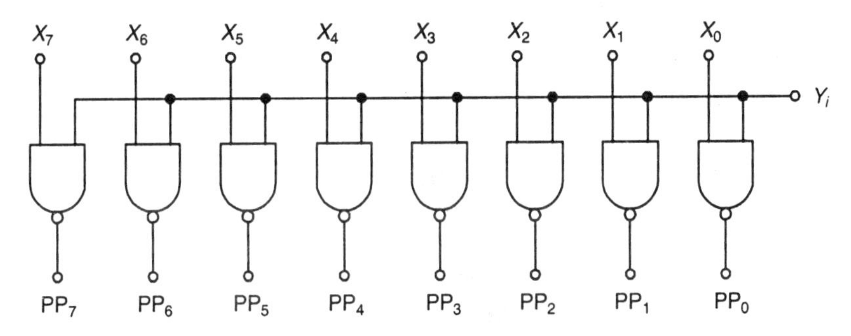

Figure 10.25 Partial-product generation logic.

Assume, for example, an 8-bit multiplier of the form 01111110, which produces six non-zero partial-product rows. One can substantially reduce the number of non-zero rows by recoding this number $(2^7 + 2^6 + 2^5 + 2^4 + 2^3 + 2^2)$ into a different format. The reader can verify that the form $1000001\bar{1}0$, with $\bar{1}$ a shorthand notation for -1, represents the same number. Using this format, we have to add only two partial products, but the final adder has to be able to perform subtraction as well. This type of transformation is called *Booth's recoding*, and it reduces the number of partial products to, at most, one half. It ensures that for every two consecutive bits, at most one bit will be 1 or -1. Reducing the number of partial products is equivalent to reducing the number of additions, which leads to a speedup as well as an area reduction. Formally, this transformation is equivalent to formatting the multiplier word into a base-4 scheme, instead of the usual binary format:

$$Y = \sum_{j=0}^{(N-1)/2} Y_j 4^j \text{ with } (Y_j \in \{-2, -1, 0, 1, 2,\}) \tag{10.23}$$

Note that $1010\cdots10$ represents the worst-case multiplier input because it generates the most partial products (one half). While the multiplication with (0, 1) is equivalent to an AND operation, multiplying with $\{-2, -1, 0, 1, 2\}$ requires a combination of inversion and shift logic. The encoding can be performed on the fly and requires some simple logic gates.

Having a variable-size partial-product array is not practical for multiplier design, and a *modified Booth's recoding* is most often used. The multiplier is partitioned into 3-bit groups that overlap by one bit. Each group of three is recoded, as shown in Table 10.2, and forms one partial

Table 10.2 Modified Booth's recoding

Partial Product Selection	
Multiplier bits	*Recoded bits*
000	0
001	+ Multiplicand
010	+ Multiplicand
011	+2 × Multiplicand
100	−2 × Multiplicand
101	− Multiplicand
110	− Multiplicand
111	0

product. The resulting number of partial products equals half of the multiplier width. The input bits to the recoding process are the two current bits, combined with the upper bit from the next group, moving from MSB to LSB.

In simple terms, the modified Booth's recoding essentially examines the multiplier for strings of ones from MSB to LSB and replaces them with a leading 1, and a −1 at the end of the string. For example, 011 is understood as the beginning of a string of ones and is therefore replaced by a leading 1 (or 100), while 110 is seen as the end of a string and is replaced by a −1 at the least significant position (or $0\bar{1}0$).

Example 10.1 Modified Booth's Recoding
Consider the 8-bit binary number 01111110. This can be divided into four overlapping groups of three bits, going from MSB to LSB; 00(1), 11(1), 11(1), 10(0). Recoding by using Table 10.2 yields: 10 (2 ×), 00 (0 ×), 00 (0 ×), $\bar{1}0$ (−2 ×), or, in combined format, $100000\bar{1}0$. This is equivalent to the result we obtained before.

10.4.3 Partial-Product Accumulation

After the partial products are generated, they must be summed. This accumulation is essentially a multioperand addition. A straightforward way to accumulate partial products is by using a number of adders that will form an array—hence, the name, *array multiplier*. A more sophisticated procedure performs the addition in a tree format.

The array multiplier. The composition of an array multiplier is shown in Figure 10.26. There is a one-to-one topological correspondence between this hardware structure and the manual multiplication shown in Figure 10.24. The generation of N partial products requires $N \times M$ 2-bit AND gates. Most of the area of the multiplier is devoted to the adding of the N partial products, which requires $N - 1$, M-bit adders. The shifting of the partial products for their proper alignment is performed by simple routing and does not require any logic. The overall structure can easily be compacted into a rectangle, resulting in a very efficient layout.

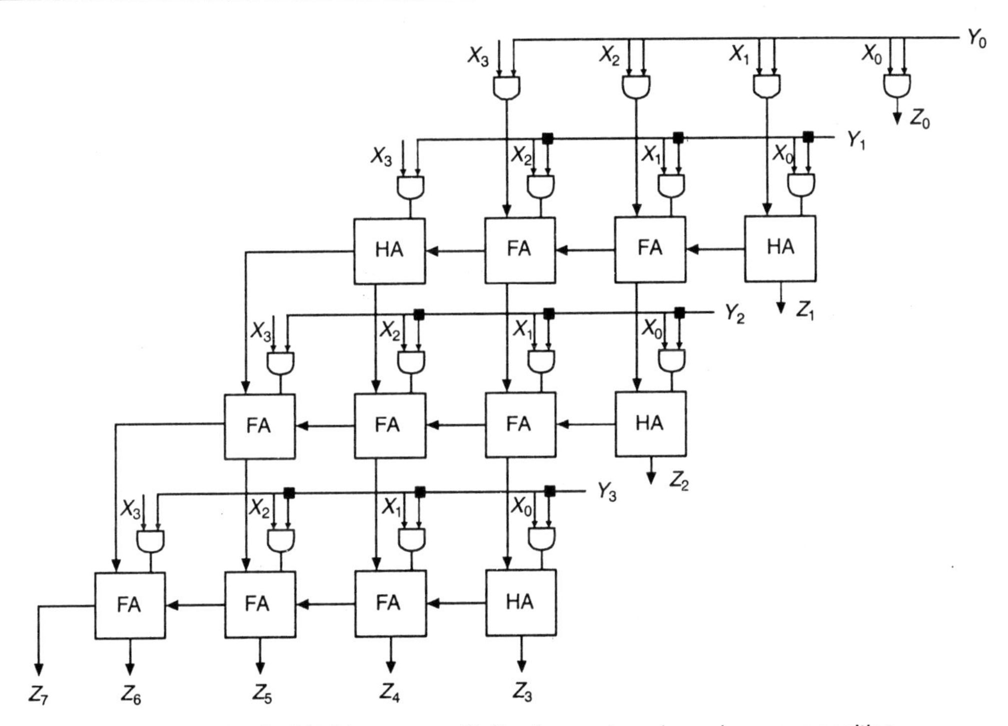

Figure 10.26 A 4X4 bit-array multiplier for unsigned numbers-composition.

Due to the array organization, determining the propagation delay of this circuit is not straightforward. Consider the implementation of Figure 10.26. The partial sum adders are implemented as ripple-carry structures. Performance optimization requires that the critical timing path be identified first. This turns out to be nontrivial. In fact, a large number of paths of almost identical length can be identified. Two of those are highlighted in Figure 10.27. A closer look at those critical paths yields an approximate expression for the propagation delay. We write this as

$$t_{\text{mult}} \approx [(M - 1) + (N - 2)]\ t_{\text{carry}} + (N - 1)\ t_{\text{sum}} + t_{\text{and}} \tag{10.24}$$

where t_{carry} is the propagation delay between input and output carry, t_{sum} is the delay between the input carry and sum bit of the full adder, and t_{and} is the delay of the AND gate.

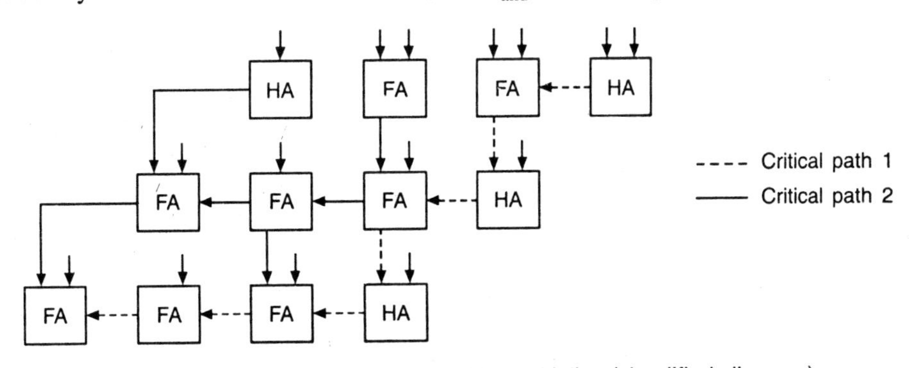

Figure 10.27 Ripple-carry based 4X4 multiplier (simplified diagram).

Since all critical paths have the same length, speeding up just one of them—for instance, by replacing one adder by a faster one such as a carry select adder—does not make much sense from a design standpoint. All critical paths have to be attacked at the same time. From Eq. (10.24), it can be deduced that the minimization of t_{mult} requires the minimization of both t_{carry} and t_{sum}. In this case, it could be beneficial for t_{carry} to equal t_{sum}. This contrasts with the requirements for adder cells, where a minimal t_{carry} was of prime importance. An example of a full adder circuit with comparable t_{sum} and t_{carry} delays was shown in Figure 10.7.

Carry-save multiplier. Due to the a large number of almost identical critical paths, increasing the performance of the structure of Figure 10.27 through transistor sizing yields marginal benefits. A more efficient realization can be obtained by noticing that the multiplication result does not change when the output carry bits are passed diagonally downwards instead of only to the right, as shown in Figure 10.28. We include an extra adder called a *vector-merging* adder to generate the final result. The resulting multiplier is called a *carry-save multiplier*, because the carry bits are not immediately added, but rather are 'saved' for the next adder stage. In the final stage, carries and sums are merged in a fast carry-propagate (e.g. carry-lookahead) adder stage. While this structure has a slightly increased area cost (one extra adder), it has the advantage that its worst-case critical path is shorter and uniquely defined, as highlighted in Figure 10.28 and is expressed as

$$t_{mult} = t_{and} + (N - 1)\, t_{carry} + t_{merge} \tag{10.25}$$

still assuming that $t_{add} = t_{carry}$.

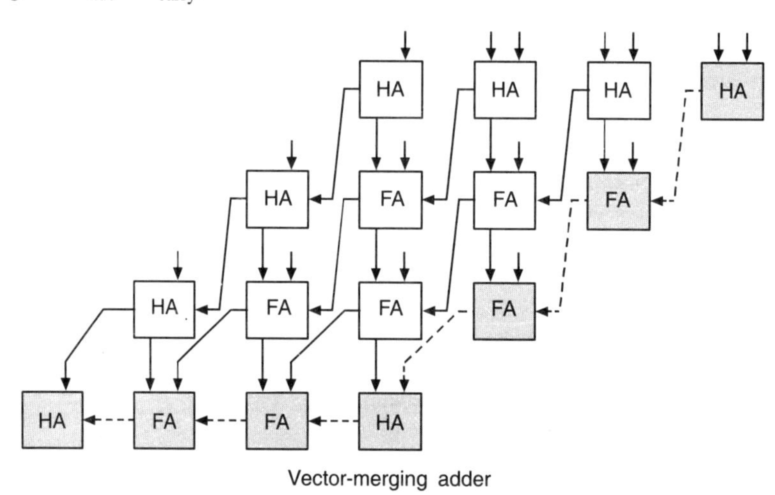

Figure 10.28 A 4X4 carry-save multiplier (the critical path is highlighted in gray).

The tree multiplier. The partial-sum adders can be rearranged in a tree-like fashion, reducing both the critical path and the number of adder cells needed. Consider the simple example of four partial products each of which is four bits wide, as shown in Figure 10.29(a). The number of full-adders needed for this operation can be reduced by observing that only column 3 in the array has to add four bits. All other columns are somewhat less complex. This is illustrated in Figure 10.29(b), where the original matrix of partial products is reorganized into a tree shape to

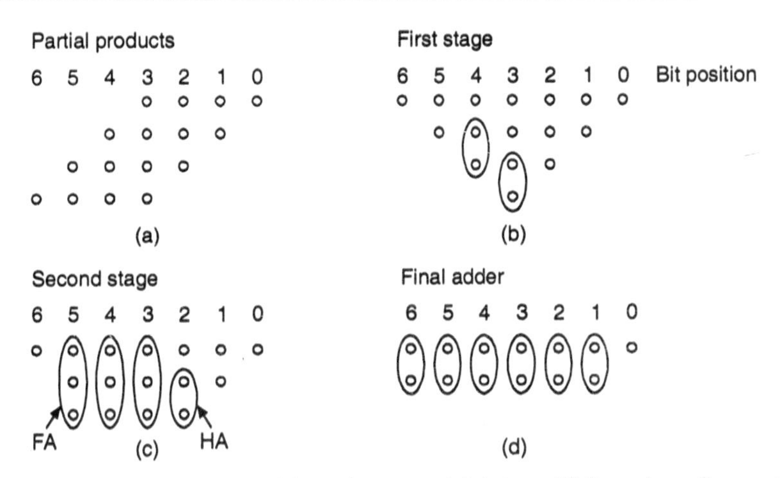

Figure 10.29 Transforming a partial-product tree (a) into a Wallace tree (b, c, c), using an iterative covering process (the example shown is for a four-bit operand).

visually illustrate its varying depth. The challenge is to realize the complete matrix with a minimum depth and a minimum number of adder elements. The first type of operator that can be used to cover the array is a full-adder, which takes three inputs and produces two outputs; the sum, located in the same column and the carry, located in the next one. For this reason, the FA is called a 3-2 *compressor*. It is denoted by a circle covering three bits. The other operator is the half-adder, which takes two input bits in a column and produces two outputs. The HA is denoted by a circle covering two bits.

To arrive at the minimal implementation, we iteratively cover the tree with FAs and HAs, starting from its densest part. In a first step, we introduce HAs in columns 4 and 3 [Figure 10.29(b)]. The reduced tree is shown in Figure 10.29(c). A second round of reductions creates a tree of depth 2 [Figure 10.29(d)]. Only three FAs and three HAs are used for the reduction process, compared with six FAs and six HAs in carry-save multiplier of Figure 10.28. The final stage (called 'Final Addition') consists of a simple two-input adder, for which any type of adder can be used.

The presented structure is called the *Wallace tree multiplier*, and its implementation is shown in Figure 10.30. The tree multiplier realizes substantial hardware savings for larger multipliers. The propagation delay is reduced as well. In fact, it can be shown that the propagation delay through the tree is equal to $O[\log_{3/2}(N)]$. While substantially faster than the carry-save structure for large multiplier word lengths, the Wallace multiplier has the disadvantage of being very irregular, which complicates the task of coming up with an efficient layout. This irregularity is visible even in the four-bit implementation of Figure 10.30.

10.4.4 Final Addition

The final step for completing the multiplication is to combine the result in the final adder. Performance of this 'vector-merging' operation is of key importance. The choice of the adder style depends on the structure of the accumulation array. A carry-lookahead adder is the preferable option if all input bits to the adder arrive at the same time, as it yields the smallest

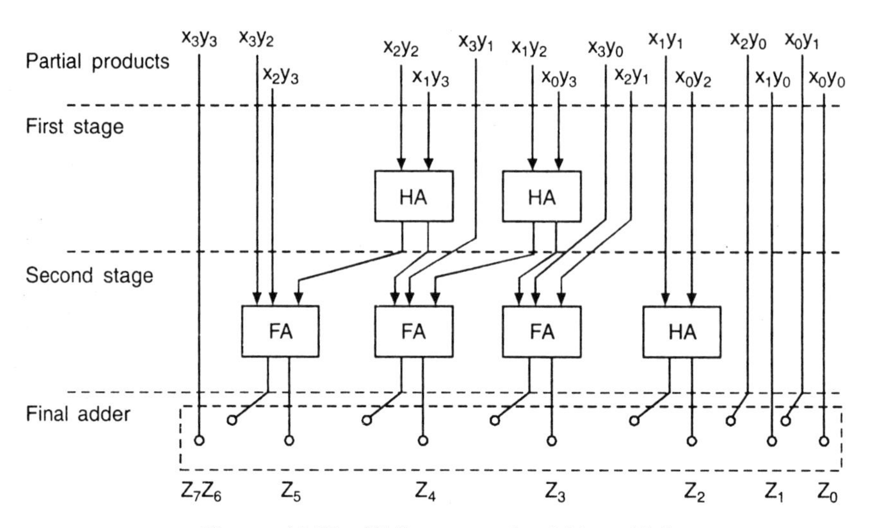

Figure 10.30 Wallace tree for 4-bit multiplier.

possible delay. This is the case if a pipeline stage is placed right before the final addition. Pipelining is a technique frequently used in high-performance multipliers. In non-pipelined multipliers, the arrival-time profile of the inputs to the final adder is quite uneven due to the varying logic depths of the multiplier tree. Under these circumstances, other adder topologies, such as carry select, often yield performance numbers similar to lookahead at a substantially reduced hardware cost.

SUMMARY

In this chapter, various arithmetic building blocks like datapath circuits, adders, multiplexers are described in detail. The implementation of arithmetic datapath operators is studied from area, performance, and power perspective. A datapath is best implemented in a bit sliced fashion. A single layout slice is used repetitively for every bit in data word. This regular approach eases the design effort and results in fast and dense layouts. A ripple carry adder has a performance that is linearly proportional to the number of bits. Circuit optimizations concentrate on reducing the delay of the carry path. A number of circuit topologies were examined, showing how careful optimization of the circuit topology and the transistor sizes helps to reduce the capacitance on the carry bit. Other adder structures use logic optimizations to increase the performance. The performance of carry-select and carry-lookahead adders depends on the number of bits in square root and logarithmic fashion, respectively. This increase in performance comes at a penalty in area. A multiplier is nothing more than a collection of cascaded adders. Its critical path is far more complex.

REVIEW QUESTIONS

1. Define full adder and explain CMOS implementation of full adder.
2. Explain monolithic lookahead adder and logarithmic lookahead adder.
3. With suitable example explain 4×4 carry save multiplier.

SHORT ANSWER QUESTIONS

1. What is a datapath?

   Ans. It is a logical and a physical structure.

2. Signal buffering inverters are used to avoid ____________ .

   Ans. propagation delay in RC networks.

3. The important property of Kogge–Stone tree is that ____________ .

   Ans. its interconnect structure is regular.

4. What is the use of Wallace tree multiplier?

   Ans. It saves hardware.

11 Programmable Logic Devices

11.1 INTRODUCTION

Programmable logic devices (PLDs) are used in many applications to replace SSI and MSI circuits. They save space and reduce the actual number and cost of devices in a given design. A PLD consists of a large array of AND gates and OR gates that can be programmed to achieve specified logic functions. The four types of devices that are classified as PLDs are the programmable read-only memory (PROM), the programmable logic array (PLA), the programmable array logic (PAL) and the generic array logic (GAL). In addition, there are much larger programmable logic devices, called complex PLDs (CPLDs) and field programmable gate arrays (FPGAs). All PLDs consist of programmable arrays. A programmable array is essentially a grid of conductors that form rows and columns with a fusible link at each cross point. Arrays can be either fixed or programmable. The PROM consists of a set of fixed (nonprogrammable) AND gates connected as a decoder and a programmable OR array. The PLA is a PLD that consists of a programmable AND array and a programmable OR array. The PLA can overcome some of the limitations imposed by the PROM. The PAL is a PLD, consisting of a programmable AND array and a fixed OR array with output logic. It was developed to overcome certain disadvantages of the PLA, such as longer delays due to the additional fusible links that result from using two programmable arrays and more circuit complexity. A more recent development in PLDs is the GAL. It has a programmable AND array and a fixed OR array with programmable output logic. The differences between GAL and PAL devices are that the GAL is reprogrammable and it has programmable output configurations.

CPLDs are essentially much larger versions of simple PLDs, with a centralized internal interconnect matrix to connect the device macrocells together. FPGAs, on the other hand, consist of a large array of simple logic cells with interconnecting horizontal and vertical routing channels. In this chapter, the programmable logic devices such as CPLD and FPGA are discussed in detail.

11.2 NMOS PLAs

The basic PLA structure consists of an AND plane driving an OR plane as shown in Figure 11.1. The terminology corresponds to a sum of products (SOP) realization of the desired function. The SOP realization converts directly into a NAND–NAND implementation. When a product of sums (POS) realization is desired, it can be implemented in OR–AND or NOR–NOR logic. In either case, the first array is referred to as the AND plane, and the second array as the OR plane. The lines connecting the AND plane to the OR plane are called the product lines.

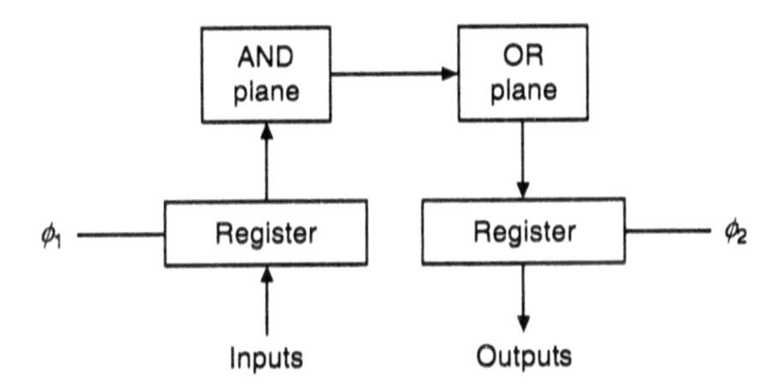

Figure 11.1 The basic PLA structure.

The OR plane matrix is identical in form to the AND plane matrix, but its layout is rotated 90 degrees with respect to the AND plane. The input and output registers need not be identical, but they are also repetitive structures. The overall size of a PLA is a function of the number of inputs, the number of product terms, the number of outputs, and the value of the parameter lambda. The PLA must be programmed by appropriately locating transistors on the array.

11.2.1 NMOS PLA Layouts

The NMOS PLA can be realized in either NAND–NAND logic or NOR–NOR logic. The NAND implementation is much smaller because it needs no metal contacts in the matrix of the AND or OR plane and is more compact. But, the NAND structure is much slower than the NOR implementation due to the series pass transistor structure. A NAND PLA with N control transistors in series is about $N \times N$ times slower than the NOR realization, while the NAND implementation is typically smaller in area than the NOR realization by a factor of about three-to-one.

Consider the PLA with 3 inputs I_0, I_1, and I_2. It is programmed to realize four product lines with three output lines. The product lines are indicated by P_0, P_1, P_2, and P_3. The outputs realized by this PLA are denoted by Y_0, Y_1, and Y_2.
The product terms are:

$$P_0 = \overline{I}_0\overline{I}_1$$

$$P_1 = \overline{I}_0 I_1$$

$$P_2 = I_0 I_1 \overline{I}_2$$

$$P_3 = I_0 I_2 \tag{11.1}$$

The outputs are:

$$Y_0 = P_1$$
$$Y_1 = P_0 + P_2 + P_3$$
$$Y_2 = P_1 + P_2 \qquad (11.2)$$

11.2.1.1 NOR–NOR realization of NMOS PLA

The NOR-NOR realization of NMOS PLA for the above example is obtained in the following way. It includes input and output buffers and two-phase clocking. First, the personality matrix, Q of the PLA is developed. In the AND plane, element $q_{ij} = 0$ if a FET is to connect product line p_i to input line I_j. The element $q_{ij} = 1$ if a FET is to connect product line p_i to input line I_i. The element q_{ij} is a **don't care** (X) if neither input is to be connected to product line p_i. In the OR plane, $q_{ij} = 1$ if product line p_i connects to output Y_j, and 0 otherwise. The personality matrix for the above example is given below:

$$Q = \begin{pmatrix} 1 & 1 & X & 0 & 1 & 0 \\ 1 & 0 & X & 1 & 0 & 1 \\ 0 & 0 & 1 & 0 & 1 & 1 \\ 0 & X & 0 & 0 & 1 & 0 \end{pmatrix} \qquad (11.3)$$

The procedure for laying out a NOR–NOR PLA is as follows:

AND plane. For each logic 1 in the input columns of the personality matrix, run a diffusion-path from the appropriate product-term line, under the corresponding inverted input line in the PLA AND plane to ground. The transistor thus created is controlled by the inverted input line. Whenever the controlling line crossing the AND plane is high, the product-term line will be low.

For each logic 0 in the input columns of the personality matrix, run a diffusion path from the appropriate product-term line, under the corresponding non-inverted input line in the PLA AND plane. The transistor thus created is controlled by the non-inverted input line. Whenever that controlling line crossing the AND plane is high, the product term line will be low. Don't care terms are connected to neither the true nor the complemented input lines.

OR plane. For each logic 1 in the output columns of the personality matrix, run a diffusion path from the next-state output line in the PLA OR plane, under the corresponding product term line, to ground. This creates a transistor controlled by the product-term line. Then, if that controlling product-term line is high, the path to the output inverter will be low, and the output will be high. The output is low unless at least one product line controlling it is high.

A NOR–NOR realization of MOS PLA for the above example with enhancement mode pull-down devices is shown in Figure 11.2, including input and output buffers and two-phase clocking (φ_1 and φ_2). The pull-up device is depletion-mode NMOS device.

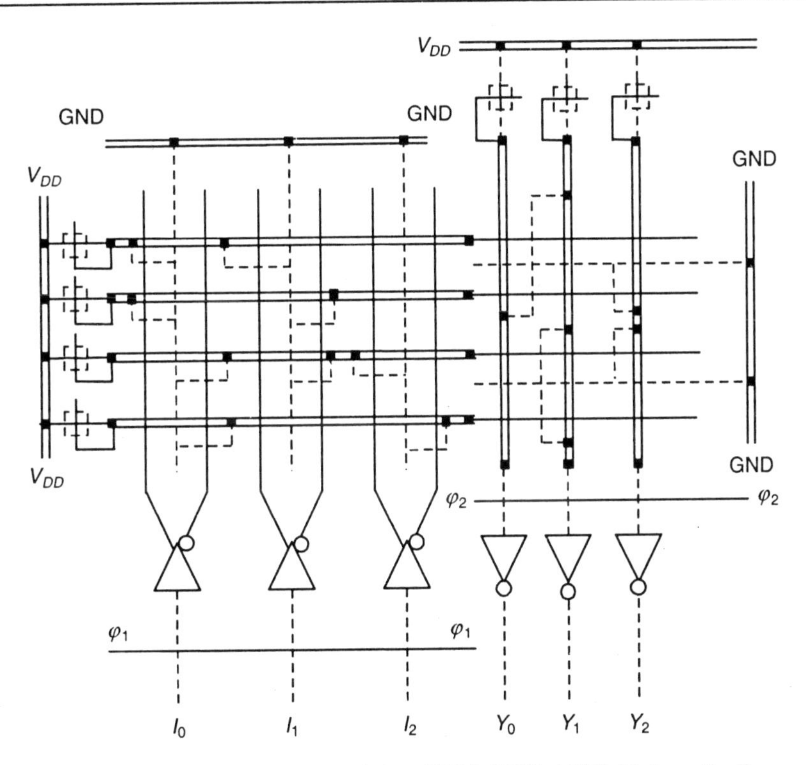

Figure 11.2 A stick drawing of an NMOS NOR–NOR PLA realization.

11.2.1.2 NAND–NAND realization of NMOS PLA

The procedure for laying out a NAND–NAND PLA is described below:

AND plane. For each logic 1 in the input columns of the personality matrix, place an ion implant under the appropriate product line where it intersects the noninverted input line in the PLA AND plane. The transistor thus created is always ON and the non-inverted input line has no control over that product line. Whenever all the controlling input lines in the AND plane are high, the product line will be low.

For each logic 0 in the input columns of the personality matrix, place an ion implant under the appropriate product line where it intersects the inverted-input line in the PLA AND plane. The transistor thus created is always ON and the inverted input line has no control over that product line. Whenever all the controlling input lines are high, the product line will be low. A don't care requires ion implants for both the true and complemented input signals.

OR plane. For each logic 0 in the output columns of the personality matrix, place an ion implant under the product line where it intersects the output line in the PLA OR plane. The transistor thus created is always ON and that product line has no control over the output line. Whenever the controlling product lines are high, the non-inverted output will be low.

A NAND–NAND realization of NMOS PLA for the above example in Eqns. (11.1) and (11.2) is shown in Figure 11.3. The pull-up device is a depletion-mode NMOS device.

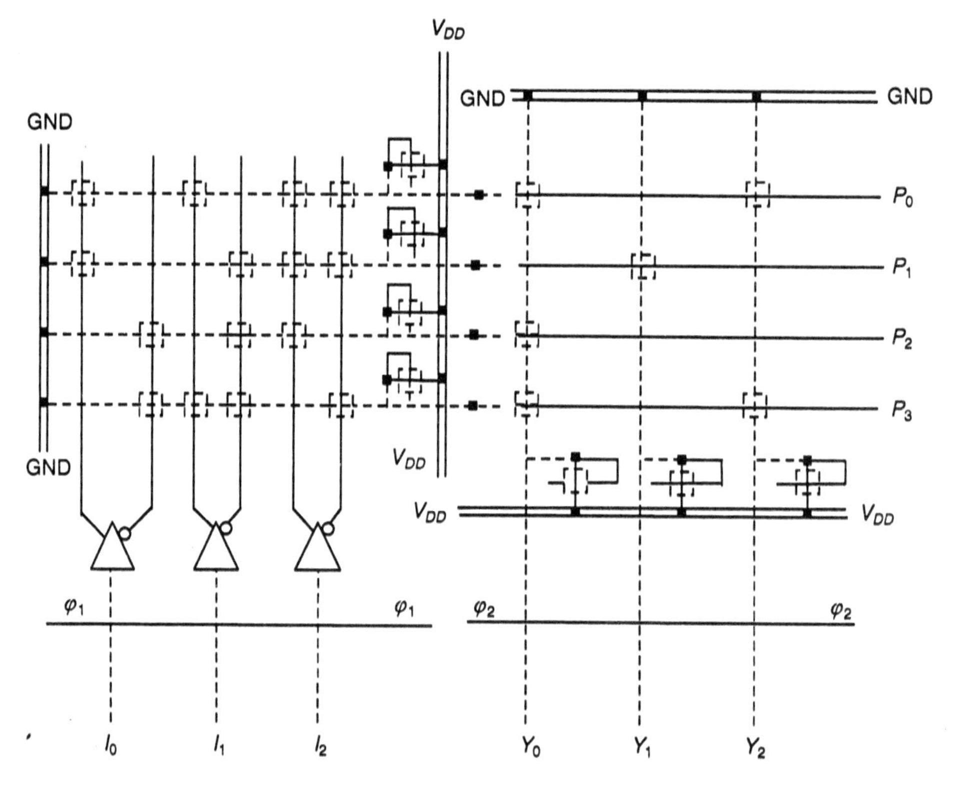

Figure 11.3 A stick drawing of an NMOS NAND–NAND PLA realization.

11.2.1.3 The precharged NMOS PLA

Depletion-mode pull-up devices are slow and precharging the output lines of both the AND plane and the OR plane avoids the slow depletion-mode pull-up. During phase I, each product line and each output line can be precharged to V_{DD}, while phase II isolates the ground and prevents the PLA from evaluating the logic. During phase II, the PLA will determine which input lines remain charged and which input lines will be discharged, while the product lines determine which output lines will be pulled low. Figures 11.4 and 11.5 show possible layouts of the NOR–NOR and NAND–NAND PLA structures given in Figures 11.2 and 11.3 respectively augmented with clock lines for precharging and evaluating the PLAs.

11.2.1.4 The CMOS PLA

The basic CMOS PLA is obtained by providing a well and replacing the pull-up devices in the NAND–NAND array or in the NOR–NOR array with enhancement mode PMOS devices. The CMOS array can be precharged or not, and can be clocked with the same two-phase clocking scheme as used for the MOS PLA. CMOS PLA design offers many more varieties of layout than does NMOS.

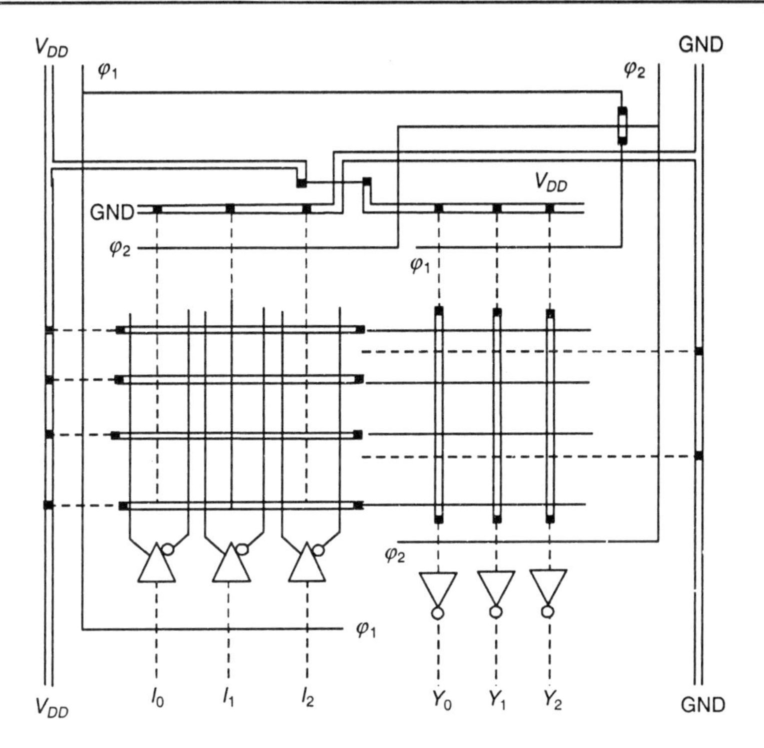

Figure 11.4 A precharged NMOS NOR–NOR PLA.

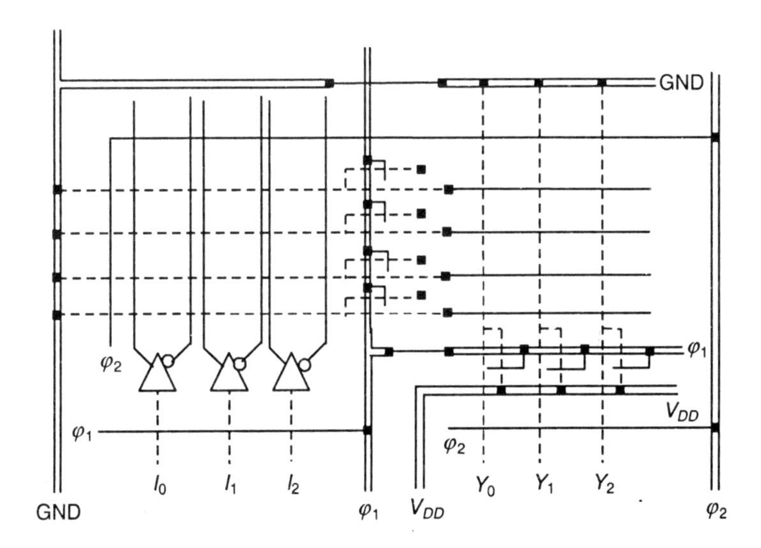

Figure 11.5 A precharged MOS NOR–NOR PLA.

11.3 OTHER PROGRAMMABLE LOGIC DEVICES

There are several close relative structures of the basic PLA. A digital, application specific IC family, widely used in VLSI design, is the programmable logic device. The AND–OR structure of the PLA is the core of all PLDs, since this structure can be used to implement any two-level Boolean function. Multilevel logic can be realized with Weinberger arrays or gate matrices.

11.3.1 Field Programmable Logic Array (FPLA)

The field-programmable logic array (FPLA) shown in Figure 11.6 has an address decoder (the AND array) and a data matrix (the OR array). In the FPLA, both the address decoder and the data matrix are programmable. This is shown by placing hollow diamonds at all the cross point sites. A cross point is the intersection of a row and a column of the PLA. When programming the PLA, the appropriate cross points can be filled into indicate connections.

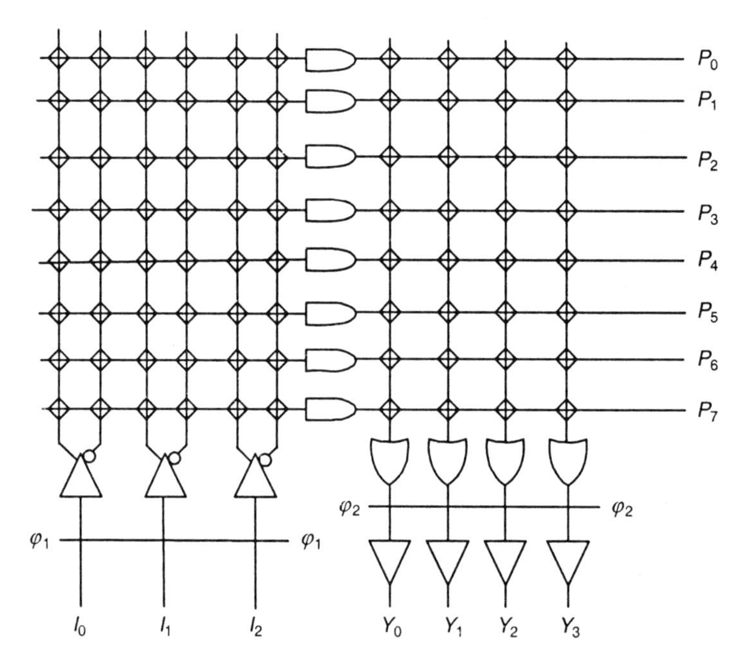

Figure 11.6 The clocked FPLA structure.

Each AND gate can have don't care inputs, which implies that multiple inputs can select the same word. Also, since multiple AND outputs can be ON simultaneously, multiple words in the OR array can be selected at the same time, thus allowing a function to be divided among multiple FPLAs.

There are two special cases of FPLA: the programmed read-only memory or PROM, the programmed array logic or PAL. In PROM, the OR matrix is programmable and the AND matrix is fixed. In PAL, the AND matrix is programmable and the OR matrix is fixed.

The PROM configuration is shown in Figure 11.7 with the AND plane cross point connections darkened. A PROM is useful for creating simple logic devices such as memory-adder decoders, but the fixed AND array limits its use in more complex applications where multiple addresses might be needed for the same word.

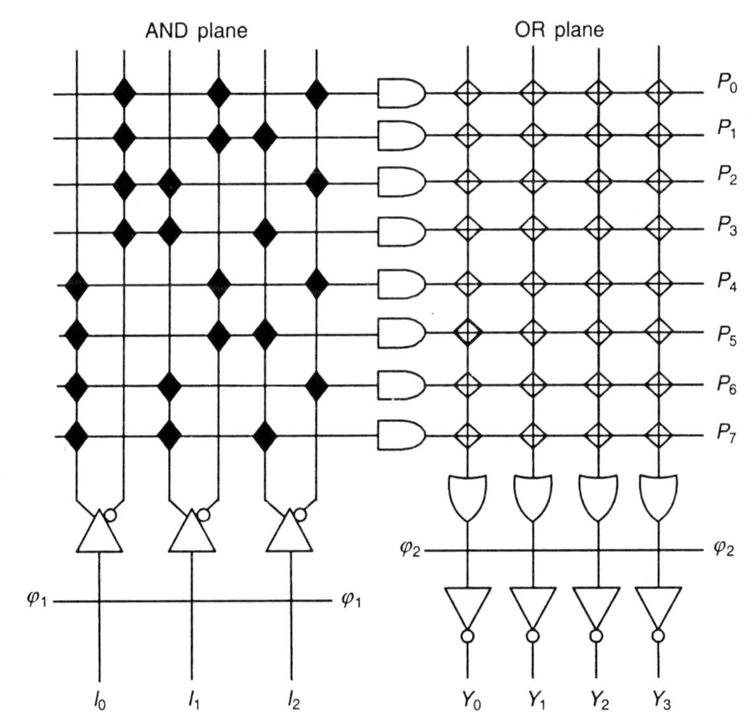

Figure 11.7 The clocked PROM structure.

11.3.2 Programmable Array Logic (PAL)

In the PAL structure, the OR plane has been preprogrammed. Since it has a programmable AND array, multiple addresses can select the same word, and multiple words in the array can be selected simultaneously (Figure 11.8).

PALs and FPLAs overcome one of the major inefficiencies of PROMs by allowing only as many inputs as necessary for a specific implementation. Some PALs do not have the fuses or the programming and testing circuitry required by the OR arrays of FPLAs, they are typically about 15% faster than FPLAs for the same power consumption. PALs used to be limited to control logic and I/O applications but recent improvements in performance make PALs suitable for data-path logic also. The PAL shown in Figure 11.8 has 2-input OR gates in the OR plane. Commercial devices typically have 8-input OR gates, and realize an 8-wide AND–OR structure.

11.3.3 Dynamic Logic Arrays (DLAs)

The dynamic logic array (DLA) resembles a PLA with the AND and OR planes merged such that the OR logic is performed in the AND plane. The DLA is ideally suited to implementations

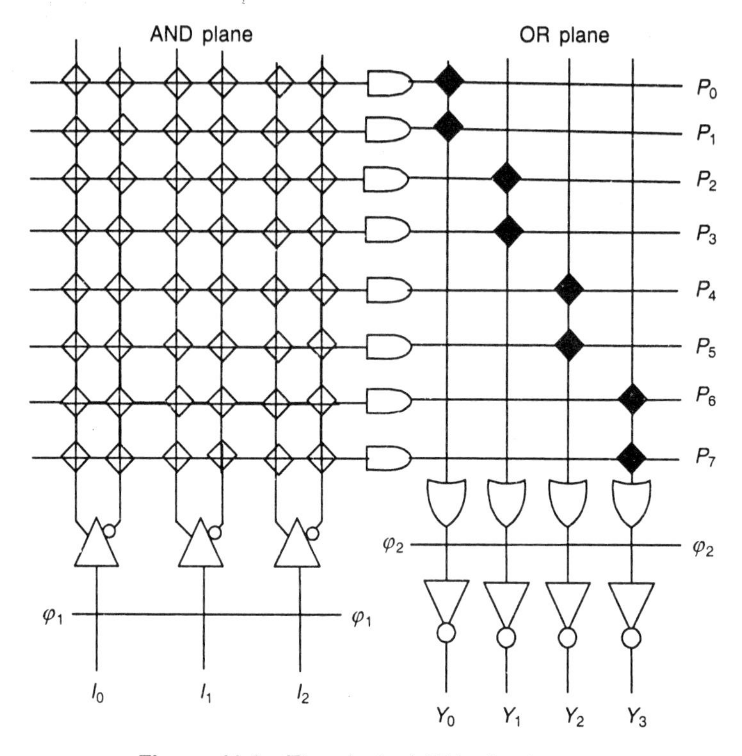

Figure 11.8 The clocked PAL structure.

with few cross point connections in the OR plane. It can realize a sum of products solution with less area than a PLA and can be clocked and precharged the same as a PLA structure. The DLA is less programmable than an equivalent PLA, but it combines the speed, simplicity and small size of dynamic logic with the programmability of a PLA.

An example of a NAND–DLA is shown in Figure 11.9. It realizes three output functions. The first output function consists of the OR of three product terms, whereas the second output function is the OR of three product terms, and the third output function consists of only one product term. The outputs are:

$$\begin{aligned} Y_0 &= \overline{I}_0\overline{I}_2 + I_1I_2 \\ Y_1 &= I_0 + I_1 + \overline{I}_2 \\ Y_2 &= I_0I_1I_2 \end{aligned} \tag{11.4}$$

The personality matrix of the DLA is:

$$Q = \begin{pmatrix} 1 & X & 1 & & 1 & 0 & 0 \\ X & 0 & 0 & & \cdot 1 & 0 & 0 \\ 0 & X & X & & 0 & 1 & 0 \\ X & 0 & X & & 0 & 1 & 0 \\ X & X & 1 & & 0 & 1 & 0 \\ 0 & 0 & 0 & & 0 & 0 & 1 \end{pmatrix} \tag{11.5}$$

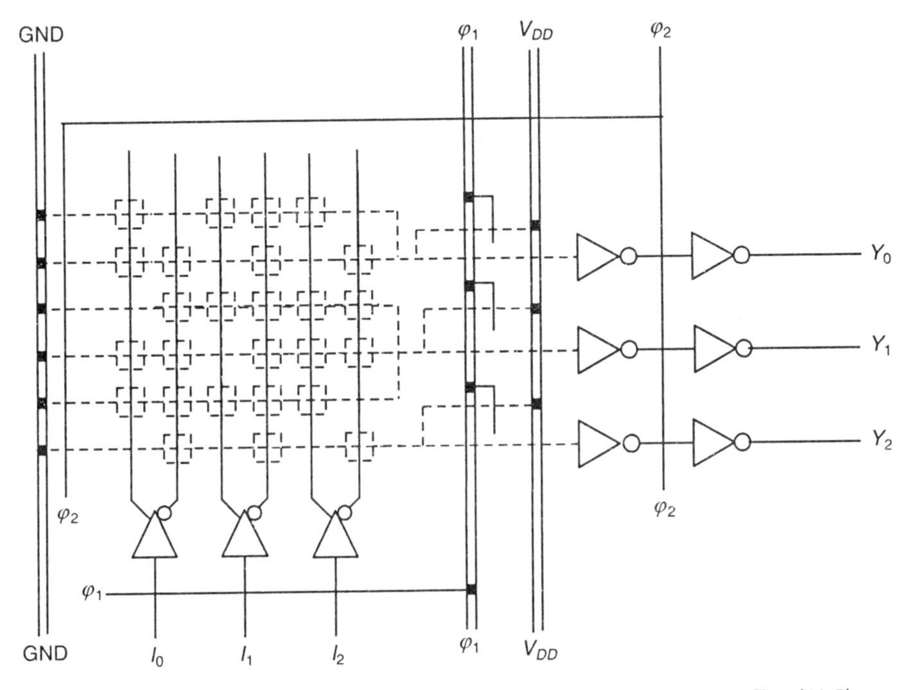

Figure 11.9 A clocked DLA realizations of the three functions in Eq. (11.5).

The four basic architectures of PLDs can be augmented by additional logic functions such as registers, latches, and feedback paths to achieve specific applications.

11.4 THE FINITE-STATE MACHINE AS A PLA STRUCTURE

When feedback is added to the AND OR PLA structure, the PLA becomes a finite state machine (FSM). If input and output buffers and two-phase clocking are added, the structure is as shown in Figure 11.10.

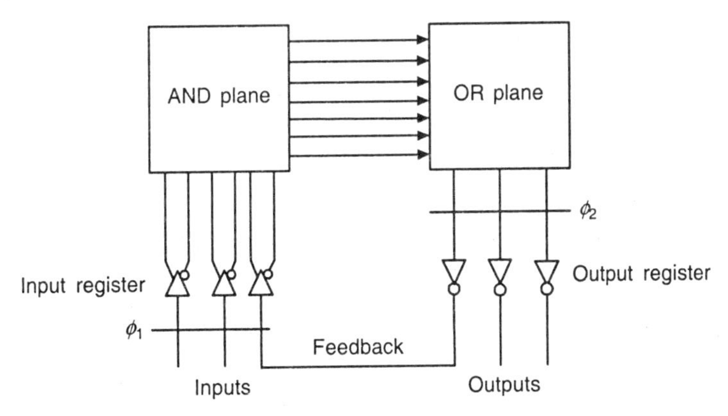

Figure 11.10 A clocked finite state machine (FSM).

Hierarchically the design procedure is as follows:

1. Plan the digital processing systems as combinations of register-to-register data transfer paths, controlled by FSMs.
2. Plan geometric shapes, relative sizes, and interconnection topologies of all subsystem modules so that all systems merge together snugly with a minimum of space and time wasted by random interconnecting wiring.
3. Dynamic storage registers are constructed using charge stored on input gates of inverting logic.
4. Combinational logic in the data-paths is implemented with steering logic composed of regular structures of pass transistors.
5. Most of the combinational logic in FSMs is implemented with PLAs.
6. All functioning is sequenced using a two-phase non-overlapping clock scheme.

An FSM can be designed as a Mealy machine or a Moore machine. The Mealy machine is shown in Figure 11.11. It has outputs, which may change with input changes in an asynchronous manner and cause erroneous behaviour. Hence, the Mealy machine should be avoided whenever possible.

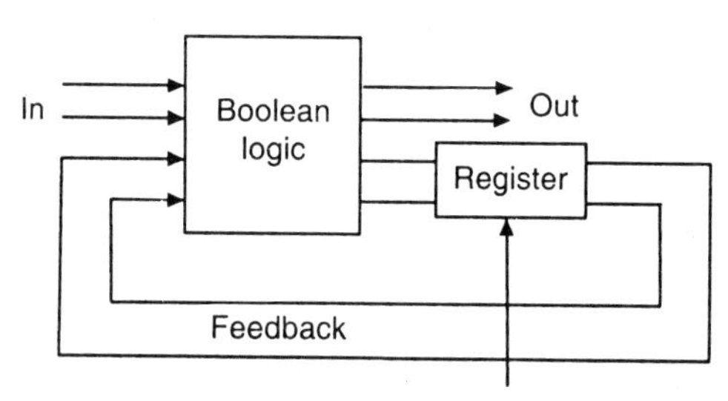

Figure 11.11 Mealy machine: some outputs are asynchronous.

The Moore machine, shown in Figure 11.12 has outputs which depend upon and change only with state changes, since all the outputs of the Boolean-logic block go through a state register, and are synchronously clocked.

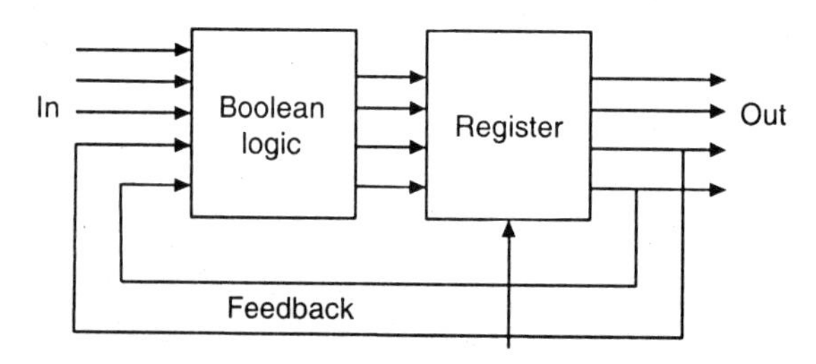

Figure 11.12 Moore machine: all outputs are clocked.

The importance of the PLA/FSM in VLSI is due to its:

(i) Regularity: It has a standard, easily expandable layout.
(ii) Convenience: Little design effort is required.
(iii) Compacted: It is efficient for small circuits.
(iv) Modularity: It makes it possible to design hierarchical PLAs and FSMs into large sequential systems.
(v) Suitability to being computer generated.

In designing PLAs, one should always be able to reset from any state: For N feedback loops, there are 2^N states. To avoid lockout, design for all $2N$ states.

Any large sequential system can be constructed from a series of FSMs as shown in Figure 11.13. If Moore machines are used, everything goes through only one combinational logic block between clock pulses, whereas Mealy machine outputs can ripple down-stream.

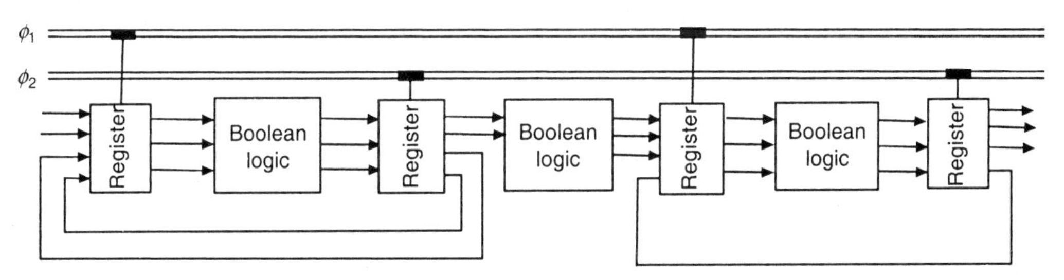

Figure 11.13 A sequential system, consisting of a series of combinational-logic modules with feedback.

11.5 COMPLEX PROGRAMMABLE LOGIC DEVICES (CPLDs)

PLAs and PALs are useful for implementing a wide variety of small digital circuits. Each device can be used to implement circuits they do not require more than the number of inputs, product terms, and outputs that are provided in the particular chip. These chips are limited to fairly modest sizes, typically supporting a combined number of inputs plus outputs of not more than 32. For implementation of circuits that require more inputs and outputs, either multiple PLAs or PALs can be employed or else a more sophisticated type of chip, called a complex programmable logic device (CPLD), can be used.

A CPLD comprises multiple circuit blocks on a single chip, with internal wiring resources to connect the circuit blocks. Each circuit block is similar to a PLA or a PAL. An example of a CPLD is shown in Figure 11.14. It includes four PAL like blocks that are connected to a set of interconnection wires. Each PAL like block is also connected to a subcircuit labelled I/O block, which is attached to a number of the chip's input and output pins.

Figure 11.15 shows an example of the wiring structure and the connections to a PAL-like block in a CPLD. The PAL-like block includes 3 macrocells (real CPLDs typically have about 16 macrocells in a PAL-like block), each consisting of a four-input OR gate (real CPLDs usually provide between 5 and 20 inputs to each OR gate). The OR gate output is connected to another type of logic gate which is not yet introduced. It is called an Exclusive-OR (XOR) gate. The behaviour of an XOR gate is the same as for an OR gate except that if both of its inputs are 1, the XOR gate produces a 0.

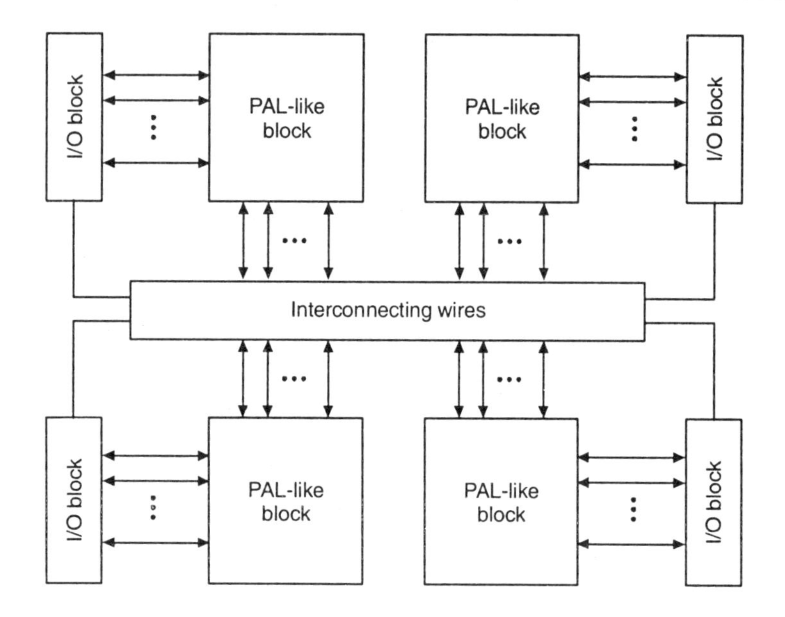

Figure 11.14 Complex Programmable Logic Device (CPLD) structure.

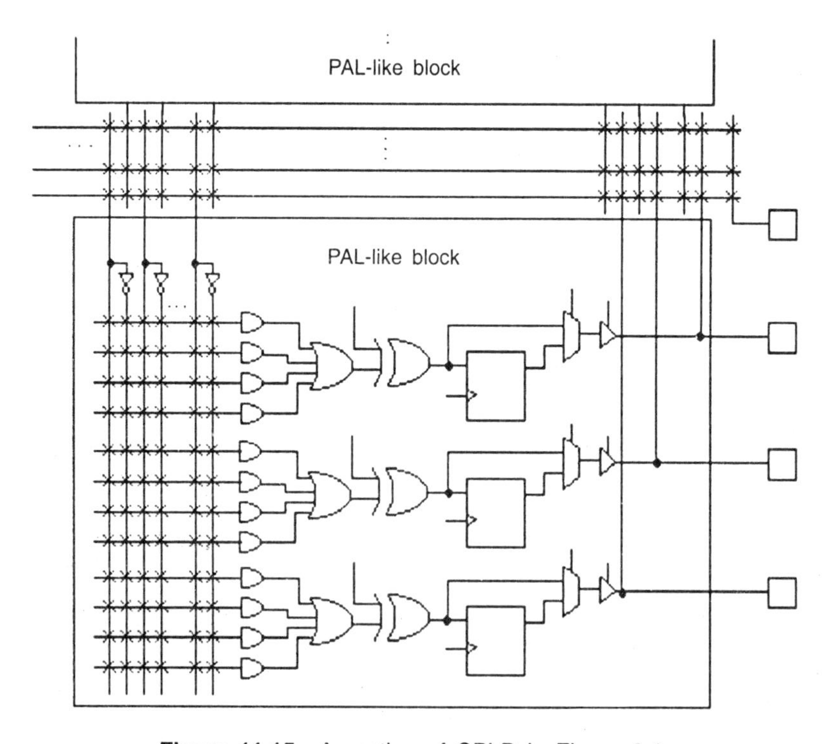

Figure 11.15 A section of CPLD in Figure 9.1.

In Figure 11.15, one input to the XOR gate can be programmably connected to 1 or 0; if 1, then the XOR gate complements the OR gate output, and if 0, then the XOR gate has no effect. The macrocell also includes a flip-flop, a multiplexer, and a tri-state buffer. The flip-flop is used to store the output value produced by the OR gate. Each tri-state buffer allows each pin to be used either as an output from the CPLD or as an input. To use a pin as an output the corresponding tri-state buffer is enabled, acting as a switch that is turned ON. If the pin is to be used as an input, then the tri-state buffer is disabled, acting as a switch that is turned OFF. In this case, an external source can drive a signal onto the pin, which can be connected to other macrocells using the interconnection wiring.

The interconnection wiring contains programmable switches that are used to connect the PAL-like blocks. Each of the horizontal wires can be connected to some of the vertical wires that it crosses, but not all of them. The number of switches is chosen to provide sufficient flexibility for typical circuits without wasting many switches in practice. Note that when a pin is used as input, the macrocell associated with that pin cannot be used and is therefore wasted. Some CPLDs include additional connections between the macrocells and the interconnection wiring that avoids wasting macrocells in such situation.

11.5.1 CPLD Packaging and Programming

Commercial CPLD's range in size from only 2 PAL like-blocks to more than 100 PAL-like blocks. They are available in a variety of packages, including the PLCC (plastic-leaded chip carrier) package that is shown in Figure 11.16. The PLCC package has pins that 'wrap around' the edges of the chip on all four of its sides. The socket that houses the PLCC is attached by solder to the circuit board, and the PLCC is held in the socket by friction.

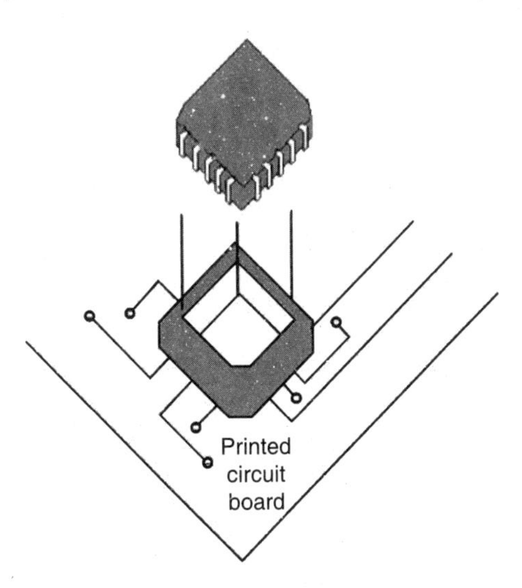

Figure 11.16 A PLCC package with socket.

Another type of package used to house CPLD chips is called a quad flat pack (QFP). The QFP package has pins on all four sides, and they extend outward from the package, with a downward-wiring shape. The QFP's pins are much thinner than those on a PLCC, which means that the package can support a larger number of pins; QFPs are available with more than 200 pins (Figure 11.17).

Figure 11.17 CPLD in quad flat pack (QFP) package.

CPLD devices usually support ISP (In-System Programming technique). A small connector is included on the PCB that houses the CPLD, and a cable is connected between that connector and a computer system. The CPLD is programmed by transferring the programming information generated by a CAD system through the cable from the computer into the CPLD. The circuitry on the CPLD that allows this type of programming has been standardized by the IEEE and is usually called a JTAG port. It uses four wires to transfer information between the computer and the device being programmed. The term JTAG stands for Joint Test Action Group. Figure 11.18 illustrates the use of JTAG port for programming two CPLDs on a circuitboard. The CPLDs are connected together so that both can be programmed using the same connection to the computer system. Once a CPLD is programmed, it retains the programmed state permanently, even when the power supply for the chip is turned OFF. This property is called nonvolatile programming.

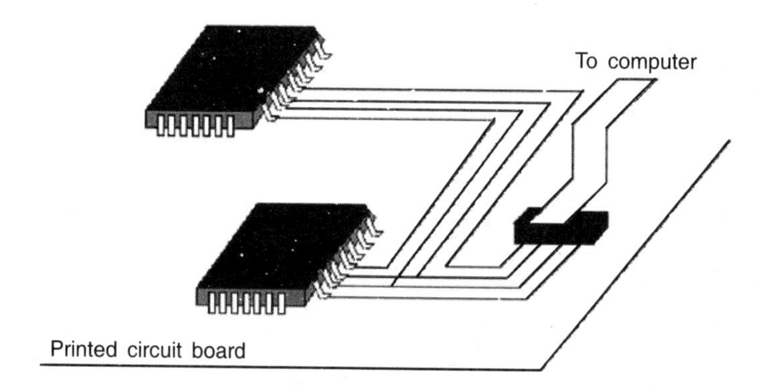

Figure 11.18 JTAG programming.

CPLDs are used for the implementation of many types of digital circuits. In industrial designs that employ some type of PLD device, CPLDs are used in about half the cases. The names of several manufacturers of Complex PLD's (CPLDs), products they offer, and the corresponding www locators are listed in Table 11.1. An example of a widely used CPLD family, the Altera MAX 7000 is described in the next section.

Table 11.1 Commercial CPLD Products

Manufacturer	*CPLD products*	*www locator*
Altera	Max 5000, 7000 and 9000	www.altera..com
Atmel	ATF, ATV	www.atmel.com
Cypress	FLASH370, Ultra 37000	www.cypress.com
Lattice	isp LSI 1000 to 8000	www.latticesemi.com
Philips	XPLA	www.philips.com
Vantis	MACH 1 to 5	www.vantis.com
Xilinx	XC 9500	www.xilinx.com

ALTERA MAX 7000. The MAX 7000 CPLD family includes chips that range in size from the 7032, which has 32 macrocells, to the 7512, which has 512 macrocells. There are two main variants of these chips, identified by the suffix, S. If this letter is present in the chip name, as in 7128S, then the chip is in-system programmable. But if the suffix is absent, as in 7128, then the chip has to be programmed in a programming unit.

The overall structure of MAX 7000 chip is illustrated in Figure 11.19. There are four dedicated output pins, two of these can be used as global clock inputs, and one can be used as a global reset for all flip-flops. Each shaded box is called a logic array block (LAB), which contains 16 macrocells. Each LAB is connected to an I/O control block, which contains tri-state buffers that are connected to pins on the chip package; each of these pins can be used as an input

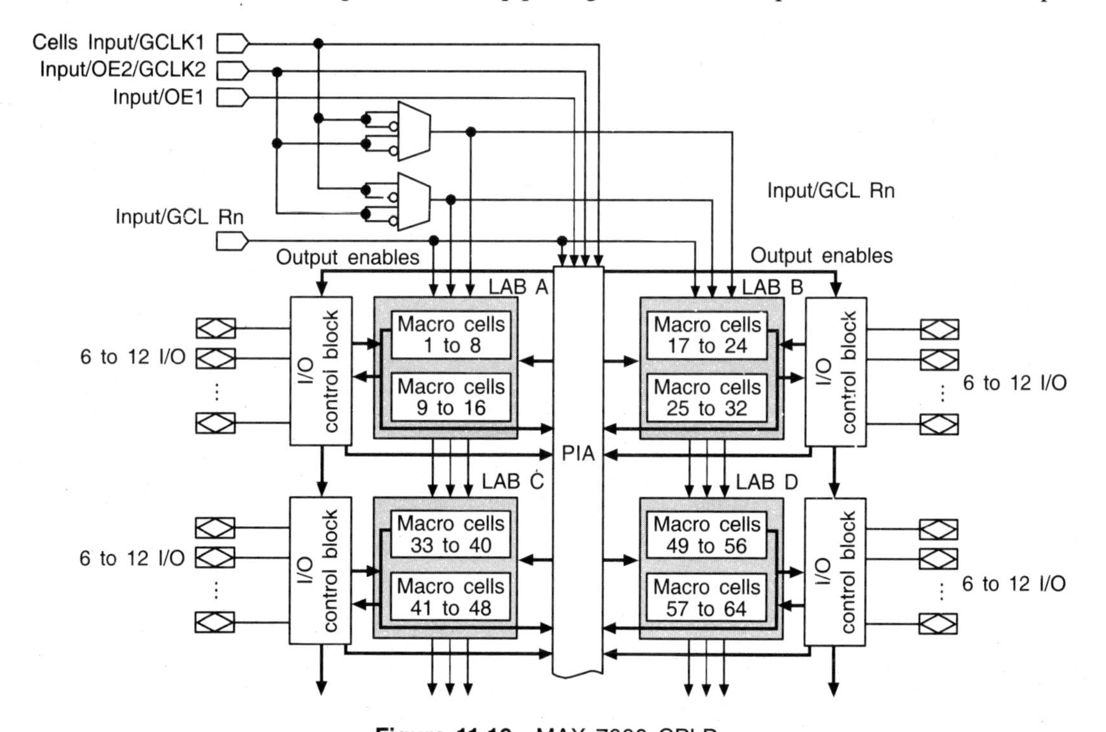

Figure 11.19 MAX 7000 CPLD.

or output pin. Each LAB is also connected to the programmable interconnect array (PIA). The PIA consists of a set of wires that spans that entire device. All interconnections between macrocells are made using the PIA.

Figure 11.20 shows the structure of a MAX 7000 macrocell. There are five product terms that can be connected through the product term select matrix to an OR gate. This OR gate can be configured to use only the product terms needed for the logic function being implemented in the macrocell. If more than five product terms are required, additional product terms can be 'shared' from other macrocells. The OR gate is connected through an XOR gate to a flip-flop, which can be bypassed.

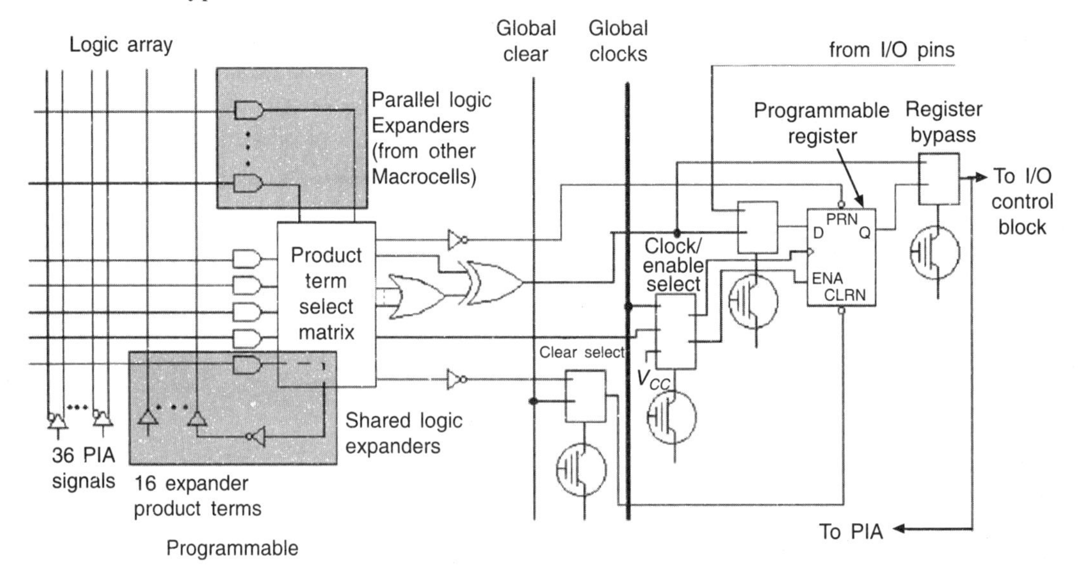

Figure 11.20 MAX 7000 macrocell.

Figure 11.21 shows how product terms can be shared between macrocells. The OR gate in a macrocell includes an extra input that can be connected to the output of the OR gate in the macrocell above it. This feature is called parallel expanders and is used for logic functions with up to 20 product terms. If even more product terms are needed, then a feature, called shared expanders is used.

Each specific MAX 7000 device is available in a range of speed grades. These grades specify the propagation delay from an input pin through the PLA and a macrocell to an output pin. For example, the chip named 7128S-7 has a propagation delay of 7.5 ns. If the logic function implemented uses parallel or shared expanders, the propagation delay is increased.

11.6 FIELD PROGRAMMABLE GATE ARRAYS (FPGAs)

A field programmable gate array (FPGA) is a programmable logic device that supports implementation of relatively large logic circuits. FPGAs can be used to implement a logic circuit with more than 20,000 gates whereas a CPLD can implement circuits of up to about 20,000 equivalent gates. FPGAs are quite different from CPLDs because FPGAs do not contain AND or OR planes. Instead, they provide logic blocks for implementation of the required functions.

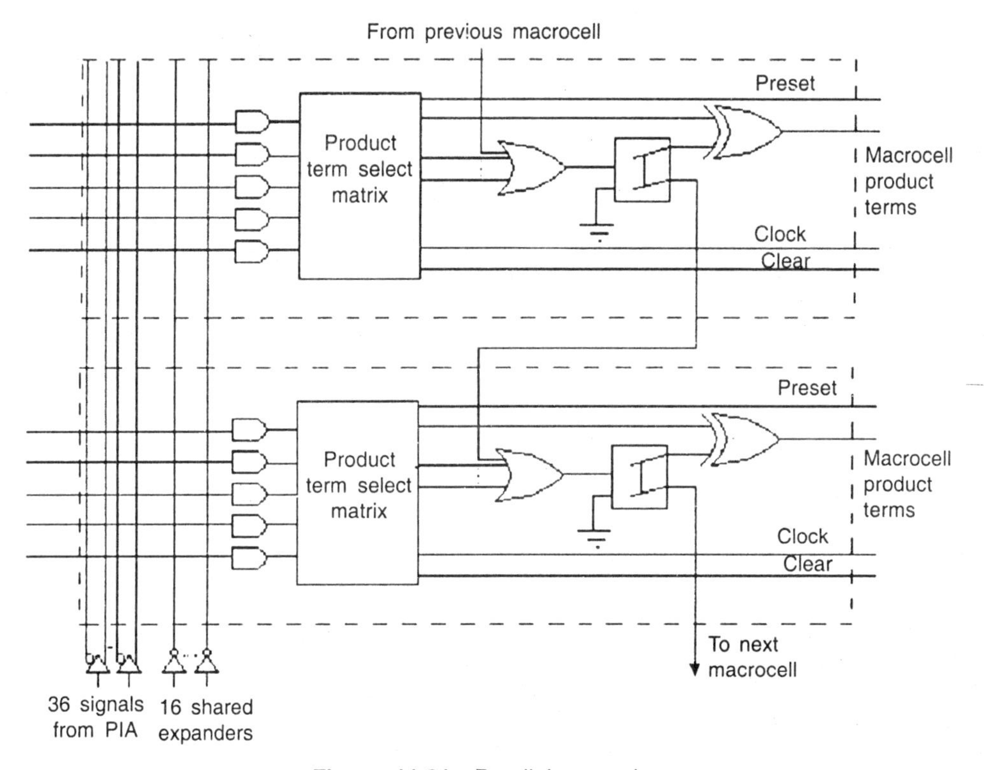

Figure 11.21 Parallel expanders.

The general structure of an FPGA is illustrated in Figure 11.22. It contains three main types of resources: logic blocks, I/O blocks for connecting to the pins of the package, and interconnection wires and switches. The logic blocks are arranged in a two-dimensional array, and the interconnection wires are organized as horizontal and vertical routing channels between rows and columns of logic blocks. The routing channels contain wires and programmable switches that allow the logic blocks to be interconnected in many ways. Figure 11.22 shows two locations for programmable switches, the blue boxes adjacent to logic block, hold switches that connect the logic block input and output terminals to the interconnection wires, and the blue boxes that are diagonally between logic blocks connect one interconnection wire to another. Programmable connections also exist between the I/O blocks and the interconnection wires. The actual number of programmable switches and wires in an FPGA varies in commercially available chips.

11.6.1 FPGA Packaging and Programming

FPGAs are available in a variety of packages, including PLCC and QFP packages. Figure 11.23 depicts another type of package, called a pin grid array (PGA). A PGA package may have up to a few hundred pins in total, which extend straight outward from the bottom of the package, in a grid pattern. Yet another packaging technology emerged known as the ball grid array

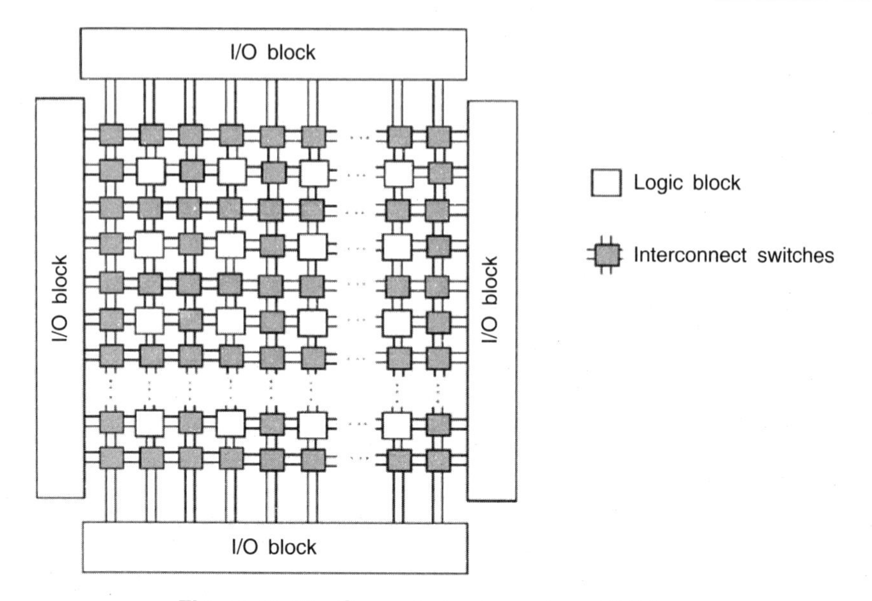

Figure 11.22 General structure of an FPGA.

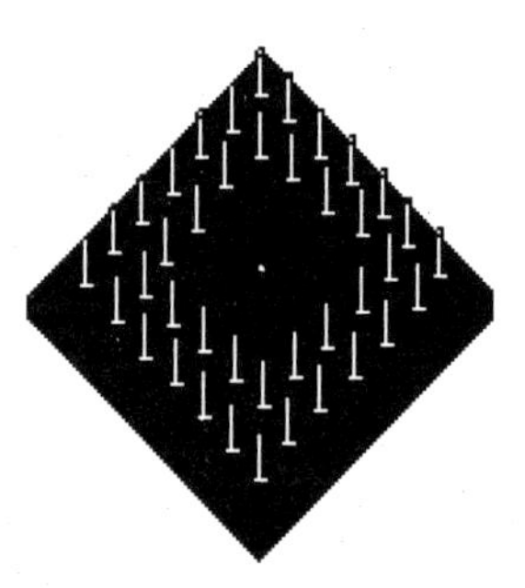

Figure 11.23 Pin grid array (PGA) package of FPGA.

(BGA), which is similar to the PGA except that the pins are small round balls, instead of posts. The advantage of BGA packages is that the pins are very small, hence more pins can be provided on the package.

Each logic block in an FPGA typically has a small number of inputs and one output. The most commonly used logic block is a lookup table (LUT), which contains storage cells that are used to implement a small logic function. Each cell is capable of holding a single logic value, either 0 or 1. The stored value is produced as the output of the storage cell. LUTs of various sizes may be created, where the size is defined by the number of inputs. Figure 11.24(a) shows the structure of a small LUT. It has two inputs x_1 and x_2, and one output, f. It is capable of implementing any logic function of two variables. This LUT has four storage cells as a two-variable truth table. The input variables x_1 and x_2 are used as the select inputs of three multiplexers, which, depending on the values of x_1 and x_2 select the content of one of the four storage cells as the output of the LUT.

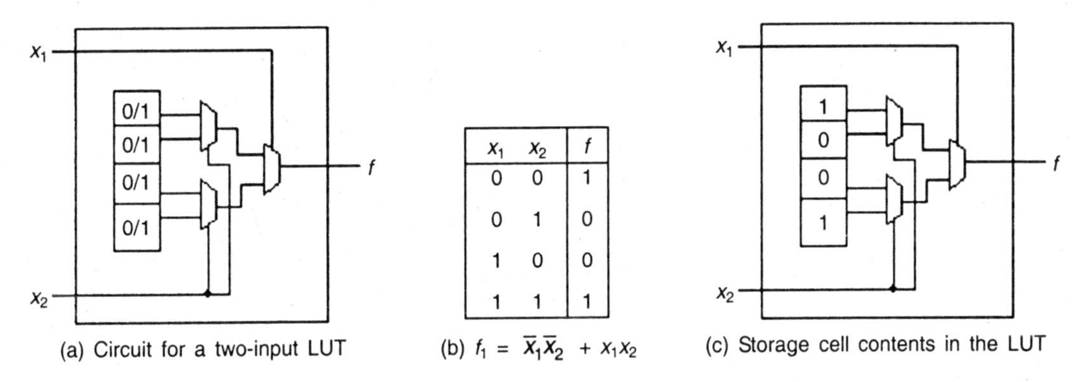

x_1	x_2	f
0	0	1
0	1	0
1	0	0
1	1	1

(a) Circuit for a two-input LUT (b) $f_1 = \bar{x}_1\bar{x}_2 + x_1x_2$ (c) Storage cell contents in the LUT

Figure 11.24 A two-input lookup table (LUT).

To see how a logic function can be realized in the two-input LUT, consider the true table in Figure 11.24(b). The function f_1 from this table can be stored in the LUT as illustrated in Figure 11.24(c). The arrangement of multiplexers in the LUT correctly realizes the function f_1. When $x_1x_2 = 0$, the output of the LUT is driven by the top storage cell, which represents the entry in the truth table for $x_1x_2 = 00$. Similarly, for all values of x_1 and x_2, the logic value stored in the storage cell corresponding to the entry in the truth table chosen by the particular valuation appears on the LUT output. Providing access to the contents of storage cells is only one way in which multiplexers can be used to implement logic functions.

Figure 11.25 shows a three-input LUT. It has eight storage cells because a three-variable truth table has eight rows. In commercial FPGA chips, LUTs usually have either four or five inputs, which require 16 and 32 storage cells, respectively.

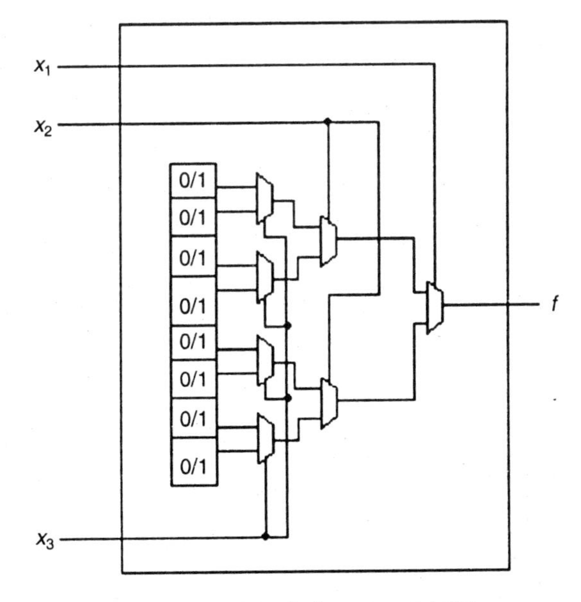

Figure 11.25 A three-input LUT.

For a logic circuit to be realized in an FPGA, each logic function in the circuit, must be small enough to fit within a single logic block. In practice, a user's circuit is automatically translated into the required form by using CAD tools. When a circuit is implemented in an FPGA, the logic blocks are programmed to realize the necessary functions and the routing channels are programmed to make the required interconnection between logic blocks. The FPGAs are configured by using the ISP (in-system programming) method. The storage cells in the LUTs in an FPGA are volatile, which means that they lose their stored elements whenever the power supply for the chip is turned OFF. Hence, the FPGA has to be programmed every time power is applied. Often a small memory chip that holds its data permanently, called a programmable read-only memory (PROM), is included on the circuitboard that houses the FPGA. The storage cells in the FPGA are loaded automatically from the PROM when power is applied to the chips.

A small FPGA that has been programmed to implement a circuit is depicted in Figure 11.26. The FPGA has two-input LUTs, and there are four wires in each routing channel. The programmed states of both the logic blocks and wiring switches in a section of FPGA are shown. Programmable wiring switches are indicated by an X. Each switch shown in blue is turned ON and makes a connection between a horizontal and a vertical wire. The switches shown in black are turned OFF. The truth tables programmed into the logic blocks in the top view of the FPGA correspond to the function $f_1 = x_1x_2$ and $f_2 = \overline{x}_2x_3$. The logic block in the bottom right is programmed to produce $f = f_1 + f_2 = x_1x_2 + \overline{x}_2x_3$.

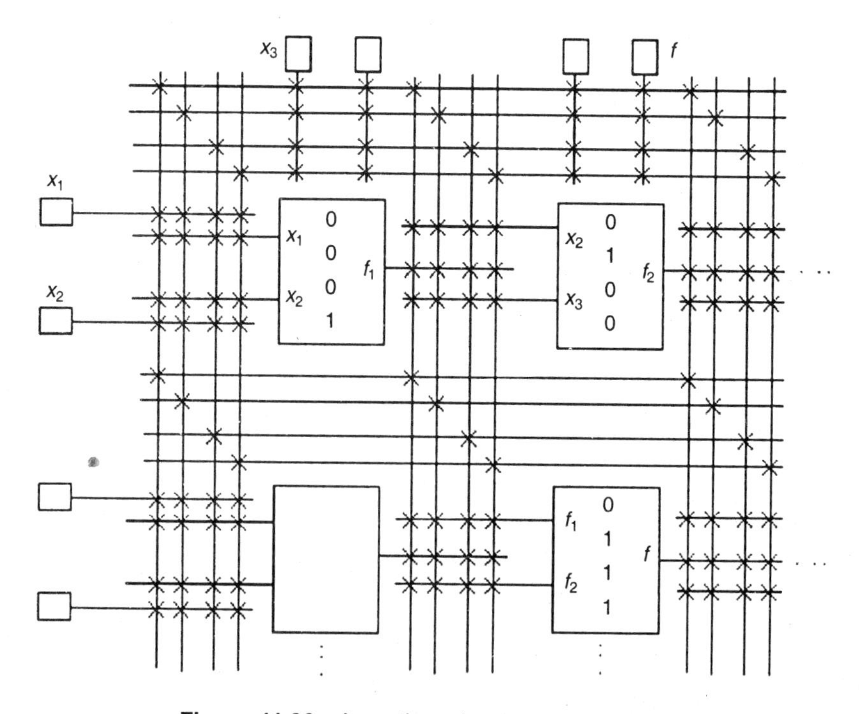

Figure 11.26 A section of programmed FPGA.

Table 11.2 lists the names of FPGA manufacturers, their products, and their www locators. Two examples of FPGAs, the Altora FLEX 10K and the XILINX XC 4000 are described here.

Table 11.2 Commercial FPGA products

Manufacturer	*FPGA products*	*www locator*
Actel	Act 1,2 and 3,MX,SX	www.actel.com
Altera	FLEX6000, 8000 and 10k APEX 20k	www.altera.com
Atmel	AT6000, AT40k	www.atmel.com
Lucent	ORCA 1, 2 and 3	www.lucent.com
QuickLogic	pASIC 1, 2 and 3	www.quicklogic.com
Vantis	VFI	www.vantis.com
Xilinx	XC3000, XC4000, XC5200, Virtex	www.xilinx.com

ALTERA FLEX 10K. Figure 11.27 shows the structure of the FLEX 10K chip. It contains a collection of logic array blocks (LABs), where each LAB comprises eight logic elements based on lookup tables (LUTs). In addition to LABs, the chip also contains embedded array blocks (EABs), which are SRAM blocks that can be configured to provide memory blocks of various

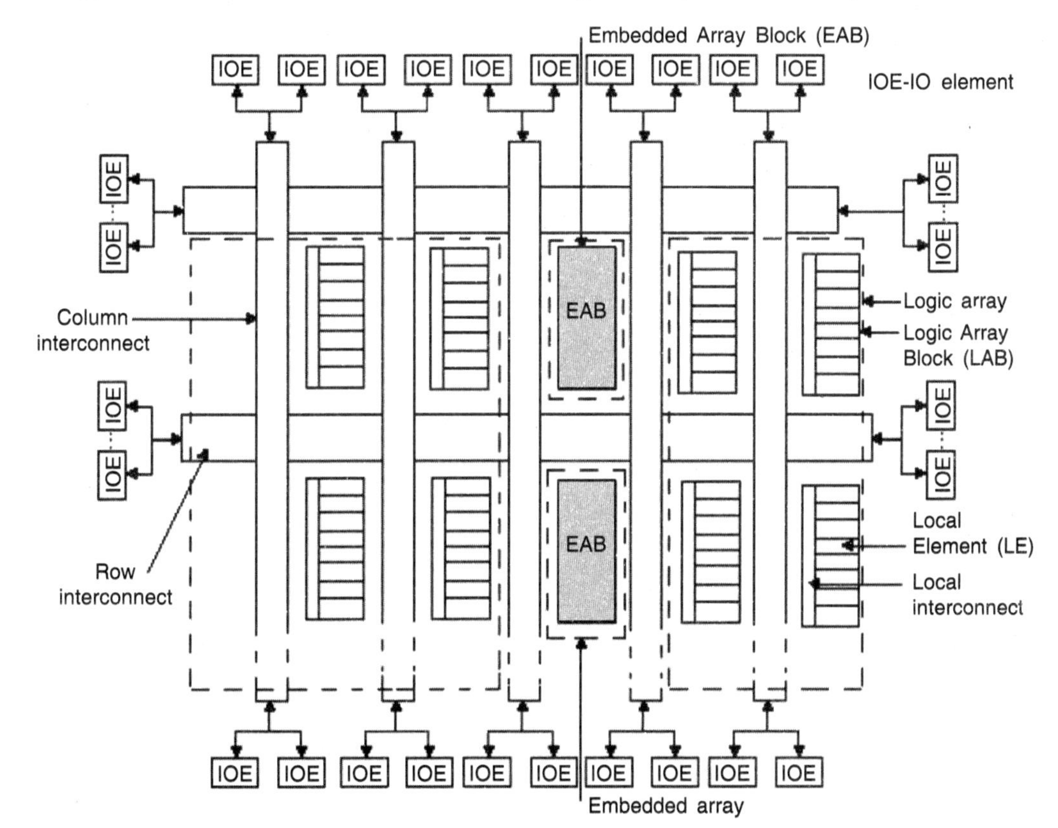

Figure 11.27 FLEX 10K FPGA.

aspect ratios. The LABs and EABs can be interconnected using the row and column interconnect wires. These wires also provide connections to the input and output pins on the chip package.

Figure 11.28 shows the contents of a LAB. It has a number of inputs that are provided from the adjacent row interconnect wires to a set of local interconnect wires inside the LAB. These local wires are used to make connections to the inputs of the logic elements, and the logic element outputs also feedback to the local wires. Local element outputs also connect to the adjacent row and column wires. The structure of a logic element is depicted in Figure 11.29. The element has a four-input LUT and a flip-flop that can be bypassed. For implementation of arithmetic adders, the four-input LUT can be used to implement 2 three-input functions, namely, the sum and carry functions in a full-adder.

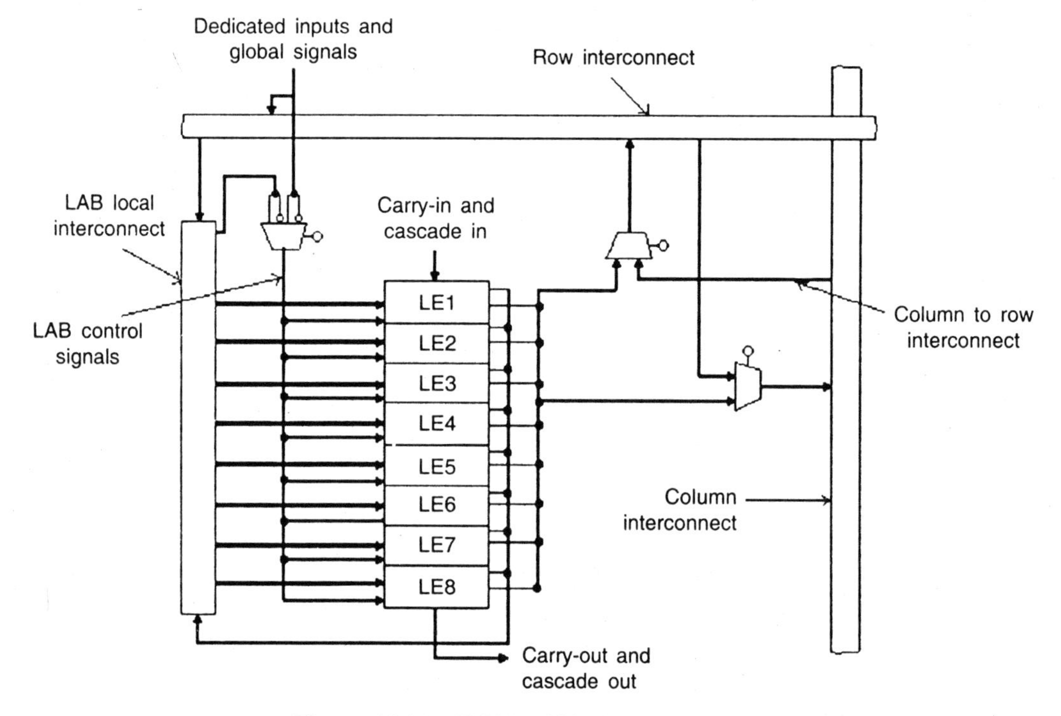

Figure 11.28 FLEX 10K logic array block.

The structure of an EAB is depicted in Figure 11.30. It contains 2048 SRAM cells, which can be used to provide memory blocks that have a range of aspect ratios: 256 × 8, 512 × 4, 1024 × 2, and 2048 × 1 bits. The address and data inputs to the memory block are provided from a set of local interconnect wires. These inputs, as well as a write enable for the memory block, can optionally be stored in flip-flops.

The number of address and data inputs connected to the memory block varies depending on the aspect ratio being used. The data outputs can also optionally be stored in flip-flops. For large memory blocks, it is possible to combine multiple EABs.

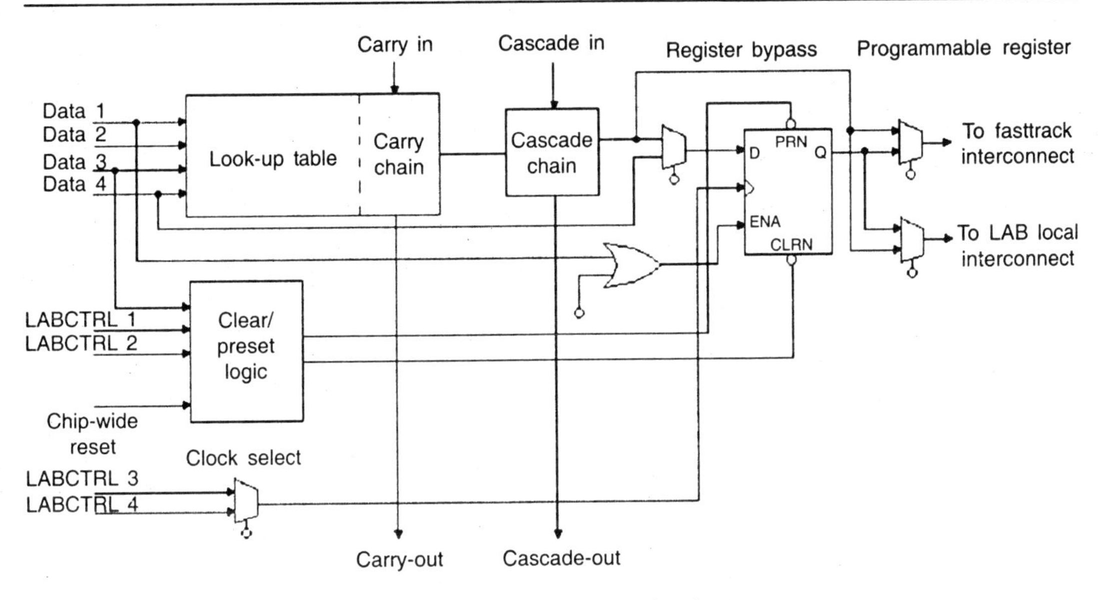

Figure 11.29 FLEX 10K logic element.

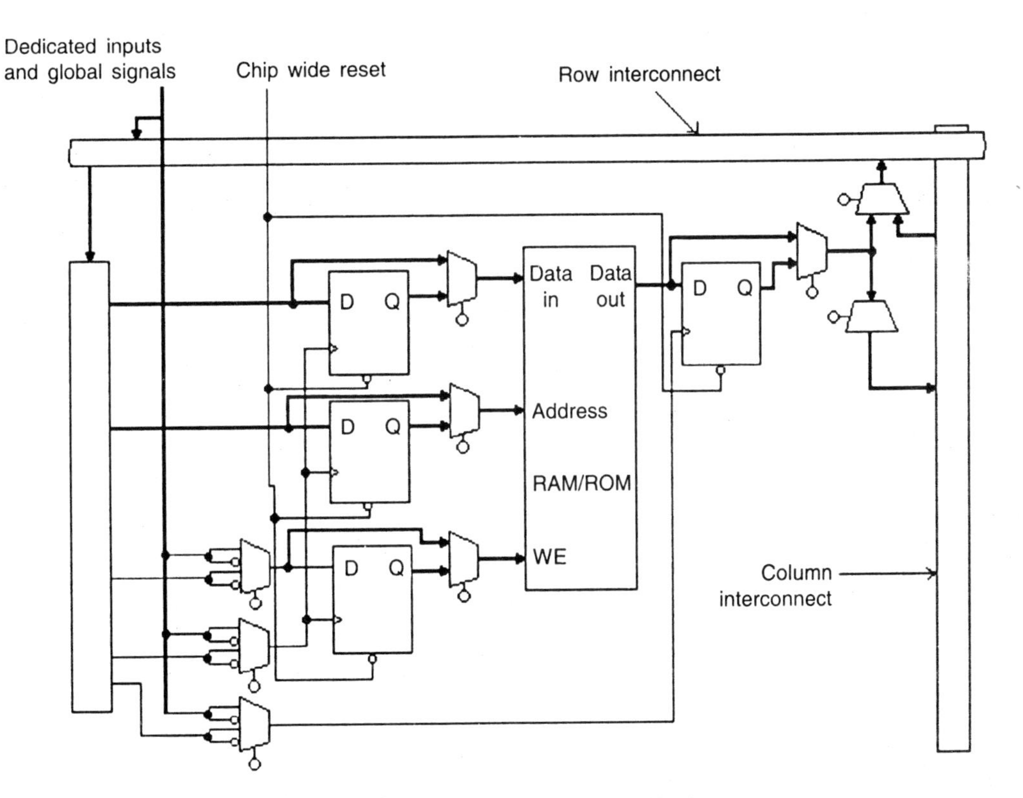

Figure 11.30 Embedded array block.

FLEX 10K chips are available in sizes ranging from the 10K10 to 10K250, which offer about 10,000 and 250,000 equivalent logic gates, respectively. Specific chips are available in various speeds, indicated using a suffix letter, such as A, as in 10K10A, and a speed grade, as in 10K10A-1. Unlike PALs and CPLDs, the speed grade for an FPGA does not specify an actual propagation delay in nanoseconds. Instead, it represents a relative speed within the device family. For instance, the 10K10-1 is a faster chip than the 10K10-2. The actual propagation delays can be examined using a timing simulator CAD tool.

11.6.2 The XILINX Programmable Gate Array

An example of an ad hoc array is a set of products from the XILINX company. The XC 3000 series and XC 4000 series are depicted in this section.

XC 3000. The architecture of the XC 3000 series is shown in Figure 11.31. An array of Configurable Logic Blocks (CLBs) is embedded within a set of horizontal and vertical channels that contain routing that can be personalized to interconnect CLBs. The configuration of the interconnect is achieved by turning ON n-channel pass transistors. The state that determines a given interconnect pattern is held in static RAM cells distributed across the chip close to the controlled elements. The CLBs and routing channels are surrounded by a set of programmable I/Os.

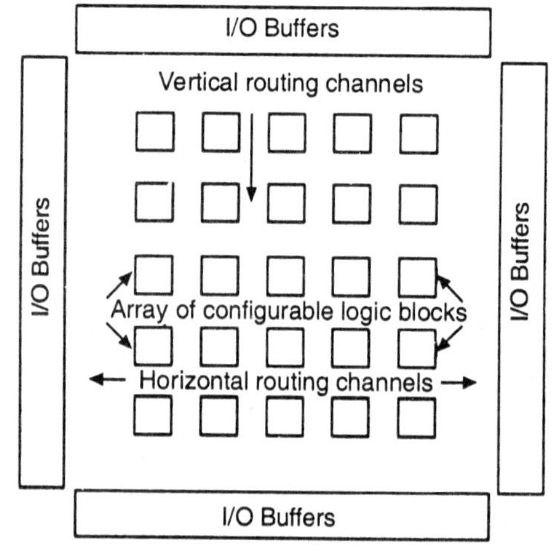

Figure 11.31 XILINX FPGA architecture (XC 3000 series).

The detailed structure of a CLB is shown in Figure 11.32. It consists of two registers, a number of multiplexers, and a combinational function unit. The latter can generate two functions of four variables, any function of five variables, or a selection between two functions of four variables. The function bit and each multiplexer is controlled by a number of RAM state bits. More recent CLBs feature enhanced table lookup function generators which can be used to build logic functions or used as register storage.

Each input and output on a CLB has a particular local interconnect pattern (called direct interconnect by XILINX), which allows most local interconnection between adjacent CLBs to take place. At the junction of the horizontal and vertical routing channels (where the general purpose interconnect runs), programmable switching matrices are employed to redirect routes.

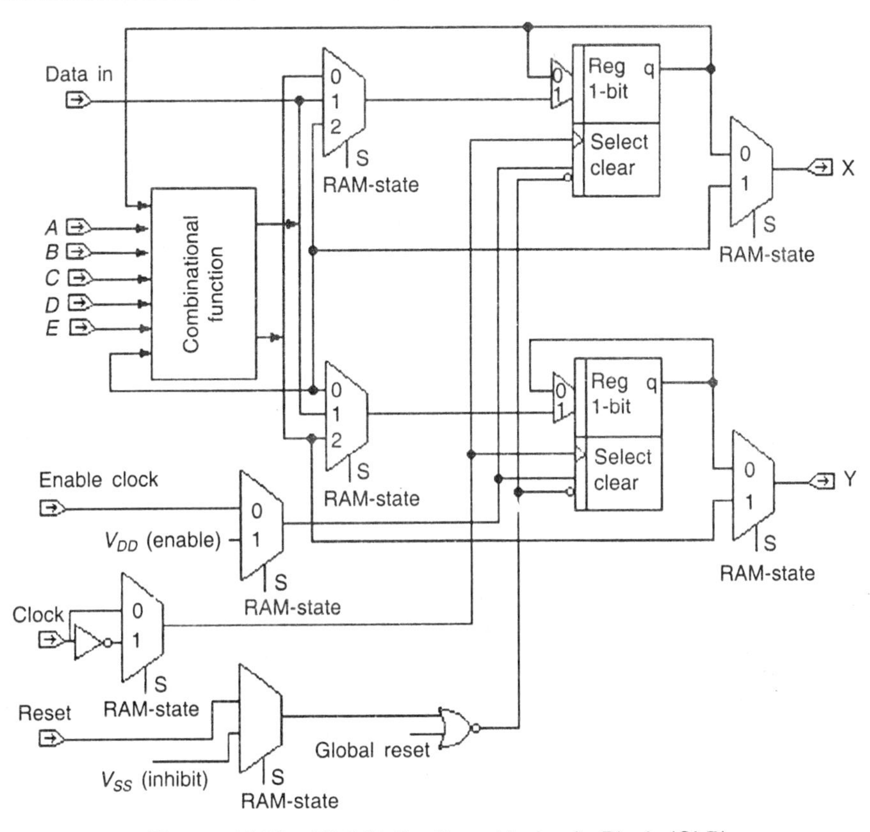

Figure 11.32 XILINX Configurable Logic Block (CLB).

Figure 11.33 shows a typical CLB surrounded by switching matrices. The switching matrices perform cross-bar switching of the global interconnect which runs both vertically and horizontally. Programmable Interconnect Points or PIPs interconnect the global routing to CLBs. Both PIPs and the switching matrices are implemented as *n*-channel pass gates controlled by 1-bit RAM cells. Extra special long-distance interconnect is used to route important timing signals with low skew.

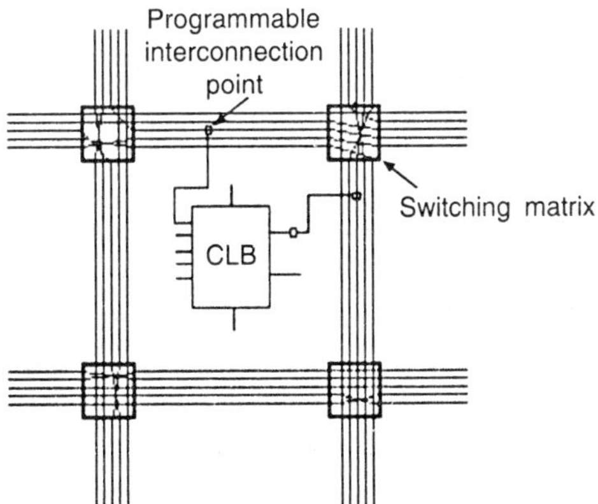

Figure 11.33 XILINX crossbar connect and CLB local connect example.

Assuming one has a board design finished, design proceeds by mapping the logic design to the CLBs and thence to one or more programmable gate arrays. Software then 'places and routes' the CLBs by loading the internal state RAM with the code needed to program the I/Os, the CLBs, and the routing. The design is then ready to be tested or used.

Currently, the largest array holds approximately 500 CLBs and has approximately 100 Kb of state RAM. In common with the Actel approach, timing is dependent on the basic CLB speed and a routing delay term. Users seem to be able to achieve system-clock rates that are 30–50% of the speed grade. Thus with 250 MHz parts an 80 MHz clock frequency is feasible.

While the XILINX arrays are standalone programmable gate arrays, the ideas may be of use to the IC-system designer who wishes to embed some reprogrammable logic within a larger system. In addition, the IC designer may find that prototyping a design in such an array might aid in system debug of a chip function. A significant advantage of the reprogrammable gate array is the ability to redesign the internals of a chip by changing software. This can be of considerable advantage in a product that has to undergo field updates.

XILINX XC4000. The structure of a XILINX XC4000 chip is similar to the FPGA structure shown in Figure 11.22. It has a two-dimensional array of configurable logic blocks (CLBs) that can be interconnected using the vertical and horizontal routing channels. Chips range in size from the XC4002 to XC40250, which have about 2000 and 250,000 equivalent logic gates, respectively. A CLB contains 2 four-input LUTs; hence it can implement any two logic functions of up to four variables. The output of each of these LUTs can optionally be stored in a flip-flop. The CLB also contains a three input LUT connected to the 2 four-input LUTs, which allows implementation of functions with four or more variables.

The CLBs are interconnected using the wires in the routing channels. Wires of various lengths are provided, from wires that span a single CLB to wires that span the entire device. The number of wires in a routing channel varies for each specific chip (Figure 11.34).

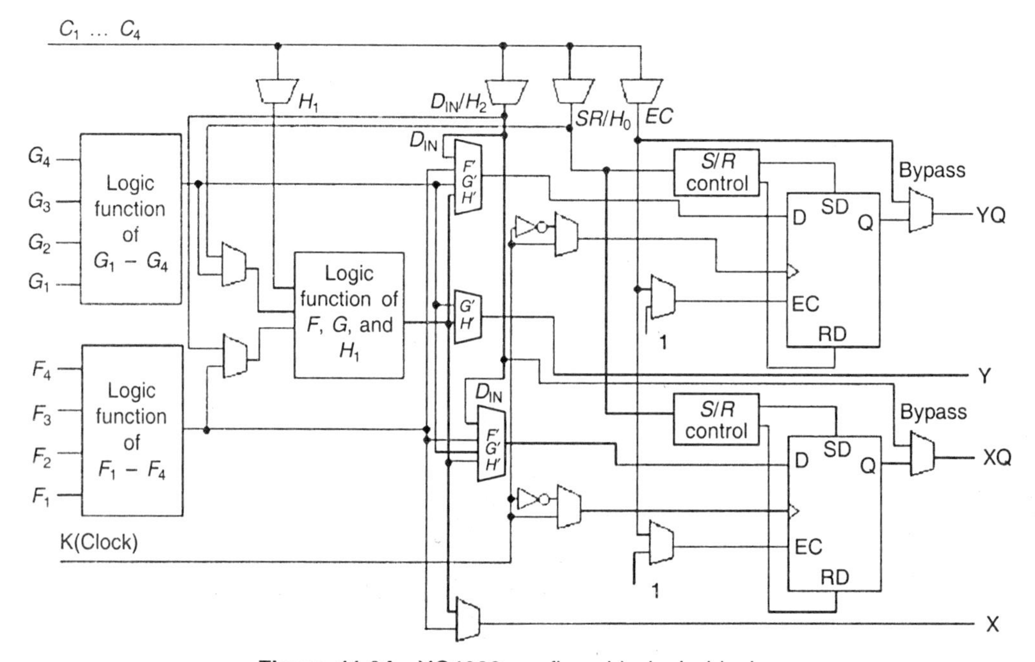

Figure 11.34 XC4000 configurable logic block.

Algotronix. An example of a regular programmable array is the CAL1024 (Configurable Array Logic) from Algotronix. This architecture contains 1024 identical logic cells arranged in a 32-by-32 matrix. At the boundary of the chip, 128 programmable I/O pin allow cascading the chips in even larger arrays. The cell interconnect is shown in Figure 11.35. Each cell is connected to the East, South, West, and North neighbour. In addition, two global interconnect signals connect to each cell. These are used to supply a low-skew signal to all cells for clocking. Each cell also receives row select lines and bit lines that are used to program RAM bits within the logic cells that dynamically customize the logic cell.

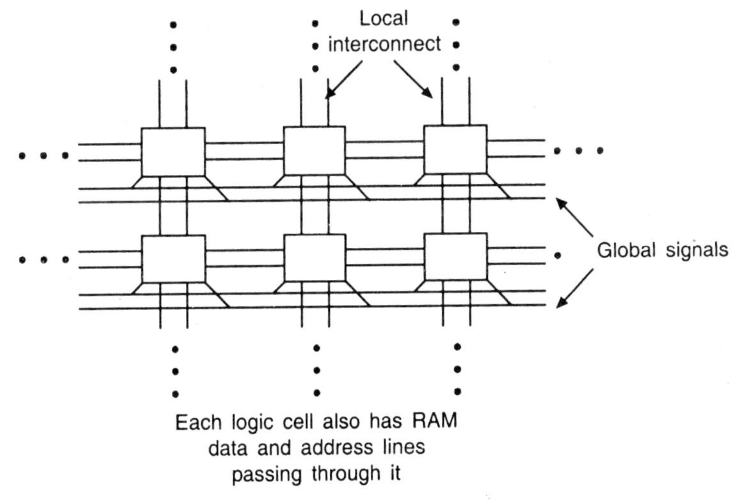

Figure 11.35 Algotronix FPGA chip architecture.

The cell design is shown in Figure 11.36. It consists of four 'through' multiplexers to route single-bit signals entering from the North, South, East, and West. In addition, two multiplexers route a selection of signals to a function unit. These signals include the signals entering on the orthogonal edges of the cell, two global 'clock' signals, and the output of the function block for feedback situations (latches). The Multiplexers are controlled by small 5-transistor static RAM cells. The functions that the logic cell can implement are listed in Table 11.3.

Table 11.3 CAL logic cell functions

Number	*Function*	*Number*	*Function*
0	ZERO	8	–X1.X2
1	ONE	9	–X1.–X2
2	X1	10	X1 + X2
3	–X1	11	X1 + –X2
4	X2	12	–X1 + X2
5	–X2	13	XNOR (X1, X2)
6	X1.X2	14	–X1 + –X2
7	X1.–X2	15	XOR (X1, X2)
16	D. Clk Latch	17	–D. Clk Latch
18	D. –Clk Latch	19	–D. –Clk Latch

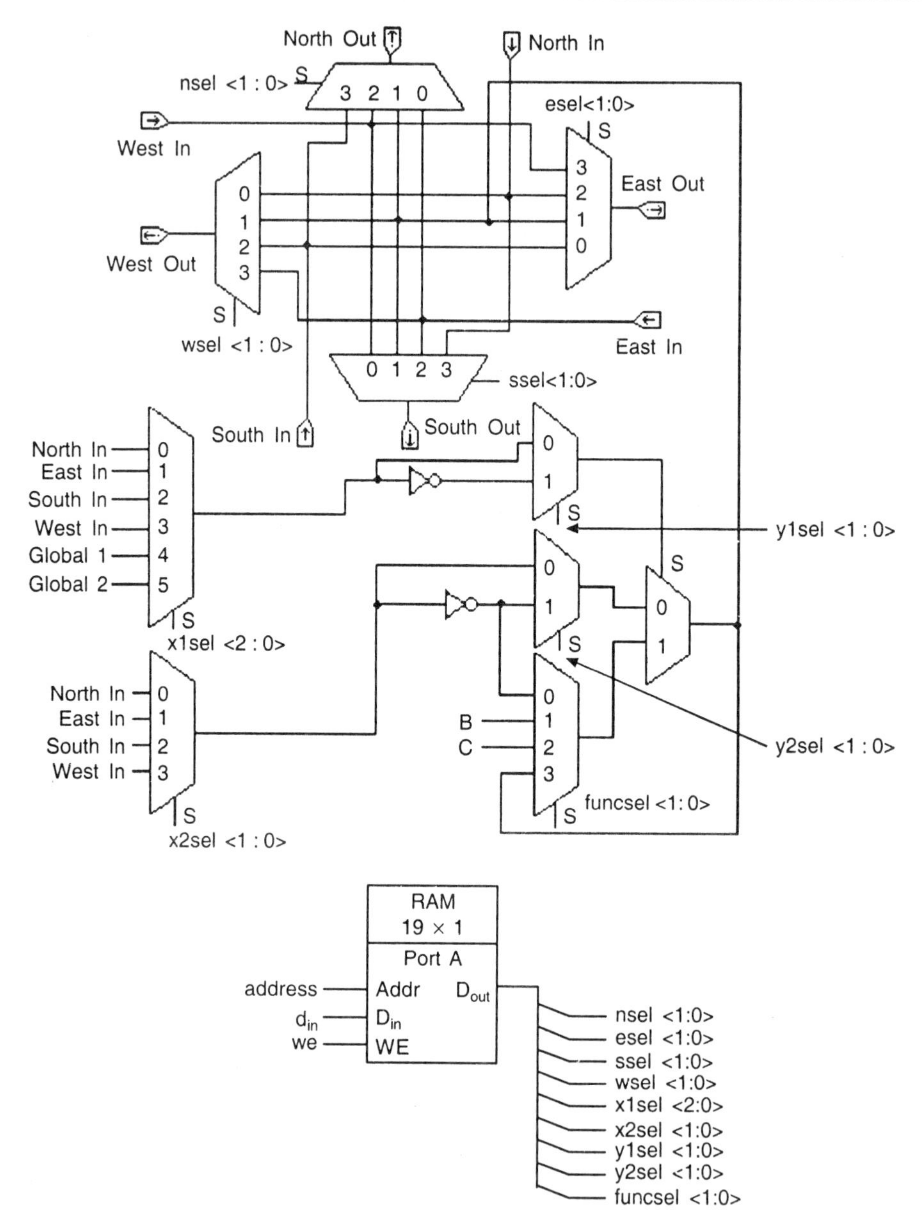

Figure 11.36 Algotronix cell design.

In the I/O pads, only one pin is used for I/O into and out of the array but have the communicating chips automatically deal with two pins that are outputs. The pads achieve this by using a ternary (three-level) logic scheme to sense when two outputs are driving each other via a contention circuit. This is then used with an XOR gate, as shown in Figure 11.37, to deduce the correct input value.

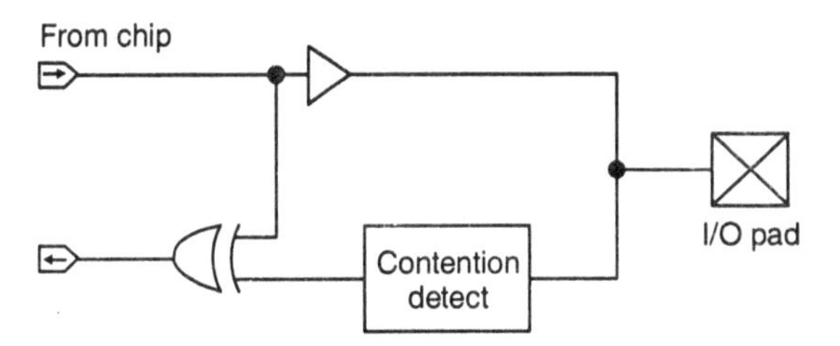

Figure 11.37 Algotronix I/O circuit.

Design is similar to both the XILINX and Actel approaches, where substantially automatic techniques can place and route a CAL chip. Unlike both other approaches, however, all routing (save the global clock lines) must pass through cells to get from one point to another. Thus, in the worst case, a signal may have to travel through 64 cells. Although implemented with fast transmission gates, this still can result in a substantial delay.

In all of these architectures, serial arithmetic may be preferable to parallel arithmetic because low-delay connections can be made between adjacent cells.

Concurrent logic. The Cli6000 series is another example of a regular array style FPGA. Concurrent designs have between 1000 and 3136 cells, with prospects of up to 10000 cells per chip in the next few years. As an example, the Cli6005 consists of a 7-by-7 array of superblocks. Each superblock has an array of 8-by-8 logic cells. Each logic cell connects to the four nearest neighbours and to a local and express bus (Figure 11.38). Compared with the Algotronix cell,

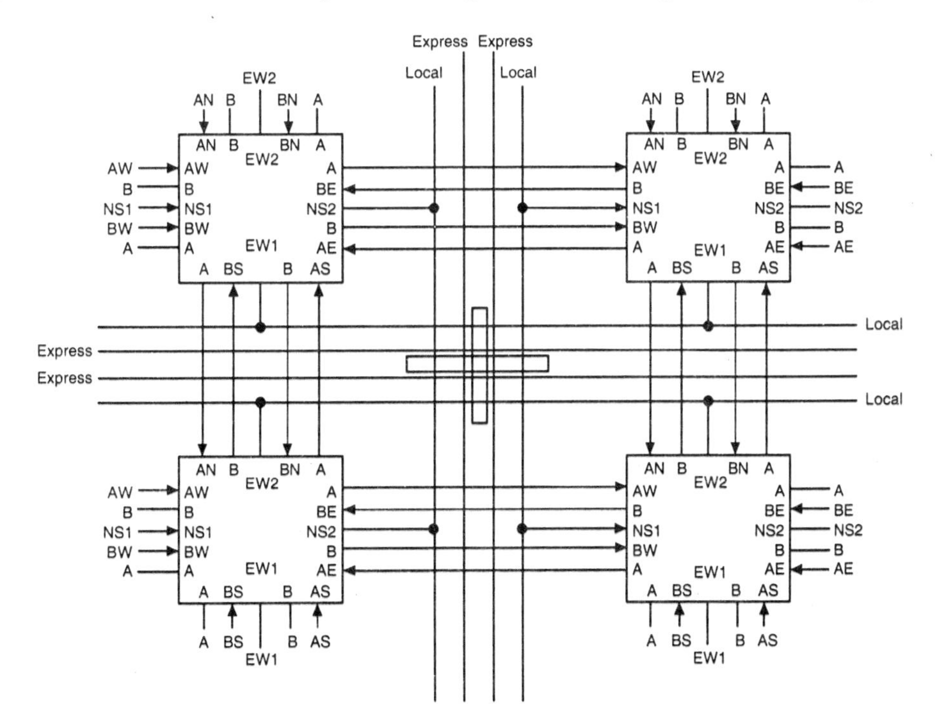

Figure 11.38 Concurrent logic array details.

cell structure of concurrent logic has considerably more functionality within a cell. A resettable register, XOR and an AND gate are included. Thus, for instance, a single-counter bit can be implemented in a single cell.

11.6.3 Implementation in FPGAs

In FPGAs, the programming information is stored in memory cells, called static random access memory (SRAM) cells. Each cell can store either a logic 0 or 1, and it provides this stored value as an output. An SRAM cell is used for each truth-table value stored in a LUT. SRAM cells are also used to configure the interconnection wires in an FPGA. In Figure 11.39, the logic block shown produces the output f_1, which is driven onto the horizontal wire (drawn in blue). This wire can be connected to some of the vertical wires that it crosses, using programmable switches. Each switch is implemented using an NMOS transistor, with its gate terminal controlled by an SRAM cell. Such a switch is known as pass-transistor switch. If a 0 is stored in an SRAM cell, then the NMOS transistor is turned OFF. But if a 1 is stored in the SRAM cell, then it is turned ON. This switch forms a connection between the two wires attached to its source and drain terminals. The number of switches are implemented using tri-state buffers, instead of pass transistors.

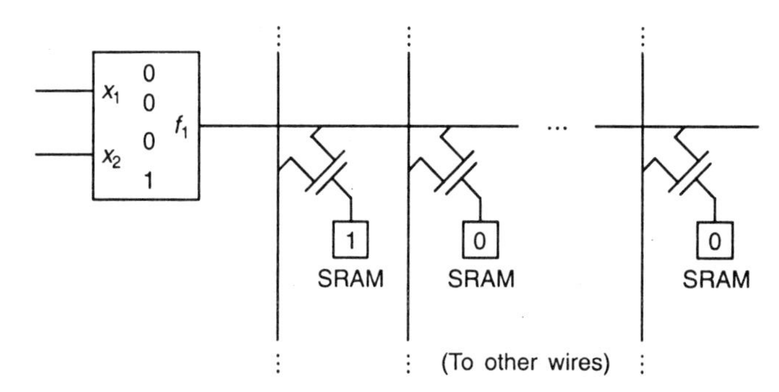

Figure 11.39 Pass-transistor switches in FPGAs.

11.6.4 Design Flow

Figure 11.40 shows the sequence of steps to design an Application Specific Integrated Circuits (ASIC); we call this a ***design flow***. The steps are listed below (numbered to correspond to the labels in Figure 11.40) with a brief description of the function of each step.

1. **Design entry:** Enter the design into an ASIC design system, either using a **hardware description language (HDL)** or schematic entry.
2. **Logic synthesis:** Use an HDL (**VHDL** or **Verilog**) and a logic synthesis tool to produce a **netlist**—a description of the logic cells and their connections.
3. **System partitioning:** Divide a logic system into ASIC-sized pieces.
4. **Prelayout simulation:** Check to see if the design functions correctly.
5. **Floorplanning:** Arrange the blocks of the netlist on the chip.
6. **Placement:** Decide the locations of cells in a block.

7. **Routing:** Make the connections between cells and blocks.
8. **Extraction:** Determine the resistance and capacitance of the interconnect.
9. **Post-layout simulation:** Check to see the design still works with the added **loads** of the interconnect.

Steps 1–4 are part of **logic design**, and Steps 5–9 are part of **physical design**.

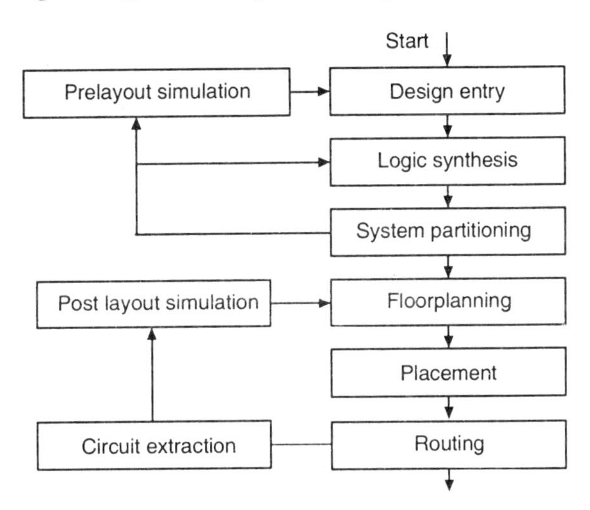

Figure 11.40 ASIC design flow.

SUMMARY

Four categories of programmable logic devices (PLDs) were investigated: the Field Programmable Gate Array (FPGA), the Programmable Read Only Memory (PROM), the Programmable Array Logic (PAL), and the Dynamic Logic Array (DLA). Both clocked and unclocked NMOS NAND–NAND and NOR–NOR PLAs were designed, after which PALs and PROMs were studied. PLDs have typical delays including external buffers, of 2 to 3 nS per gate, and power dissipations of the order of 0.5 to 1.5 mW per equivalent gate. PLDs can also be programmed very quickly as compared to gate-array prototypes. Another benefit of PLDs is that the devices can be tested very quickly in situ. Both ac and dc timing-analysis programs for semicustom IC designs are being continuously improved; but critical-path delays, switching noise and line reflections still cannot be predicted accurately until the chip is tested in the system. Adding feedback to a PLD gives a finite-state-machine (FSM). Mealy and Moore FSMs are discussed and compared. Two phase clocked Moore machines are the preferred FSM. The chapter also describes about the programmable logic devices: CPLD and FPGA. The CPLDs are larger versions of simple PLDs, it comprises multiple circuit blocks on a single chip with internal wiring resources to connect the circuit blocks. Each circuit block is similar to a PLA or a PAL. The chapter explains the architecture of CPLD and also the different packaging and programming techniques. The FPGA architecture and its different programming methods are described in this chapter. In an FPGA, more than 20,000 gates can be implemented in a logic circuit. They contain logic blocks for the implementation of the required functions. The implementation of FPGA is accomplished with SRAM cell, called pass transistor which acts as a memory cell to store information. Some FPGAs use tri-state buffers as switches instead of pass transistors. The chapter also gives a brief description of the design flow used in many Application Specific Integrated Circuits (ASICs).

REVIEW QUESTIONS

1. Explain field programmable gate array.
2. Explain how FPGA can be programmed.
3. With suitable diagram explain XILINX programmable Gate Array.
4. Explain the design flow of ASIC.

SHORT ANSWER QUESTIONS

1. What are the four types of devices that are classified as PLDs?

   Ans. (a) PROM
   (b) PLA
   (c) PAL
   (d) GAL

2. FSM can be designed as ___________ or ___________.

   Ans. (a) Mealy machine
   (b) Moore machine

3. A type of package used to house CPLD chip is called ___________.

   Ans. quad flat pack.

4. Give an example of widely used CPLD family.

   Ans. Altera MAX 7000

5. What is the most commonly used logic blocks in FPGA?

   Ans. Look up table

6. The FPGAs are configured by ___________ method.

   Ans. in system programming method

7. What is the use of programmable interconnect points?

   Ans. It is used to interconnect the global routing to CLBs.

8. Making connections between cells and blocks is called ___________.

   Ans. routing

12 CMOS Chip Design

12.1 INTRODUCTION

The design description for an integrated circuit may be described in terms of three domains, namely: behavioural, structural and physical domains. In each of these domains, there are a number of design options that may be selected to solve a particular problem. For instance, at the behavioural level, the freedom to choose, say, a sequential or a parallel algorithm is available. In the structural domain, the decision about which particular logic family, clocking strategy, or circuit style to use is initially unbound. At the physical level, how the circuit is implemented in terms of chips, boards, and cabinets also provides many options to the designer. These domains may be hierarchically divided into levels of design abstraction. Classically, these include the following:

- Architectural or Functional Level
- Register Transfer Level (RTL)
- Logic level
- Circuit level

The relationship between description domains and levels of design abstraction are elegantly shown by the Y-Chart in Figure 12.1. In this diagram, the three radial lines represent the three description domains, namely, behavioural, structural and physical domains. Along each line are enumerated types of objects in that domain. Concentric circles around the centre indicate the various levels of abstraction that are common in electronic design. The particular abstraction levels and design objects may differ slightly, depending on the design method.

In this chapter, we begin by discussing the various design strategies available to the CMOS IC designer, the CMOS chip design options: i.e., the different types of Application Specific Integrated Circuits (ASICs), programmable logic structures, ASIC design flow, etc.

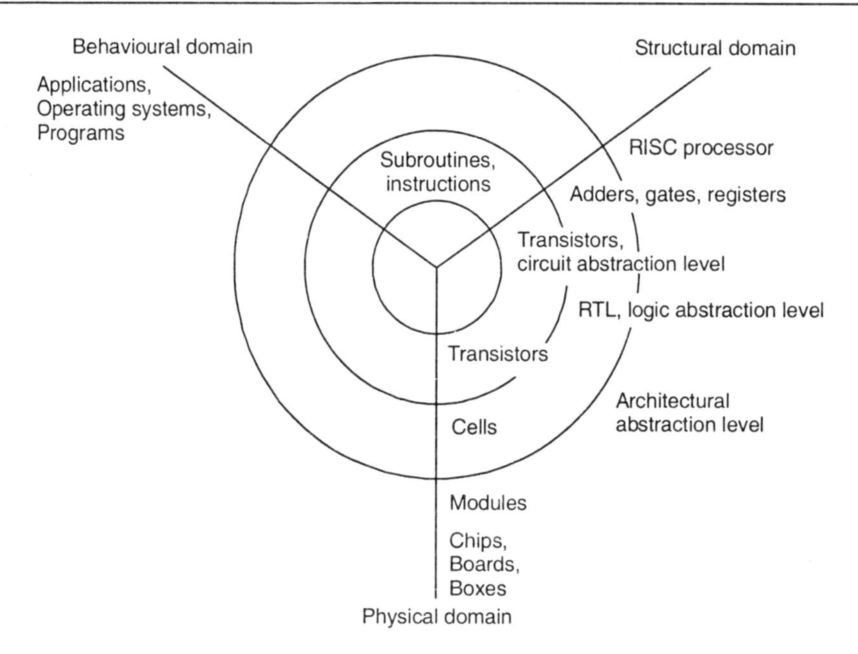

Figure 12.1 Y-Chart showing description domains and levels of design abstraction.

12.2 DESIGN STRATEGIES

A good VLSI design system should provide for consistent descriptions in all three description domains (behavioural, structural, and physical) and at all levels of abstraction. The means by which this is accomplished may be measured in terms of various design parameters as summarized below:

- Performance—speed, power, function, flexibility.
- Size of die and its cost.
- Time of design and hence cost of engineering and schedule.
- Ease of test generation and testability.

Design is a continuous tradeoff to achieve the adequate results for all levels of the above parameters. The tools and methodologies used for a particular chip will be a function of these parameters. Given that the process of designing a system on silicon is complicated, the role of good VLSI design aids is to reduce this complexity, increase productivity, and assure the designer of a working product. A good method of simplifying the approach to a design is by the use of constraints and abstractions.

12.2.1 Structured Design Strategies

The successful implementation of almost any IC requires an attention to the details of the engineering design process. A number of structured design techniques have been developed to deal with both complex hardware and software projects. Whether under consideration is a small chip designed by a single designer or a large system designed by a team of designers, the basic principles of structured design will improve the prospects of success.

12.2.2 Hierarchy

The use of hierarchy, or 'divide and conquer', involves dividing a module into submodules until the complexity of the submodules is at an appropriately comprehensive level of detail. This parallels the software case where large programs are split into smaller and smaller sections until simple subroutines, with well-defined functions and interfaces, can be written.

12.2.3 Regularity

Hierarchy involves dividing a system into a set of submodules. However, hierarchy alone does not necessarily solve the complexity problem. For instance, we could repeatedly divide the hierarchy of a design into different submodules but still end up with a large number of different submodules. With regularity, the designer attempts to divide the hierarchy into a set of similar building blocks. The use of iteration to form arrays of identical cells is an illustration of the use of regularity in an IC design.

Regularity can exist at all levels of the design hierarchy. At the circuit level, uniformly sized transistors might be used. At the logic-module level, identical gate structures might be employed. At higher levels, one can construct architectures that use a number of identical processor structures. Regularity allows an improvement in productivity by reusing specific designs in a number of places, thereby reducing the number of different designs that need to be completed.

12.2.4 Modularity

The principle of modularity adds to hierarchy and regularity the condition that submodules have well-defined functions and interfaces. If modules are 'well-formed', the interaction with other modules may be well characterized. A good starting point is the criteria placed on a well-formed software subroutine. A well-defined interface is required which in the case of software is an argument list with typed variables. In the IC case, this corresponds to a well-defined behavioural, structural, and physical interface that indicates the position, name, layer type, size and signal type of external interconnections, along with logic function and electrical characteristics. Modularity helps the designer to clarify and document an approach to a problem, and also allows a design system to more easily check the attributes of a module as it is constructed. The ability to divide the task into a set of well-defined modules aids in a team design where each of a number of designers has a portion of a complete chip to design.

12.2.5 Locality

By defining well-characterized interfaces for a module, the other internals of the module become unimportant to any exterior interface. In this way, by performing a form of 'information hiding', the apparent complexity of the module is reduced. In software, this is accomplished by reducing the global variables to a minimum (to zero).

12.3 CMOS CHIP DESIGN OPTIONS

In this section, a range of design options that may be used to implement a CMOS system design will be examined. The design options include different types of Application Specific Integrated Circuits (ASICs) like full custom ASICs, Standard cell based ASICs, Gate array based ASICs, Channelled, Channelless and Structured gate array, and programmable logic such as programmable logic structures, programmable interconnect and reprogrammable gate arrays.

12.3.1 Application Specific Integrated Circuits (ASICs)

An Application Specific Integrated Circuit (ASIC) is an integrated circuit (IC) customised for a particular use, rather than intended for general-purpose use. For example, a chip designed solely to run a cell phone is an ASIC. In contrast, the 7400 series and 4000 series integrated circuits are logic building blocks that can be wired together to perform many different applications. Intermediate between ASICs and standard products are Application Specific Standard Products (ASSPs).

The maximum complexity and hence functionality possible in an ASIC has grown from 5000 gates to over 100 million. Modern ASICs often include entire 32-bit processors, memory blocks including ROM, RAM, EEPROM, Flash and other logic building blocks. Such an ASIC is often termed as System on a Chip (SoC). Designers of digital ASICs use a Hardware Description Language (HDL), such as Verilog or VHDL, to describe the functionality of ASICs. ASICs are commonly used in networking devices to maximize performance.

12.3.2 Types of ASICs

ICs are made on a thin (a few hundred microns thick), circular silicon wafer, with each wafer holding hundreds of dice (sometimes people use dies or dice for the plural of die). The transistors and wiring are made from many layers (usually between 10 and 15 distinct layers) built on top of one another. Each successive mask layer has a pattern that is defined using a mask similar to a glass photographic slide. The first half-dozen or so layers define the transistors. The last half-dozen or so layers define the metal wires between the transistors (the interconnect).

A **full-custom IC** includes some (possibly all) logic cells that are customized and all mask layers that are customized, e.g. a microprocessor—designers spend many hours squeezing the most out of every last square micron of microprocessor chip space by hand. Customizing all of the IC features in this way allows designers to include analog circuits, optimized memory cells, or mechanical structures on an IC. Full-custom ICs are the most expensive to manufacture and to design. The manufacturing lead time (the time it takes just to make an IC—not including design time) is typically eight weeks for a full-custom IC.

We shall discuss the different full-custom ASICs briefly next, mainly about **semicustom ASICs** and **programmable ASICs**. In **semicustom ASICs**, all of the logic cells are predesigned and some (possibly all) of the mask layers are customized. Using predesigned cells from a cell library makes our lives as designers much, much easier. There are two types of semicustom ASICs: standard-cell based ASICs and gate-array based ASICs. Following this, the

programmable ASICs are described. In this, all of the logic cells are predesigned and none of the mask layers are customized. There are two types of programmable ASICs: the programmable logic device (PLD) and the field-programmable gate array (FPGA).

12.3.2.1 Full-custom ASICs

In a full-custom ASIC, some or all of the logic cells, circuits, or layout are designed specifically for one ASIC. This means the designer abandons the approach of using pretested and precharacterized cells for all or part of that design. It makes sense to take this approach only if there are no suitable existing cell libraries available that can be used for the entire design. This might be because existing cell libraries are not fast enough, or the logic cells are not small enough or consume too much power.

Bipolar technology has historically been used for precision analog functions. There are some fundamental reasons for this. In all integrated circuits, the matching of component characteristics between chips is very poor, while the matching of characteristics between components on the same chip is excellent. Suppose we have transistors T1, T2, and T3 on an analog/digital ASIC. The three transistors are all the same size and are constructed in an identical fashion. Transistors T1 and T2 are located adjacent to each other and have the same orientation. Transistor T3 is the same size as T1 and T2 but is located on the other side of the chip from T1 and T2 and has a different orientation. ICs are made in batches called wafer lots. A wafer lot is a group of silicon wafers that are all processed together. Usually there are between 5 and 30 wafers in a lot. Each wafer can contain tens or hundreds of chips depending on the size of the IC and the wafer.

If we were to make measurements of the characteristics of transistors T1, T2, and T3, we would find the following:

- Transistors T1 will have virtually identical characteristics to T2 on the same IC. We say that the transistors match well or the tracking between devices is excellent.
- Transistor T3 will match transistors T1 and T2 on the same IC very well, but not as closely as T1 matches T2 on the same IC.
- Transistors T1, T2, and T3 will match fairly well with transistors T1, T2, and T3 on a different IC on the same wafer. The matching will depend on how far apart the two ICs are on the wafer.
- Transistors on ICs from different wafers in the same wafer lot will not match very well.
- Transistors on ICs from different wafer lots will match very poorly.

Device physics dictates that a pair of bipolar transistors will always match more precisely than CMOS transistors of a comparable size. Bipolar technology has historically been more widely used for full-custom analog design because of its improved precision. Despite its poorer analog properties, the use of CMOS technology for analog functions is increasing. There are two reasons for this. The first reason is that CMOS is now by far the most widely available IC technology. Many more CMOS ASICs and CMOS standard products are now being manufactured than bipolar ICs. The second reason is that increased levels of integration require mixing analog and digital functions on the same IC: this has forced designers to find ways to use CMOS technology to implement analog functions.

12.3.2.2 *Standard-cell based ASICs*

A **cell-based ASIC** (or **CBIC**) uses predesigned logic cells (for example, AND gates, OR gates, multiplexers, and flip-flops) known as **standard cells**. The standard-cell areas (also called flexible blocks) in a CBIC are built of rows of standard cells. The standard-cell areas may be used in combination with larger predesigned cells, perhaps microcontrollers or even microprocessors, known as **megacells**. Megacells are also called megafunctions, full-custom blocks, system-level macros (SLMs), fixed blocks, cores, or Functional Standard Blocks (FSBs). The ASIC designer defines only the placement of the standard cells and the interconnect in a CBIC. However, the standard cells can be placed anywhere on the silicon; this means that all the mask layers of a CBIC are customized and are unique to a particular customer.

The advantage of CBICs is that designers save time, money, and reduce risk by using a predesigned, pretested, and precharacterized **standard-cell library**. In addition, each standard cell can be optimized individually. For example, during the design of the cell library, each and every transistor in every standard cell can be chosen to maximize speed or minimize area. The disadvantages are the time or expense of designing or buying the standard-cell library and the time needed to fabricate all layers of the ASIC for each new design.

Figure 12.2 shows a CBIC. The important features of this type of ASIC are as follows:

- All mask layers are customized—transistors and interconnect.
- Custom blocks can be embedded.
- Manufacturing lead time is about eight weeks.

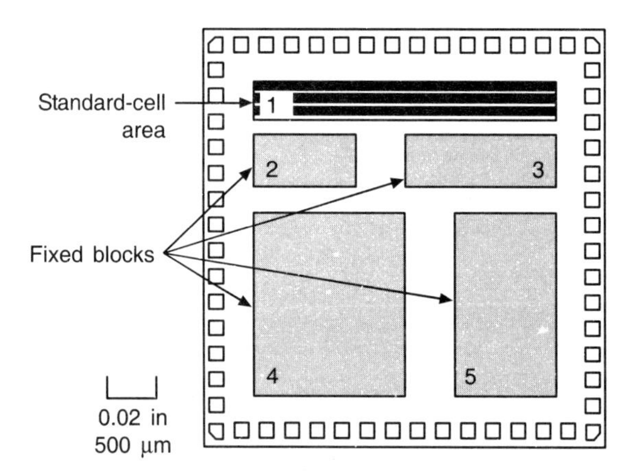

Figure 12.2 A cell-based ASIC (CBIC) die with a single standard-cell area (a flexible block) together with four fixed blocks. The flexible block contains rows of standard cells. The small squares around the edge of the die are bonding pads that are connected to the pins of the ASIC package.

Each standard cell in the library is constructed using full-custom design methods, but these predesigned and precharacterized circuits can be used without having to do any full-custom design. This design style gives the same performance and flexibility advantages of a full-custom ASIC but reduces design time and risk.

Standard cells are designed to fit together like bricks in a wall. Figure 12.3 shows an example of a simple standard cell. Power and ground buses (V_{DD} and GND or V_{SS}) run horizontally on metal lines inside the cells.

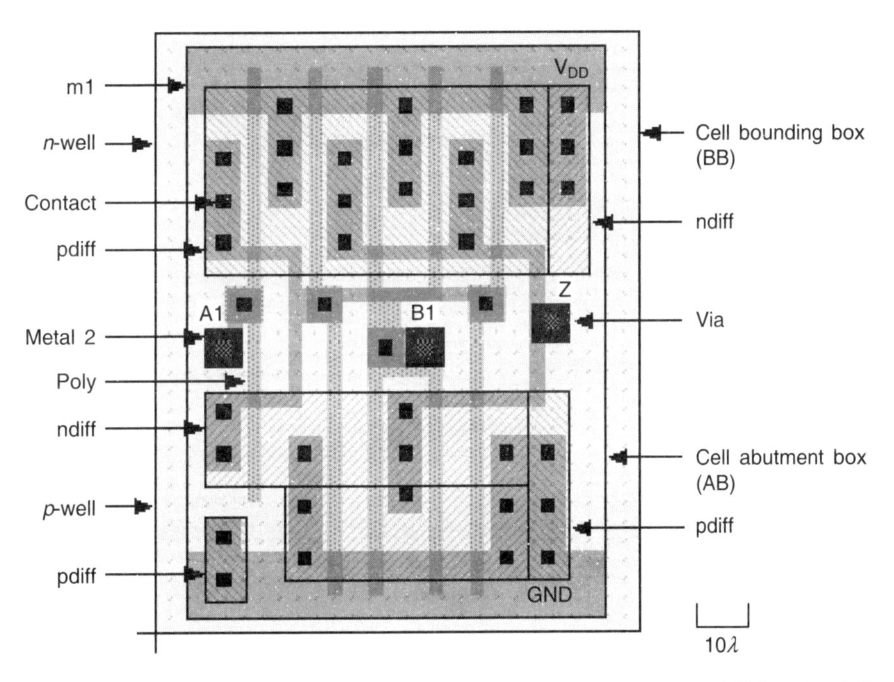

Figure 12.3 This cell would be approximately 25 microns wide on an ASIC with λ (lambda) = 0.25 microns (a micron is 10^{-6} m). Standard cells are stacked like bricks in a wall; the abutment box (AB) defines the edges of the brick. The difference between the bounding box (BB) and the AB is the area of overlap between the bricks. Power supplies (labelled V_{DD} and GND) run horizontally inside a standard cell on a metal layer that lies above the transistor layers. Each different shaded and labelled pattern represents a different layer. This standard cell has centre connectors (the three squares, labelled A1, B1, and Z) that allow the cell to connect to others.

Standard-cell design allows the automation of the process of assembling an ASIC. Groups of standard cells fit horizontally together to form rows. The rows stack vertically to form flexible rectangular blocks (which you can reshape during design). A flexible block built from several rows of standard cells may then be connected to other standard-cell blocks or other full-custom logic blocks.

Modern CMOS ASICs use two, three, or more levels (or layers) of metal for interconnect. This allows wires to cross over different layers in the same way that we use copper traces on different layers on a printed-circuit board. In a two-level metal CMOS technology, connections to the standard-cell inputs and outputs are usually made using the second level of metal (**metal 2**, the upper level of metal) at the tops and bottoms of the cells. In a three-level metal technology, connections may be internal to the logic cell (Figure 12.3). This allows for more sophisticated routing programs to take advantage of the extra metal layer to route interconnect over the top of the logic cells.

A connection that needs to cross over a row of standard cells uses a **feedthrough**. The term feedthrough can refer either to the piece of metal that is used to pass a signal through a cell or to a space in a cell waiting to be used as a feedthrough. Figure 12.4 shows two feedthroughs: one in cell A.14 and one in cell A.23.

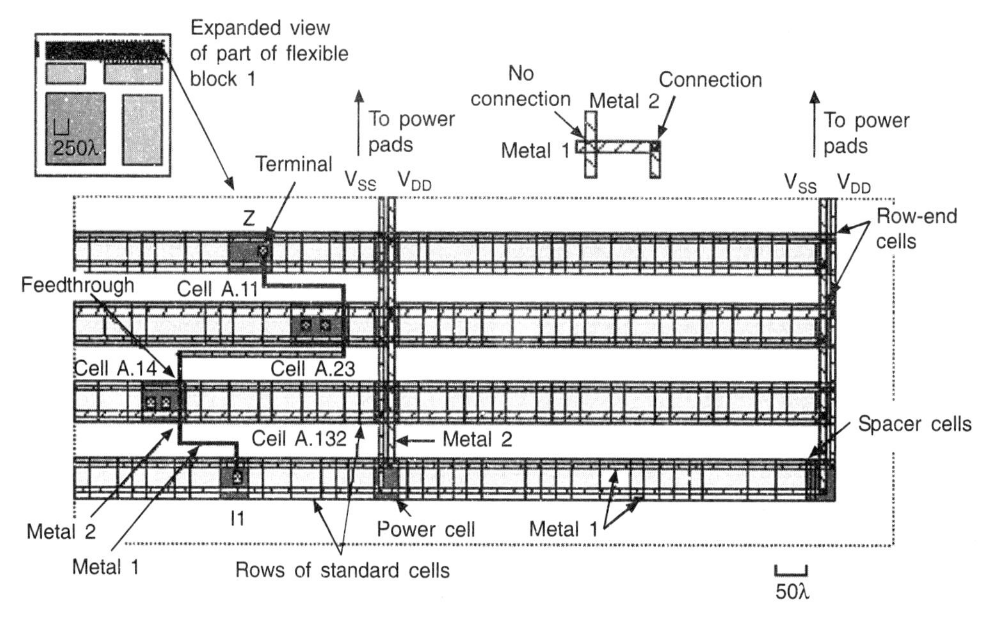

Figure 12.4 Routing the CBIC (cell-based IC) shown in Figure 12.2. The use of regularly shaped standard cells, such as the one in Figure 12.3, from a library allows ASICs like this to be designed automatically. This ASIC uses two separate layers of metal interconnect (metal 1 and metal 2) running at right angles to each other (like traces on a printed-circuit board). Interconnections between logic cells use spaces (called channels) between the rows of cells. ASICs may have three (or more) layers of metal allowing the cell rows to touch with the interconnect running over the top of the cells.

In both two-level and three-level metal technology, the power buses (V_{DD} and GND) inside the standard cells normally use the lowest (closest to the transistors) layer of metal (**metal 1**). The width of each row of standard cells is adjusted so that they may be aligned using **spacer cells**. The power buses, or rails, are then connected to additional vertical power rails using **row-end cells** at the aligned ends of each standard-cell block. If the rows of standard cells are long, then vertical power rails can also be run in metal 2 through the cell rows using special **power cells** that just connect to V_{DD} and GND. Usually the designer manually controls the number and width of the vertical power rails connected to the standard-cell blocks during physical design. A diagram of the power distribution scheme for a CBIC is shown in Figure 12.4.

All the mask layers of a CBIC are customized. This allows megacells (SRAM, a SCSI controller, or an MPEG decoder, for example) to be placed on the same IC with standard cells. Megacells are usually supplied by an ASIC or library company complete with behavioural models and some way to test them (a test strategy). ASIC library companies also supply compilers to generate flexible DRAM, SRAM, and ROM blocks. Since all mask layers on a standard-cell design are customized, memory design is more efficient and denser than for gate arrays.

For logic that operates on multiple signals across a data bus—**a datapath** (DP)—the use of standard cells may not be the most efficient ASIC design style. Some ASIC library companies provide a datapath compiler that automatically generates **datapath logic**. A **datapath library** typically contains cells such as adders, subtracters, multipliers, and simple **arithmetic and logical units (ALUs)**. The connectors of datapath library cells are **pitch-matched** to each other so that they fit together. Connecting datapath cells to form a datapath usually, but not always, results in faster and denser layout than using standard cells or a gate array.

Standard-cell and gate-array libraries may contain hundreds of different logic cells, including combinational functions (NAND, NOR, AND, OR gates) with multiple inputs, as well as latches and flip-flops with different combinations of reset, preset and clocking options. The ASIC library company provides designers with a data book in paper or electronic form with all of the functional descriptions and timing information for each library element.

12.3.2.3 Gate-array based ASICs

In a **gate array** (GA) or gate-array based ASIC the transistors are predefined on the silicon wafer. The predefined pattern of transistors on a gate array is the **base array**, and the smallest element that is replicated to make the base array is the **base cell** (sometimes called a **primitive cell**). Only the top few layers of metal, which define the interconnect between transistors, are defined by the designer using custom masks. To distinguish this type of gate array from other types of gate array, it is often called a **masked gate array (MGA)**. The designer chooses from a gate-array library of predesigned and precharacterized logic cells. The logic cells in a gate-array library are often called **macros**. The reason for this is that the base-cell layout is the same for each logic cell, and only the interconnect (inside cells and between cells) is customized, so that there is a similarity between gate-array macros and a software macro. Inside IBM, gate-array macros are known as **books**.

Both cell-based and gate-array ASICs use predefined cells, but there is a difference—we can change the transistor sizes in a standard cell to optimize speed and performance, but the device sizes in a gate array are fixed. This results in a trade-off in performance and area in a gate array at the silicon level. The trade-off between area and performance is made at the library level for a standard-cell ASIC.

We can complete the diffusion steps that form the transistors and then stockpile wafers (sometimes we call a gate array a **prediffused array** for this reason). Since only the metal interconnections are unique to an MGA, we can use the stockpiled wafers for different customers as needed. Using wafers prefabricated up to the metallization steps, reduces the time needed to make an MGA, the **turnaround time**, to a few days or at most a couple of weeks. The costs for all the initial fabrication steps for an MGA are shared for each customer and this reduces the cost of an MGA compared to a full-custom or standard-cell ASIC design.

There are the following different types of MGA or gate-array based ASICs:

- Channelled gate arrays.
- Channelless gate arrays.
- Structured gate arrays.

In the channelled gate-array architecture, the gate array is channelled. There are two common ways of arranging (or arraying) the transistors on a MGA: in a channelled gate array we leave space between the rows of transistors for wiring; the routing on a channelless gate array uses rows of unused transistors. The channelled gate array was the first to be developed, but the channelless gate-array architecture is now more widely used. A structured (or embedded) gate array can be either channelled or channelless but it includes (or embeds) a custom block.

12.3.2.4 *Channelled gate array*

Figure 12.5 shows a channelled gate array. The important features of this type of MGA are:

- Only the interconnect is customized.
- The interconnect uses predefined spaces between rows of base cells.
- Manufacturing lead time is between two days and two weeks.

A channelled gate array is similar to a CBIC—both use rows of cells separated by channels used for interconnect. One difference is that the space for interconnect between rows of cells are fixed in height in a channelled gate array, whereas the space between rows of cells may be adjusted in a CBIC.

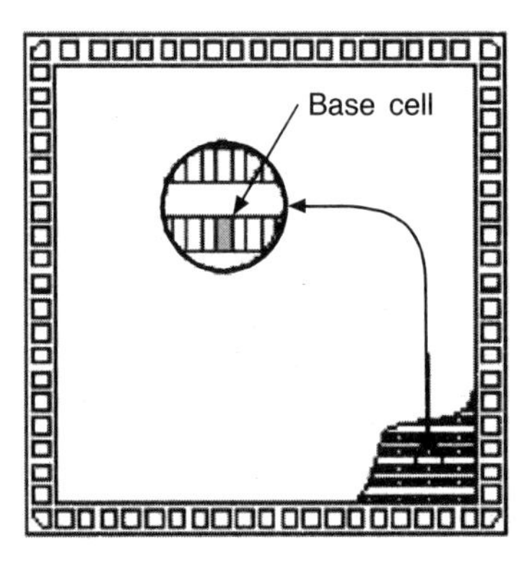

Figure 12.5 A channelled gate-array die. The spaces between rows of the base cells are set aside for interconnect.

12.3.2.5 *Channelless gate array*

Figure 12.6 shows a channelless gate array (also known as a channel-free gate array, sea-of-gates array, or SOG array). The important features of this type of MGA are as follows:

- Only some (the top few) mask layers are customized—the interconnect.
- Manufacturing lead time is between two days and two weeks.

The key difference between a channelless gate array and channelled gate array is that there are no predefined areas set aside for routing between cells on a channelless gate array. Instead we route over the top of the gate array devices. We can do this because we customize the contact layer that defines the connections between metal 1, the first layer of metal, and the transistors. When we use an area of transistors for routing in a channelless array, we do not make any contacts to the devices lying underneath; we simply leave the transistors unused.

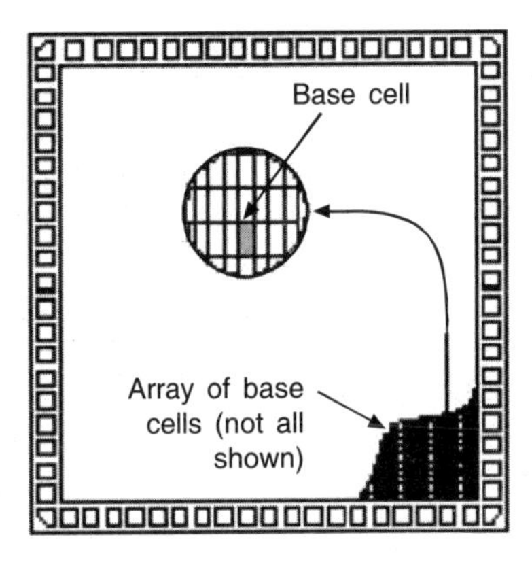

Figure 12.6 A channelless gate-array or sea-of-gates (SOG) array die. The core area of the die is completely filled with an array of base cells (the base array).

12.3.2.6 Structured gate array

An **embedded gate array** or **structured gate array** (also known as **masterslice** or **masterimage**) combines some of the features of CBICs and MGAs. One of the disadvantages of the MGA is the fixed gate-array base cell. This makes the implementation of memory, for example, difficult and inefficient. In an embedded gate array, we set aside some of the IC area and dedicate it to a specific function. This embedded area either can contain a different base cell that is more suitable for building memory cells, or it can contain a complete circuit block, such as a microcontroller.

Figure 12.7 shows an embedded gate array. The important features of this type of MGA are the following:

- Only the interconnect is customized.
- Custom blocks (the same for each design) can be embedded.
- Manufacturing lead time is between two days and two weeks.

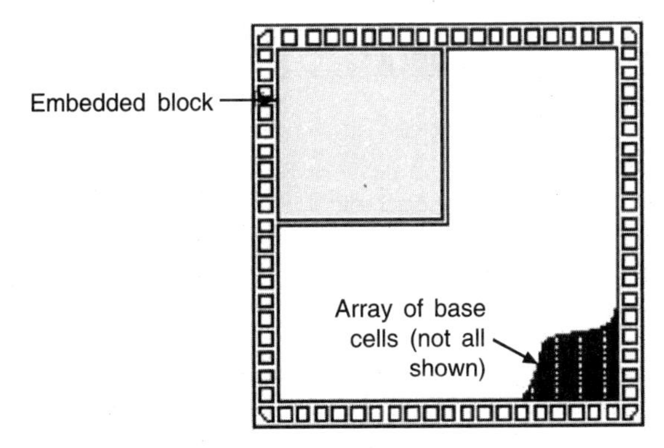

Figure 12.7 A structured or embedded gate-array die showing an embedded block in the upper left corner (e.g., a static random-access memory). The rest of the die is filled with an array of base cells.

An embedded gate array gives the improved area efficiency and increased performance of a CBIC but with the lower cost and faster turnaround of an MGA. One disadvantage of an embedded gate array is that the embedded function is fixed. For example, if an embedded gate array contains an area set aside for a 32 Kb memory, but we only need a 16 Kb memory, then we may have to waste half of the embedded memory function. However, this may still be more efficient and cheaper than implementing a 32 Kb memory using macros on a SOG array.

12.3.2.7 *Programmable Logic Devices (PLDs)*

Programmable logic devices (PLDs) are standard ICs that are available in standard configurations. However, PLDs may be configured or programmed to create a part customized to a specific application, and so they also belong to the family of ASICs. PLDs use different technologies to allow programming of the device. Figure 12.8 shows a PLD and the following important features that all PLDs have in common:

- No customized mask layers or logic cells
- Fast design turnaround
- A single large block of programmable interconnect
- A matrix of logic macrocells that usually consist of programmable array logic followed by a flip-flop or latch

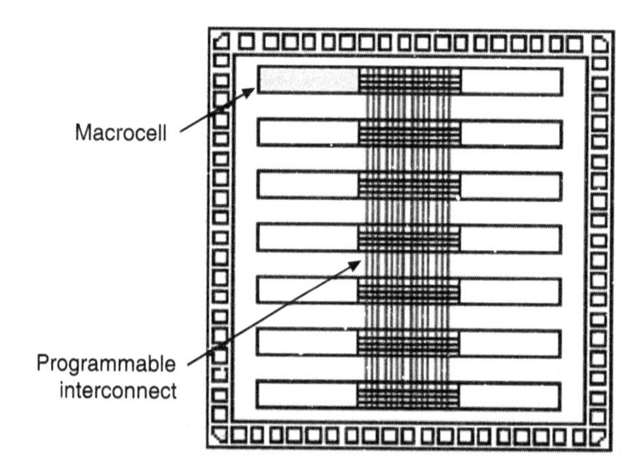

Figure 12.8 A programmable logic device (PLD) die. The macrocells typically consist of programmable array logic followed by a flip-flop or latch. The macrocells are connected using a large programmable interconnect block.

The simplest type of programmable IC is a **read-only memory (ROM)**. The most common types of ROM use a metal fuse that can be blown permanently (a **programmable ROM or PROM**). An electrically programmable ROM, or EPROM, uses programmable MOS transistors whose characteristics are altered by applying a high voltage. One can erase an EPROM either by using another high voltage (an **electrically erasable PROM**, or **EEPROM**) or by exposing the device to ultraviolet light (**UV-erasable PROM**, or **UVPROM**).

There is another type of ROM that can be placed on any ASIC—a **mask-programmable ROM** (mask-programmed ROM or masked ROM). A masked ROM is a regular array of transistors permanently programmed using custom mask patterns. An embedded masked ROM is thus a large, specialized, logic cell.

The same programmable technologies used to make ROMs can be applied to more flexible logic structures. By using the programmable devices in a large array of AND gates and an array of OR gates, we create a family of flexible and programmable logic devices called **logic arrays**. The company Monolithic Memories was the first to produce **Programmable Array Logic (PAL)** devices. A PAL can also include registers (flip-flops) to store the current state information so that you can use a PAL to make a complete state machine.

Just as we have a mask-programmable ROM, we could place a logic array as a cell on a custom ASIC. This type of logic array is called a **programmable logic array (PLA)**. There is a difference between a PAL and a PLA: a PLA has a programmable AND logic array, or AND plane, followed by a programmable OR logic array, or OR plane; a PAL has a programmable **AND plane** and, in contrast to a PLA, a fixed **OR plane**.

Depending on how the PLD is programmed, we can have an **erasable PLD (EPLD)**, or **mask-programmed PLD** (sometimes called a **masked PLD** but usually just PLD). The first PALs, PLAs, and PLDs were based on bipolar technology and used programmable fuses or links. CMOS PLDs usually employ floating-gate transistors.

12.3.2.8 Field-programmable gate arrays (FPGAs)

A step above the PLD in complexity is the field-programmable gate array (FPGA). There is very little difference between an FPGA and a PLD—an FPGA is usually just larger and more complex than a PLD. In fact, some companies that manufacture programmable ASICs call their products FPGAs and some call them complex PLDs. FPGAs are the newest member of the ASIC family and are rapidly growing in importance, replacing TTL in microelectronic systems. Even though an FPGA is a type of gate array, we do not consider the term gate-array based ASICs to include FPGAs. This may change as FPGAs and MGAs start to look more alike.

Figure 12.9 illustrates the essential characteristics of an FPGA:

- None of the mask layers are customized.
- A method for programming the basic logic cells and the interconnect.
- The core is a regular array of programmable basic logic cells that can implement combinational as well as sequential logic (flip-flops).
- A matrix of programmable interconnect surrounds the basic logic cells.
- Programmable I/O cells surround the core.
- Design turnaround is a few hours.

The complex programmable logic device (CPLD) and Field Programmable Gate Array (FPGA) and their programming techniques are described in detail in Chapter 11.

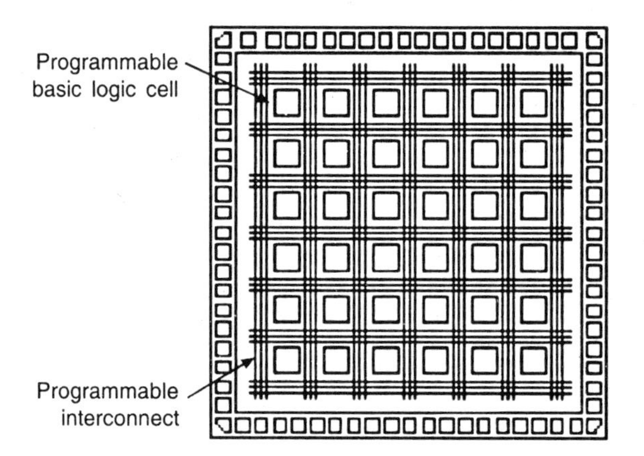

Figure 12.9 A field-programmable gate array (FPGA) die. All FPGAs contain a regular structure of programmable basic logic cells surrounded by programmable interconnect. The exact type, size, and the number of the programmable basic logic cells varies tremendously.

12.3.3 Economics of ASICs

In this section, the economics of using ASICs in a product is discussed and compared for the most popular types of ASICs: an FPGA, an MGA, and a CBIC. To make an economic comparison between these alternatives, the ASIC itself is considered as a product and examined to determine the components of product cost: fixed costs and variable costs. The most obvious economic factor in making a choice between the different ASIC types is the part cost. Part costs vary enormously—one can pay anywhere from a few dollars to several hundreds of dollars for an ASIC. In general, however, FPGAs are more expensive per gate than MGAs, which are, in turn, more expensive than CBICs. The price per gate for an FPGA to implement the same function is typically 2–5 times the cost of an MGA or CBIC.

12.3.3.1 Product cost

The total cost of any product can be separated into **fixed costs** and **variable costs**:

$$\text{Total cost of product} = \text{Fixed cost of product} + \text{variable cost of product} \times \text{numbers sold} \quad (12.1)$$

Fixed costs are independent of **sales volume**—the number of products sold. However, the fixed costs amortized per product sold (fixed costs divided by products sold) decrease as sales volume increases. Variable costs include the cost of the parts used in the product, assembly costs, and other manufacturing costs. The total part cost can be obtained as follows:

$$\text{Total cost of parts} = \text{Fixed cost of parts} + \text{variable cost of parts} \times \text{volume of parts} \quad (12.2)$$

The fixed cost when an FPGA is used can be low—i.e., we just have to buy the software and any programming equipment. The fixed part costs for an MGA or CBIC are higher and include the costs of the masks, simulation, and test program development.

Figure 12.10 shows a **break-even graph** that compares the total part cost for an FPGA, MGA, and a CBIC with the following assumptions:

- FPGA fixed cost is $21,800, part cost is $39.
- MGA fixed cost is $86,000, part cost is $10.
- CBIC fixed cost is $146,000, part cost is $8.

At low volumes, the MGA and the CBIC are more expensive because of their higher fixed costs. The total part costs of two alternative types of ASIC are equal at the break-even volume. In Figure 12.10 the **break-even volume** for the FPGA and the MGA is about 2000 parts. The break-even volume between the FPGA and the CBIC is about 4000 parts. The break-even volume between the MGA and the CBIC is higher—at about 20,000 parts.

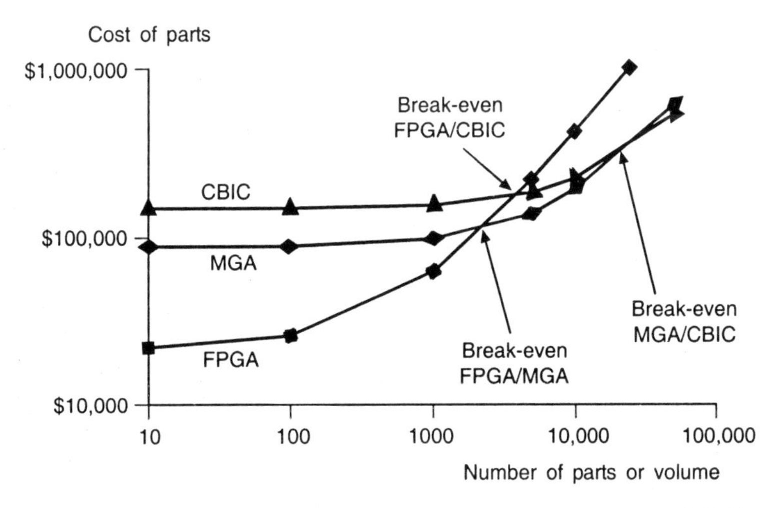

Figure 12.10 A break-even analysis for an FPGA, a masked gate array (MGA) and a custom cell-based ASIC (CBIC).The break-even volume between two technologies is the point at which the total cost of parts are equal. These numbers are very approximate.

12.3.3.2 ASIC fixed costs

Figure 12.11 shows a spreadsheet, 'Fixed Costs,' that calculates the fixed part costs associated with ASIC design.

The **training cost** includes the cost of the time to learn any new electronic **design automation (EDA)** system. For example, a new FPGA design system might require a few days to learn; a new gate-array or cell-based design system might require taking a course. Figure 12.11 assumes that the cost of an engineer (including overhead, benefits, infrastructure, and so on) is between $100,000 and $200,000 per year or $2000 to $4000 per week.

Next, the **hardware and software cost** for ASIC design are considered. Figure 12.11 shows some typical figures, but one can spend anywhere from $1000 to $1 million (and more) on ASIC design software and the necessary infrastructure.

	FPGA		MGA		CBIC	
Training:	$800		$2,000		$2,000	
Days		2		5		5
Cost/day		$400		$400		$400
Hardware	$10,000		$10,000		$10,000	
Software	$1,000		$20,000		$40,000	
Design:	$8,000		$20,000		$20,000	
Size (gates)		10,000		10,000		10,000
Gates/day		500		200		200
Days		20		50		50
Cost/day		$400		$400		$400
Design for test:			$2,000		$2,000	
Days				5		5
Cost/day				$400		$400
NRE:			$30,000		$70,000	
Masks				$10,000		$50,000
Simulation				$10,000		$10,000
Test prcgram				$10,000		$10,000
Second source:	$2,000		$2,000		$2,000	
Days		5		5		5
Cost/day		$400		$400		$400
Total fixed costs	$21,800		$86,000		$146,000	

Figure 12.11 A spreadsheet, "Fixed Costs," for a field-programmable gate array (FPGA), a masked gate array (MGA), and a cell-based ASIC (CBIC). These costs can vary widely.

The **productivity** of an ASIC designer in gates (or transistors) per day is measured. This is like trying to predict how long it takes to dig a hole, and the number of gates per day an engineer averages varies wildly. ASIC design productivity must increase as ASIC sizes increase and will depend on experience, design tools, and the ASIC complexity. If we are using similar design methods, design productivity ought to be independent of the type of ASIC, but FPGA design software is usually available as a complete bundle on a PC. This means that it is often easier to learn and use than semicustom ASIC design tools.

Every ASIC has to pass a **production test** to make sure that it works. With modern test tools the generation of any test circuits on each ASIC that are needed for production testing can be automatic, but it still involves a cost for **design for test**. An FPGA is tested by the manufacturer before it is sold to you and before you program it. You are still paying for testing an FPGA, but it is a hidden cost folded into the part cost of the FPGA. You do have to pay for any **programming costs** for an FPGA, but we can include these in the hardware and software cost.

The **nonrecurring-engineering (NRE)** charge includes the cost of work done by the ASIC vendor and the cost of the masks. The production test uses sets of test inputs called test vectors, often many thousands of them. Most ASIC vendors require simulation to generate test vectors and test programs for production testing, and will charge for a test-program development cost.

The number of masks required by an ASIC during fabrication can range from three or four (for a gate array) to 15 or more (for a CBIC). Total mask costs can range from \$5000 to \$50,000 or more. The total NRE charge can range from \$10,000 to \$300,000 or more and will vary with volume and the size of the ASIC. If you commit to high volumes (above 100,000 parts), the vendor may waive the NRE charge. The NRE charge may also include the costs of software tools, design verification, and prototype samples.

If a design does not work the first time, one has to complete a further **design pass** (**turn** or **spin**) that requires additional NRE charges. Normally one signs a contract (sign off a design) with an ASIC vendor that guarantees first-pass success—this means that if one designed the ASIC according to rules specified by the vendor, then the vendor guarantees that the silicon will perform according to the simulation or one gets his money back.

Nowadays it is almost routine to have an ASIC work on the first pass. However, if a design fails, it is little consolation to have a second pass for free if the company goes bankrupt in the meantime. Figure 12.12 shows a **profit model** that represents the **profit flow** during the **product lifetime**. Using this model, we can estimate the lost profit due to any delay.

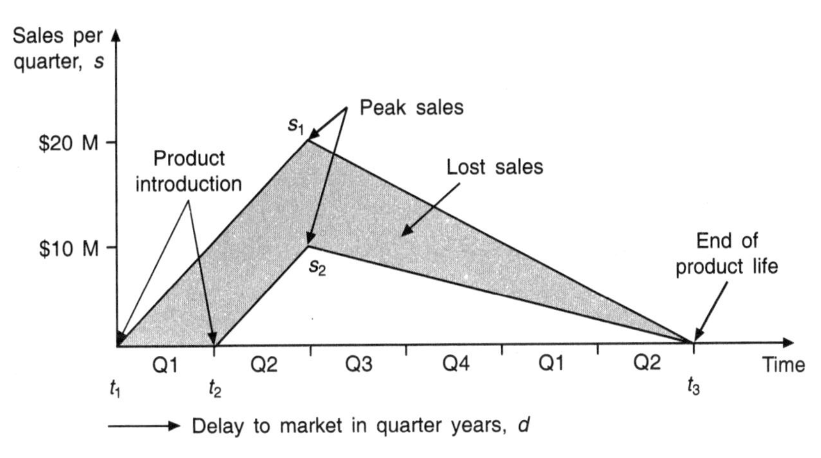

Figure 12.12 A profit model. If a product is introduced on time, the total sales are \$60 million (the area of the higher triangle). With a three-month (one fiscal quarter) delay the sales decline to \$25 million. The difference is shown as the shaded area between the two triangles and amounts to a lost revenue of \$35 million.

The last fixed cost shown in Figure 12.11 corresponds to an 'insurance policy.' When a company buys an ASIC part, it needs to be assured that it will always have a back-up source, or second source, in case something happens to its first or primary source. Established FPGA companies have a second source that produces equivalent parts. With a custom ASIC you may have to do some redesign to transfer your ASIC to the second source. However, for all ASIC types, switching production to a second source will involve some cost. Figure 12.11 assumes a second-source cost of \$2000 for all types of ASIC (the amount may be substantially more than this).

12.3.3.3 ASIC variable costs

Figure 12.13 shows a spreadsheet, 'Variable Costs,' that calculates some example part costs. This spreadsheet uses the terms and parameters defined below the figure.

	FPGA	MGA	CBIC	Units
Wafer size	6	6	6	inches
Wafer cost	1,400	1,300	1,500	$
Design	10,000	10,000	10,000	gates
Density	10,000	20,000	25,000	gates/sq.cm
Utilization	60	85	100	%
Die size	1.67	0.59	0.40	sq.cm
Die/wafer	88	248	365	
Defect density	1.10	0.90	1.00	defects/sq.cm
Yield	65	72	80	%
Die cost	25	7	5	$
Profit margin	60	45	50	%
Price/gate	0.39	0.10	0.08	cents
Part cost	$39	$10	$8	

Figure 12.13 A spreadsheet, 'Variable Costs,' to calculate the part cost (that is the variable cost for a product using ASICs) for different ASIC technologies.

- The **wafer size** increases every few years. From 1985 to 1990, 4-inch to 6-inch diameter wafers were common; equipment using 6-inch to 8-inch wafers was introduced between 1990 and 1995; the present size is the 300 cm or 12-inch wafer.
- The **wafer cost** depends on the equipment costs, process costs, and overhead in the fabrication line. A typical wafer cost is between $1000 and $5000, with $2000 being average; the cost declines slightly during the life of a process and increases only slightly from one process generation to the next.
- **Moore's Law** models the observation that the number of transistors on a chip roughly doubles every 18 months. Not all designs follow this law, but a 'large' ASIC design seems to grow by a factor of 10 every 5 years (close to Moore's Law). In 1990, a large ASIC design size was 10 k-gate, in 1995 a large design was about 100 k-gate, in 2000 it was 1 M-gate, in 2005 it was 10 M-gate.
- The **gate density** is the number of gate equivalents per unit area (a gate equivalent, or gate, corresponds to a two-input NAND gate).
- The **gate utilization** is the percentage of gates that are on a die that we can use (on a gate array, some gate space is wasted for interconnect).
- The **die size** is determined by the design size (in gates), the gate density, and the utilization of the die.
- The number of **dies per wafer** depends on the die size and the wafer size.
- The **defect density** is a measure of the quality of the fabrication process. The smaller the defect density, the less likely there is to be a flaw on any one die. A single defect

on a die is almost always fatal for that die. Defect density usually increases with the number of steps in a process. A defect density of less than 1 cm^{-2} is typical and required for a submicron CMOS process.

- The **yield** is the fraction of die on a wafer that is good (expressed as a percentage). The yield of a process is the key to a profitable ASIC company. Yield depends on the complexity and maturity of a process. A process may start out with a yield close to zero for complex chips, which then climbs to above 50 percent within the first few months of production. Within a year the yield has to be brought to around 80 percent for the average complexity ASIC for the process to be profitable. Yields of 90 percent or more are not uncommon.
- The **die cost** is determined by wafer cost, number of die per wafer, and the yield. Of these parameters, the most variable and the most critical to control is the yield.
- The **profit margin** (what you sell a product for, less what it costs you to make it, divided by the cost) is determined by the ASIC company's fixed and variable costs. ASIC vendors that make and sell custom ASICs have huge fixed and variable costs associated with building and running fabrication facilities (a fabrication plant is a **fab**). FPGA companies are typically **fabless**—they do not own a fab—they must pass on the costs of the chip manufacture (plus the profit margin of the chip manufacturer) and the development cost of the FPGA structure in the FPGA part cost. The profitability of any company in the ASIC business varies greatly.
- The **price per gate** is determined by die costs and design size. It varies with design size and declines over time.
- The **part cost** is determined by all of the preceding factors. It will vary widely with time, process, yield, economic climate, ASIC size and complexity, and many other factors.

As an estimate, the price per gate for any process technology falls at about 20% per year during its life (the average life of a CMOS process is 2-4 years, and can vary widely). Beyond the life of a process, prices can increase as demand falls and the fabrication equipment becomes harder to maintain.

12.3.4 CMOS Chip Design with Programmable Logic

In this section, a range of CMOS chip design options using programmable logic are examined. These are arranged in order of increased design investment. The sequence is also some what in order of complexity of device that may be implemented. In CMOS, the programmable devices may be divided into three areas:

- Chips with programmable logic structures
- Chips with programmable inter connect
- Chips with reprogrammable gate arrays

The CMOS system designer should be familiar with these options for two reasons:

(i) First, it allows the designer to competently assess a particular system requirement for an IC and recommend a solution, given the system complexity, the speed-of-operation, cost goals, time-to-market goals, and other top-level concerns.

(ii) Second, it familiarizes the IC system designer with methods of making any chip design re-programmable and hence more useful and of widespread use.

12.3.4.1 Programmable logic structures

The first broad class of programmable CMOS devices are represented by the programmable logic devices referred to as PALs (Programmable Array Logics), Generally, these devices are implemented as AND–OR plane devices. In the design shown, a number of inputs feed vertical wires, which are selectively connected to an AND–OR gate. Each AND–OR gate has a variable number of product terms that feed the gate. This gate in turn feeds an I/O cell, which allows registering of the registered result into the AND–OR plane. PAL devices come in a large range of sizes with a variable number of inputs, outputs, product terms, and I/O cell complexity.

The 22V10 is an industry standard for devices with the following characteristics:

- 12 inputs
- 10 I/Os
- #product terms 9 10 12 14 16 14 12 10 8
- 24 pins

12.3.4.2 Programming of PALs

The programming of PALs is done in three main ways:

- Fusible links
- UV—erasable EPROM
- EEPROM (E^2PROM)—Electrically Erasable Programmable ROM

Fusible links. Fusible links use a metal such as platinum silicide or titanium tungsten to form links that are blown when a certain current is exceeded in the fuse. This is normally accomplished by using a higher than normal programming voltage applied to the device. This technology is normally used in conjunction with a bipolar process (as opposed to a CMOS process) where the small devices can readily sink the current needed to blow the fuses. Programming is a one-time operation. As an alternative to current, a laser can be used to out aluminium fuses in normal CMOS technologies. Often this is used in redundant memory techniques where a spare column may be switched in to replace a failing one.

UV-erasable EPROM. UV—erasable memories typically use a floating gate structure as shown in Figure 12.14. Here, a floating gate is interposed between the regular MOS transistor gate and the channel. To program the cell, a voltage around 13–14 V is applied to the control gate while the drain of the transistor to be programmed is held at around 12 V.This results in the floating gate becoming charged negatively. This increases the threshold of the transistor (to around 7 V), this rendering it permanently 'off' for all normal circuit voltages (maximum 5–6 V). The process can be reversed by illuminating the gate with UV light, 'permanently' means at least 10 years at 125°C. At elevated temperatures, the storage time will be reduced and programming may be completed numerous times. The chips are usually housed in glass-lidded packages to allow illumination by UV light.

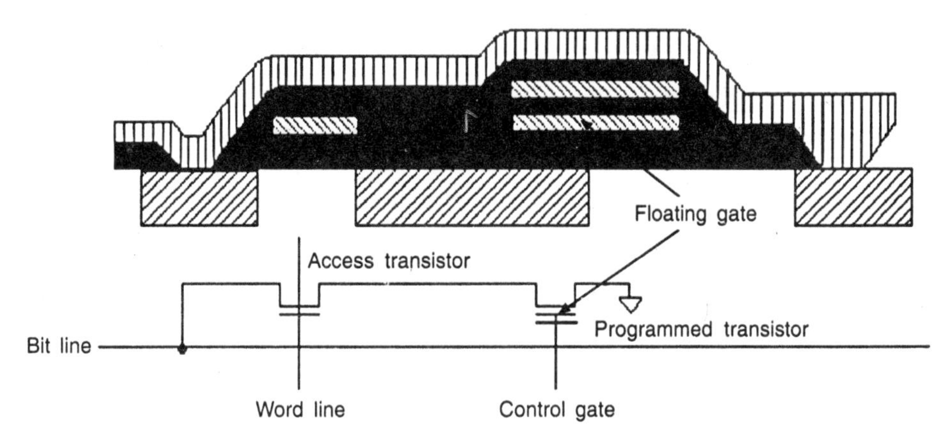

Figure 12.14 UV-erasable EPROM structure.

Electrically erasable programmable ROM. The electrically erasable programmable ROM (EEPROM) technology allows the electrical programming and erasure of CMOS ROM cells. This type of programming is most popularly used today for CMOS. A typical structure is shown in Figure 12.15.

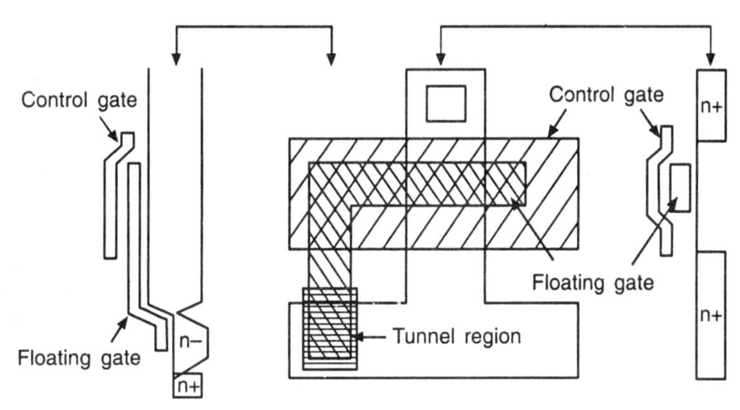

Figure 12.15 EEPROM structure.

Two transistors are typically used in a ROM cell. One is an access transistor while the other is the programmed transistor. A two-poly sandwich is again used in the programmed transistor with the control gate on the top. A very thin oxide layer between the floating gate and the drain of the device allows the electrons to 'tunnel' to or from the floating gate (thus charging the gate oxide) to turn the cell off or on respectively. The series—access transistor allows programming of cells. EEPROM has a testability advantage over fused technologies. Each device can be fully tested before shipment. A range of ROM architectures have been used, including the normal NOR ROM structure and NAND structures.

12.3.4.3 Programmable interconnect

In a PAL, the device is programmed by changing the characteristics of the switching element. An alternative would be to program the routing. This has been demonstrated via a number of techniques including Laser Pantography, where a laser lays down paths of metal under computer control. Commercially programmable routing approaches are represented by products from Actel, Quick Logic, and other companies.

The Actel field programmable arrays are based on an element called a PLICE (Programmable Low-Impedance Circuit Element) or antifuse. An antifuse is normally high resistance (>100 MΩ). On application of appropriate programming voltages, the antifuse is changed permanently to a low-resistance structure (200–500 Ω). The structure of an antifuse is shown in Figure 12.16(a). It consists of an ONO (oxide–nitride–oxide) layer sandwiched between a polysilicon layer on top and an n^+ diffusion on the bottom. The Quicklogic array is based on a structure called a Via Link,® which consists of a sandwich of material between metal 1 and metal 2. This is illustrated in Figure 12.16(b). The 'ON' resistance of this structure is somewhat lower than that in Figure 12.16(a).

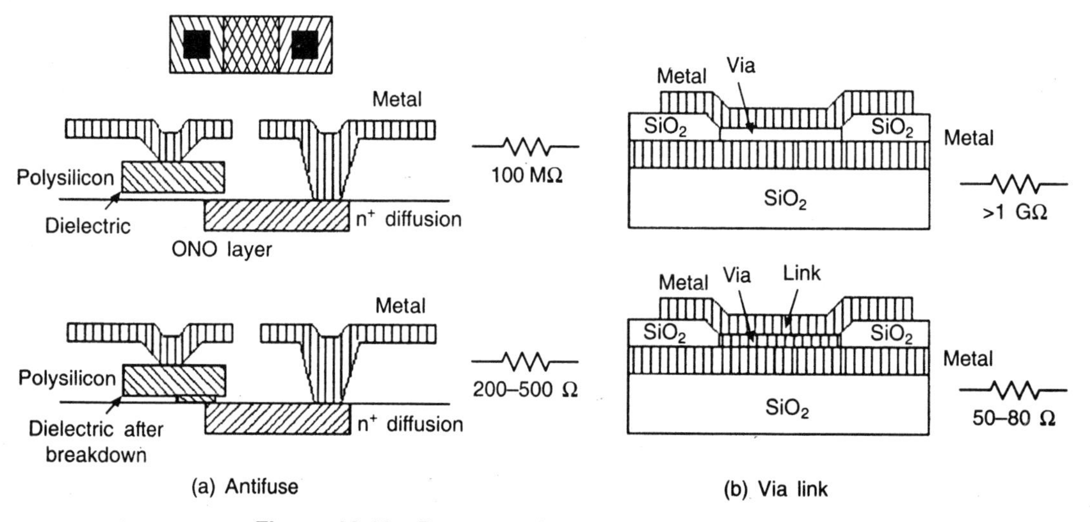

Figure 12.16 Programmable interconnect structures.

One chip architecture that uses the antifuse is shown in Figure 12.17. Logic elements are arranged in rows separated by horizontal interconnect. Interconnect permanently connected to the logic elements passes vertically. Both horizontal and vertical segments are segmented into a variety of lengths. Segments may be joined by programming antifuses. Certain special signals such as power and a clock line are routed globally to all logic. The logic elements are surrounded by I/O pads and programming and diagnostic logic.

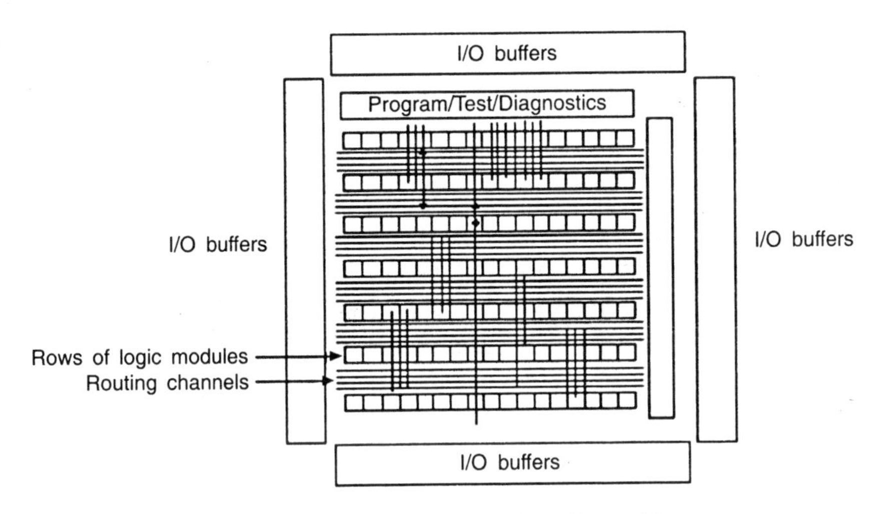

Figure 12.17 Actel FPGA chip architecture.

The structure of an Actel logic element shown in Figure 12.18. It consists of three input MUXes and a NOR gate. This structure can implement all 2- and 3-input logic functions and some 4-input functions. A latch may be implemented with one logic element, while a register requires two elements. The Quick Logic cell is shown in Figure 12.19. In addition to the structure shown in Figure 12.18, it includes a resettable register and numerous logic gates. An interesting trade-off in these types of arrays is the granularity of the logic cell verses the amount of routing.

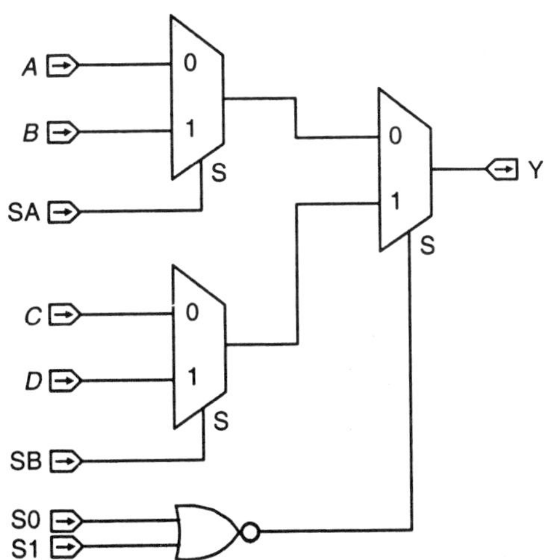

Figure 12.18 Actel logic cell.

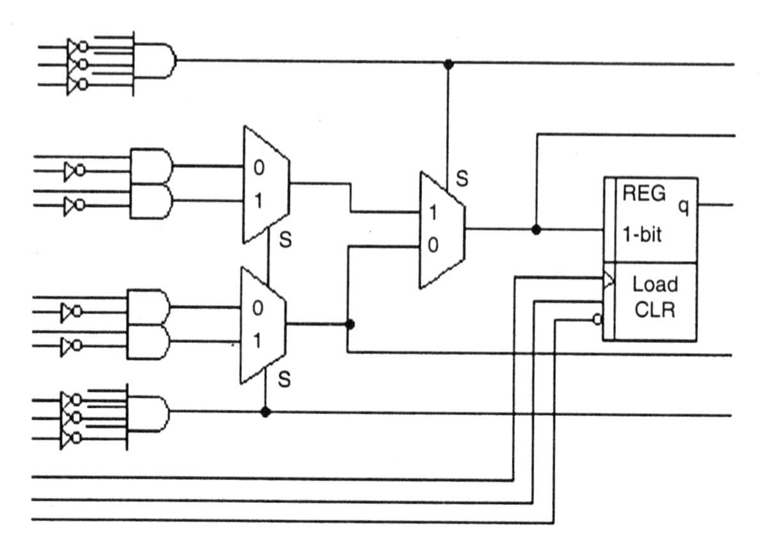

Figure 12.19 Quick logic cell.

The Actel programmable I/O pad is shown in Figure 12.20. Two antifuses allow the configuration to operate as an input pad, output pad, or bidirectional pad. If the ENABLE pin is not programmed, then the pad is bidirectional. If the ENABLE antifuse to V_{DD} is blown, the pad is an output, whereas if the V_{SS} antifuse is blown, the pad is an input. The isolation devices isolate the pad if necessary during programming and testing. A highly desirable feature of the Actel architecture is the ability to observe any node in the chip using the series—pass transistors that are used for programming.

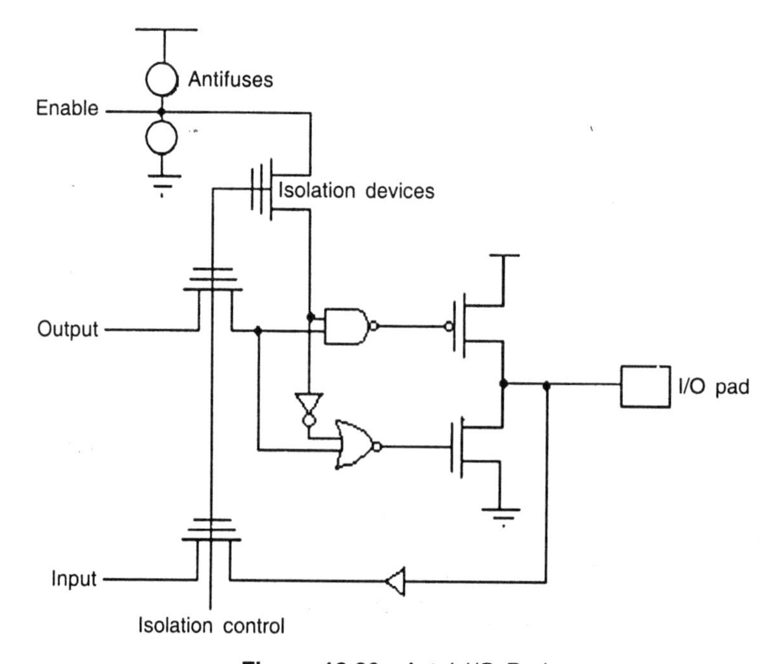

Figure 12.20 Actel I/O Pad.

At the time of writing, these arrays would implement 550 logic modules and 70 I/O modules. The speed of a particular circuit depends on the logic element speed and the delay through antifuse elements in any routing. A single logic module exhibits a delay from 7 ns to 14 ns (5 V and 25°C) depending on fan-out in a 2 μm technology. Long route delays through many antifuses can range from 15 ns to 35 ns. With smaller technologies, the logic module delays would decrease while the routing delays might decrease somewhat. More drastic reductions in the routing delays would come with lower 'on-resistance' antifuses.

If a 32-bit adder is implemented in an Actel array, 160 logic modules would be needed, and it would add in approximately 65 ns. Thus, roughly 3.5 32-bit adders would fit in a single FPGA chip.

12.3.4.4 Reprogrammable gate arrays

A further class of programmable device is the programmable (or reprogrammable) gate array. These may be further categorized into ad-hoc and structured arrays. The different versions of reprogrammable gate arrays are: The Xilinx programmable Gate Array, Algotronix, Concurrent logic, etc. Xilinx Gate array is an example of an ad-hoc array, e.g., XC 3000 series. An example of a regular programmable array is the CAL 1024 from Algotronix. The CLI 6000 series is another example of a regular array style FPGA. A detailed description about these versions is given in Chapter 11.

SUMMARY

This chapter has covered a broad spectrum of design issues that may be encountered when designing CMOS chips. The structured design strategies are useful for any kind of CMOS-chip design method. The Application Specific Integrated Circuits (ASICs) is also described in a detailed manner. A range of implementation options are given to give the reader an appreciation for the wide spectrum of solutions that are available today.

REVIEW QUESTIONS

1. Explain the different design strategies required for designing a VLSI system.
2. Explain in detail the different types of Application Specific Integrated Circuits and the design constraints.
3. Explain in detail the CMOS chip design using programmable logic structures.
4. Explain in detail the different ways of programming PLAs.
5. Explain with example the CMOS chip design using programmable interconnect.
6. What are reprogrammable gate arrays? Explain the use of reprogrammable gate arrays for the design of CMOS chip design.

SHORT ANSWER QUESTIONS

1. A good VLSI design system should provide for consistent descriptions in ____________, ____________, and ____________ description domains.
   Ans. behavioural; structural; physical.
2. ASIC is often termed as ____________.
   Ans. System on a Chip (SoC)
3. What are the different types of ASICs?
   Ans. (a) Full-custom ASIC
   (b) Semicustom ASIC
   (c) Programmable ASIC.
4. What are the two types of programmable ASICs?
   Ans. (a) Programmable logic device
   (b) Field programmable gate array
5. ICs are made in batches called ____________.
   Ans. wafer lots.
6. A cell-based ASIC uses predesigned logic cells known as ____________.
   Ans. standard cells.
7. In routing, a connection that needs to cross over a row of standard cells uses a ____________.
   Ans. feedthrough.
8. The predefined pattern of transistors on a gate array is ____________.
   Ans. base array.
9. What are the different types of gate array based ASICs?
   Ans. (i) Channelled gate arrays.
   (ii) Channelless gate arrays.
   (iii) Structured gate arrays.
10. What are the different names of channelless gate array?
    Ans. (i) Channel free gate array
    (ii) Sea of gates array
11. Structured gate array is otherwise called ____________.
    Ans. masterslice
12. What is the total product cost?
    Ans. Total cost of product = Fixed cost of product + variable cost of product × numbers sold
13. ____________ is determined by the design size (in gates), gate density, and the utilization of the die.
    Ans. Die size
14. How the programming of PLAs is done?
    Ans. (a) Fusible links
    (b) UV-erasable EPROM
    (c) EEPROM.

13 Routing Procedures

13.1 INTRODUCTION

Once the designer has floorplanned a chip and the logic cells within the flexible blocks have been placed, it is time to make the connections by routing the chip. This is still a hard problem that is made easier by dividing it into smaller problems. Routing is usually split into global routing followed by detailed routing.

13.2 GLOBAL ROUTING

The details of global routing differ slightly between cell-based ASICs, gate arrays, and FPGAs, but the principles are the same in each case. A global router does not make any connections, it just plans them. We typically global route the whole chip (or large pieces if it is a large chip) before detailed routing the whole chip (or the pieces). There are two types of areas to global route: inside the flexible blocks and between blocks.

13.2.1 Goals and Objectives

The input to the global router is a floorplan that includes the locations of all the fixed and flexible blocks: the placement information for flexible blocks and the location of all the logic cells. The goal of global routing is to provide complete instructions to the detailed router on where to route every net. The objectives of global routing are one or more of the following:

- Minimize the total interconnect length
- Maximize the probability that the detailed router can complete the routing
- Minimize the critical path delay

In both floorplanning and placement, with minimum interconnect length as an objective, it is necessary to find the shortest total path length connecting a set of terminals. This path is the MRST, which is hard to find. The alternative, for both the floorplanning and the placement, is to use simple approximations to the length of the MRST (usually the half-perimeter measure). Floorplanning and placement both assume that interconnect may be put anywhere on a rectangular grid, since at this point nets have not been assigned to the channels, but the global router must use the wiring channels and find the actual path. Often the global router needs to find a path that minimizes the delay between two terminals—this is not necessarily the same as finding the shortest total path length for a set of terminals.

13.2.2 Measurement of Interconnect Delay

Floorplanning and placement need a fast and easy way to estimate the interconnect delay in order to evaluate each trial placement; often this is a predefined look-up table. After placement, the logic cell positions are fixed and the global router can afford to use better estimates of the interconnect delay. To illustrate one method, we shall use the Elmore constant to estimate the interconnect delay for the circuit shown in Figure 13.1.

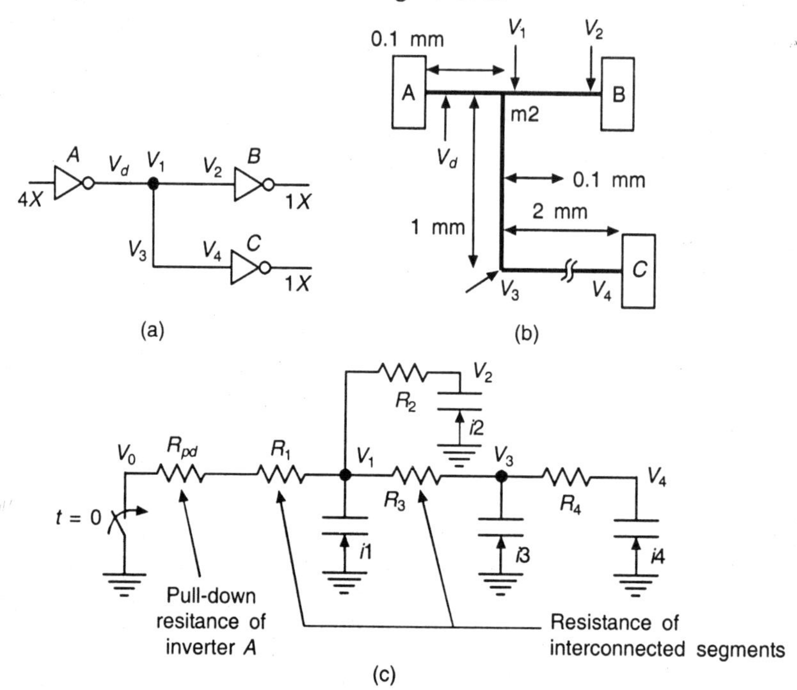

Figure 13.1 Measuring the delay of a net. (a) A simple circuit with an inverter A driving a net with a fan-out of two. Voltages V_1, V_2, V_3, and V_4 are the voltages at intermediate points along the net. (b) The layout showing the net segments (pieces of interconnect). (c) The RC model with each segment replaced by a capacitance and resistance. The ideal switch and pull-down resistance R_{pd} model the inverter A.

The problem is to find the voltages at the inputs to logic cells B and C taking into account the parasitic resistance and capacitance of the metal interconnect. Figure 13.1(c) models logic cell A as an ideal switch with a pull-down resistance equal to R_{pd} and models the metal interconnect using resistors and capacitors for each segment of the interconnect.

The Elmore constant for node 4 (labelled V_4) in the network shown in Figure 13.1(c) is

$$\tau_{D4} = \sum_{k=1}^{4} R_{kr} C_k = R_{14}C_1 + R_{24}C_2 + R_{34}C_3 + R_{44}C_4 \tag{13.1}$$

where,

$$R_{14} = R_{pd} + R_1 \qquad R_{24} = R_{pd} + R_1$$

$$R_{34} = R_{pd} + R_1 + R_3 \qquad R_{44} = R_{pd} + R_1 + R_3 + R_4 \tag{13.2}$$

In Eq. (13.2) notice that $R_{24} = R_{pd} + R_1$ (and not $R_{pd} + R_1 + R_2$) because R_1 is the resistance to V_0 (ground) shared by node 2 and node 4.

Suppose we have the following parameters (from the generic 0.5 μm CMOS process, G5) for the layout shown in Figure 13.1(b):

- m2 resistance is 50 mΩ/square.
- m2 capacitance (for a minimum-width line) is 0.2. pF-min^{-1}.
- 4X inverter delay is 0.02 ns + $0.5C_L$ ns (C_L is in picofarads).
- Delay is measured using 0.35/0.65 output trip points.
- m2 minimum width is $3\lambda = 0.9$ μm.
- IX inverter input capacitance is 0.02 pF (a standard load).

First we need to find the pull-down resistance R_{pd}, of the 4X inverter. If we model the gate with a linear pull-down resistor R_{pd}, driving a load C_L, the output waveform is exp $-t/(C_L R_{pd})$ (normalized to 1 V). The output reaches 63% of its final value when $t = C_L R_{pd}$, because exp (–1) = 0.63. Then, because the delay is measured with a 0.65 trip point, the constant 0.5 nspF^{-1} = 0.5 kΩ is very close to the equivalent pull-down resistance. Thus R_{pd} = 500 Ω.

From the given data, we can calculate the R's and C's:

$$R_1 = R_2 = \frac{(0.1\text{ mm})(50\times 10^{-3}\ \Omega)}{0.9\ \mu\text{m}} = 6\ \Omega$$

$$R_3 = \frac{(1\text{ mm})(50\times 10^{-3}\ \Omega)}{0.9\ \mu\text{m}} = 56\ \Omega \tag{13.3}$$

$$R_4 = \frac{(2\text{ mm})(50\times 10^{-3}\ \Omega)}{0.9\ \mu\text{m}} = 112\ \Omega$$

$$C_1 = (0.1\text{ mm})\ (0.2\text{ pF mm}^{-1}) = 0.02\text{ pF}$$

$$C_2 = (0.1\text{ mm})\ (0.2\text{ pF mm}^{-1}) + 0.02\text{ pF} = 0.04\text{ pF}$$

$$C_3 = (1\text{ mm})\ (0.2\text{ pF mm}^{-1}) = 2\text{ pF} \tag{13.4}$$

$$C_4 = (2\text{ mm})\ (0.2\text{ pF mm}^{-1}) + 0.02\text{ pF} = 0.42\text{ pF}$$

Now we can calculate the path resistance, R_{ki} values (notice that $R_{ki} = R_{ik}$):

$$R_{14} = 500\ \Omega + 6\ \Omega = 506\ \Omega$$

$$R_{24} = 500\ \Omega + 6\ \Omega = 506\ \Omega$$

$$R_{34} = 500\ \Omega + 6\ \Omega + 56\ \Omega = 562\ \Omega \tag{13.5}$$

$$R_{44} = 500\ \Omega + 6\ \Omega + 56\ \Omega + 112\ \Omega = 674\ \Omega$$

Finally, we can calculate Elmore's constants for node 4 and node 2 as follows:

$$\tau_{D4} = R_{14}C_1 + R_{24}C_2 + R_{34}C_3 + R_{44}C_4 \quad (13.6)$$
$$= (506)(0.02) + (506)(0.04) + (562)(0.2) + (674)(0.42)$$
$$= 425 \text{ ps}$$

$$\tau_{D2} = R_{12}C_1 + R_{22}C_2 + R_{32}C_3 + R_{42}C_4$$
$$= (R_{pd} + R_1 + R_2)C_2 + (R_{pd} + R_1)(C_1 + C_3 + C_4) \quad (13.7)$$
$$= (500 + 6 + 6)(0.04) + (500 + 6)(0.02 + 0.2 + 0.42)$$
$$= 344 \text{ ps}$$

and $$\tau_{D4} - \tau_{D2} = (425 - 344) = 81 \text{ ps}$$

A **lumped-delay model** neglects the effects of interconnect resistance and simply sums all the node capacitances (the lumped capacitance) as follows:

$$t_D = R_{pd}(C_1 + C_2 + C_3 + C_4)$$
$$= (500)(0.02 + 0.04 + 0.2 + 0.42) \quad (13.8)$$
$$= 340 \text{ ps}$$

Comparing Eqs. (13.6)–(13.8), we can see that the delay of the inverter can be assigned as follows: 20 ps (the intrinsic delay, 0.2 ns, due to the cell output capacitance), 340 ps (due to the pull-down resistance and the output capacitance), 4 ps (due to the interconnect from A to B), and 65 ps (due to the interconnect from A to C). We can see that the error from neglecting interconnect resistance can be important.

Even using the Elmore constant we still made the following assumptions in estimating the path delays:

- A step-function waveform drives the net.
- The delay is measured from when the gate input changes.
- The delay is equal to time constant of an exponential waveform that approximates the actual output waveform.
- The interconnect is modelled by discrete resistance and capacitance elements.

The global router could use more sophisticated estimates that remove some of these assumptions, but there is a limit to the accuracy with which delay can be estimated during global routing. For example, the global router does not know how much of the routing is on which of the layers, or how many vias will be used and of which type, or how wide the metal lines will be. It may be possible to estimate how much interconnect will be horizontal and how much is vertical. Unfortunately, this knowledge does not help much if horizontal interconnect may be completed in either m1 or m3 and there is a large difference in parasitic capacitance between m1 and m3. For example, when the global router attempts to minimize interconnect delay, there is an important difference between a path and a net. The path that minimizes the delay between two terminals on a net is not necessarily the same as the path that minimizes the total path length of the net. For example, to minimize the path delay (using the Elmore constant as a measure) from the output of inverter A in Figure 13.1(a) to the input of inverter B requires a rather complicated algorithm to construct the best path.

13.2.3 Global Routing Methods

Global routing cannot use the interconnect length approximations, such as the half perimeter measure, that were used in placement. What is needed now is the actual path and not an approximation to the path length. However, many of the methods used in global routing are still based on the solutions to the tree on a graph problem.

One approach to global routing takes each net in turn and calculates the shortest path using tree on graph algorithms—with the added restriction of using the available channels. This process is known as **sequential routing**. As a sequential routing algorithm proceeds, some channels will become more congested since they hold more interconnects than others. In the case of FPGAs and channelled gate arrays, the channels have a fixed channel capacity and can only hold a certain number of interconnects. There are two different ways that a global router normally handles this problem. Using order-independent routing a global router proceeds by routing each net, ignoring how crowded the channels are. Whether a particular net is processed first or last does not matter, the channel assignment will be the same. In order-independent routing, after all the interconnects are assigned to channels, the global router returns to those channels that are the most crowded and reassigns some interconnects to other, less crowded channels. Alternatively, a global router can consider the number of interconnects already placed in various channels as it proceeds. In this case, the global routing is order dependent—the routing is still sequential, but now the order of processing the nets will affect the results. Iterative improvement or simulated annealing may be applied to the solution found from both order-dependent and order-independent algorithms. This is implemented in the same way as for system partitioning and placement: A constructed solution is successively changed, one interconnect path at a time, in a series of random moves.

In contrast to sequential global-routing methods, which handle nets one at a time, hierarchical routing handles all nets at a particular level at once. Rather than handling all of the nets on the chip at the same time, the global-routing problem is made more tractable by dividing the chip area into levels of hierarchy. By considering only one level of hierarchy at a time, the size of the problem is reduced at each level. There are two ways to traverse the levels of hierarchy. Starting at the whole chip, or highest level, and proceeding down to the logic cells is the ***top-down approach***. The ***bottom-up approach*** starts at the lowest level of hierarchy and globally routes the smallest areas first.

13.2.4 Global Routing Between Blocks

Figures 13.2(a), (b) and (c) illustrate the global-routing problem for a cell-based ASIC. Each edge in the channel-intersection graph in Figure 13.2(c) represents a channel. The global router is restricted to using these channels. The weight of each edge in the graph corresponds to the length of the channel. The global router plans a path for each interconnect using this graph.

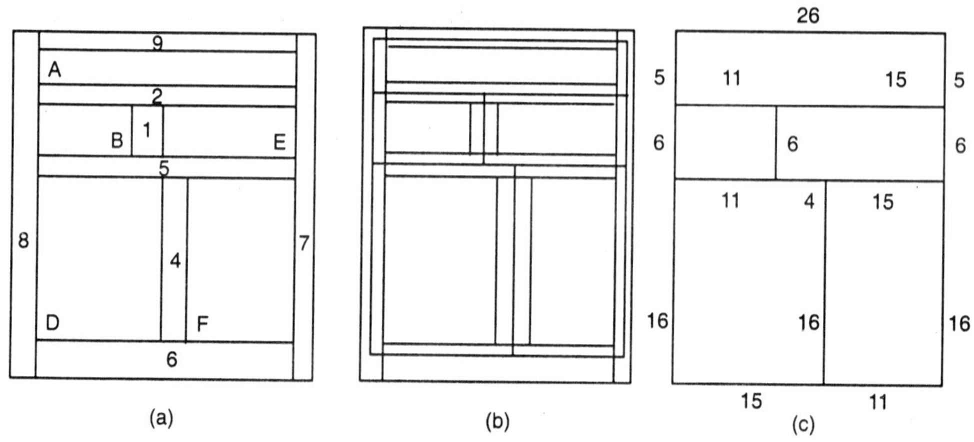

Figure 13.2 Global routing for a cell-based ASIC formulated as a graph problem. (a) A cell-based ASIC with numbered channels. (b) The channels from the edges of a graph. (c) The channel-intersection graph. Each channel corresponds to an edge on a graph whose weight corresponds to the channel length.

Figure 13.3(a) shows an example of global routing for a net with five terminals, labelled A1 through F1, for the cell-based ASIC shown in Figure 13.2. If a designer wishes to use minimum total interconnect path length as an objective, the global router finds the minimum length tree shown in Figure 13.3(b). This tree determines the channels the interconnects will use. For example, the shorter connection from A1 to B1 uses channels 2, 1, and 5 (in that order). This is the information the global router passes to the detailed router. Figure 13.3(c) shows that minimizing the total path length may not correspond to minimizing the path delay between two points.

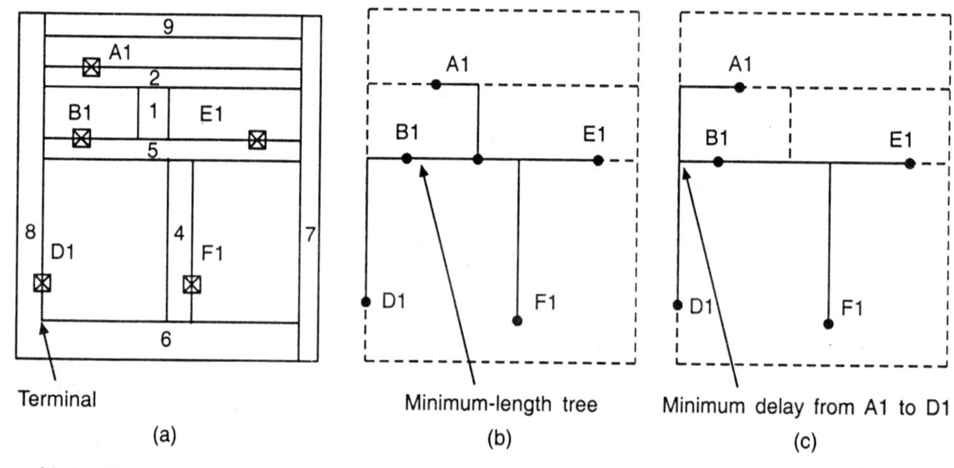

Figure 13.3 Finding paths in global routing. (a) A cell-based ASIC (from Figure 13.2) showing a single net with a fan-out of four (five terminals). We have to order the numbered channels to complete the interconnect path for terminals A1 through F1. (b) The terminals are projected to the centre of the nearest channel, forming a graph. A minimum-length tree for the net that uses the channels and takes into account the channel capacities. (c) The minimum-length tree does not necessarily correspond to minimum delay. If we wish to minimize the delay from terminal A1 to D1, a different tree might be better.

Global routing is very similar for cell-based ASICs and gate arrays, but there is a very important difference between the types of channels in these ASICs. The size of the channels in sea-of-gates arrays, channelless gate arrays, and cell-based ASICs can be varied to make sure there is enough space to complete the wiring. In channelled gate-arrays and FPGAs, the size, number, and location of channels are fixed. The good news is that the global router can allocate as many interconnects to each channel as it likes, since that space is committed anyway. The bad news is that there is a maximum number of interconnects that each channel can hold. If the global router needs more room, even in just one channel on the whole chip, the designer has to repeat the placement and routing steps and try again (or use a bigger chip).

13.2.5 Global Routing Inside Flexible Blocks

We shall illustrate global routing using a gate array. Figure 13.4(a) shows the routing resources on a sea-of-gates or channelless gate array. The gate array base cells are arranged in 36 blocks, each block containing an array of 8-bý gate-array base cells, making a total of 4068 base cells.

The horizontal interconnect resources are the routing channels that are formed from unused rows of the gate-array base cells, as shown in Figures 13.4(b) and 13.4(c). The vertical resources are feedthroughs. For example, the logic cell shown in Figure 13.4(d) is an inverter that contains two types of feedthroughs. The inverter logic cell uses a single gate-array base cell with terminals (or connectors) located at the top and bottom of the logic cell. The inverter input pin has two electrically equivalent terminals that the global router can use as a feedthrough. The output of the inverter is connected to only one terminal. The remaining vertical track is unused by the inverter logic cell, so this track forms an uncommitted feedthrough.

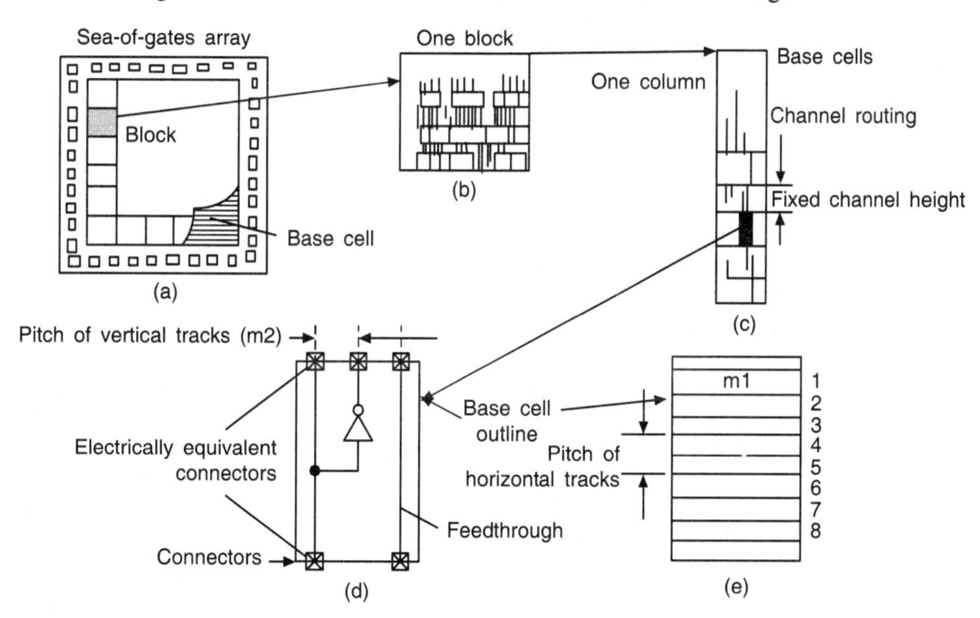

Figure 13.4 Gate-array global routing: (a) A small gate array, (b) An enlarged view of the routing. The top channel uses three rows of gate-array base cells; the other channels use only one, (c) A further enlarged view showing how the routing in the channels connects to the logic cells, (d) One of the logic cells, an inverter, (e) There are seven horizontal wiring tracks available in one row of gate-array base cells—the channel capacity is thus 7.

You may see any of the terms ***landing pad*** (because we say that we 'drop' a via to a landing pad), ***pick-up point, connector, terminal, pin,*** or ***port*** used for the connection to a logic cell. The term ***pick-up point*** refers to the physical pieces of metal (or sometimes polysilicon) in the logic cell to which the router connects. In a three-level metal process, the global router may be able to connect to anywhere in an area—an ***area pick-up point***. In this book, we use the term connector to refer to the physical pick-up point. The term pin more often refers to the connection on a logic schematic icon (a dot, square box, or whatever symbol is used), rather than layout. Thus the difference between a pin and a connector is that we can have multiple connectors for one pin. Terminal is often used when we talk about routing. The term port is used when we are using text (EDIF netlists or HDLs, for example) to describe circuits.

In a gate array, the channel capacity must be a multiple of the number of horizontal tracks in the gate array base cell. Figure 13.4(e) shows a gate-array base cell with seven horizontal tracks. Thus, in this gate array, we can have a channel with a capacity of 7, 14, 21, ... horizontal tracks—but not between these values.

Figure 13.5 shows the inverter macro for the sea-of-gates array shown in Figure 13.4. Figure 13.5(a) shows the base cell. Figure 13.5(b) shows how the internal inverter wiring on m1 leaves one vertical track free as a feedthrough in a two level metal process (connectors placed at the top and bottom of the cell). In a three level metal process, the connectors may be placed inside the cell abutment box [Figure 13.5(c)]. Figure 13.6 shows the global routing for the sea-of-gates array. We divide the array into non-overlapping routing bins (or just bins, also called global routing cells or GRCs), each containing a number of gate array base cells.

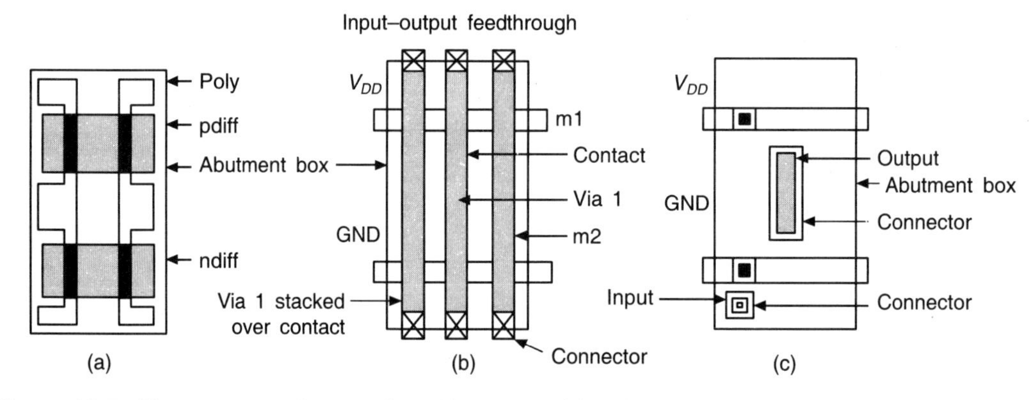

Figure 13.5 The gate-array inverter from Figure 13.4(d): (a) An oxide-isolated gate-array base cell, showing the diffusion and polysilicon layers, (b) The metal and contact layers for the inverter in a 2LM (two-level metal) process, (c) The router's view of the cell in a 3LM process.

We need an aside to discuss our use of the term cell. Be careful not to confuse the global routing cells with gate-array base cells (the smallest element of a gate array, consisting of a small number of n-type and p-type transistors), or with logic cells (which are NAND gates, NOR gates, and so on).

A large routing bin reduces the size of the routing problem, and a small routing bin allows the router to calculate the wiring capacities more accurately. Some tools permit routing bins of different sizes in different areas of the chip (with smaller routing bins helping in areas of dense routing). Figure 13.6(a) shows a routing bin that is 2-by-4 gate-array base cells. The logic cells occupy the lower half of the routing bin. The upper half of the routing bin is the channel area reserved for wiring. The global router calculates the edge capacities for this routing bin, including the vertical feedthroughs. The global router then determines the shortest path for each net considering these edge capacities. An example of a global-routing calculation is shown in Figure 13.6(b). The path, described by a series of adjacent routing bins, is passed to the detailed router.

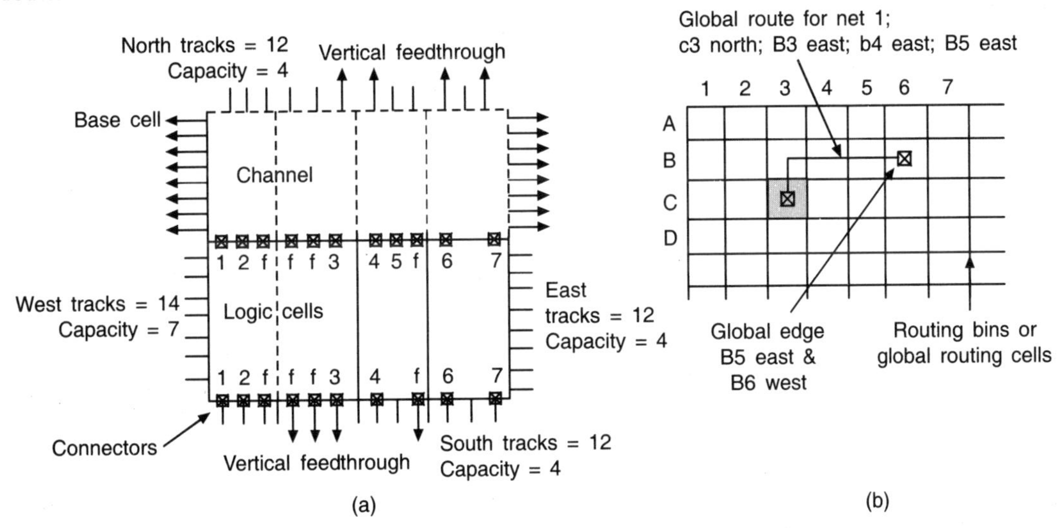

Figure 13.6 Global routing a gate array: (a) A single global-routing cell (GRC or routing bin) containing 2-by-4 gate-array base cells. For this choice of routing bin, the maximum horizontal track capacity is 14, the maximum vertical track capacity is 12. The routing bin labelled C3 contains three logic cells, two of which have feedthroughs marked ‘f’. This results in the edge capacities shown, (b) A view of the top left-hand corner of the gate array showing 28 routing bins. The global router uses the edge capacities to find a sequence of routing bins to connect the nets.

13.2.6 Timing-Driven Methods

Minimizing the total pathlength using a Steiner tree does not necessarily minimize the interconnect delay of a path. Alternative tree algorithms apply in this situation, most using the Elmore constant as a method to estimate the delay of a path. As in timing-driven placement, there are two main approaches to timing driven routing:

Path-based methods are more sophisticated. For example, if there is a critical path from logic cell A to B to C, the global router may increase the delay due to the interconnect between logic cells A and B if it can reduce the delay between logic cells B and C. Placement and global routing tools may or may not use the same algorithm to estimate net delay. If these tools are from different companies, the algorithms are probably different. The algorithms must be compatible, however. There is no use performing placement to minimize predicted delay if the global router uses completely different measurement methods. Companies that produce floorplanning and placement tools make sure that the output is compatible with different routing tools—often to the extent of using different algorithms to target different routers.

13.2.7 Back-Annotation

After global routing is complete, it is possible to accurately predict what the length of each interconnect in every net will be after detailed routing, probably to within 5%. The global router can give us not just an estimate of the total net length (which was all we knew at the placement stage), but the resistance and capacitance of each path in each net. This RC information is used to calculate net delays. We can back-annotate this net delay information to the synthesis tool for in-place optimization or to a timing verifier to make sure there are no timing surprises. Differences in timing predictions at this point arise due to the different ways in which the placement algorithms estimate the paths and the way the global router actually builds the paths.

13.3 DETAILED ROUTING

The global routing step determines the channels to be used for each interconnect. Using this information, the detailed router decides the exact location and layers for each interconnect. Figures 13.7(a), (b), (c) and (d) show typical metal rules. These rules determine the m1 routing pitch (track pitch, track spacing, or just pitch). We can set the m1 pitch to one of these values:

1. via-to-via (VTV) pitch (or spacing)
2. via-to-line (VTL or line-to-via) pitch, or
3. line-to-line (LTL) pitch.

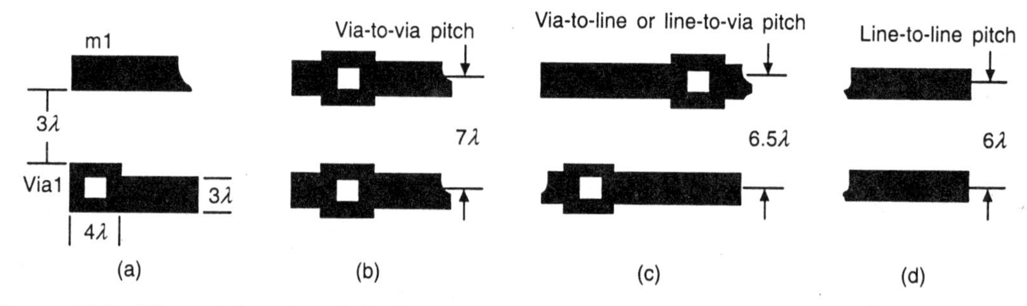

Figure 13.7 The metal routing pitch: (a) An example of λ-based metal design rules for m1 and via1 (m1/m2 via), (b) Via-to-via pitch for adjacent vias, (c) Via-to-line (or line-to-via) pitch for nonadjacent vias, (d) Line-to-line pitch with no vias.

The same choices apply to the m2 and other metal layers if they are present. Via-to-via spacing allows the router to place vias adjacent to each other. Via-to-line spacing is hard to use in practice because it restricts the router to nonadjacent vias. Using line-to-line spacing prevents the router from placing a via at all without using jogs and is rarely used. Via-to-via spacing is the easiest for a router to use and the most common. Using either via-to-line or via-to-via spacing means that the routing pitch is larger than the minimum metal pitch.

Sometimes people draw a distinction between a cut and a via when they talk about large connections such as shown in Figure 13.8(a). We split or **stitch** a large via into identically sized cuts (sometimes called a **waffle via**). Because of the profile of the metal in a contact and the way current flows into a contact, often the total resistance of several small cuts is less than that

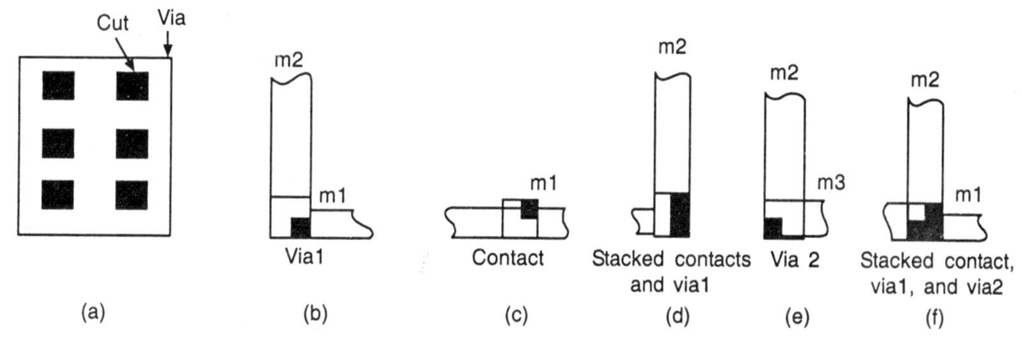

Figure 13.8 (a) A large m1 to m2 via. The black squares represent the holes (or cuts) that are etched in the insulating material between the m1 and m2 layers, (b) A m1 to m2 via (a via1), (c) A contact from m1 to diffusion or polysilicon (a contact), (d) A via 1 placed over (or stacked over) a contact, (e) A m2 to m3 via (a via2), (f) A via 2 stacked over a contact. Notice that the black squares in parts b-c do not represent the actual location of the cuts. The black squares are offset so you can recognize stacked vias and contacts.

of one large cut. Using identically sized cuts also means the processing conditions during contact etching, which may vary with the area and perimeter of a contact, are the same for every cut on the chip.

In a **stacked via** the contact cuts all overlap in a layout plot and it is impossible to tell just how many vias on which layers are present. Figures 13.8(b) to (f) show an alternative way to draw contacts and vias. Though this is not a standard, using the diagonal box convention makes it possible to recognize stacked vias and contacts on a layout (in any orientation). We shall use these conventions when it is necessary.

In a two-level metal CMOS ASIC technology, we complete the wiring using the two different metal layers for the horizontal and vertical directions, one layer for each direction. This is ***Manhattan routing***, because the results look similar to the rectangular north-south and east-west layout of streets in New York City. Thus, for example, if terminals are on the m2 layer, then we route the horizontal branches in a channel using m2 and the vertical trucks using m1. Figure 13.9 shows that, although we may choose a **preferred direction** for each metal layer (for example, m1 for horizontal routing and m2 for vertical routing), this may lead to problems in cases that have both horizontal and vertical channels. In these cases, we define a **preferred metal layer** in the direction of the channel spine. In Figures 13.9(a) and (b), because the logic cell connectors are on m2, any vertical channel has to use vias at every logic cell location. By changing the orientation of the metal directions in vertical channels, we can avoid this, and instead we only need to place vias at the intersection of horizontal and vertical channels.

Figure 13.10 shows an imaginary logic cell with connectors. Double-entry logic cells intended for two-level metal routing have connectors at the top and bottom of the logic cell, usually in m2. Logic cells intended for processes with three or more levels of metal have connectors in the centre of the cell, again usually on m2. Logic cells may use both m1 and m2 internally, but the use of m2 is usually minimized. The router normally uses a simplified view of the logic cell called a ***phantom***. The phantom contains only the logic cell information that the

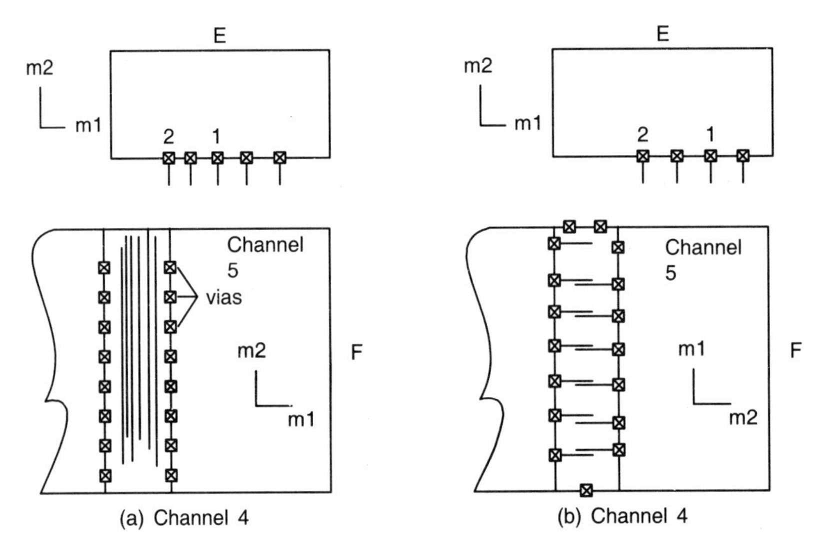

Figure 13.9 An expanded view of part of a cell-based ASIC: (a) Both channels 4 and 5 use m1 in the horizontal direction and m2 in the vertical direction. If the logic cell connectors are on m2, this requires vias to be placed at every logic cell connector in channel 4; (b) Channels 4 and 5 are routed with m1 along the direction of the channel spine (the long direction of the channel). Now vias are required only for nets 1 and 2, at the intersection of the channels.

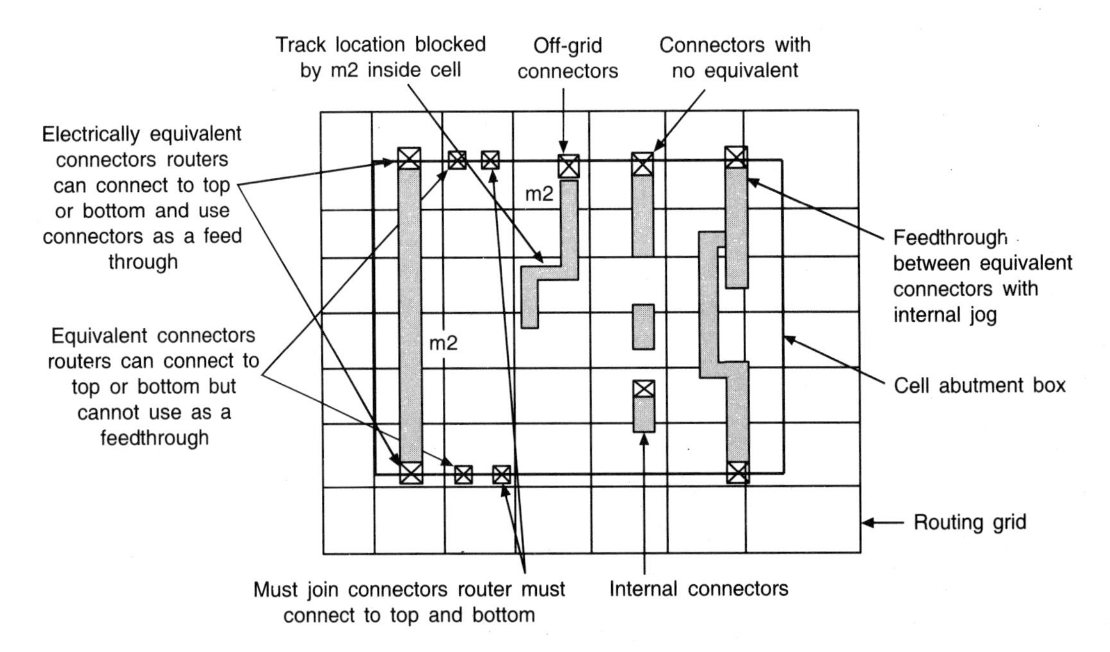

Figure 13.10 The different types of connections that can be made to a cell. This cell has connectors at the top and bottom of the cell and internal connectors.

router needs; the connector locations, types, and names; the abutment and bounding boxes; enough layer information to be able to place cells without violating design rules; and a ***blockage map***—the locations of any metal inside the cell that blocks routing.

Figure 13.11 illustrates some terms used in the detailed routing of a channel. The channel spine in Figure 13.11 is horizontal with terminals at the top and the bottom, but a channel can also be vertical. In either case, terminals are spaced along the longest edges of the channel at the given, fixed locations. Terminals are usually located on a grid defined by the routing pitch on that layer (we say terminals are either ***on-grid*** or ***off-grid***). We make connections between terminals using interconnects that consist of one or more trunks running parallel to the length of the channel and branches that connect the trunk to the terminals. If more than one trunk is used, the trunks are connected by ***doglegs***. Connections exit the channel at ***pseudoterminals***.

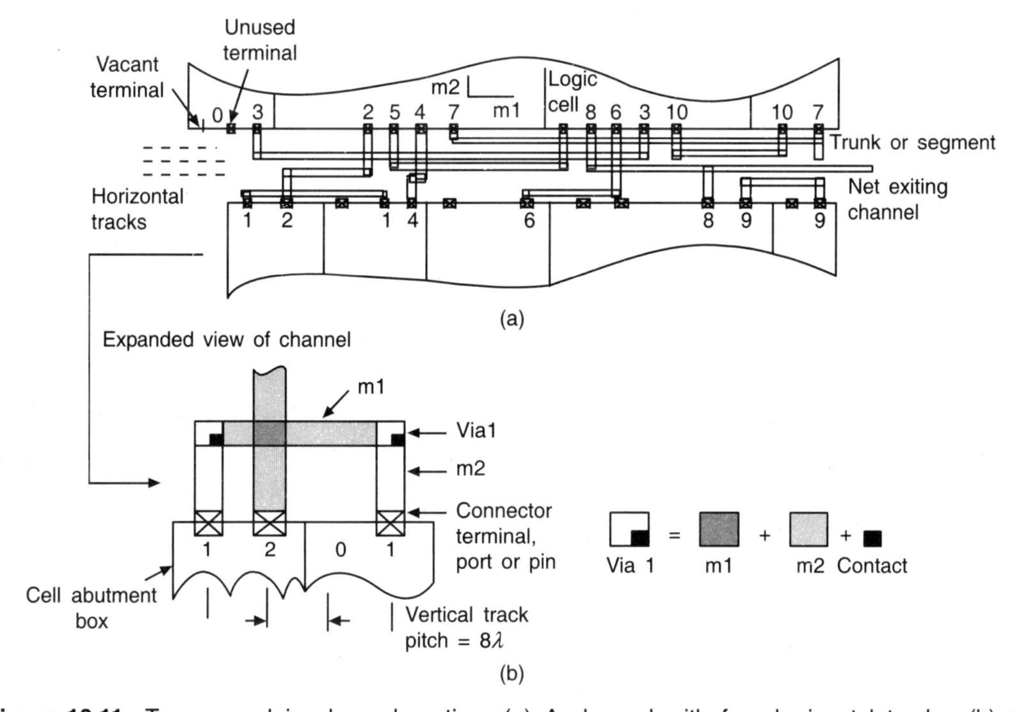

Figure 13.11 Terms used in channel routing: (a) A channel with four horizontal tracks, (b) An expanded view of the left-hand portion of the channel showing (approximately to scale) now the m1 and m2 layers connect to the logic cells on either side of the channel, (c) The construction of a via 1 (m1/m2 via).

The trunk and branch connections run in ***tracks*** (equispaced, like railway tracks). If the trunk connections use m1, the horizontal trunk spacing (usually just called the ***track spacing*** for channel routing) is equal to the m1 routing pitch. The maximum number of interconnects we need in a channel multiplied by the horizontal track spacing gives the minimum height of a channel. Each terminal occupies a column. If the branches use m2, the column spacing (or vertical track spacing) is equal to the m2 routing pitch.

13.3.1 Goals and Objectives

The goal of detailed routing is to complete all the connections between logic cells. The most common objective is to minimize one or more of the following:

- The total interconnect length and area.
- The number of layer changes that the connections have to make.
- The delay of critical paths.

Minimizing the number of layer changes corresponds to minimizing the number of vias that add parasitic resistance and capacitance to a connection.

In some cases, the detailed router may not be able to complete the routing in the area provided. In the case of a cell-based ASIC or sea-of-gates array, it is possible to increase the channel size and try the routing steps again. A channelled gate array or FPGA has fixed routing resources and in these cases we must start all over again with floorplanning and placement, or use a larger chip.

13.3.2 Measurement of Channel Density

We can describe a channel routing problem by specifying two list of nets; one for the top edge of the channel and one for the bottom edge. The position of the net number in the list gives the column position. The net number zero represents a vacant or unused terminal. Figure 13.12 shows a channel with the numbered terminals to be connected along the top and the bottom of the channel.

We call the number of nets that cross a line drawn vertically anywhere in a channel the ***local density***. We call the maximum local density of the channel, the global density or sometimes just ***channel density***. Figure 13.12 has a channel density of 4, channel density is an important measure in routing, it tells a router the absolute fewest number of horizontal interconnects that

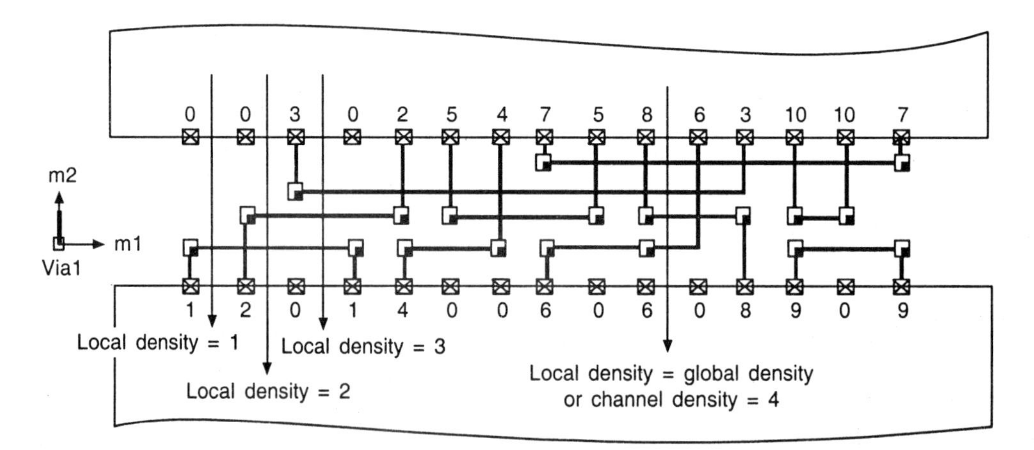

Figure 13.12 The definitions of local channel density and global channel density. Lines represent the m1 and m2 interconnect in the channel to simplify the drawing.

it needs at the point where the local density is highest. In two-level routing (all the horizontal interconnects run on one routing layer) the channel density determines the minimum height of the channel. The channel capacity is the maximum number of interconnects that a channel can hold. If the channel density is greater than the channel capacity, then that channel definitely cannot be routed.

13.3.3 Algorithms

We start discussion of routing methods by simplifying the general channel routing problem. The restricted channel routing problem limits each net in the channel to use only one horizontal segment. In other words, the channel router uses only one trunk for each net. This restriction has the effect of minimizing the number of connections between the routing layers. This is equivalent to minimizing the number of vias used by the channel router in a two-layer metal technology. Minimizing the number of vias is an important objective in routing a channel, but it is not always practical. Sometimes constraints will force a channel router to use jogs or other methods to complete the routing. Next, we shall study an algorithm that solves the restricted channel-routing problem.

13.3.4 Left-Edge Algorithm

The left-edge algorithm (LEA) is the basis for several routing algorithms. The LEA applies to two-layer channel routing, using one layer for the trunks and the other layer for the branches. For example, m1 may be used in the horizontal direction and m2 in the vertical direction. The LEA proceeds as follows.

1. Sort the nets according to the leftmost edges of the net's horizontal segment.
2. Assign the first net on the list to the first free track.
3. Assign the next net on the list, which will fit, to the track.
4. Repeat this process from step 3 until no more nets will fit in the current track.
5. Repeat steps 2–4 until all nets have been assigned to tracks.
6. Connect the net segments to the top and bottom of the channel.

Figure 13.13 illustrates the LEA. The algorithm works as long as none of the branches touch—which may occur if there are terminals in the same column belonging to different nets. In this situation, we have to make sure that the trunk that connects to the top of the channel is placed above the lower trunk. Otherwise two branches will overlap and short the nets together. In the next section we shall examine this situation more closely.

13.3.5 Constraints and Routing Graphs

Two terminals that are in the same column in a channel create a ***vertical constraint***. We say that the terminal at the top of the column imposes a vertical constraint on the lower terminal. We can draw a graph showing the vertical constraints imposed by terminals. The nodes in a ***vertical constraint graph*** represent terminals. A vertical constraint between two terminals is shown by

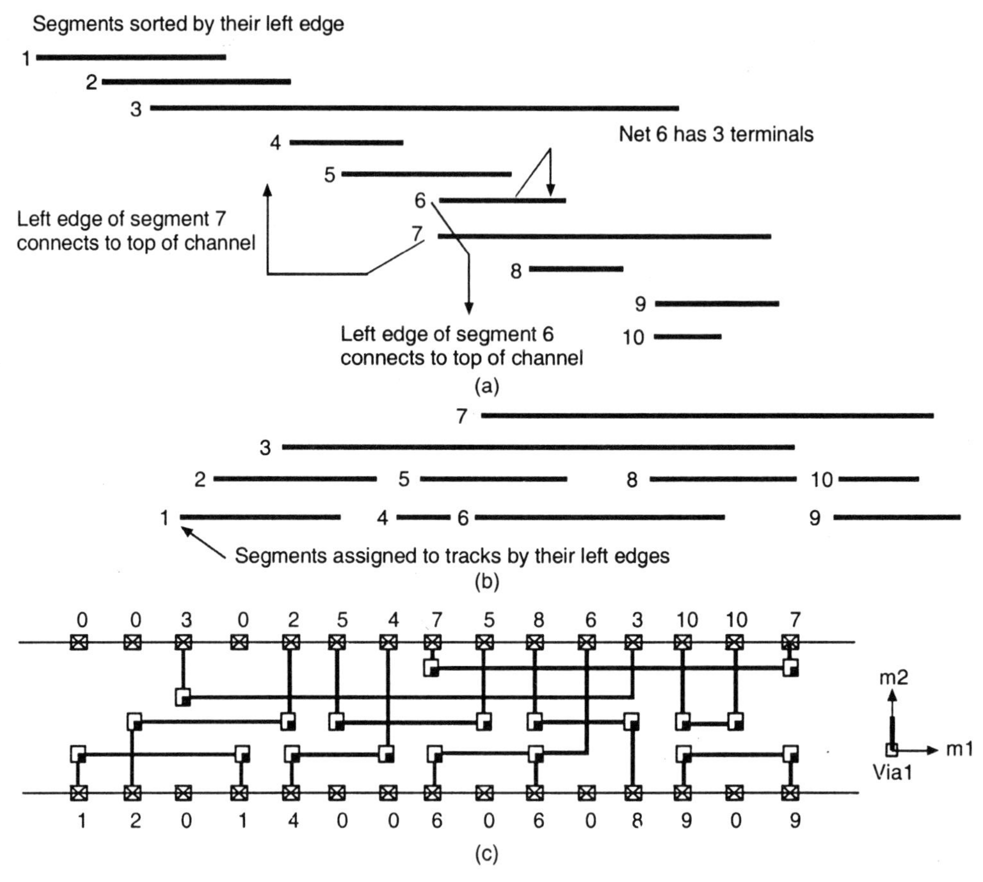

Figure 13.13 Left-edge algorithm. (a) Sorted list of segments. (b) Assignment to tracks, (c) Completed channel route (with m1 and m2 interconnect represented by lines.)

an edge of the graph connecting the two terminals. A graph that contains information in the direction of an edge is a directed graph. The arrow on the graph edge shows the direction of the constraint pointing to the lower terminal, which is constrained. Figure 13.14(a) shows an example of a channel, and Figure 13.14(b) shows its vertical constraint graph.

We can also define a ***horizontal constraint*** and a corresponding ***horizontal constraint graph***. If the trunk for net 1 overlaps the trunk of net 2, then we say there is a horizontal constraint between net 1 and 2. Unlike a vertical constraint, a horizontal constraint has no direction. Figure 13.14(c) shows an example of a horizontal constraint graph and shows a group of 4-terminal (numbered 3, 5, 6, and 7) that must all overlap. Since this is the largest such group, the global channel density is 4.

If there are no vertical constraints at all in a channel, we can guarantee that the LEA will find the minimum number of routing tracks. The addition of vertical constraints transforms the restricted routing problem into an NP-complete problem. There is also an arrangement of vertical constraints that none of the algorithms based on the LEA can cope with. In Figure 13.15(a), net 1 is above net 2 in the first column of the channel. Thus net 1 imposes a vertical constraint on net 2. Net 2 is above net 1 in the last column of the channel. Then net 2 also imposes a vertical

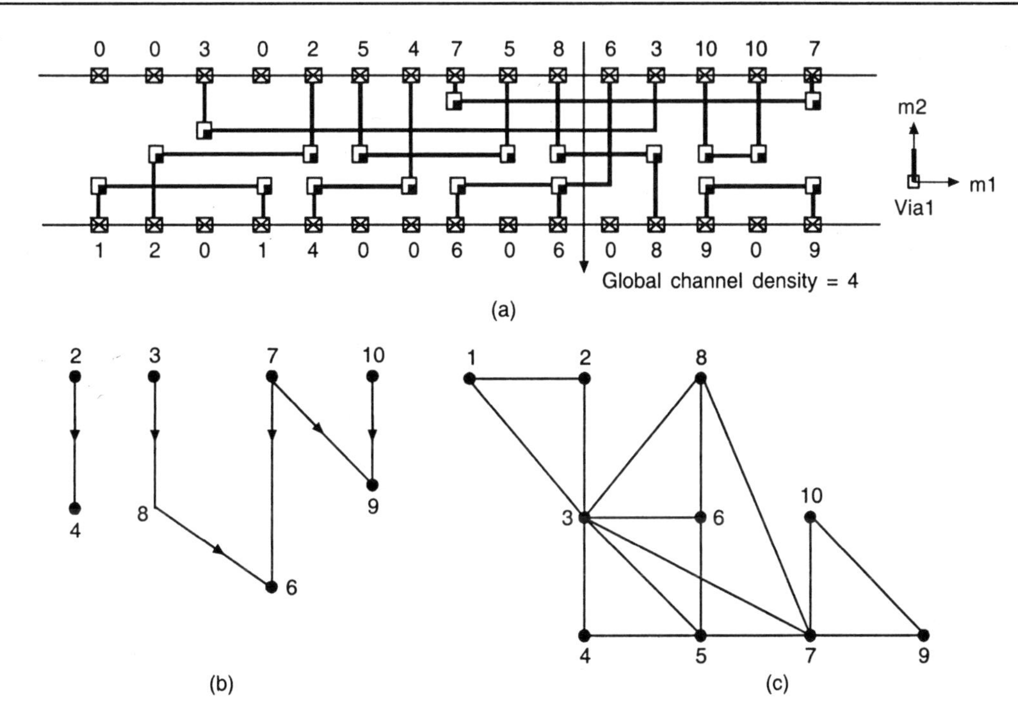

Figure 13.14 Routing graphs: (a) Channel with a global density of 4, (b) The vertical constraint graph. If two nets occupy the same column, the net at the top of the channel imposes a vertical constraint on the net at the bottom, (c) Horizontal-constraint graph. If the segments of two nets overlap, they are connected in the horizontal-constraint graph.

constraint on net 1. It is impossible to route this arrangement using two routing layers with the restriction of using only one trunk for each net. If we construct the vertical-constraint routing graph for this situation [Figure 13.15(b)], there is a loop or cycle between nets 1 and 2. If there is any vertical-constraint cycle (or cyclic constraint) between two or more nets, the LEA will fail. A ***dogleg router*** removes the restriction that each net can use only one track or trunk. Figure 13.15(c) shows how adding a dogleg permits a channel with a cyclic constraint to be routed.

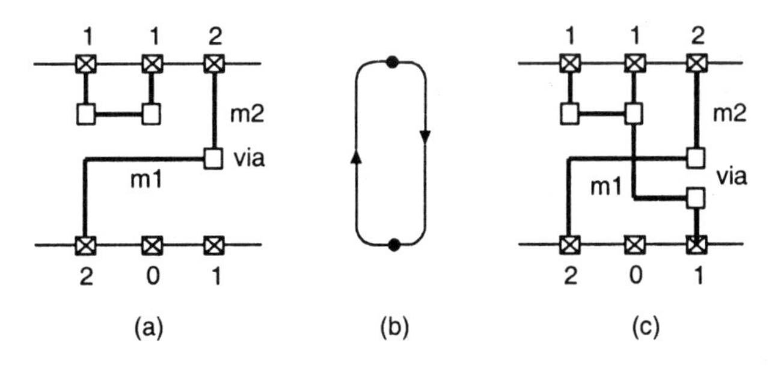

Figure 13.15 The addition of a dogleg, an extra trunk in the wiring of a net can resolve cyclic vertical constraints.

The channel-routing algorithms interconnects on one layer to run on top of other interconnects on a different layer. These algorithms allow interconnects to cross at right angles to each other on different layers, but not to overlap. When we remove the restriction that horizontal and vertical routing must use different layers, the density of a channel is no longer the lower bound for the number of tracks required. For two routing layers, the ultimate lower bound becomes half of the channel density. The practical reasoning for restricting overlap is the parasitic **overlap capacitance** between signal interconnects. As the dimensions of the metal interconnect are reduced, the capacitance between adjacent interconnects on the same layer (**coupling capacitance**) is comparable to the capacitance of interconnects that overlap on different layers (**overlap capacitance**).

The channel height is fixed for channelled gate arrays; it is variable in discrete steps for channelless gate arrays; it is continuously variable for cell-based ASICs. The use of channel routing compaction for a two-layer channel can reduce the channel height by 15% to 20%.

13.3.6 Area-Routing Algorithms

There are many algorithms used for the detailed routing of general-shaped areas. The first group are the grid-expansion or maze-running algorithms. A second group of methods, which are more efficient, are the line-searching algorithms.

Figure 13.16 illustrates the Lee-maze running algorithm. The goal is to find a path from X to Y—from the start (or source) to the finish (or target)—avoiding the obstacles. The algorithm is often called wave propagation because it sends out waves, which spread out like those created by dropping a stone into a pond.

5	4	3	2	3	4			
4	3	2	1	2	3	4		
3	2	1	X	1	2	3	4	4
4	3					4	4	
	4					4		
				4	3	2	1	Y
					4	3	2	1

Figure 13.16 The Lee maze running algorithm.

Algorithms that use lines rather than waves to search for connections are more efficient than algorithms based on the Lee algorithm. Figure 13.17 shows a **Hightower algorithm**—a **line-search algorithm** (or **line-probe algorithm**):

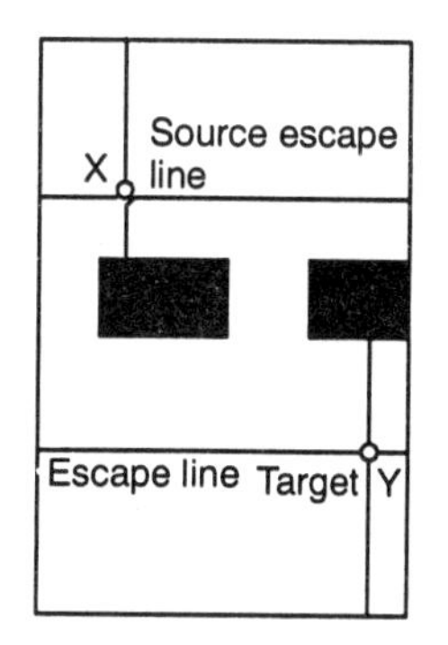

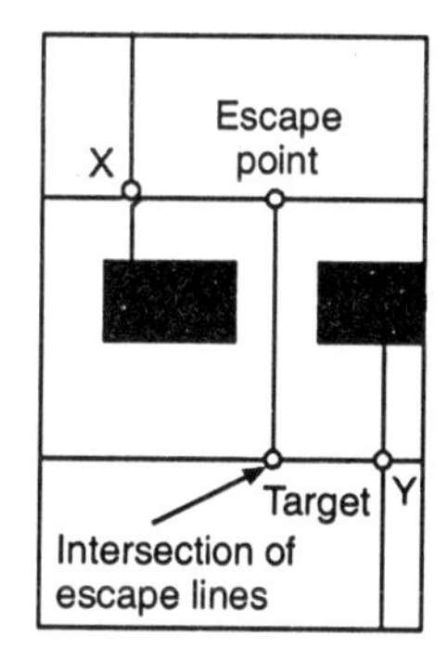

Figure 13.17 Hightower area routing algorithm.

1. Extend lines from both the source and target towards each other.
2. When an extended line, known as escape line, meets an obstacle, choose a point on the escape line from which to project another escape line at right angles to the old one. This point is called the **escape point.**
3. Place an escape point on the line so that the next escape line just misses the edge of the obstacle. Escape lines emanating from the source and target intersect to form the path.

The Hightower algorithm is faster and requires less memory that methods based on the Lee algorithm.

13.3.7 Multilevel Routing

Using two-layer routing, if the logic cells do not contain any m2, it is possible to complete some routing in m2 using over-to-cell (OTC) routing. Sometimes poly is used for short connections in the channel in a two-level metal technology. This is known as **2.5 layer routing**. Using a third level of metal in **three-layer routing**, there is a choice of approaches. **Reserved layer routing** restricts all the interconnect on each layer to flow in one direction in a given routing area. **Unreserved-layer routing** moves in both horizontal and vertical directions on a given layer. Most routers use reserved routing. Reserved-three level metal routing offers another choice: either use m1 and m3 for horizontal routing, with m2 for vertical routing (**HVH routing**) or use **VHV routing**. Since the logic cell interconnect usually blocks most of the area on the m1 layer, HVH routing is normally used. Some processes have more than three levels of metal. Sometimes the upper one or two metal layers have a coarser pitch than the lower layers and are used in **multilevel routing**.

13.3.8 Timing-driven Detailed Routing

Before detailed routing, the global router has already set the path the interconnect will flow. At this point little can be done to improve timing except to reduce the number of vias, alter the interconnect width to optimize delay, and minimize overlap capacitance. The gains here are relatively small, but for very long branching nets, even small gains may be important. For high frequency clock nets, it may be important to shape and chamfer (round) the interconnect to match impedences at branches and control reflections at corners.

13.3.9 Final Routing Steps

If the algorithms to estimate congestion in the floorplanning tool accurately perfectly reflected the algorithms used by the global router and detailed router, routing completion should be guaranteed. Often, however, the detailed router will not be able to route all the nets. These problematic routes are known as **unroutes**. Routers handle this situation in one of two ways. The first method leaves the problematic nets unconnected. The second one completes all interconnects anyway but with some design-rule violations. **Routing inspection** can then be performed as a final step. Routing of nets that require special attention, clock and power nets for example, are normally done before detailed routing of signal nets.

13.4 SPECIAL ROUTING

Routing of nets that require special attention, clock and power nets, for example, normally done before detailed routing of signal nets. The architecture and structure of these nets is performed as part of floorplanning, but the sizing and topology of the nets is finalized as part of the routing step.

13.4.1 Clock Routing

Arrays normally use a clock spine (a regular grid), eliminating the need for speed routing. The clock distribution grid is signed at the same time as the gate-array base to ensure a minimum clock skew minimum clock latency-given power dissipation and clock buffer area limitation cell-based ASICs may use either a clock spine, a clock tree, or a hybrid each. Figure 13.18(a), (b) shows how a clock router may minimize clock skew in a spine by making the path lengths, and thus net delays, to every leaf node using jogs in the interconnect paths if necessary. More sophisticated clocks perform clock-tree synthesis (automatically choosing the depth and structure of the clock tree) and clock-buffer insertion (equalizing the delay to the leaf by balancing interconnect delays and buffer delays).

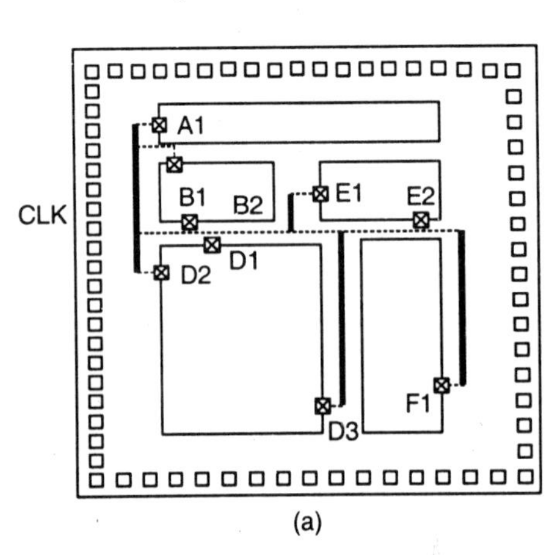

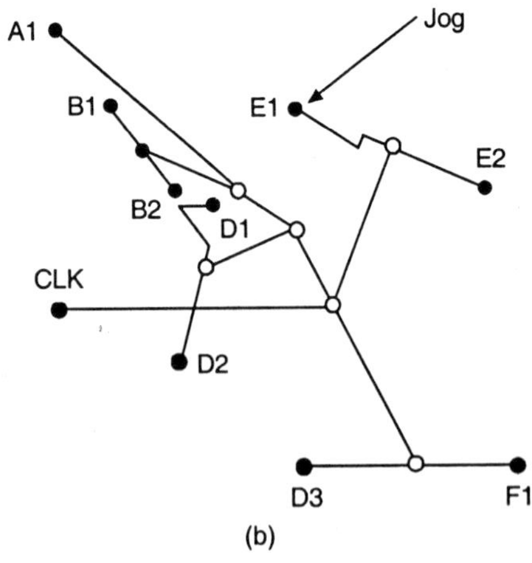

Figure 13.18 Clock routing: (a) A clock network for the cell-based ASIC, (b) Equalizing the interconnect segments between CLK and all destinations (by including jogs if necessary) minimizes clock skew.

The clock tree may contain multiply-driven nodes (more than one active element driving a net). The net delay models that we have used breakdown in this case may have to extract the clock network and perform circuit simulation, fold by back-annotation of the clock delays to the netlist and the bus currents to the clock router. The sizes of the clock depend on the current they must carry. The limits are set by reliability issues discussed next.

The factor contributing to unpredictable clock skew is changes in clock–buffer delays with variations in power-supply voltage due to data dependent activity. This activity induced clock skew can easily be larger than the skew achievable using a clock router. For example, there is little point in using software capable of reducing clock skew to less than 100 ps if, due to fluctuations in power-supply voltage when part of the chip becomes active, the clock-network delays change by 200 ps.

The power buses supplying the buffers driving the clock spine carry direct current (unidirectional current or DC), but the clock spine itself carries alternating current (bidirectional current or AC). The difference between electro migration failure rates due to AC and DC leads to different rules for sizing clock buses. "Clock Planning", the fastest way to drive a large load in CMOS is to taper successive stages by approximately $e = 3$. This is not necessarily the smallest-area or lowest-power approach.

13.4.2 Power Routing

Each of the power buses has to be sized according to the current it will carry. To much current in a power bus can lead to a failure through a mechanism known at electromigration [Young and Christou, 1994]. The required power-bus widths can be estimated automatically from library information, from a separate power simulation tool, or by entering the power-bus widths to the routing software by hand. Many routers use a default power-bus width so that it is quite easy to complete routing of an ASIC without even knowing about this problem.

For a direct current (DC) the mean time to failure (MTTF) due to electromigration is experimentally found to obey the following equation:

$$\text{MTTE} = AJ^{-2} \exp \frac{-E}{KT}$$

Where J is the current density; E is approximately 0.5 eV; k, Boltzmann's constant, is 8.62×10^{-5} eVK^{-1}; and T is absolute temperature in Kelvins.

There are a number of different approaches to model the effect of an AC component. A typical expression is

$$\text{MTTF} = \frac{A \exp(-E/KT)}{\bar{J}\,|\bar{J}| + k_{\text{AC/DC}}\,|\bar{J}^2|}$$

where $\bar{J}$ is the average of $J(t)$, and $|\bar{J}|$ is the average of $|j|$. The constant $k_{\text{AC/DC}}$ plates the relative effects of AC and DC and is typically between 0.01 and 0.0001. Electromigration problems become serious with a MTTF of less than 10^5 hours (approximately 10 years) for current densities (DC) greater than 0.5 GAm^{-2} at temperatures above 150°C.

SUMMARY

The completion of routing finishes the ASIC physical design process. Routing is a complicated problem best divided into two steps: Global and Detailed routing. Global routing plans the wiring by finding the channels to be used for each path. There are differences between global routing for different types of ASiCs, but the algorithms to find the shortest path are similar. Two main approaches to global routing are one net at a time or all nets at once. With the inclusion of timing driven routing objectives of the routing problem become much harder and require understanding of the differences between finding the shortest net and finding the net with the shortest delay. Different types of detailed routing include channel routing and area based or maze routing.

The most important points in this chapter are:

- Routing is divided into global and detail routing
- Routing algorithm should match with the placement algorithm
- Routing is not complete if there are unroutes
- Clock and power are handled as special cases.

REVIEW QUESTIONS

1. Explain the different kinds of routing in detail.
2. Give detailed explanation about left edge algorithm.
3. Explain different kinds of routing algorithm.
4. Explain different routing steps and in particular explain about the clock routing.
5. What is power routing and explain.

SHORT ANSWER QUESTIONS

1. Routing is usually split into ____________ routing followed by ____________ routing.

   Ans. global; detailed
2. What a lumped-delay model neglects?

   Ans. Sums of all the node capacitances, i.e. the lumped capacitance.
3. How the bottom-up approach reacts?

   Ans. Starts at the lowest level of hierarchy.
4. In a logic cell array, the vertical resources are ____________.

   Ans. feedthroughs
5. Explain the term pick-up point.

   Ans. It refers to the physical pieces of metal (or sometimes polysilicon) in the logic cell to which the router connects.
6. We can set the m1 pitch to one of these values:

   Ans. (a) via-to-via (VTV) pitch (or spacing)

   (b) via-to-line (VTL or line-to-via) pitch, or

   (c) line-to-line (LTL) pitch.

7. Why Manhattan routing is called so?

   Ans. Because the results look similar to the rectangular north-south and east-west layout of streets in New York City.

8. If more than one trunk is used, the trunks are connected by ___________. Connections exit the channel at ___________.

   Ans. dogleg routers; pseudo terminals

9. Where exactly the trunk and branch connections run?

   Ans. In tracks.

10. The number of nets that cross a line drawn vertically anywhere in a channel are called ___________.

    Ans. local density

11. The maximum local density of the channel is called ___________.

    Ans. channel density.

12. Two terminals that are in the same column in a channel create a ___________.

    Ans. vertical constraint

13. What is the advantage of a dogleg router?

    Ans. It can remove the restriction that each net can use only one track or trunk.

14. The use of channel routing compaction for a two-layer channel can reduce the channel height by ___________.

    Ans. 15% to 20%

15. Algorithms that use lines rather than waves to search for connections are called

    Ans. Lee algorithms.

14 CMOS Testing

14.1 INTRODUCTION

Once a logic function has been designed, it must be tested and the faulty chips should be identified from good chips. For its checking, we must develop tests methodology. VLSI testing deals with techniques that are used to determine if a die (chip) behaves properly after the fabrication sequence is completed. If a die passes the testing phase, it is packaged and sold. So, fault may be understood to be a manufacturing defect. A fault may be caused by mechanisms ranging from crystalline dislocations to lithographic errors to bad electing of vias. In this chapter, the various testing schemes and fault diagnosis for both combinational as well as sequential logic gates, logic networks are described.

14.2 NEED FOR TESTING

Due to the complexity of manufacturing process, not all dice on a wafer correctly operate. Small imperfections in starting material, processing steps, or in photomasking may result in bridged connections or missing features. It is the aim of a test procedure to determine which dice are good and should be used in end systems. Testing a die (chip) can occur:

- at the wafer level
- at the packaged-chip level
- at the board level
- at the system level
- in the field

By detecting a malfunctioning chip at an earlier stage, the manufacturing cost may be kept low. There are two types of tests: functionality tests and manufacturing tests.

14.2.1 Functionality Tests

The functionality tests verify that the chip performs its intended function, e.g., it performs a digital filtering, acts as a microprocessor, or communicates using a particular protocol. These tests assert that all the gates in the chip achieve a desired function. These tests are usually used early in the design cycle to verify the functionality of the circuit.

For most systems, functionality test involves proving that the circuit is functionally equivalent to some specification. The specification might be a verbal description, a plain-language textual specification, a description in some high-level computer language such as C, FORTRAN, Pascal or LISP or in a hardware description language such as VHDL, ELLA or Verilog®, or simply a table of inputs and required outputs. Functional equivalence involves running a simulator at some level on the two descriptions of the chip, one at the gate level and one at a functional level, and ensuring for all inputs applied, the outputs are equivalent at some convenient check-points in time. The functional equivalence may be carried out at various levels of the design hierarchy. If the description is in a behavioural language, the behaviour at a system level may be verifiable.

There is no good theory on how to ensure that good functional tests be written. The best way is to simulate the chip or system as closely as possible to the way it will be used in the real world.

14.2.2 Manufacturing Tests

These tests verify that every gate and register in the chip functions correctly. These are used after the chip is manufactured to verify that the silicon is intact. The functionality tests verify the function of a chip as a whole whereas manufacturing tests are used to verify that every gate operates as expected. This is necessary due to a number of manufacturing defects that might occur during chip fabrication or during accelerated life testing. Typical defects include the following:

- Layer-to-layer shorts (i.e., metal-to-metal)
- Discontinuous wires
- Thinoxide shorts to substrate or well

These in turn lead to particular circuit difficulties, including the following:

- Nodes shorted to power or ground
- Nodes shorted to each other
- Inputs floating/outputs disconnected

Apart from the verification of internal gates, I/O integrity is also tested through completing the following tests:

(i) I/O level test
(ii) Speed test
(iii) I_{DD} test

The I/O level test checks for the noise margin for TTL, ECL or CMOS I/O pads. These three tests check the leakage if the circuit is composed of complementary logic. Any value markedly above the expected value for a given wafer normally indicates an internal shorting failure (or very bad leakage). Wafer tests may be done at high speed or low speed (1 MHz) due to possible power and ground bounce effects.

In general, manufacturing test generation assumes that the circuit/chip functions correctly and ways of exercising all gate inputs and of monitoring all gate outputs are required.

14.2.3 Test Process

Depending on whether a wafer or a packaged part is tested, a probe card or Device-Under-Test (DUT) board would be needed to connect the tester outputs and inputs to the die I/O pads or chip packages' pins. The next requirement for a chip tester is the existence of 'test program'. This is a file with a format of inputs and outputs that suit the chip tester to be used to test the chip. When the stimulus is applied to a circuit via a simulator, the output of the simulator may be dumped to a file (often called an activity file). If this output is filtered so that only the chip inputs and chip outputs are retained and further filtered so that only the quiescent signal values are kept after an input or inputs change, then the resulting file may be used to generate a 'test program'.

Normally, a mapping file is required that maps a given input or output in the test program to a physical connection (pin) in the tester. This pin may be programmed to be an input, output, tristate, bidirect, or, in some cases, a multiplexed data pin. Each pin on the tester is driven by a function memory that is used either to assert a value or to check a value at a DUT pin. In addition, various control memories may be present to control the drive on the tester pin (i.e., to control a tristate pin) or to mask data from the chip (i.e., to ignore certain pins at certain times).

The clock speed is specified by specifying a test cycle time, T_c. The time at which outputs are asserted or inputs are sampled is also specified on a pin-by-pin basis (T_s). The format of the test data may usually be chosen from Non Return to Zero (NRZ), Return To Zero (RTZ), or other formats, such as Surround By Zero (SBZ).

The probe card or DUT board is connected to the tester. The test program is compiled and downloaded into the tester, and the tests are applied to the circuit. The tester samples the chip outputs and compares the values with those provided by the test program. If there are any differences, the chip is marked as faulty. In the case of a probe card, the card is raised, moved to the next die on the wafer and lowered, and the test procedure is repeated. In the case of a DUT board with automatic part handling, the tested part is binned into a good or bad bin and a new part is fed to the DUT board, and the test is repeated. In most cases, these procedures take a few seconds for each part tested.

14.3 GENERAL CONCEPTS OF TESTING

The wafer testing procedure is illustrated schematically in Figure 14.1. A test probe head allows electrical contact to the inputs and the response is taken from the outputs. The system is programmed to accept or reject the die based on this set of measurements. A bad die is marked for future reference. After every die has been tested, the wafer is scored along lines that run in

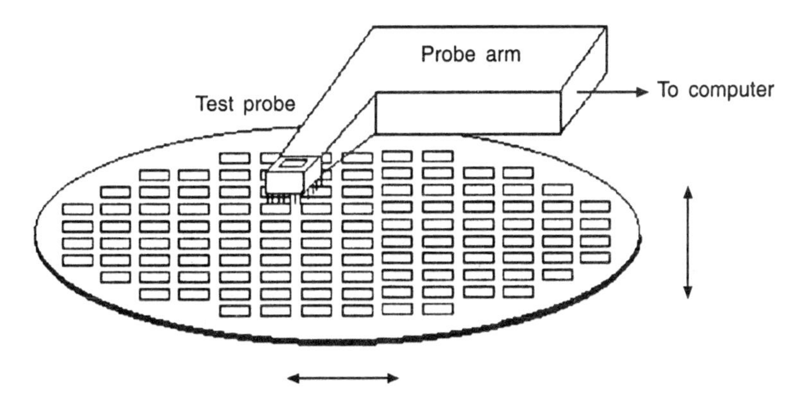

Figure 14.1 Visualization of wafer testing.

between the individual sites. Applying a little pressure to the wafer induces cracks along the score lines, resulting in individual die without ruining the circuitry. Good circuits enter the final assembly phase where robotic equipment is used to place the die into a package, connect the die to the package electrodes, and then seal the package.

14.3.1 Reliability

Reliability is concerned with projecting the lifetime of a component once it is placed into operation. The projected lifetime is the number of hours of operation that can be expected before a failure occurs. This is described graphically using a bathtub curve as shown in Figure 14.2, which plots the number of failures of a given system as a function of time.

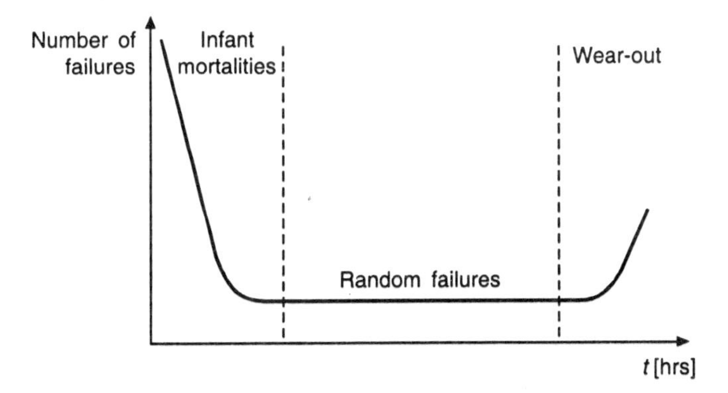

Figure 14.2 Bathtub reliability curve.

Bathtub curves are semilog plots with time t in hours, and points plotted as $\log_{10}(t)$. There are three general regions shown in the curve. Infant mortalities are failures that occur after a very short period of time, i.e., early in the life of the system. These tend to arise from manufacturing defects that manifest themselves after a few hours of operation. The central portion of the curve represents random failures during normal operation, while wear-out describes the end of life.

The number of infant mortalities can be large for any device, especially during the initial month of production where the design and manufacturing are being refined. Once a circuit is mounted on a board, the cost of repair exceeds the cost of the IC itself. So, it is desirable to perform burn-in operation. During a burn-in, the circuits are operated under stressed conditions with higher than normal voltage levels, high temperatures, and high humidity environments. The idea is to induce potential failures to occur during burn-in and avoid using the devices at the board level. This increases the system reliability considerably. Electronic parts vendors, including VLSI chip manufacturers, usually provide some type of written warranty for their products. It is to both sides advantage to produce reliable components.

14.3.2 Reliability Modelling

Reliability data can be used to construct mathematical models that are useful in predigesting failure rates. Suppose that we test a group of devices starting at time $t = 0$. As time progresses, a device fails at time t_1, then end at time t_2, and so on. At the end of the test period, we find that N devices fail. The total number of operational hours for the group is the sum

$$T = \sum_{i=1}^{N} t_i \tag{14.1}$$

where T is measured in hours.

The average failure rate of the test group is defined as the number of failures divided by the total number of operational hours.

$$\lambda_{\text{av}} = \frac{N}{T} \tag{14.2}$$

The mean time to failure (MTTF) is given by

$$\text{MTTF} = \frac{T}{N} = \frac{1}{\lambda_{\text{av}}} \tag{14.3}$$

MTTF represents an average lifetime for the test group.

One of the most commonly used metric to describe failure rates is the failure in time (FIT). It is defined as

$$1 \text{ FIT} = 1 \text{ failure in 1 million parts over 1000 hours} \tag{14.4}$$

or

$$1 \text{ FIT} = 1 \text{ ppm/k} \tag{14.5}$$

where ppm stands for parts per million. A failure rate of 1 FIT (10^9 operating hours) corresponds to a lifetime of about 125000 years for one device out of a large sample. Accelerated stress life testing where extreme conditions are used to simulate aging is often employed to determine reasonable FIT values. This simple approach to failure analysis provides valuable insight, but is not sufficient to yield accurate estimates. More sophisticated mathematical models provide higher confidence levels. One general approach uses probability density function (PDF) $f(t)$ as a function of time.

The cumulative distribution function (CDF) $F(t)$ is related to the PDF by

$$F(t) = \int f(n)\, dn \tag{14.6}$$

where n is a dummy variable. The device lifetimes are described by the models known as life distributions. The CDF is a life distribution that can be interpreted as below:

- $F(t)$ is the probability that a random unit of the test group will fail by t hours.
- Alternately, $F(t)$ is the fraction of units in the test group that fail by t hours.

The reliability function $R(t)$ is defined by

$$R(t) = 1 - F(t) \tag{14.7}$$

This function describes the units that have not failed, i.e., the ones that are still functioning after t hours.

The hazard rate $h(t)$ can be interpreted as the probability of a unit failing in a time between t and $(t + dt)$. The hazard rate can be defined as

$$h(t) = \left(\frac{1}{R(t)}\right)\left(\frac{dF}{dt}\right) \tag{14.8}$$

or

$$h(t) = \frac{f(t)}{R(t)} \tag{14.9}$$

The hazard rate is also called the instantaneous failure rate or just the failure rate. The bathtub curve in Figure 14.2 is an example of a hazard rate plot.

The cumulative hazard function $H(t)$ is obtained by integrating the hazard rate.

$$H(t) = \int_o^t h(n)\,dn \tag{14.10}$$

The relationship between $F(t)$ and $H(t)$ is given by

$$F(t) = 1 - e^{-H(t)} \tag{14.11}$$

The average failure rate (AFR) over a time T hours is given by

$$\text{AFR}(T) = \frac{H(T)}{T} = \frac{\ln R(T)}{T} \tag{14.12}$$

The AFR is very useful to specify the failure rate of a device that is operated for T hours.

Reliability modelling is a fascinating field of study that employs statistical analysis of data to determine projected lifetimes and failure rates. As the complexity of VLSI processing equipment increases, reliability issues become more critical.

Example 14.1 Consider a small chip that has about 200,000 FETs. What is the FIT value needed to achieve an average reliability of no more than 1 transistor failure over 1 year?

Assuming 8760 hours per year, we see that the FETs represent a total of 200,000 × 8760 = 1.752×10^9 device hours/year. To find the FITs needed to obtain 1 failure per year, we write

$$\left(\frac{x}{10^9}\right)(1.752 \times 10^9) = 1 \tag{14.13}$$

where x is the FIT value.
solving gives x = 0.67 FITs as the required rate.

14.4 MANUFACTURING TEST PRINCIPLES

A critical factor in all LSI and VLSI designs is the need to incorporate methods of testing circuits. Figure 14.3(a) shows a combinational circuit with n-inputs. To test this circuit exhaustively, a sequence of 2^n-inputs (or test vectors) must be applied and observed to fully exercise the circuit. This combinational circuit is converted to a sequential circuit with the addition of m-storage elements as shown in Figure 14.3(b). The state of the circuit is determined by the inputs and the previous state. A minimum of $2^{(n+m)}$ test vectors must be applied to exhaustively test the circuit or $2^{(n+m)}$ inputs are required to exhaustively test circuit for $n = 25$, $m = 50$, 1 μS/test, the test time is over 1 billion years.

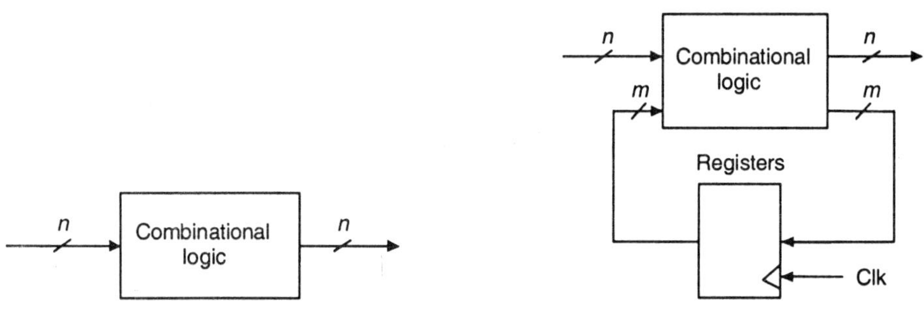

(a) 2^n-inputs required to exhaustively test circuit (b) The combinational explosion in test vectors

Figure 14.3 Testing the circuits.

14.4.1 Fault Models

14.4.1.1 Stuck-at faults

In order to deal with the existence of good and bad parts, it is necessary to propose a 'fault model', i.e., a model for how faults occur and their impact on circuits. The most popular model is called the 'stuck-at' model. These are logic-based faults: stuck-at-0 (SA0) and stuck-at-1 (SA1). These faults occur when a node is accidentally connected to the power supply (SA1) or ground (SA0).

The effect of a stuck-at-fault varies with the location. For example, an SA0 fault at the gate of an NFET implies that it can never turn ON, while an SA0 fault at the gate of a PFET implies that the transistor cannot be shut off. Obviously, these affect the operation of the logic network. Figure 14.4(a) and (b) illustrates how an SA0 or SA1 fault might occur.

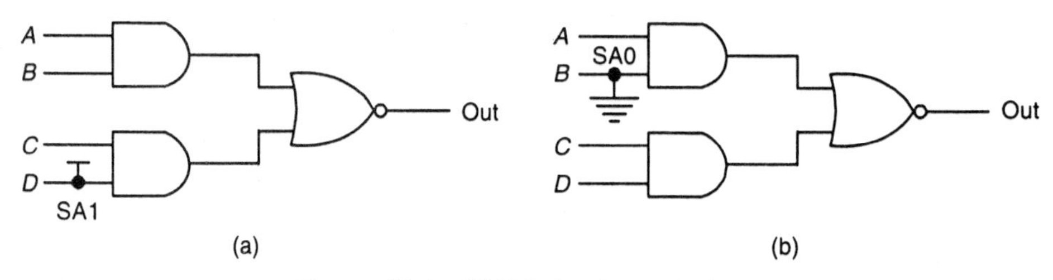

Figure 14.4 CMOS Stuck—at faults.

A related set of faults are gate–source and gate–drain shorts as shown in Figure 14.5(a) and (b) fault models for these defects respectively.

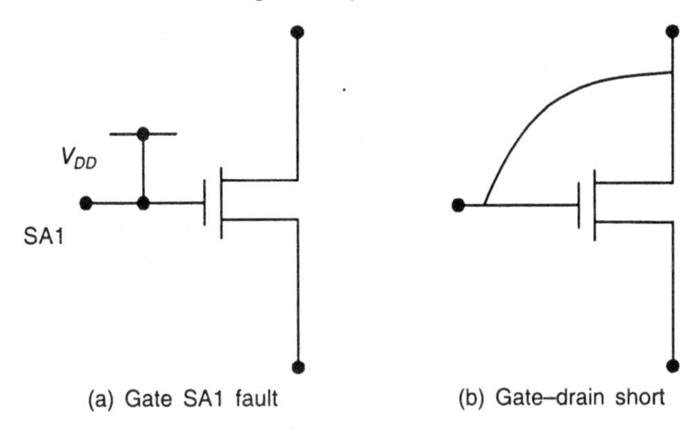

(a) Gate SA1 fault (b) Gate–drain short

Figure 14.5 Fault models for the above defects.

14.4.1.2 Short-Circuit and Open-Circuit Faults

A short-circuited FET is one that always conducts drain–source current with an applied drain–source voltage, V_{DS}; the gate has no control over the operation. This is also called a stuck-on fault as shown in Figure 14.6(a). An open-circuit or stuck-off fault is exactly opposite: Current never flows regardless of V_{GS} or V_{DS} (as shown in Figure 14.6(b)). The circuit models for these two faults are shown in Figure 14.6. Physically, these problems tend to be due to normalization or etching problems or mask registration errors.

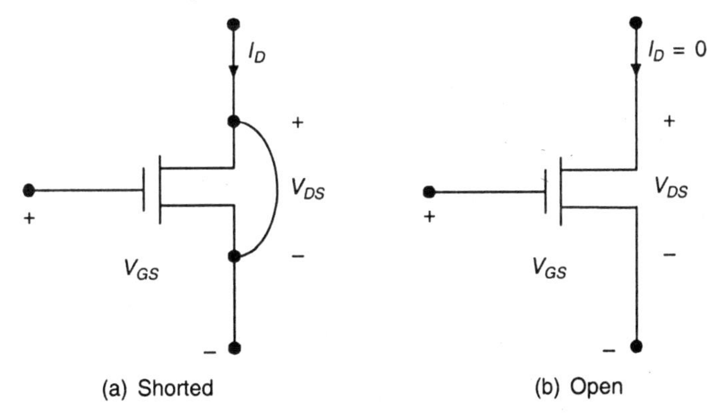

(a) Shorted (b) Open

Figure 14.6 Circuit models for the two faults.

Two shorted faults are shown in Figure 14.7. Considering the faults, the short S1 is modelled by an SA0 fault at input A, while short S2 modifies the function of the gate. To ensure the most accurate modelling, faults should be modelled at the transistor level, because it is only at this level that the complete circuit structure is known.

A particular problem that arises with CMOS is that it is possible for a fault to convert a combinational circuit into a sequential circuit. This is illustrated for the case of a 2-input NOR

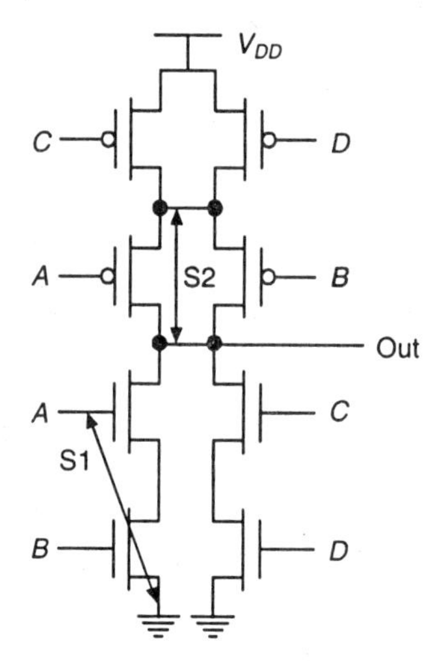

Figure 14.7 Two-shorted faults.

gate in which one of the transistors is rendered ineffective (stuck open or stuck closed) in Figure 14.8. This might be due to the missing source, drain, or gate connection. If one of the n-transistors (A connected to gate) is stuck open, then the function displayed by the gate will be

$$F = \overline{A + B} + A \cdot \overline{B} \cdot F_n \tag{14.14}$$

where F_n is the previous state of the gate. Similarly, if the B and n-transistor drain connection is missing, the function is

$$F = \overline{A + B} + \overline{A} \cdot B \cdot F_n \tag{14.15}$$

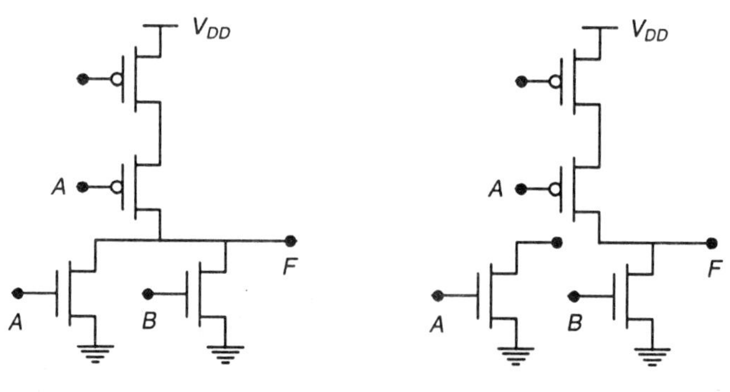

(a) 2-input CMOS NOR gate $F = \overline{A + B}$

(b) 2-input CMOS NOR gate with one of n-transistors stuck-open $F = \overline{A + B} + A \cdot \overline{B} \cdot F_n$

Figure 14.8 A case in which one of the transistors is made ineffective.

If either of the *p*-transistors is open, the node would be arbitrarily charged (i.e., it might be high due to some weird charging sequence) until one of the *n*-transistors discharged the node. Thereafter it would remain at zero, bar-charge leakage effects. This problem has caused researchers to search for new methods of test generation to detect such behaviour.

Gate-oxide shorts (GOSs) are specific to MOSFETs. They occur when the insulating gate oxide has a defect and the gate material contacts the substrate as shown in Figure 14.9(a) for an NFET. Assuming an *n*-doped poly gate, the GOS creates a parasitic *pn*-junction diode between the gate and the *p*-type substrate. The gate voltage has no control over the drain current, which makes the circuit non-functional. Gate-oxide shorts are found in both NFETs and PFETs and tend to originate from non-uniform growth of the gate–oxide due to defects on the wafer. In this case, the affected die(s) tend to have many GOS-defective FETs, which make them easier to find. Figure 14.9(b) shows the symbolic circuit designation.

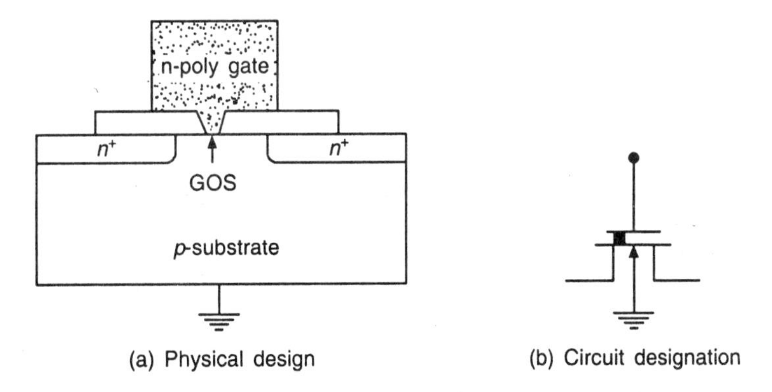

Figure 14.9 Gate-oxide short in an NFET.

Currently, debate ranges over whether an SA0/SA1 approach to testing is adequate for testing CMOS. It is also possible to have switches (transistors) exhibit a 'stuck-open' or 'stuck-closed' state. Stuck-closed states can be detected by observing the static V_{DD} current (I_{DD}) while applying the test vectors.

14.4.2 Gate Level Testing

Fault models are used to characterize failures in logic gates. Creating different fault circuits allows one to find a set of test vectors that can be applied to the circuit. We examine a simple 2-input NAND gate with inputs *A* and *B* as an example. In Figure 14.10(a), a stuck-at-1 fault at the gate of PFET (connected to *A* input) keeps the transistor in cut off, while the corresponding NFET (connected to *A* input) is always conducting.

The circuit in Figure 14.10(b) has a stuck-at-0 fault at the *B*-input. This prevents the NFET (connected to *B* input) from turning ON, and keeps the NFET (connected to *B* input) in a conducting state.

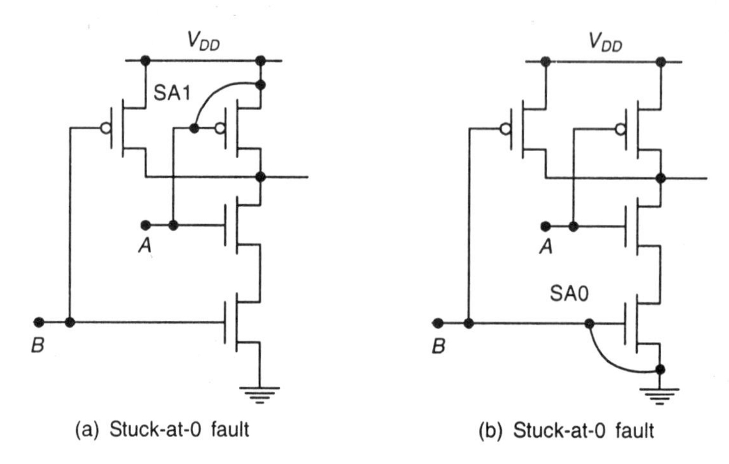

(a) Stuck-at-0 fault (b) Stuck-at-0 fault

Figure 14.10 2-input NAND gate with stuck-at faults.

Both circuits represent distinct cases. The circuits can be used to derive the test vectors needed to find each problem. Table 14.1 represents the function tables. The normal response of NAND gate is shown in F. The response for the SA1 fault in Figure 14.10(a) is denoted by F_{SA1}. Since the PFET connected to A-input never conducts, the output of the gate cannot be pulled to a logic 1 with an input of $(A,B) = (0,1)$. This vector can be used to test for this problem since it should produce logic 1 output. The SA0 fault in Figure 14.10(b) causes the output to behave as summarized in the F_{SA0} column of Table 14.1. In this case, the PFET connected to B input is always ON so that the output is stuck at a 1. Using an input vector of $(A,B) = (1,1)$ would find this fault.

Table 14.1 Function Table

A	B	F	F_{SA1}	F_{SA0}
0	0	1	1	1
0	1	1	0	1
1	0	1	1	1
1	1	0	0	1

The logic gate testing requires a fault model. In the simplest fault model, the entire logic gate is to be considered as a single unit. In SA0/SA1 fault model, the output of a faulty logic gate is 0 or 1, which is independent of its input values.

In SA0 fault, one should search for a set of inputs that sets a fault free gate's output to 1. So, for these inputs, the output of the gate is tested whether it has true or faulty value. Tables 14.2 and 14.3 shows the behaviour of two-input NAND and NOR gates with SA0 and SA1 faults.

Table 14.2 Behavior of Two Input NAND Gate with SA0 and SA1 Faults

Inputs		*NAND outputs*		
A	*B*	*Fault free output*	*SA0 output*	*SA1 output*
0	0	1	0	1
0	1	1	0	1
1	0	1	0	1
1	1	0	0	1

Table 14.3 Behavior of Two Input NOR Gate with SA0 and SA1 Faults

Inputs		*NOR outputs*		
A	*B*	*Fault free output*	*SA0 output*	*SA1 output*
0	0	1	0	1
0	1	0	0	1
1	0	0	0	1
1	1	0	0	1

For the NAND gate, if the output is tested for SA0 fault, the three input sets (0,0), (0,1), (1,0) should result a logic 1 at output. But since it is stuck-at 0 (SA0) fault, the output remains 0. In the case of SA1 fault, there is only one way to test for it by setting both inputs to 1 [i.e., (1,1) input set]. Fault free output for this input combination is logic 0, but due to SA1 fault, the output remains 1 (Table 14.2).

For the NOR gate, the input combination (0,0) gives a fault free output of 1. Due to SA0 fault, the output is logic 0 for this input combination. For checking SA1 fault, three input combinations of (0,1), (1,0), (1,1) are used. These sets provide a fault-free output of 0, but due to SA1 fault, the output remains at 1 (Table 14.3).

Thus, in a 2-input NAND gate, the input vectors (0,0), (0,1), (1,0) are used for testing SA0 fault and the vector (1,1), for SA1 fault. Similarly, in a 2-input NOR gate, the input vector (0,0) is used for testing SA0 fault, and the vectors (0,1), (1,0) and (1,1) are used for SA1 fault testing.

CMOS testing is complicated by the fact that every circuit node is capacitive and has the ability to store charge for a short period of time. If we apply a set of input test vectors to the gate, the response may be affected by this characteristic. Consider an open fault in the 2-input NAND gate as shown in Figure 14.11. This prevents the PFET connected to A from conducting and should be detected by the input combination $(A,B) = (0,1)$. However, note that the output node has a capacitance C_{out} that can't be eliminated. If we cycle through the input sequence $(A,B) = (0,0), (0,1), (1,0), (1,1)$, the stored charge may make it appear that the gate is operating properly. This is justified by the Table 14.4. The first input $(A,B) = (0,0)$ gives a logic 1 output, and the capacitor C_{out} has a voltage $V = V_{DD}$ across it. If we quickly apply the next input $(A,B) = (0,1)$ to insure a short testing cycle, then the output will still look like a logic 1 since C_{out} can hold the charge. Cycling through the remaining inputs gives normal results, so we have missed the fault entirely.

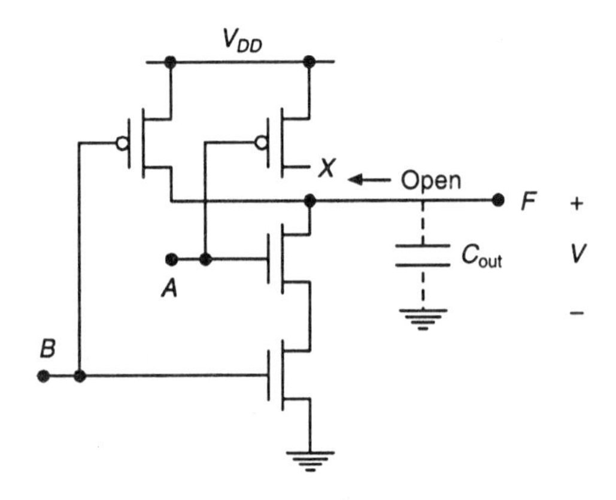

Figure 14.11 Charge storage effects on testing.

Table 14.4 Function table for charge-storage problem

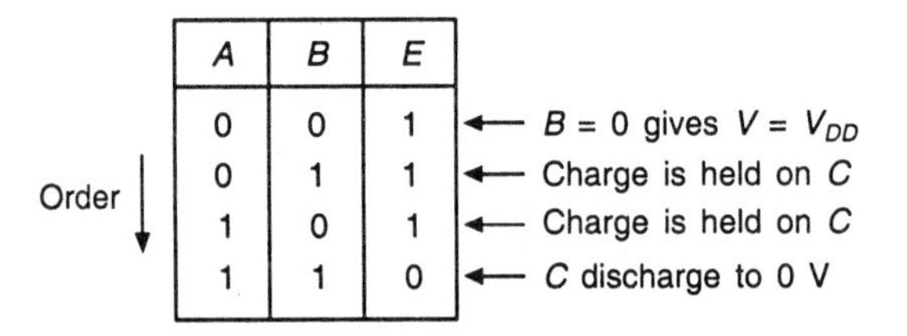

Order	A	B	E	
↓	0	0	1	← $B = 0$ gives $V = V_{DD}$
	0	1	1	← Charge is held on C
	1	0	1	← Charge is held on C
	1	1	0	← C discharge to 0 V

To compensate for this problem, we use an initialization vector that "prepares" the gate for the actual test vector. In the present case, the sequence $(A,B) = (1,1)$, $(0,1)$ would find the fault since the initialization vector $(A,B) = (1,1)$ discharges the output to 0 V, and the fault prevents $(A,B) = (0,1)$ from producing a logic 1 output.

Another type of problem arises with stuck-on and stuck-off faults. Consider the circuit shown in Figure 14.12(a) where the PFET connected to A input has a stuck-on (shorted) fault. If we apply an input vector of $(A,B) = (1,1)$, then the NFETs connected to inputs A and B

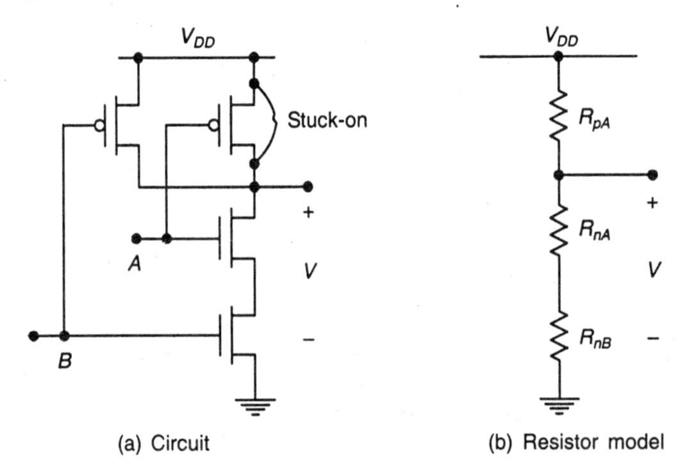

Figure 14.12 Stuck-on fault in a NAND gate.

conduct along with the PFET connected to A input. This leads to the resistor equivalent model shown in Figure 14.12(b). The output voltage is given by the voltage-divider rule as

$$V = \left[\frac{R_{nA} + R_{nB}}{R_{nA} + R_{nB} + R_{pA}} \right] V_{DD} \tag{14.16}$$

Since the NFET resistances depend on the aspect ratio while R_{pA} is due to the short, the voltage may give a low value of V which would make it appear that the gate is operating properly. This would be the case if the sum $(R_{nA} + R_{nB})$ is small compared to R_{pA}. If $R_{pA} \approx (R_{nA} + R_{nB})$, then N would be around one-half of V_{DD} which may or may not be detected as an incorrect value.

14.4.3 Observability

The observability of a particular internal circuit node is the degree to which one can observe that node at the outputs of an integrated circuit (i.e., the pins). This measure is of importance when a designer/tester desires to measure the output of a gate within a larger circuit to check that it operates correctly. Given a limited number of nodes that may be directly observed, it is the aim of well-designed chips to have easily observable gate outputs, and the adoption of some basic test design techniques can aid tremendously in this respect. Ideally, one should be able to observe directly or somewhat indirectly (i.e., one may have to wait a few cycles) every gate output within an IC.

14.4.4 Controllability

The controllability of an internal circuit node within a chip is a measure of the ease of setting the node to a 1 or 0 state. This measure is of importance when assessing the degree of difficulty of testing a particular signal within a circuit. An easily controllable node would be directly settable via an input pad. A node with little controllability might require many hundreds or thousands of cycles to get it to the right state. Often one finds it impossible to generate a test sequence to set a number of poorly controllable nodes into the right state. It should be the aim of a well-designed circuit to have all nodes easily controllable. In common with observability, the adoption of some simple for test techniques can aid tremendously in this respect.

14.4.5 Fault Coverage

A measure of goodness of a test program is the amount of fault coverage it achieves, i.e., for the vectors applied, what percentage of the chip's internal nodes were checked. For calculating the fault coverage, each circuit node is taken in sequence and held to 0 (SA0), and the circuit is simulated, comparing the chip outputs with a known 'good machine'. When a discrepancy is detected between the 'faulty machine' and the good machine, the fault is marked as detected and the simulation is stopped. This is repeated for setting the node to 1 (SA1). In turn, every node is stuck at 1 and 0 sequentially.

The total number of nodes that, when set to 0 or 1, result in the detection of the fault, divided by the total number of nodes in the circuit, is called the percentage-fault coverage. The above method of fault analysis is called sequential fault grading. The time to complete the fault grading may be very long. On average *KN* cycles need to be simulated, where *K* is the number of nodes in the circuit and *N* is the length of the test sequence. To overcome these long simulation times, many techniques have been invented to deal with fault simulation.

14.4.6 Automatic Test Pattern Generation (ATPG)

A physical fault can be transformed into a logical fault model that allows one to develop sets of test vectors. Many techniques have been developed for testing CMOS VLSI chips that use common circuit design styles. Most Automatic Test Pattern Generation (ATPG) approaches have been based on simulation. A five-valued logic form is commonly used to implement test generation algorithms (more advanced algorithms use up to 10 level logic). This consists of the states 1, 0, D, $\bar{D}$ and X, where 0 and 1 represent logical zero and logical one respectively, D represents a logic 1 in a good machine and a logic 0 in a faulty machine, $\bar{D}$ represents a logic 0 in a good machine and a logic 1 in a faulty machine. X represents the don't-care state.

The truth tables for NOT, AND and OR gates for these states are shown in Tables 14.5, 14.6 and 14.7.

Table 14.5 Inverter truth table ($Z = \bar{A}$)

A	$Z = \bar{A}$
0	1
1	0
X	X
D	$\bar{D}$
$\bar{D}$	D

Table 14.6 2-input AND gate truth table ($Z = A \cdot B$)

A \ B	0	1	X	D	$\bar{D}$
0	0	0	0	0	0
1	0	1	X	D	$\bar{D}$
X	0	X	X	X	X
D	0	D	X	D	0
$\bar{D}$	0	$\bar{D}$	X	0	$\bar{D}$

Table 14.7 2-input OR gate truth table ($Z = A + B$)

A \ B	0	1	X	D	$\bar{D}$
0	0	1	X	D	D
1	1	1	1	1	1
X	X	1	X	X	X
D	D	1	X	D	1
$\bar{D}$	$\bar{D}$	1	X	1	D

Examples

Consider the circuit shown in Figure 14.13, to examine the use of five-valued logic. Assume that an SA0 fault is to be detected at node *h*. This node *h* would have value *D*. There are two steps to be performed:

- Propagate the *D* on node *h* to one or more Primary Outputs (POs).
  A primary output is a directly observable signal. This path to the primary output (or outputs) is called the sensitized path.
- Set node *h* to state *D* via a set of Primary Inputs (PIs).
  A primary input is one that can be directly set via a pad or some other means. The gate driving node *h* is the Gate Under Test (GUT).

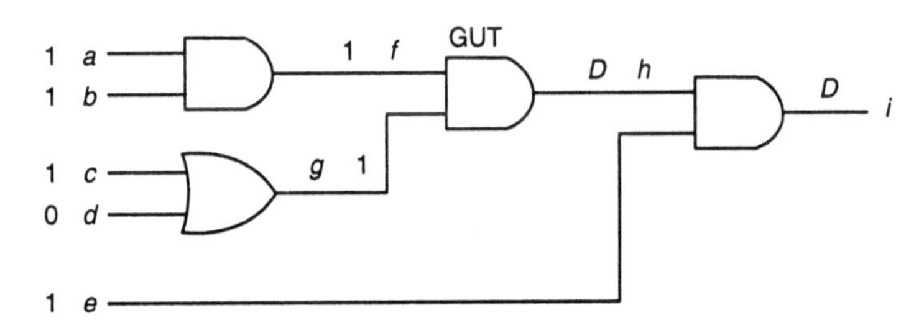

Figure 14.13 The *D*-algorithm-Sensitization step.

When the gate to be tested is embedded within a larger logic network, we can use the existing circuitry to create a specific path from the location of the fault to an observable output point. This technique is called path sensitization. The process of creating a path is called propagation since the fault is viewed as being propagated through the logic network.

From node *h*, we track back to the primary inputs (a,b,c,d,e) to find the necessary input vector required to set node *h* to a 1. Since the gate driving node *h* is an AND gate, both inputs (f,g) have to be set to 1 to get 1 in node *h*. Proceeding further towards the inputs, to assert node *f* as a 1, both inputs *a* and *b* have to be set to a 1.

Similarly, since node *g* is driven by an OR gate, either input *c* or input *d* need to be at logic 1 to assert node *g*. Thus, a vector $\{a,b,c,d\}$ of $\{1,1,1,0\}$ or $\{1,1,0,1\}$ is required to control node *h*. To observe that the output node has been set to a *D*, the input *e* has to be set to a 1. Thus, the resultant test vector is $\{a,b,c,d,e\} = \{1,1,0,1,1\}$ or $\{1,1,1,0,1\}$.

Now, assume that the fault at node *h* is SA1. In order to check for an SA1 fault, the node must be set to a 0. For this case, the test vector would be $\{a,b,c,d,e\} = \{0,1,X,X,1\}$ or $\{1,0,X,X,1\}$ or $\{0,0,X,X,1\}$ or $\{1,1,0,0,1\}$. Similarly, for other nodes, a summary of the vectors is listed in Table 14.8.

Table 14.8 Node vector summary of D-algorithm for node in Figure 14.13

Node	*Test*	*Vector {a,b,c,d,e}*
h	SA0	{1,1,0,1,1}, {1,1,1,0,1}
h	SA1	{0,1,*X*,*X*,1}, {1,0,*X*,*X*,1}, {0,0,*X*,*X*,1}, {1,1,0,0,1}
f	SA0	{1,1,0,1,1}, {1,1,1,0,1}
f	SA1	{0,0,0,1,1}, {0,0,1,0,1}
g	SA0	{1,1,0,1,1}, {1,1,1,0,1}, {1,1,1,1,1}
g	SA1	{1,1,0,0,1}

The next step is to collapse the vectors into the least set that covers all nodes. A possible set is {1,1,0,1,1}, {0,0,1,0,1}, {1,1,0,0,1}.

The reason for using a five-valued logic is shown in Figure 14.14. Here an additional AND gate and INVERT gate have been added to the circuit. We can see that a fault at node *h* is essentially unobservable (due here to the non-sensical logic). This circuit suffers from what is called reconvergent fan-out.

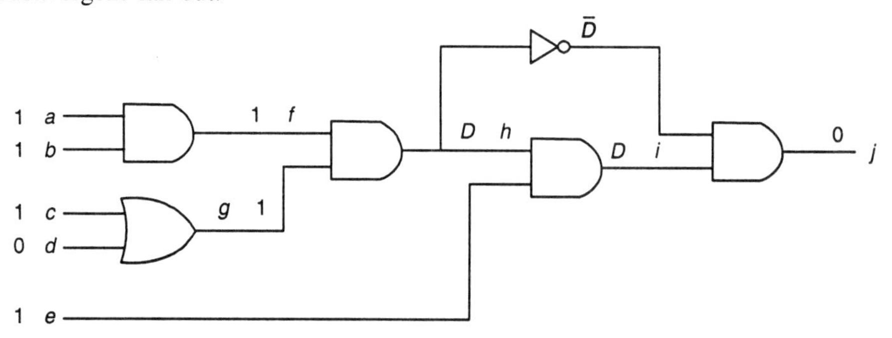

Figure 14.14 Reconvergent fan-out with *D* notation.

The usual basis for manual generation of tests by test engineers and many current automatic test-pattern generation programs is the D-algorithm (DALG). PODEM and PODEM-X are improved algorithms that are more efficient than the original DALG and in addition treat error-correcting circuits composed of XOR gates with reconvergent fan-out. Another ATPG algorithm is called FAN and an improved efficiency algorithm dealing with tristate drivers called ZALG has been developed. Other work has concentrated on dealing at a module level rather than at a gate level. Basically, these algorithms start by propagating the *D* value on an internal node to a primary output. This is called the *D*-propagation phase. The selection of which gates to pass through to the output is guided by observability indexes assigned to gates. At any particular gate input, the gate with the highest observability is selected. Once the *D* value is observable at a primary output, the next step is to determine the primary input values that are required to enable the fault to be observed and tested. This proceeds by backtracking from the faulted signal and sensitized path enables toward the primary inputs. The selection of which path to proceed along toward the inputs is aided by controllability indices assigned to nodes. This is known as the backtrace step.

Controllabilities and observabilities can be assigned statically (that is, without regard to the logic state of the network) or dynamically (that is, according to the current state of the network). The SCOAP algorithm is one method of assigning controllabilities and observabilities. In the SCOAP system the following six testability measures (TMs) are defined for each circuit node:

- CC0(n)—combinatorial 0 controllability of node n (i.e., the extent to which a combinatorial node can be forced to a zero).
- CC1(n)—combinatorial 1 controllability of node n.
- CO(n)—combinatorial observability of node n.
- SC0(n)—sequential 0 controllability of node n.
- SC1(n)—sequential 1 controllabillity of node n.
- SO(n)—sequential observability of node n.

The combinatorial measures are applied to the outputs of logic gates, while the sequential measures apply to registers and other 'sequential' modules. As an example, for the AND gate shown in Figure 14.15, the CC1 value is

$$\mathrm{CC1}(z) = \mathrm{CC1}(a) + \mathrm{CC1}(b) + 1$$

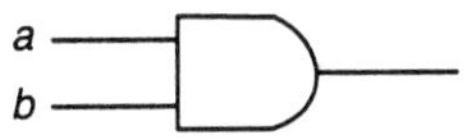

Figure 14.15 AND gate.

That is, the 1-controllability of the output of the AND gate is the sum of the 1-controllabilities of each input because each input has to be set to 1 to set the output to 1. The 1 is added at the end because the AND gate represents one stage of combinatorial logic. The sequential 1-controllability is given by

$$\mathrm{SC1}(z) = \mathrm{SC1}(a) + \mathrm{SC1}(b)$$

The combinatorial 0-controllability is given by

$$\mathrm{CC0}(z) = \min[\mathrm{CC0}(a), \mathrm{CC0}(b)] + 1$$

This arises due to the fact that either a 0 on a or b forces a 0 at the output. Therefore the easiest controllable input may be used (the lowest combinatorial controllability). The sequential controllability is given by

$$\mathrm{SC0}(z) = \min[\mathrm{SC0}(a), \mathrm{SC0}(b)]$$

The combinatorial observability of a is given by

$$\mathrm{CO}(a) = \mathrm{CO}(z) + \mathrm{CC1}(b) + 1$$

that is, the observability of z added to the combinatorial 1-controllability of b. This occurs because b has to be forced to a 1 to make a observable. The sequential observability of a is given by

$$\mathrm{SO}(a) = \mathrm{SO}(z) + \mathrm{SO}(b)$$

Similar equations may be derived for other gate types. The SCOAP algorithm proceeds by first calculating the circuit controllabilities by propagating controllabilities from the logic inputs. Following this, the observabilities are propagated from the logic outputs. Figure 14.16(a) shows a logic circuit with the 1-controllabities annotated. Figure 14.6(b) shows the observabilities.

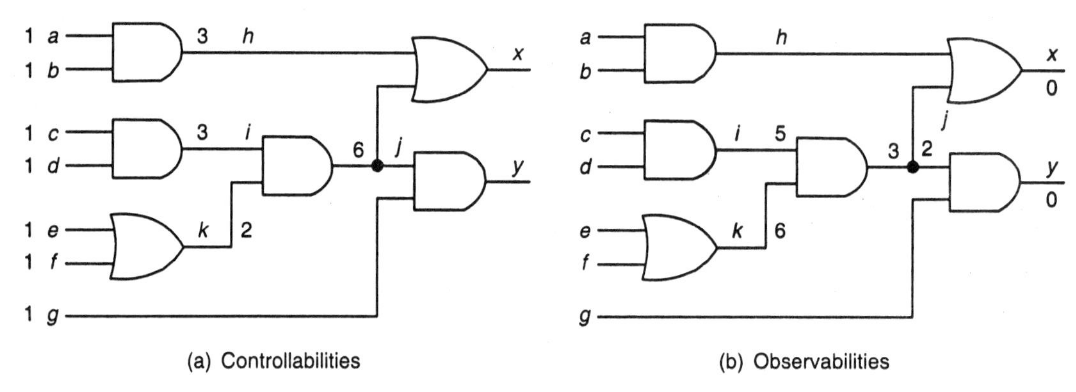

Figure 14.16 SCOAP testability measure example.

In cases of multiple fan-out, the minimum observability measure is used. The presence of high controllability numbers indicates a node that is difficult to control, while the presence of high observability numbers indicates nodes that are difficult to observe. The testability measures are used to guide the selection of paths in the *D*-propagation and backtrace phase of the *D*-algorithm-based ATPG procedures.

Other testability measures, such as COP and LEVEL are also used. COP testability measures are probabilistic in nature.

More recently authors have proposed the use of massively parallel methods for ATPG. Methods have also been developed that model faults as changes to a Boolean network. Equivalence checking is used to prove that the two networks are not equivalent. These methods, when combined with random-fault generation and fault simulation, have demonstrated a great deal of success.

14.4.7 Fault Grading and Fault Simulation

Fault grading consists of two steps. First, the node to be faulted is selected. Normally global nodes such as reset lines and clock lines are excluded because faulting them can lead to unnecessary simulation (i.e., if the reset or clock line is stuck, then not much is going to happen in the circuit). A simulation is run with no faults inserted, and the results of this simulation (that is, the primary output responses for each input test vector) are saved. Following this process, in principle, each node or line to be faulted is set to 0 and then 1 and the test vector set is applied. If, and when, a discrepancy is detected between the faulted circuit response and the good circuit response, the fault is said to be detected and the simulation is stopped, and the process is repeated for the next node to be faulted. If the number of nodes to be faulted is K, and the average number of test vectors is N, the number of simulation cycles, S_K, is approximately given by

$$S_K = 2(N/2)\ K + N$$

$$= K(N + 1) = KN \tag{14.17}$$

This serial fault simulation process is therefore running K sets of the test vector set. With a small vector set, simple circuit, or very fast simulator, this approach is feasible. However, for large test sets and circuits, it is highly impractical.

To deal with this problem, a number of ideas have been developed to increase the speed of fault simulation. Parallel simulation is one method for speeding up simulation of multiple machines. In this method m words in an n-bit computer are used to encode the state of n 'machines' for a 2^m-state simulator. Two n-bit words may be used to encode n machines for a three-state simulator. More computer words may be used to encode simulator with more states. Moreover this principle has been extended to special-purpose hardware where the computer word length could be optimized to deal with substantially more circuits in parallel. Now if M circuits can be simulated in parallel, then

$$S_K = (KN)/M \tag{14.18}$$

Concurrent Simulation is currently the most popular method for software-based fault simulators. The technique uses a nonfaulted version of the circuit to create a 'good' machine model. Each fault creates a new faulty machine that is simulated in parallel with the good machine. Thus $N + 1$ simulations may have to be completed, where N is the number of faults. Concurrent simulators rely on a number of heuristics to reduce the amount of simulation. For instance, when a difference is noted between a faulted machine and a good machine at an externally observable point (i.e., the pads), the faulty machine is dropped from the simulation queue and the fault is "detected". If the bad machine has a X or Z compared to a 1 or 0 for the good machine, the fault is a 'possible detect'. Obviously, the more externally observable nodes a circuit has, the quicker the bad machines get dropped from the simulation. Normally, only the good machine state is stored, with each node listing the fault machines that differ with the good machine. The different state is often small, which implies that there is a small amount of extra simulation to be done. In other words, most simulation for a faulty machine is exactly the same as the good machine. This is what concurrent simulation exploits. Fault collapsing occurs when two different faults result in the same faulty machine. This is noted, and one of the faulty machines may be dropped. Some machines perform static fault collapsing prior to simulation. For instance, an SA0 fault on the input of an inverter is the same as an SA1 fault at the output of the same inverter. With some fault simulators it is possible to create a fault dictionary. This is a cross reference that maps an observed fault to a set of possible internal faults. It is of use when the tester wishes to track down the actual internal failure (such as to perform yield improvement) rather than just cull the part.

Apart from software-based simulations, hardware-fault simulation accelerators that can provide a speedup over software-based simulators are also available.

14.4.8 Delay Fault Testing

The fault models we have dealt with to this point have neglected timing. Failures that occur in CMOS could leave the functionality of the circuit untouched, but affect the timing. For instance, consider the layout of circuit shown in Figure 14.17 for a high-power NAND gate composed of paralleled *n*-and *p*-transistors. If the link was opened, the gate would still function, but with increased pull-down time. In addition, the fault now becomes sequential because the detection of the fault depends on the previous state of the gate and the simulation clock speed.

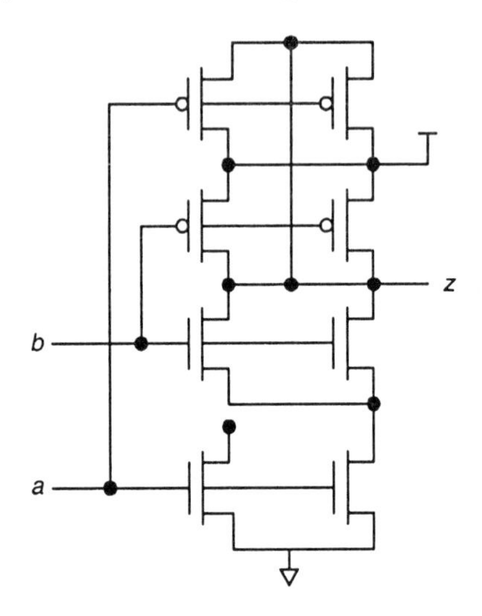

Figure 14.17 An example of a delay fault.

14.4.9 Statistical Fault Analysis

Conventional fault analysis can consume large CPU resources and take a long time. An alternative to this is what is called statistical fault analysis (STAFAN). This method of fault analysis relies on estimating the probability that a fault will be detected. In summary, a fault free simulation is performed on a circuit in which some extra statistics are gathered by a modified simulator on a per-input vector bases. These are as follows:

- Zero-counter: The 0 count on each gate input when a $1 \rightarrow 0$ change of the output is detected.
- One-counter: The 1 count on each gate input when a $0 \rightarrow 1$ change of the output is detected.
- Sensitization-counter: Incremented if the input change causes the output to be sensitized.
- Loop-counter: Used to detect and deal with feedback.

The one-controllability of line l is given by

$$\mathrm{CI}(l) = \text{one-count}/N,$$

where N is the number of vectors.

The zero-controllability is given by

$$C0(l) = \text{zero-count}/N$$

The one-level sensitization probability is

$$S(l) = (\text{sensitization} - \text{count})/N$$

The observabilities are calculated by propagating from gate outputs to gate inputs. For common gates, Jain and Agrawal derive the one-observabilities (B1) and zero-observabilities (B0) for common gates, as shown in Table 14.9.

Table 14.9 Statistical fault analysis 1 and 0 observabilities

Gate Type	*B1(l)*	*B0(l)*
AND	$B1(m) \cdot [C1(m)/C1(l)]$	$B0(m) \cdot \{[S(l) - C1(m)]/C0(l)\}$
OR	$B1(m) \cdot \{[S(l) - C0(m)]/C1(l)\}$	$B0(m) \cdot [C0(m)/C0(l)]$
NAND	$B0(m) \cdot [C0(m)/C1(l)]$	$B1(m) \cdot \{[S(l) - C1(m)]/C0(l)\}$
NOR	$B0(m) \cdot \{[S(l) - C1(m)]/C1(l)\}$	$B1(m) \cdot [C1(m)/C0(l)]$
NOT	$B0(m)$	$B1(m)$

Methods also exist to deal with fan-out where two observabilities must be combined. Once these observability and controllability measures have been determined, the probability of fault detection may be calculated as follows:

$$D1(l) = B0(l) \cdot C0(l),$$

where $D1(l)$ is the probability of detection that line l is SA1.

$$D0(l) = B1(l) \cdot C1(l)$$

where $D0(l)$ is the probability of detection that line l is SA0.

From these values the fault coverage of the circuit may be calculated. The results of using this technique follow very closely the results generated by conventional fault simulation.

14.4.10 Fault Sampling

Another approach to fault analysis is known as fault sampling. This is used in circuits where it is impossible to fault every node in the circuit. Nodes are randomly selected and faulted. The resulting fault-detection rate may be statistically inferred from the number of faults that are detected in the fault set and the size of the set. As with all probabilistic methods, it is important that the randomly selected faults be unbiased. Although this approach does not yield a specific level of fault coverage, it will determine whether the fault coverage exceeds a desired level. The level of confidence may be increased by increasing the number of samples.

14.5 DESIGN STRATEGIES FOR TEST

14.5.1 Design for Testability

There are two key concepts to designing circuits that are testable. They are:

1. Controllability
2. Observability

Controllability is the ability to set (to 1) and reset (to 0) every node internal to the circuit. Observability is the ability to observe either directly or indirectly the state of any node in the circuit. Quite simply, these concepts ensure that the designer considers the provision of means of setting or resetting key nodes in the system and of observing the response at key points.

Three main approaches to design for testability are:

(i) Ad hoc testing
(ii) Scan-based approaches
(iii) Self-test and built-in testing

14.5.2 Ad hoc Testing

Ad hoc techniques are collections of ideas aimed at reducing the combinational explosion of testing. Common techniques involve:

- Partitioning large sequential circuits
- Adding test points
- Adding multiplexers
- Providing for.easy state reset

Ad hoc testing increases the observability and controllability of a design.

14.5.2.1 Ad hoc test techniques applied to a counter

Long counters can be tested by ad hoc techniques. We want to test an 8-bit counter. Figure 14.18(a) shows an implementation in which the counter only has a RESET and a CLOCK input, with the terminal count (TC) being observable. The designer probably thought that a reset and 256 clock cycles, followed by the observation of TC, would be adequate for testing purposes. Apart from the non-observability of the count value (Q <7:0>), the main problem is the number of cycles required to test a single counter. Possible ad hoc test techniques are shown in Figures 14.18(b) and 14.18(c).

In Figure 14.18(b), a parallel-load feature is added to the counter. This enables the counter to be preloaded with appropriate values to check the carry propagation within the counter. Another technique is to reduce the length of each counter to, say, 4-bits, as shown in Figure 14.18(c). This is achieved by having the test signal block the carry propagate at every 4-bit section. The carry propagation between 4-bit sections may be tested with a few additional vectors.

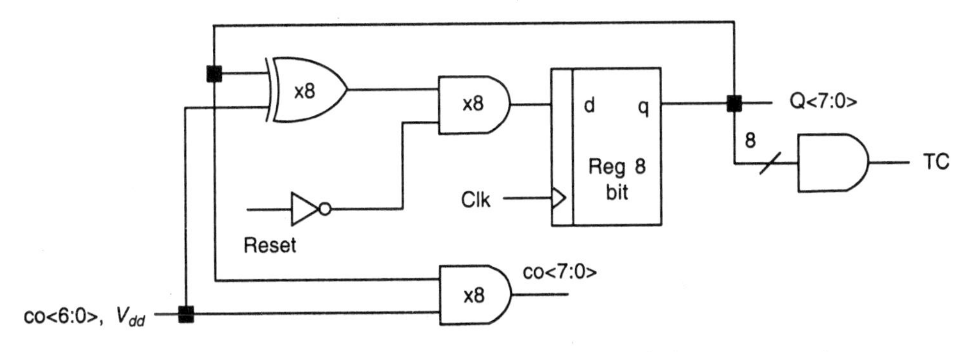

(a) Implementation in which the counter only has a reset and clock input, with terminal count (TC) being observable

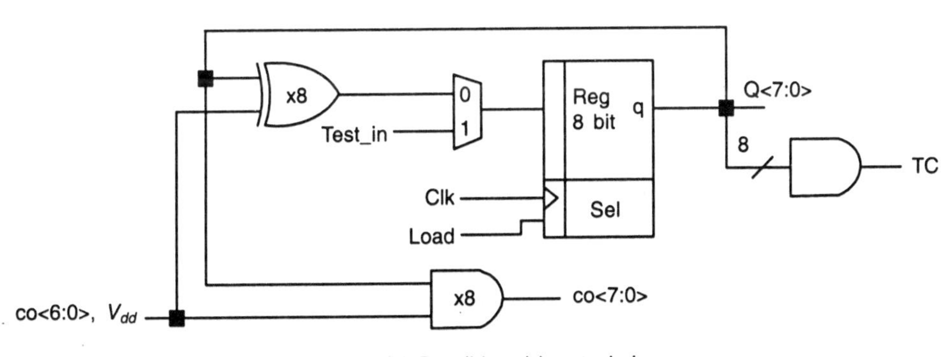

(b) Possible ad hoc technique

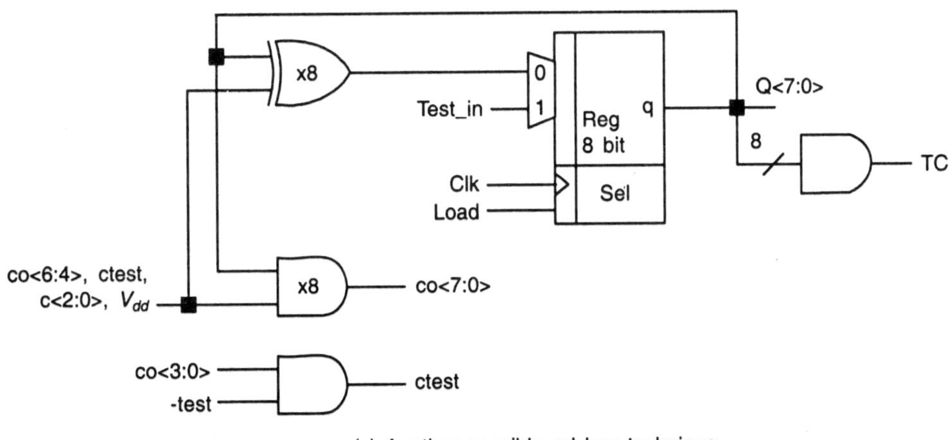

(c) Another possible ad hoc technique

Figure 14.18 Ad hoc test techniques applied to a counter.

14.5.2.2 Bus-oriented test techniques

Another technique is the use of bus in a bus-oriented system for test purposes. This is shown in Figure 14.19(a) for a very simple accumulator. Each register has been made loadable from

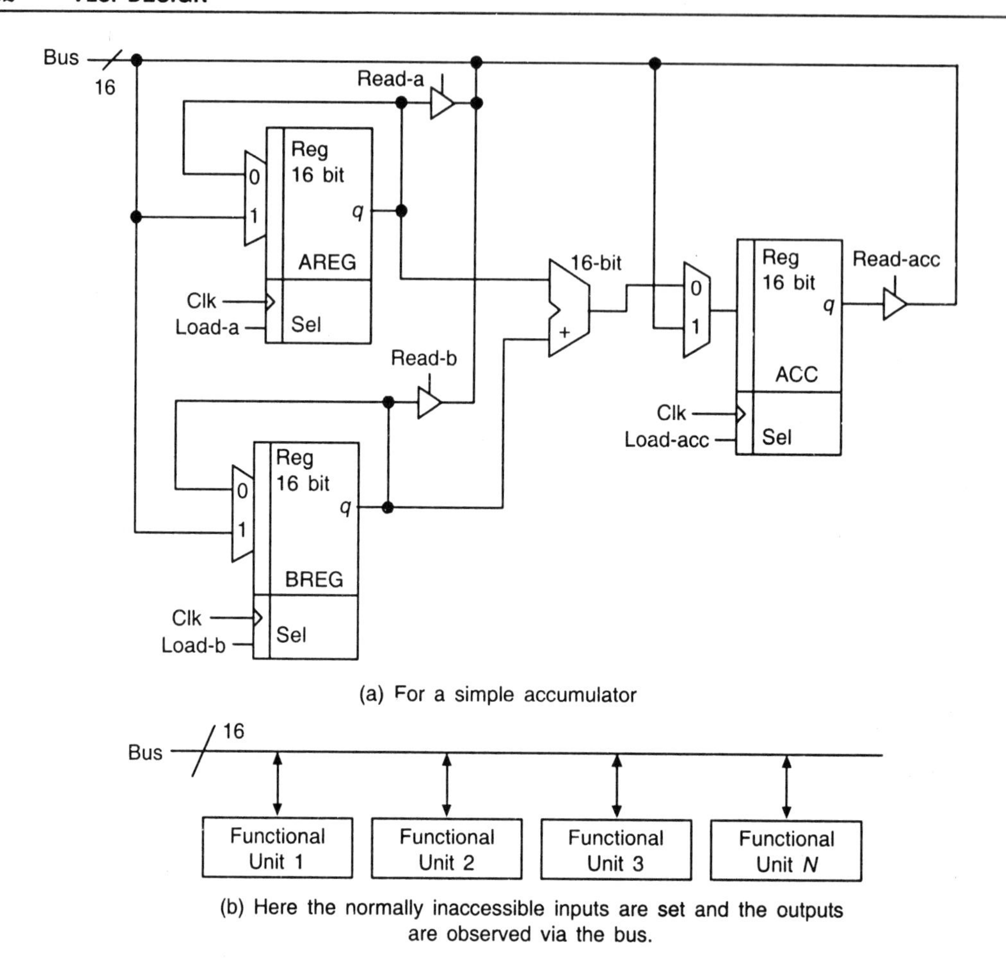

(a) For a simple accumulator

(b) Here the normally inaccessible inputs are set and the outputs are observed via the bus.

Figure 14.19 Bus-oriented test techniques.

the bus and capable of being driven onto the bus. Here the internal logic values that exist on a data bus are enabled onto the bus for test purposes. A more general scheme is shown in Figure 14.19(b), where the normally inaccessible inputs are set and the outputs are observed via the bus.

14.5.2.3 *Multiplexer-based testing*

The controllability and observability of a memory can be improved dramatically by adding multiplexers on the data and address buses. Figure 14.20(a) shows a simple processor with its data memory. Under normal configuration, the memory is only accessible through the processor. Writing and reading a data value into and out of a single memory position requires a number of clock cycles. Figure 14.20(a) shows the arrangement of Figure 14.20(b) with a multiplexer. During normal operation mode, these selectors direct the memory ports to the processor. During test, the data and address ports are connected directly to the I/O pins, and testing the memory can proceed more efficiently.

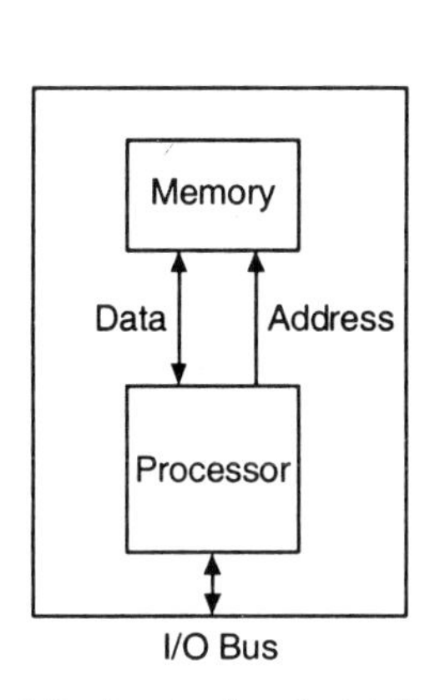

(a) Design for low testability

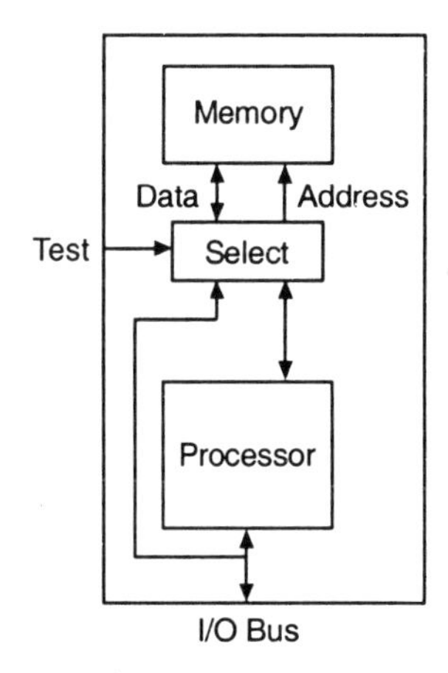

(b) Adding a selector improves testability

Figure 14.20 Improving testability by inserting multiplexers.

Any design should always have a method of resetting the internal state of the chip within a single cycle or almost a few cycles. Apart from making testing easier, this also makes simulation faster because a few cycles are required to initialize the chip.

An extensive collection of ad hoc test approaches has been devised. The applicability of most of these techniques depends upon the application and architecture at hand. Their insertion into a given design requires expert knowledge and is difficult to automate. Structured and automatable approaches are more desirable.

14.5.3 Scan-Based Test Techniques

A collection of approaches have evolved for testing that lead to a structured approach to testability. The approaches stem from the basic principles of controllability and observability.

14.5.3.1 Level Sensitive Scan Design (LSSD)

This approach is based on two concepts. First, that the circuit is level sensitive. The second is that each register may be converted to a serial shift register. A logic system is level-sensitive if, and only if, the steady state response to any allowed input state change is independent of the circuit and wire delays within the system. Also, if an input state change involves the changing of more than one input signal, then the response must be independent of the order in which they change. Steady state response is the final value of all logic gate outputs after all change activity has terminated.

The basic building block in LSSD is the shift register latch, or SRL. A block-level implementation of a polarity-hold SRL is shown in Figure 14.21(a). It consists of two latches L1 and L2. L1 has a serial data port, I, and an enable, A. It also has a data port D, and an enable C. When A is high, the value of L1 (T1) is set by the value of I, while when C is high, L1 is set by D. A and C cannot be simultaneously high. When signal B in L2 is high, T1 is passed to T2. A gate level implementation of the SRL is shown in Figure 14.21(b) and Figure 14.21(c).

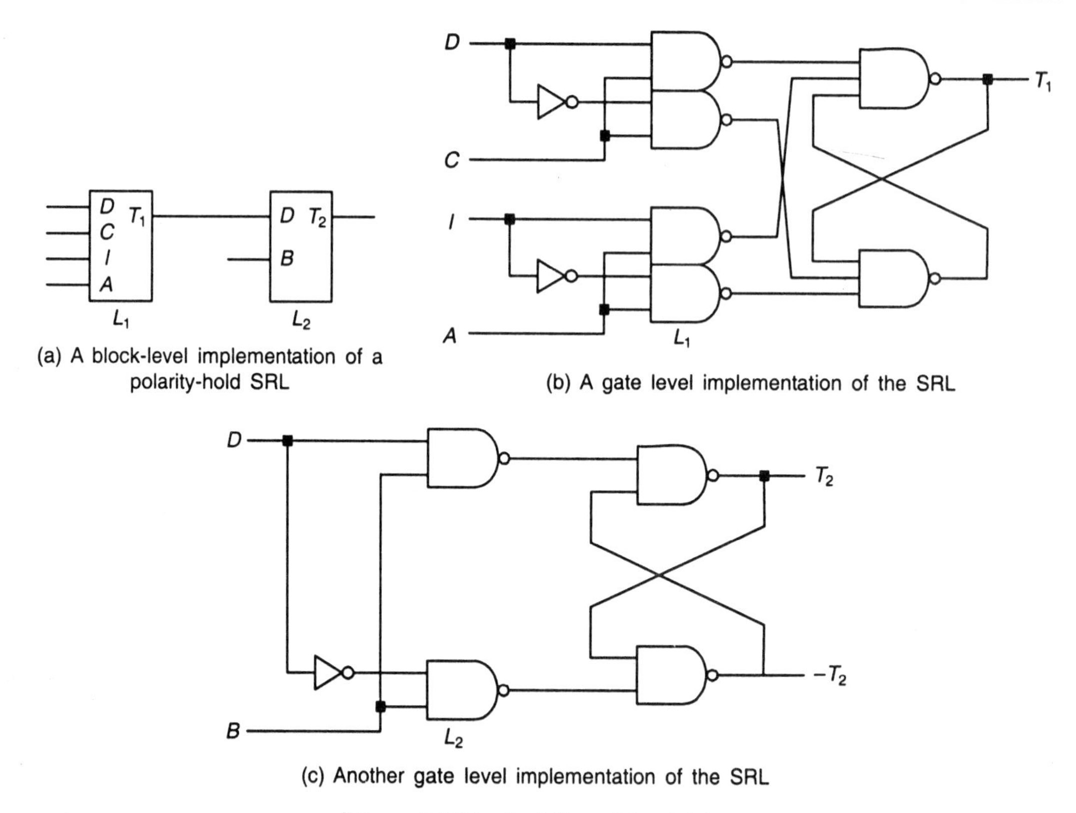

Figure 14.21 A shift register latch.

In normal operation, the D input is the normal input to the register, while the T2 signal is the output. L1 is the master while L2 is the slave. SRLs may be connected in series by using the T2 output and the I input of successive latches. During normal system operation, A is held low and C and B may be thought of as a two phase non-overlapping clock. When data is to be loaded into the SRLs or dumped out of the SRLs, A and B are used as a two-phase shift clock.

Figure 14.22(a) shows a typical LSSD scan system. An expanded view is show in Figure 14.22(b). The first rank of SRLs have inputs driven from a preceding stage and have outputs QA1, QA2 and QA3. These outputs feed a block of combinational logic. The output of this logic block feeds a second rank of SRLs with outputs QB1, QB2 and QB3.

Figure 14.22(c) shows a typical clocking sequence. Initially, the shift-clock and C2 are clocked three times to shift data into the first rank of SRLs (QA1 to QA3). C1 is asserted and then C2 is asserted, clocking the output of the logic block into the second rank of SRLs (QB1 to QB3). Shift-clock and C2 are then clocked three times to shift QB1, QB2 and QB3 out via the serial-data out line. Testing proceeds in this manner of serially clocking the data through the SRLs to the right point in the circuit, running a single 'system' clock cycle and serially clocking the data out for observation. In this scheme, every input to the combinational block may be controlled and every output may be observed. In addition, running a serial sequence of 1s and 0s (such as 110010) through the SRLs can test them.

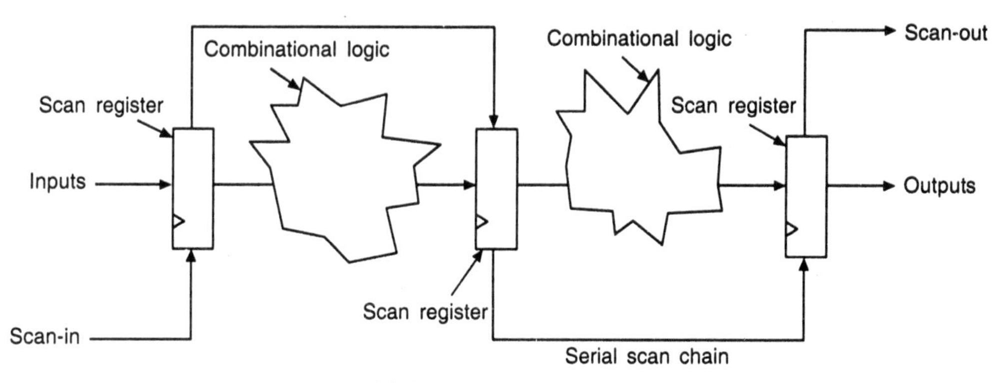

(a) A typical LSSD scan.

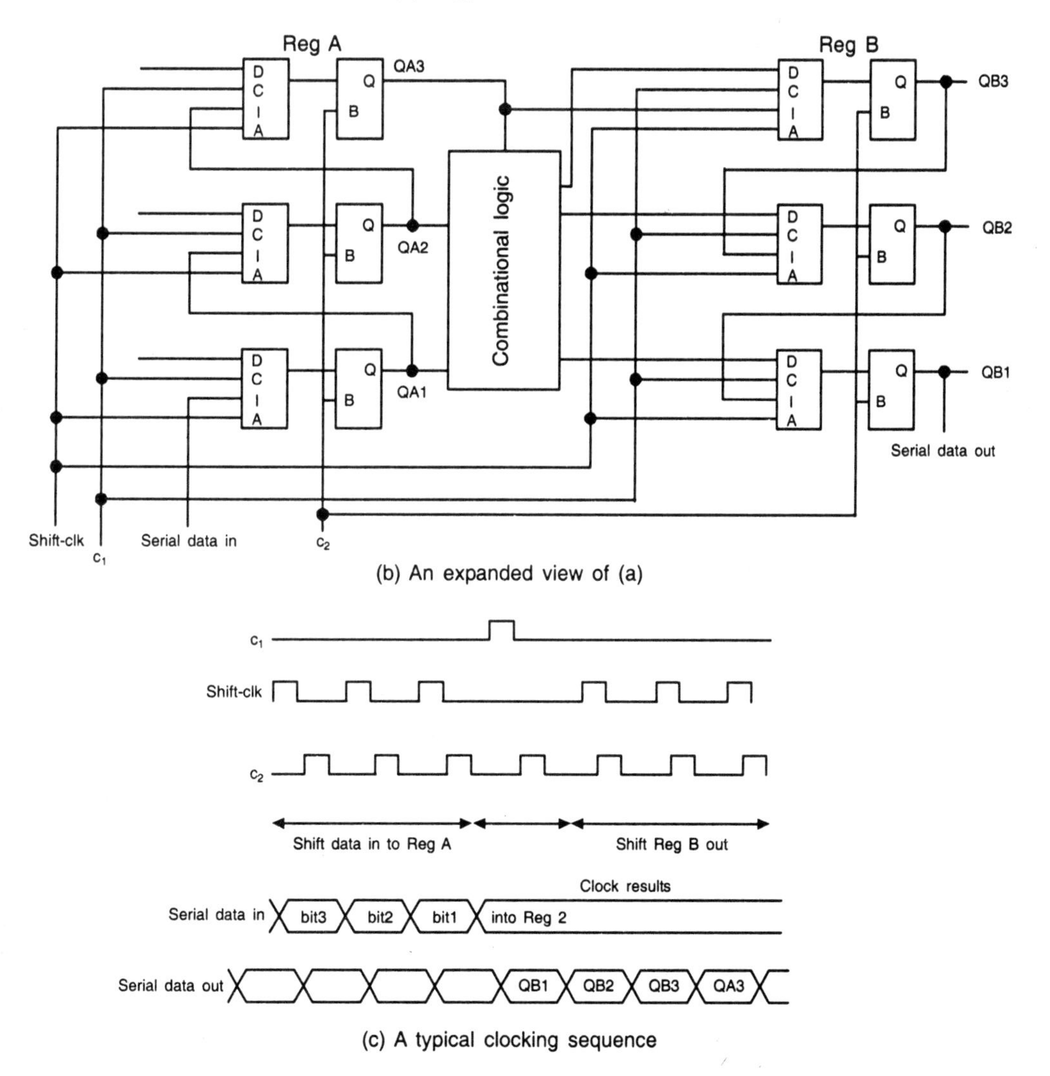

(b) An expanded view of (a)

(c) A typical clocking sequence

Figure 14.22 An LSSD scan chain.

Test generation for this type of test architecture may be highly automated. ATPG techniques may be used for the combinational blocks, and the SRLs are easily tested. The main disadvantage is the complexity of the SRLs (i.e., impacting density and speed).

14.5.3.2 Serial scan

It has become difficult to provide a hazard-free latching scheme with level sensitive scan. Faster clock speeds and design for smaller overhead in the registers has led to simplification in the SRL that give up a little on the hazard front but retain the same principles mentioned above.

A schematic for a commonly used CMOS edge-sensitive scan-register is shown in Figure 14.23. A MUX is added before the master latch in a conventional D register. TE is the test enable pin, and TI is the test input pin. When TE is enabled, TI is clocked into the register

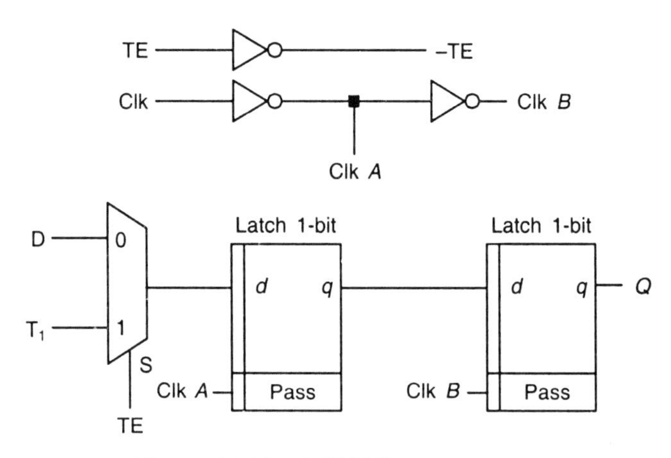

Figure 14.23 A CMOS scan register.

by the rising edge of clock. Figure 14.24 shows some circuit-level diagrams of CMOS SRL implementations. Figure 14.24(a) shows a frequently used implementation, which uses transmission gates to implement the multiplexers. The layout density overhead for this latch is minimal. In addition, because the addition of the testability MUX places two transmission gates in series, the increase in delay is minimized. Two further implementations of the input MUX are shown in Figures 14.24(b) and 14.24(c). Figure 14.24(b) shows the addition of only two transistors and a single control line. A register so implemented does have the normal problems associated with used single-polarity transmission gates. Alternatively, the clocks may be gated, as in Figure 14.24(c). While this minimizes transistors, it may lead to unacceptable hold-time constraints on the register. Because the signals applied to master latch are delayed with respect to the main clock, the date has to be held, for a long time at the input.

14.5.3.3 Partial serial scan

Quit often in a design, one may not find it area- and speed-efficient to implement scan registers in every location where a register is used. This occurs, for example, in signal-processing circuits where many pipeline registers might be used to achieve high-speed. If these are in the data flow

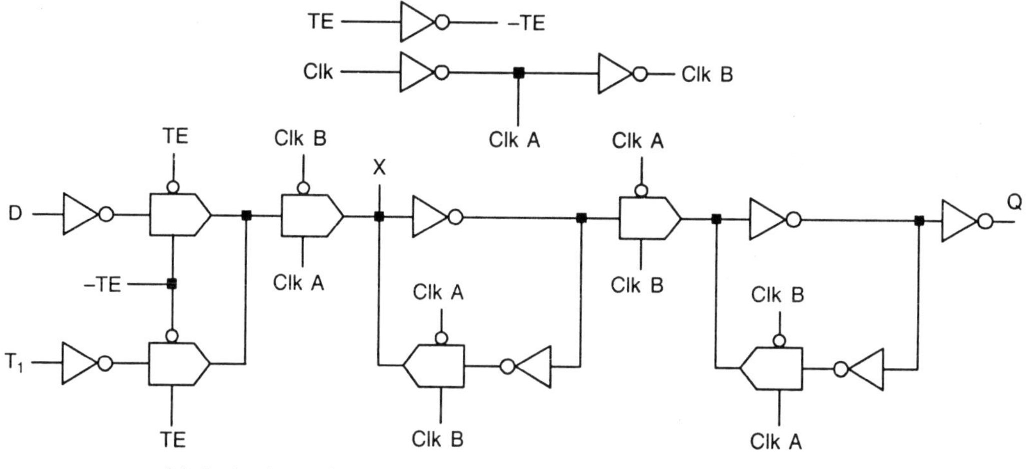

(a) An implemention using transmission gates to implement multiplexers.

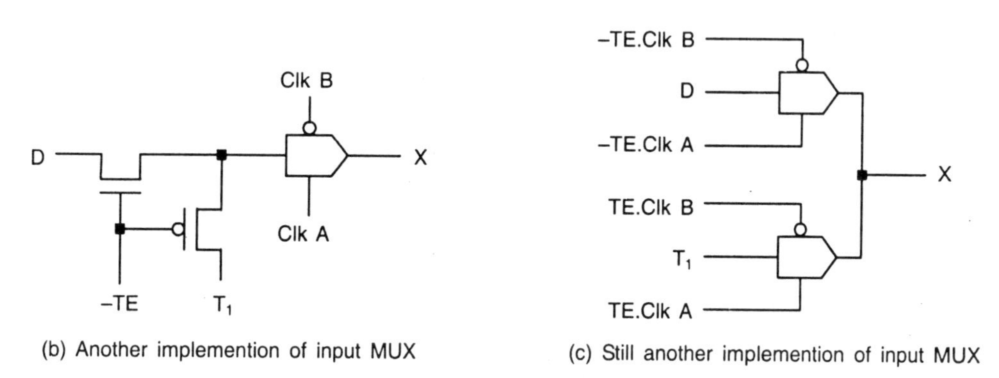

(b) Another implemention of input MUX

(c) Still another implemention of input MUX

Figure 14.24 CMOS scan-latch options.

section of the chip, then one can think of the logic that has to be tested as the logic with the pipeline registers removed. In this case, only the input and the output registers need be made scannable. This technique of testing is known as partial scan, and depends on the design making decisions about which registers need to be made scannable.

Consider the design shown in Figure 14.25. In a full scan test strategy, all registers would have to be scannable. A partial-scan design is shown in Figure 14.25(a) where only two registers have been made scannable (R6 and R3). In addition, these registers have the ability to hold their state dependent on a HOLD control. The part of the circuit that is being tested and monitored by the scan registers (known as the kernel) is shown in Figure 14.25(b). It may be proven that, by holding the vectors at the input of the kernel for three clock cycles, the kernel may be (represented by the combinational-equivalent circuit shown in Figure 14.25(c). This circuit may be used by an ATPG program to generate test vectors.

The serial-scan chains can become quite long, and the loading and unloading sequence can dominate testing sequence.

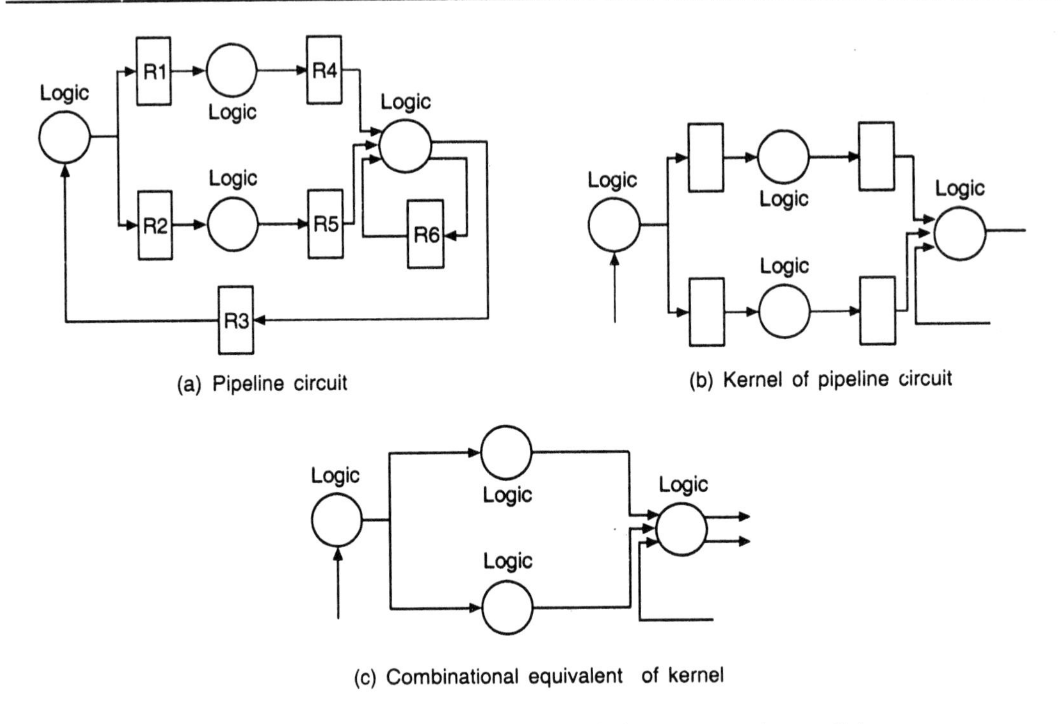

Figure 14.25 The application of scan techniques to employ partial scan.

14.5.3.4 Parallel scan

An extension of serial scan is called random-access or parallel scan. The basic structure of parallel scan is shown in Figure 14.26. Each register in the design is arranged on an imaginary (or real) grid where registers on common rows receive common data lines and registers in

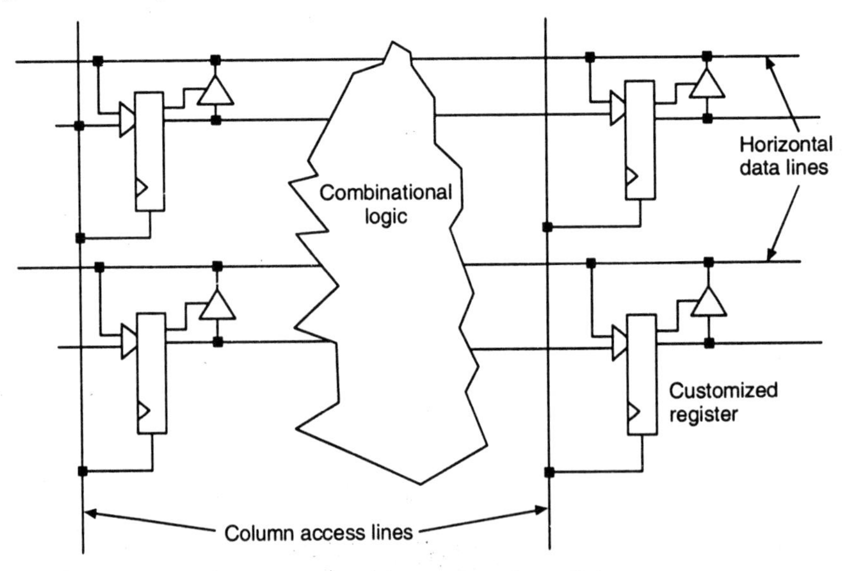

Figure 14.26 Basic structure of parallel scan.

common columns receive common read- and write-control signals. In the figure, and array of 2-by-2 registers is shown. The D and Q signals of the registers are connected to the normal circuit connections. Any register output may be observed by enabling the appropriate column read line and setting the appropriate address on an output data multiplexer. Similarly, data may be written to any register.

Figure 14.27 shows a D-register implementation called a cross-coupled latch. It consists of a normal CMOS master–slave edge-triggered register augmented by two small n-transistors, N1 and N2. When test–write–enable is high, Probe (j) is high, and clock is low, the value of node Y, i.e., the value D through transmission gate may be sensed on Sense (i) via transistor N2. When test–write–enable is low, Probe (j) is high, and clock is high, the value on sense (i) can be driven onto node Y. This is seen immediately at the output of the register. The net effect on the register-timing parameters of the extra transistors is to slightly increase the minimum clock-pulse width. The area impact for an ASIC based register is around 3%.

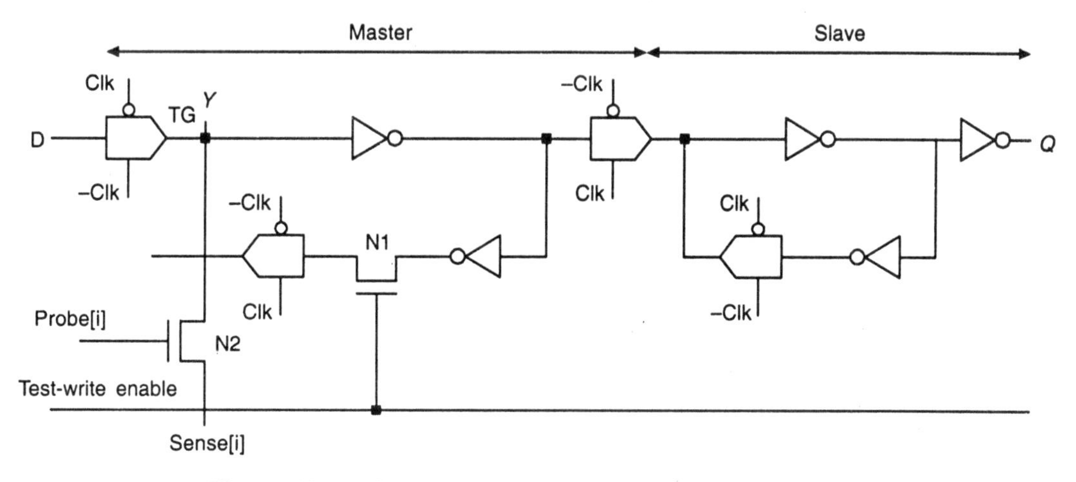

Figure 14.27 Parallel scan registers (a cross-coupled latch).

14.5.4 Self-Test Techniques

Self-test and built-in test techniques rely on augmenting circuits to allow them to perform operations on themselves that prove correct operation.

14.5.4.1 Signature analysis and BILBO

One method of incorporating built-in test module is to use signature analysis or cyclic-redundancy checking. This involves the use of pseudo-random sequence generator (PRSG) to generate the input signals for a section of combinational circuitry and then using a signature analyzer to observe the output signals.

A PRSG implements a polynomial of some length N. It is constructed from a linear feedback shift register (LFSR), which is constructed, in turn, from a number of 1-bit registers connected in a serial fashion, a shown in Figure 14.28. The outputs of certain shift bits are XORed and fed back to the input of the LFSR to calculate the required polynomial. For instance, in Figure 14.28,

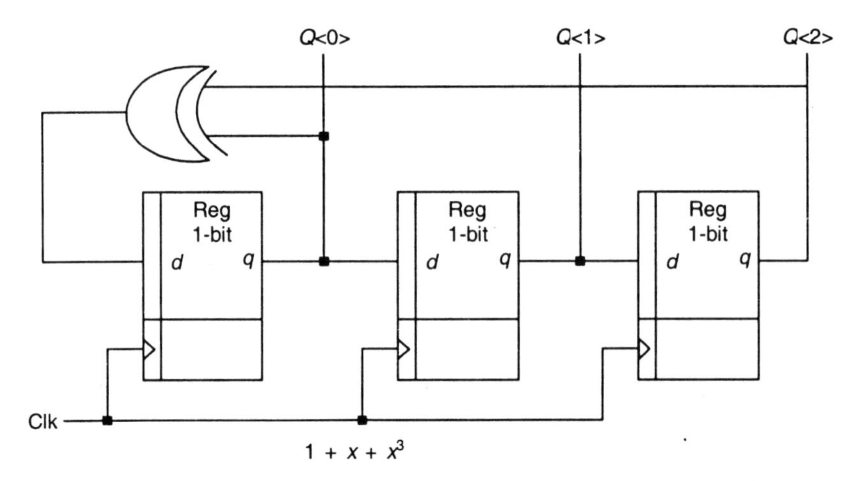

Figure 14.28 Pseudo-Random Sequence Generator (PRSG).

the 3-bit shift register is computing the polynomial $f(x) = 1 + x + x^3$. For an n-bit LFSR, the output will cycle through $2^n - 1$ states before repeating the sequence. A complete feedback shift register (CFSR) includes the zero state, which may be required in some test situations.

A signature analyzer is constructed by cyclically adding the outputs of a circuit to a shift register or an LFSR if successive logic blocks are to be tested in a like manner. A typical circuit is shown in Figure 14.29(a). As each test vector is run, the incoming data is XORed with the contents of the LFSR. At the end of a test sequence, the LFSR contains a number, known as the syndrome, which is a function of the current output and all previous outputs. This can be compared with the correct syndrome (derived by running a test program on the good logic) to determine whether the circuit is good or bad.

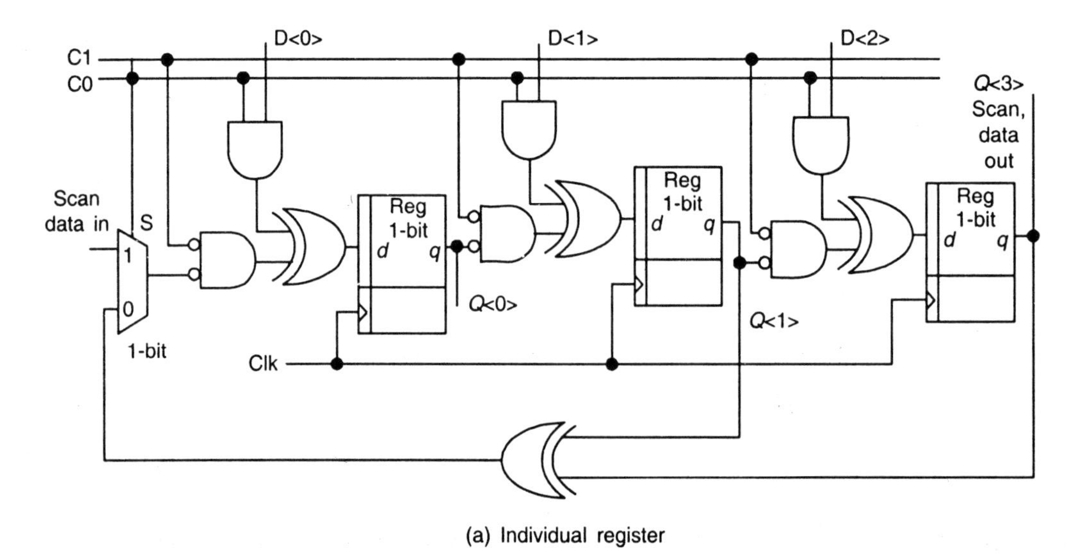

(a) Individual register

Figure 14.29 (contd.)

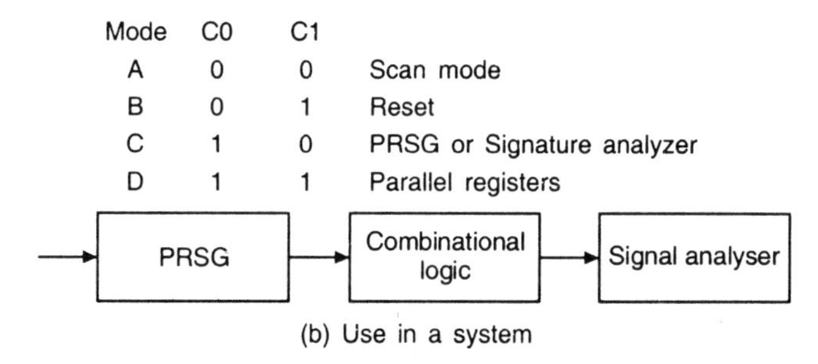

(b) Use in a system

Figure 14.29 Built-in logic block observation (BILBO).

Signature analysis can be merged with the scan technique to create a structure known as BILBO for Built-In Logic Block Observation.

A 3-bit register is shown with the associated circuitry. In mode D (C0 = C1 = 1), the registers act as conventional parallel registers. In mode A (C0 = C1 = 0), the registers act as scan registers. In mode C (C0 = 1, C1 = 0), the registers act as a signature analyzer or pseudo-random sequence generator (PRSG). The registers are reset if C0 = 0 and C1 = 1. Thus a complete test-generation and observation arrangement can be implemented, as shown in Figure 14.29(b). In this case, two sets of registers have been added in addition to some random logic to effect the test structure.

14.5.4.2 Memory self-test

Embedding self-test circuits for memories in higher-speed circuits not only may be the way of testing the structures at speed but can save on the number of external test vectors that have to be run. A typical read/write memory (RAM) test program for an M-bit address memory might be as follows:

FOR i = 0 to M - 1 write ($\overline{\text{data}}$)

FOR i = 0 to M - 1 read ($\overline{\text{data}}$) then write (data)

FOR i = 0 to M - 1 read (data) then write ($\overline{\text{data}}$)

FOR i = M - 1 to 0 read ($\overline{\text{data}}$) then write (data)

FOR i = M - 1 to 0 read (data) then write ($\overline{\text{data}}$)

data is 1 and $\overline{\text{data}}$ is 0 for a 1-bit memory or a selected set of patterns for an n-bit word. For an 8-bit memory data, might be x00, x55, x33, and x0F. An address counter, some multiplexers, and a simple-state machine result in a fairly low overhead self-test structure for read/write memories. Oshawa et al. describe a 4-Mbit RAM with self-test. The self-test consists of 256K cycles that input a checkerboard pattern to test for cell-to-cell interference.

This is followed by 256K cycles in which the data is read out. Then a complemented checkerboard is written and read. A total of 1 million cycles provide a test sufficient for system maintenance.

ROM memories may be tested by placing a signature analyzer at the output of the ROM and incorporating a test mode that cycles through the contents of the ROM. A significant advantage of all self-test methods is that testing may be completed when the part is in the field. With care, self-test may even be performed during normal system operation.

14.5.4.3 *Iterative logic array testing*

Arrays of logic present an interesting problem to the test architect because the replication can be used to advantage in reducing the number of tests. In addition, by augmenting the logic, extremely high fault coverage rates are possible. An iterative logic array (ILA) is a collection of identical logic modules (such as an n-bit adder). An ILA is C-testable if it can be tested with a constant number of input vectors independent of the iteration count. An ILA is I-testable if a particular fault that occurs in any module as a result of an applied input vector is identical for all modules in the ILA. Assuming that only one module is faulty, the detection of a fault may be made by using an equality test on the ILA outputs.

14.5.5 I_{DDQ} Testing

Applying a power supply voltage to a CMOS chip causes a current I_{DD} to flow. When the signal inputs are stable (not switching), the quiescent leakage current I_{DDQ} can be measured. This is illustrated in Figure 14.30. Every chip design is found to have a range of 'normal' levels. I_{DDQ} testing is based on the assumption that an abnormal reading of the leakage current indicates a problem on the chip. I_{DDQ} testing is usually performed at the beginning of the testing cycle. If a die fails, it is reflected and no further tests are performed.

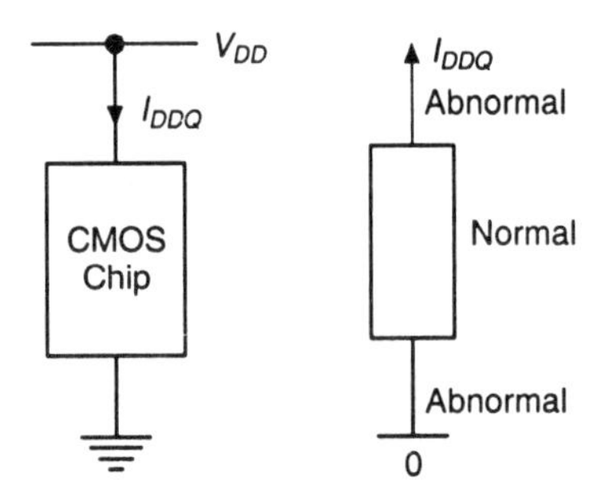

Figure 14.30 Basic I_{DDQ} test.

The components of a basic I_{DDQ} measurement system is shown in Figure 14.31. The test chip is modelled as being in parallel with the testing capacitance, C_{test}. A power supply with a value V_{DD} is connected to the chip by a switch that is momentarily closed at time $t = 0$. The current I_{DD} is monitored by a buffer (a unity-gain amplifier) and gives the output voltage.

$$I_{DD} = C[\Delta V_o/\Delta t]$$

where the voltage falls by an amount ΔV_o in a time Δt. The total capacitance C in the equation is the sum

$$C = C_{test} + C_{chip}$$

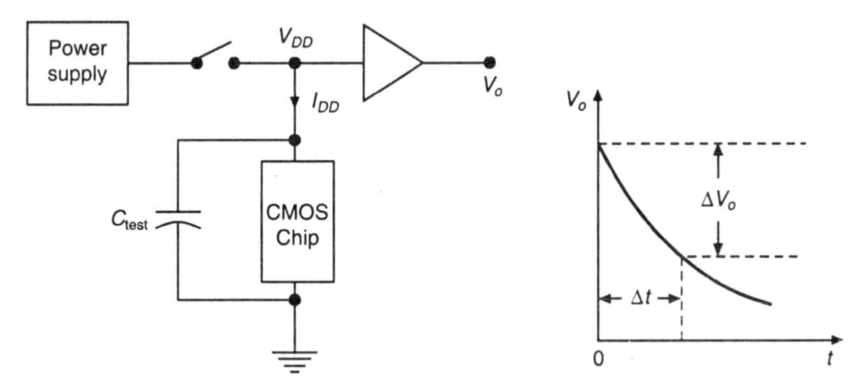

Figure 14.31 Components of an I_{DDQ} measurement system.

14.6 CHIP-LEVEL TEST TECHNIQUES

Here, we look at the application of the design strategies to the following types of circuits.

1. Regular logic arrays (datapaths)
2. Memories (RAM, ROM)
3. Random logic (multilevel standard-cell, two-level PLA)

In the past, the design process was frequently divided between a designer who designed the circuit and a test engineer who designed the test to apply to that circuit. The advent of the ASIC, small design teams, the desire for reliable ICs, and rapid times to market have all forced the 'test problem' earlier in the design cycle. In fact, the designer who is only thinking about what functionality has to be implemented and not about how to test the circuit will quite likely cause product deadlines to be missed and in extreme cases, products to be stillborn. In this section, we will examine some practical methods of incorporating test requirements into a design.

14.6.1 Regular Logic Arrays

Partial serial scan or parallel scan is probably the best approach for structures such as datapaths. One approach is shown in Figure 14.32. Here the input buses may be driven by a serially loaded register. These in turn may be used to load the internal datapath registers. The datapath registers may be sourced onto a bus, and this bus may be loaded into a register that may be serially accessed. All of the control signals to the datapath are also made scannable.

14.6.2 Memories

Memories may use the self-testing techniques. Alternatively, the provision of multiplexers on data inputs and addresses, and convenient external access to data outputs enables the testing of embedded memories. It is a mistake to have memories indirectly accessible (i.e., data is written by passing through logic, data is observed after passing through logic, addresses cannot be conveniently sequenced). Because the memories have to be tested exhaustively, any overhead on writing and reading the memories can substantially increase the test time and, probably more significantly, turn the testing task into an effort in inscrutability.

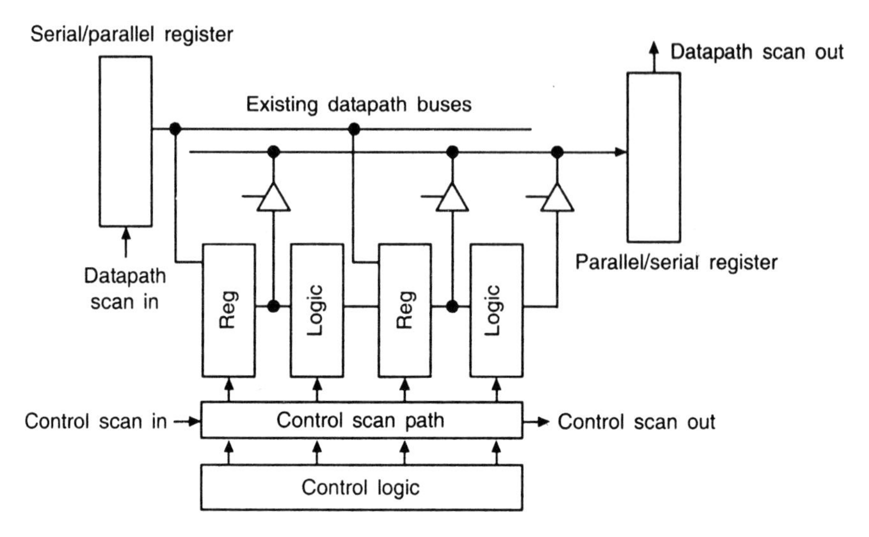

Figure 14.32 Datapath test scheme.

14.6.3 Random Logic

The random logic is probably best tested via full serial scan or parallel scan.

14.7 SYSTEM-LEVEL TEST TECHNIQUES

In the previous section, methods of testing individual chips were discussed. At the board level, 'bed-of-nails' testers have been used to test boards. In this type of tester, the board under test is lowered onto a set of test points (nails) that probe points of interest on the board. These may be sensed (the observable points) and driven (the controllable points) to test the complete board. At the chassis level, software programs are frequently used to test a complete board set. For example, when a computer boots, it might run a memory test on the installed memory, to detect possible faults.

The increasing complexity of boards and the movement to technologies like Multichip Modules (MCMs) and surface-mount technologies resulted in system designers using a scan-based methodology for testing chips at the board and system level. This is called boundary scan.

14.7.1 Boundary Scan

The IEEE 1149 Boundary scan architecture is shown in Figure 14.33. It provides a standardized serial scan path through the I/O pins of an IC. At the board level, ICs obeying the standard may be connected in a variety of series and parallel combinations to enable testing of a complete board or, possibly, collection of boards. The standard allows for the following types of tests to be run in a certified testing framework:

- Connectivity tests between components.
- Sampling and setting chip I/Os.
- Distribution and collection of self-test or built-in test results

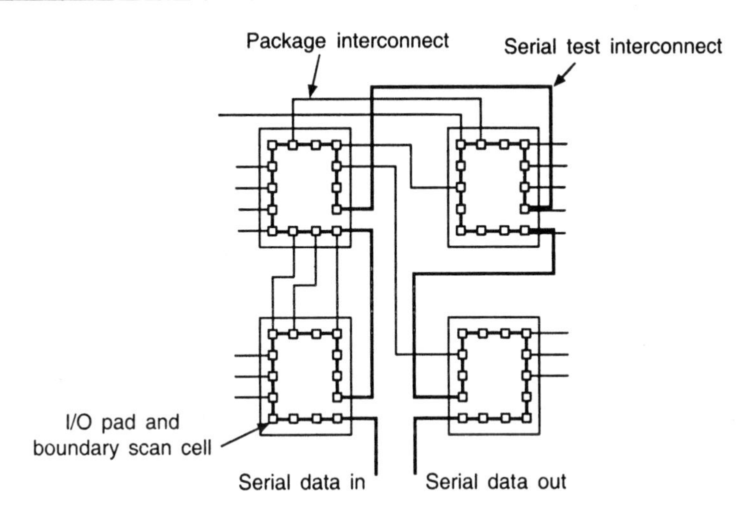

Figure 14.33 Boundary scan architecture.

14.7.1.1 Test Access Port (TAP)

It represents the interface that needs to be included in an IC to make it capable of being included in a Boundary Scan Architecture. The port has four or five single-bit connections, as follows:

- TCK (The Test Clock Input)
  - used to clock tests into and out of chips.
- TMS (The Test Mode Select)
  - used to control test operations.
- TDI (The Test Data Input)
  - used to input test data to a chip.
- TDO (The Test Data Output)
  - used to output test data from a chip.
- TRST (The Test Reset signal)
  - an optional signal
  - used to asynchronously reset the TAP controller
  - also used if a power-up reset signal is not available in the chip being tested.

The TDO signal is defined as a tristate signal that is only driven when the TAP controller is outputting test data.

14.7.1.2 Test architecture

The basic test architecture that must be implemented on a chip is shown in Figure 14.34. The architecture in Figure 14.34 consists of:

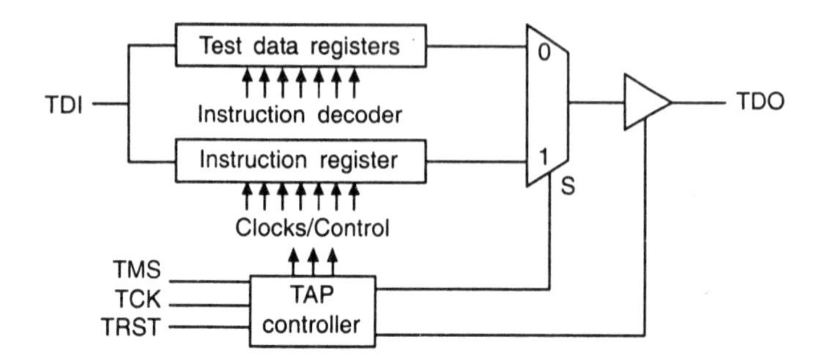

Figure 14.34 TAP architecture.

- the TAP interface pins
- a set of test-data registers to collect data from the chip
- an instruction register to enable test inputs to be applied to the chip
- a TAP controller, which interprets test instructions and controls the flow of data into and out of the TAP

The data that is input via the TDI port may be fed to one or more test data registers or an instruction register. An output MUX selects between the instruction register and the data registers to be output to the tristate TDO pin.

14.7.1.3 TAP controller

The TAP Controller is a 16-state FSM that proceeds from state to state based on the TCK and TMS signals. It provides signals that control the test data registers, and the instruction register. These include serial-shift clocks and update clocks. The state diagram is shown in Figure 14.35. The state adjacent to each state transition is that of the TMS signal at the rising edge of TCK.

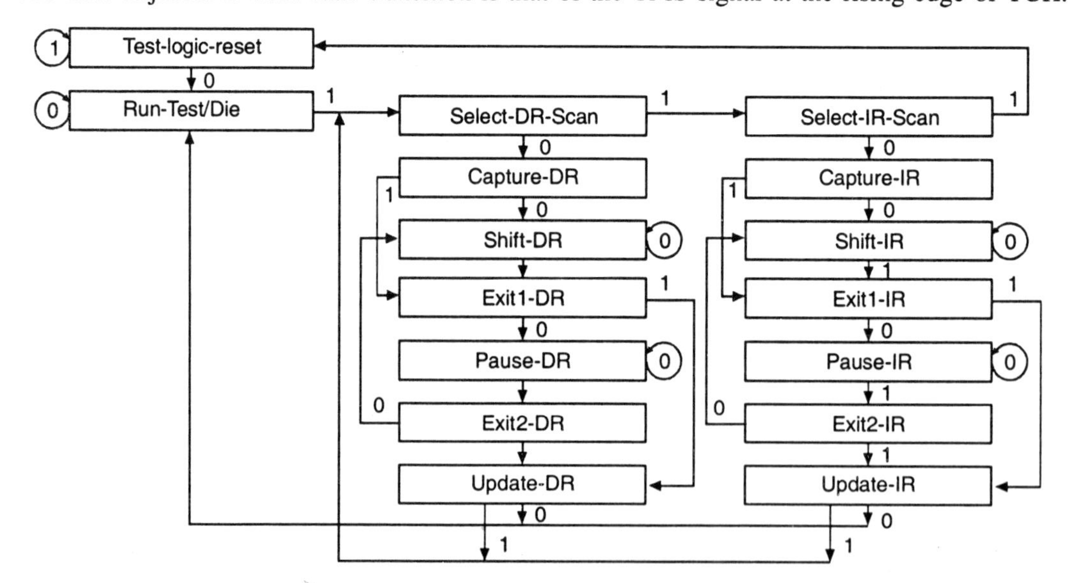

Figure 14.35 TAP controller state diagram.

Starting initially in the Test–logic–Reset state, a low on TMS transitions the FSM to the Run-Test/Idle Mode. Holding TMS high for the next three TCK cycle places the FSM in the select-DR-scan, select-IR-scan, and finally capture-IR mode. In this mode, two bits are input to the TDI port and shifted into the instruction register. Asserting TMS for a cycle allows the instruction register to pause while serially loading to allow tests to be carried out. Asserting TMS for two cycles allows the FSM to enter the Exit-2-IR mode on exit from the pause-IR state and then to enter the Update-IR mode where the Instruction Register is updated with the new IR value. Similar sequencing is used to load the data registers.

14.7.1.4 Instruction Register (IR)

The Instruction Register has to be at least two bits long, and logic detecting the state of the instruction register has to decode at least three instructions which are as follows:

- BYPASS
  - This instruction is represented by an IR having all 0s in it.
  - used to bypass any serial-data registers in a chip with a 1-bit register
  - This allows specific chips to be tested in a serial-scan chain without having to shift through the accumulated SR stages in all the chips.
- EXTEST
  - This instruction allows for the testing of off-the-chip circuitry and is represented by all 1s in the IR.
- SAMPLE/PRELOAD
  - This instruction places the boundary-scan registers (i.e., at the chip's I/O pins) in the DR chain, and samples or preloads chips I/Os.

In addition to these instructions, the following are also recommended:

- INTEST
  - This instruction allows for single-step testing of internal circuitry via the boundary-scan registers.
- RUNBIST
  - This instruction is used to run internal self-testing procedures within a chip.

A typical IR bit is shown in Figure 14.36.

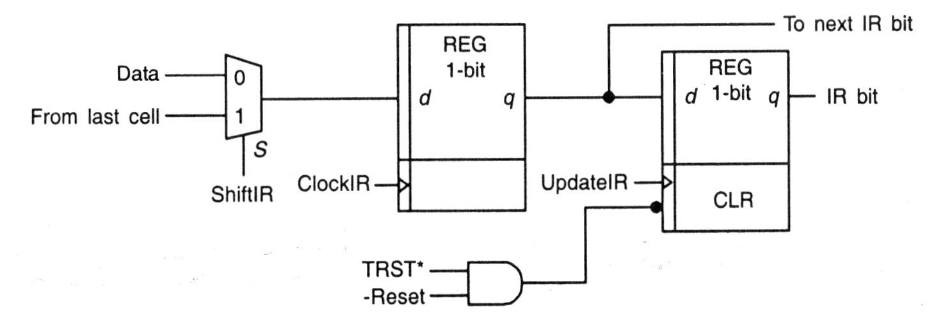

Figure 14.36 IR bit implementation.

14.7.1.5 *Test-Data Registers (DRs)*

The test-data registers are used to set the inputs of modules to be tested, and to collect the results of running tests. The simplest data-register configuration would be a boundary-scan register (passing through all I/O pads) and a bypass register (1-bit long). Figure 14.37 shows a generalized view of the data registers where one internal data register has been added. A multiplexer under the control of the TAP controller selects which particular data register is routed to the TDO pin.

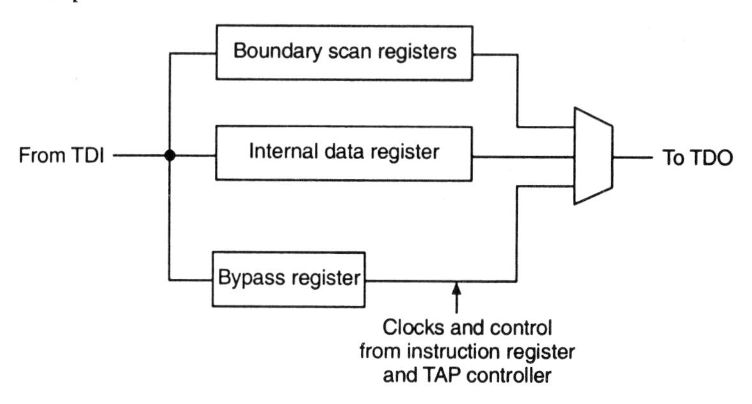

Figure 14.37 TAP data registers.

14.7.1.6 *Boundary Scan Registers*

The boundary scan register is a special case of a data register. It allows circuit-board interconnections to be tested, external components tested, and the state of chip digital I/Os to be sampled. Apart from the bypass register, it is the only data register required in a Boundary Scan compliant part.

A single structure in addition to the existing I/O circuitry can be used for all I/O pad types, depending on the connections made on the cell. It consists of two multiplexers and two edge-triggered registers. Figure 14.38(a) shows this cell used as an input pad. Two register bits allow the serial shifting of data through the boundary-scan chain and the local storage of a data bit. This data bit may be directed to internal circuitry in the INTEST or RUNBIST modes (mode = 1). When mode = 0, the cell is in EXTEST or SAMPLE/PRELOAD mode. A further multiplexer under the control of shiftDR controls the serial/parallel nature of the cell. The signal clock DR and update DR generated by the TAP controller load the serial and parallel register respectively.

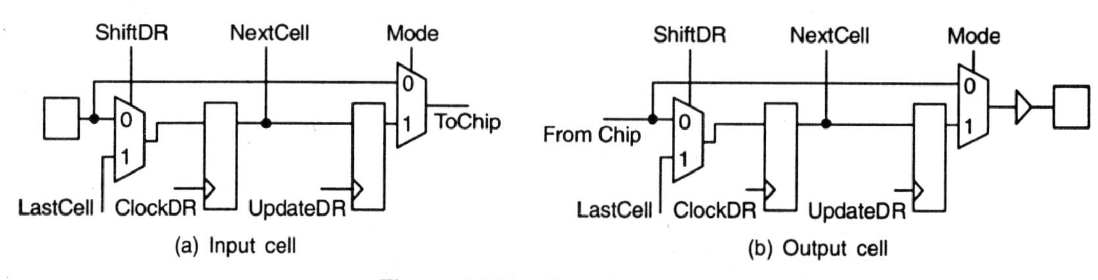

Figure 14.38 Boundary scan.

An output cell is shown in Figure 14.38(b). When mode = 1, the cell is in EXTEST, INTEST, or RUNBIST modes, communicating the internal data to the output pad. When mode = 0, the cell is in the SAMPLE/PRELOAD mode.

Two output cells may be combined to form a tristate boundary-scan cell, as in Figure 14.39. The output signal and tristate-enable each have their own MUXes and registers. The Mode control is the same for the output-cell example. Finally, a bidirectional pin combines an input and tristate cell as in Figure 14.40.

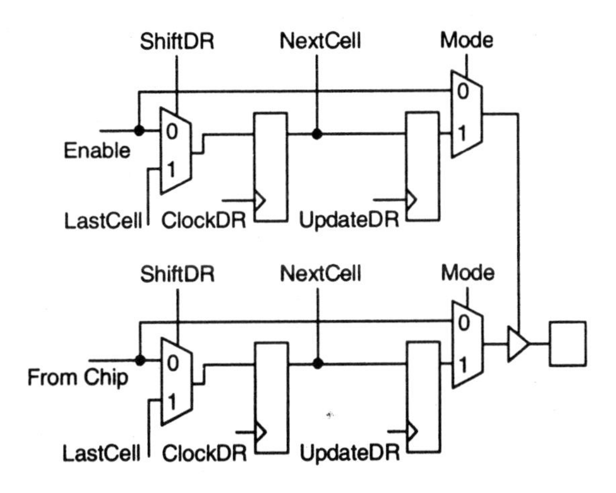

Figure 14.39 Boundary-scan tristate cell.

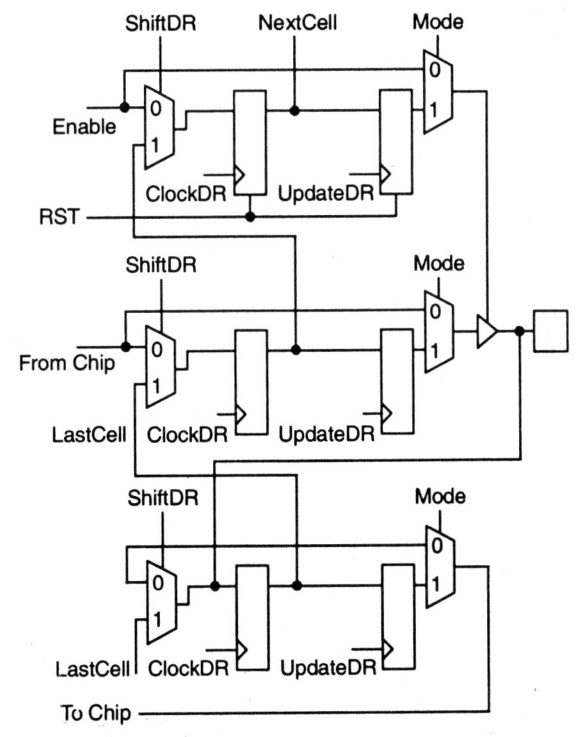

Figure 14.40 Boundary-scan bidirectional cell.

14.8 LAYOUT DESIGN FOR IMPROVED TESTABILITY

In order to predict layout styles and improve testability, a designer has to have some idea of the nature and frequency of defects for a particular process. The types of defects that commonly occur may be divided into those that short together conductors and those that create open-circuits. Shorts are possible interlayer in all layers used for connections, i.e., diffusion, polysilicon, metal 1, metal 2, and metal 3, if used. The gate onside may short to the substrate or to either the source or drain. The source and drain regions may also short. Similarly, for open-circuits, all conducting layers might have open-circuits. In addition, contacts may be misaligned, missing, or badly etched, leading to interlayer opens. For open-circuits, the ideas proposed in the literature to increase the immunity to open-circuit faults usually involve incorporating connection redundancy.

SUMMARY

This chapter has summarized the important issues in CMOS chip testing and has provided some methods for incorporating test conditions into chips from the start of the design. The importance of writing adequate tests for both functional verification and manufacturing verification cannot be understated. It is probably the single most important activity in any CMOS chip design cycle and usually takes the longest time no matter what design methodology is used. A chip designer should be absolutely rigorous about the testing activity surrounding the chip project and that testing should rank first in any design trade-offs.

REVIEW QUESTIONS

1. Explain what is meant by a Stuck-At-1 (SA1) fault and a Stuck-At-0 (SA0) fault.
2. How are sequential faults caused in CMOS? Give an example.
3. Explain the different kinds of physical faults that can occur on a CMOS chip, and relate them to typical circuit failures.
4. Explain the need for CMOS testing.
5. Explain the terms controllability, observability and fault coverage.
6. Explain how serial scan testing is implemented.
7. Explain how a pseudo-random sequence generator (PRSG) may be used to test a 16-bit datapath. How would the outputs be collected and checked?
8. Design a block diagram of a test generator for an 8x4 K static RAM.

SHORT ANSWER QUESTIONS

1. The functionality tests verify ____________ as a whole.
   Ans. function of a chip.
2. Manufacturing tests are used to verify that ____________ as expected.
   Ans. every gate operates

3. When do the manufacturing defects occur?

   Ans. This occurs during chip fabrication or during accelerated life testing.

4. What is mean time to failure?

   Ans. $\text{MTTF} = \dfrac{T}{N} = \dfrac{1}{\lambda_{av}}$

5. What are the different stuck at faults?

   Ans. (a) Stuck at 0
   (b) Stuck at 1.

6. Mention a few testing algorithms.

   Ans. D-algorithm,
   PODEM, and
   PODEM-X.

7. What are the two key concepts for designing circuits that are testable?

   Ans. Controllability, Observability.

8. Three main approaches to design for testability are ____________, ____________, ____________.

   Ans. Ad hoc testing; scan-based approaches; and self-test

9. What is RUNBIST.

   Ans. This instruction is used to run internal self testing procedure within a chip.

15 Verilog HDL

15.1 INTRODUCTION

Hardware Description Languages (HDLs) are ideal vehicles for hierarchical design. A system can be specified from the highest abstract architectural level down to primitive logic gates and switches.

Two HDLs dominate the field: VHDL (VHSIC HDL) and Verilog HDL. VHDL started as government effort to unify projects from different contractors, while Verilog was the result or private development. Both are now standardized and widely used in industry, so either one could be presented here. Verilog was chosen because of its popularity in VLSI design. Compared to VHDL, it is a relatively loose and free-flowing language, and most chip designers feel that it adheres to their way of thinking. Verilog is structured after the C programming language and uses similar procedures and constructs. We should node, however, that C or C++ themselves can be used as an HDL, and several companies develop their own language.

15.2 BASIC CONCEPTS

A hardware description language allows us to specify the components that make up a digital system using words and symbols instead of having to use a pictorial representation like a block or logic diagram. Every component is defined by its input and output ports, the logic function it performs, and timing characteristics such as delays and clocking. An entire digital system can be described in text format using a prescribed set of rules and keywords (reserved words). The file is then processed with the language complier, and the output can be analyzed for proper operations. This can be applied to simple logic gates or to an entire microprocessor design. Logic verification using an HDL is usually considered mandatory to validate the design.

A typical design hierarchy is portrayed in Figure 15.1. At the highest level is a behavioural description that describes the system in terms of its architectural features. This is generally quite abstract in that it does not contain any details on how to implement the design. Once the behavioural model is simulated and refined, the design moves down to the register-transfer level

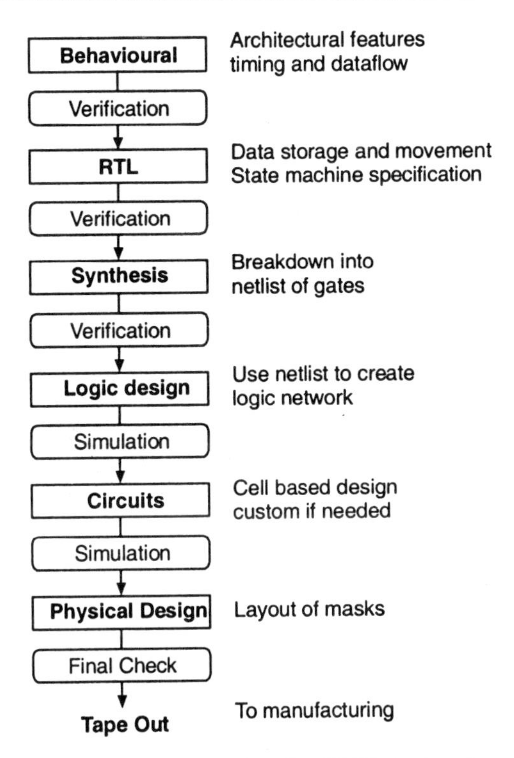

Figure 15.1 Example of a VLSI design flow

(RTL). An RTL description of a digital network concentrates on how the data moves about the system from unit to unit and the main operations. State machines and sequential circuits can be introduced at this level. Timing windows are checked and rechecked and validation of the design is again a primary objective.

The next level in the design process is called synthesis. In fully automated design, the RTL description is sent through a synthesis tool that produces a netlist of the hardware components needed to actually build the system. The success or failure of the synthesis process often depends upon the skill of the code writer. Not all HDL constructs can be synthesized, with a typical estimate hovering somewhere around 50%.

After the synthesis step, the netlist is used to design the logic network. Verification at this level consists of simulations to ensure that the logic is correct. Once the logic is validated, the cell library can be used to design the circuits. Components are wired together, and both the electrical characteristics and the logic are verified using simulation. The cell instances and wirings are translated into silicon patterns in the physical design phase. After verifying the layout, the design is (at last) complete and sent to manufacturing for the first silicon test chip.

Verilog HDL provides for descriptions of a digital system at all of the levels listed above. Every level is related to every other level, and the hierarchical design philosophy is linked by different types of code. Each level has its own coding style using certain sets of commands and constructs. Verilog even provides for switch modelling of MOSFETs, although it is not as robust and sensitive to the CMOS processing variables as a circuit simulator such as SPICE. Verilog-A is an extension of an intrinsically digital language to the analog world.

The concept that links the various levels is that of a module. A Verilog module is the description of a unit that performs some function. It may be as simple as a basic FET switch, or as complex as a 64-bit ALU. Instantiations of simple modules are used to create more complex modules. The hierarchical structure is analogous to that used in the design of cells in a layout editor that was discussed earlier in the book.

Our treatment of Verilog will start at the digital logic level where simple gates are used to build more complex logic units. Once the structure of the language is understood, higher levels of abstraction are introduced.

15.3 STRUCTURAL GATE-LEVEL MODELLING

Structural modelling describes digital logic networks in terms of the components that make up the systems. Gate-level modelling is based on using primitive logic gates and specifying how they are wired together. It is the easiest to learn since it parallels the ideas developed in elementary logic.

Verilog is built using certain keywords that are understood by the compiler. Included in the group are primitives (such as logic gates), signal types, and commands. At the structural modelling level, the keywords are often primitive logic operations (gates) which result in a very readable coding style. A straightforward approach to learning Verilog is to study how a logic network is translated into a Verilog description using a line-by-line analysis. This will illustrate the ideas and syntax in a direct manner.

15.3.1 Verilog by Example

Consider the 4-input AOI circuit shown in Figure 15.2. The logic is constructed using primitive AND and NOR gates that take the inputs *a*, *b*, *c*, *d* and produce an output of

$$f = \text{NOT}(a.b + c.d)$$

```
module AO14 (f, a, b, c, d)
input a, b, c, d
output f;
wire w1, w2;
and G1 (w1, a, b);
and G2 (w2, c, d);
nor G3 (f, w1, w2);
endmodule
```

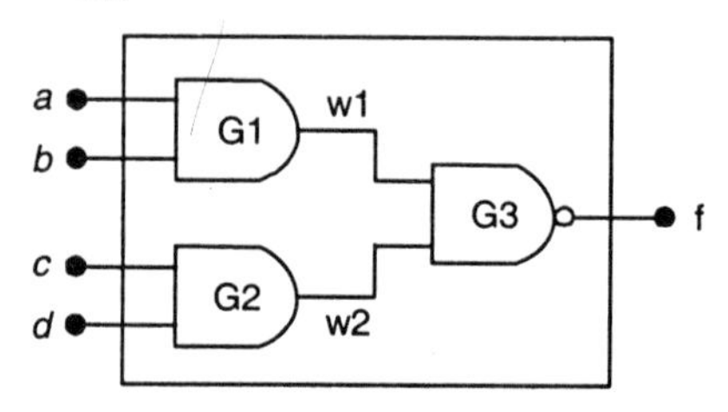

Figure 15.2 AOI module example.

The next group of lines are the port keywords input and output that identify the input and output variables. The wire keyword identifies *w*1 and *w*2 as internal values that are needed to describe the network, but are not input and output ports. A wire declaration is a data type called net. A net value is determined by the output of the driving gate. In this case, *w*1 and *w*2 are the outputs of AND2 gates, which are in turn determined by the input values.

Identifiers. Identifiers are names of modules, variables, and other objects that we can reference in the design. Examples of identifiers used so far include `AO14`, a, b, `in_0`, and `s_out`. Identifiers consist of upper- and lowercase letters, digits 0 through 9, the underscore `character (_)`, and the dollar sign ($). The first character must be a letter or the underscore in normal usage. An identifier must be a single group of characters. For example, `input_control_A` is a single object, but input control A is not allowed as a single identifier.

It is important to point out that the Verilog language is **case sensitive**. One must be careful not to mix upper and lowercase letters, as they will mean different things. For example, `in_0`, `In_0`, and `IN_0` are all distinct and are not interchangeable. Listings are insensitive to white space, so you may insert as many spaces or blank lines to help readability.

Value set. The value set refers to the specific values that a binary variable can have. Verilog provides four levels for the values needed to describe hardware: 0, 1, *x*, and *z*. The 0 and 1 levels are the usual binary values. A 0 is either a logic 0 or a FALSE statement, while a 1 indicates either a logic 1 or a TRUE statement. The context determines which interpretation is valid. An *x* represents an unknown value, *x* is important as there are many situations where there is insufficient information. For example, when we first power up a circuit, the outputs of logic gates are unknown; we must wait for an input set to establish a value.

In addition to the four levels, 0 and 1 values can be subdivided into eight 'strengths.' These are used to model various physical phenomena that degrade the signals that contend for control of a line. Strengths will be discussed in detail later.

Gate primitives. Primitive logic function keywords provide the basis for structural modelling at this level. The important operations in Verilog are `and`, `nand`, `or`, `nor`, `xor`, `xnor`, `not`, and `buf`, where `buf` is a non-inverting drive buffer. All gates except `not` and `buf` can have 2 or more inputs.

The truth tables for 0 and 1 inputs are defined in the usual manner. However, since *x* and *z* levels are allowed, we must define how a gate reacts to an expanded set of input stimuli. The `buf` and `not` gates are defined by the tables presented in Figure 15.3. The input values on the top row produce the outputs on the second row, making these self-explanatory.

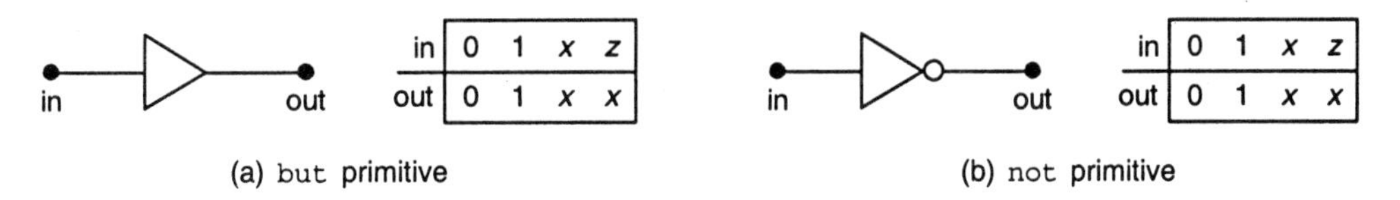

(a) `but` primitive (b) `not` primitive

Figure 15.3 Function maps for `buf` and `not` gates.

Figure 15.4 provides the truth table for the multiple-input gates and `nand`, `or`, `nor`, `xor`, and `xnor`. The tables themselves are for two inputs and must be extrapolated for 3 or more inputs. The format of the tables is standard in Verilog, and has the structure of a Karnaugh map. The top row gives the values for one input, while the left column is the other. The output value out for each possibility is read from the matrix contained within the box by aligning a row with a column. The 4X4 sub-matrix in the upper left-hand corner is easily recognized as the standard K-map for 0 and 1 inputs.

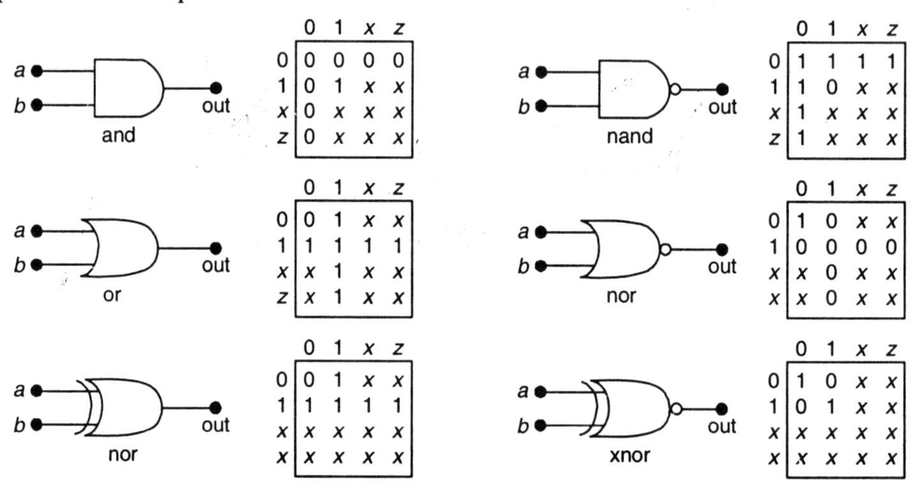

Figure 15.4 Multiple-input gate maps.

Tri-state primitives are `bufif0`, `bufif1`, `notif0`, and `notif1`. The names help remember the operation. The `bufif0` gate is a buffer if the control is 0; if the control is 1, then it is tri-stated with a Hi-Z output. Similarly, `notif1` acts as a not if the control is 1, while a control of 0 gives a Hi-Z output. Tri-state gates have one input, but can have more than one output corresponding to their usages as drivers. To describe them we use the form

```
tristate_name instance_name (out_0, out_1, out_2, ..., input, control);
```

where `instance_name` is the optional name of the instance. The logic maps for these primitives are summarized in Figure 15.5.

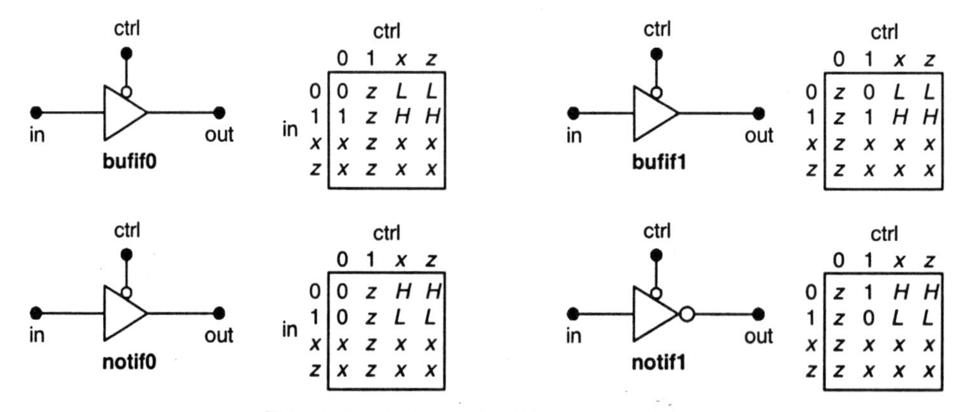

Figure 15.5 Maps for tri-state primitives.

An example of a tri-state circuit is the 2:1 MUX shown in Figure 15.6. The logic for this network is out = p0.s̄ + p1.s̄
and is described by the Verilog listing

```
module 2_1_mux (out, p0,p1, s);
input p0, p1, s;
output out;
bufif0 (mux_out, p0,s);
bufif1 (mux_out, p1,s);
endmodule
```

Ports. Ports are interface terminals that allow a module to communicate with the other modules. These correspond to the input and output points on a library cell. All ports must be declared within a module listing. The examples thus far have been of the form

```
input in_0, in_1;
output s_out, c_out;
```

A bidirectional port is declared with the syntax

```
inout IO_0, IO_1;
```

where the identifiers `IO_0` and `IO_1` can be used as either inputs or outputs to the module. Example for `nor` based `sr_latch`

```
module sr_latch (q, q_bar, s, r);
input s, r;
output q, q_bar;
reg q, q_bar;
nor (q_bar, s,q), (q,r, q_bar)
endmodule
```

Gate Delays. A hardware description language must use modelling that allows the simulation to include time delays. Verilog provides several techniques for introducing delay at the gate level.

The logic delay through a gate is sometimes modelled using a single delay time (propagation delay) from the input to the output. Delays are specified in instantiations using the pound `sign(#)` as in

```
nand # (prop_delay) G1 (output, in_a, in_b);
```

where `prop_delay` is the value of the delay.

Numerical values of gate delay values are specified as integer values of an internal time step unit. For example

```
and #(4,2) A1 (out, A_in, B_in);
```

assigns t_rise = 4 units and t_fall = 2 units. Relative units are sufficient for a broad class of simulations, so it is not necessary to use absolute time values (i.e., seconds).

If numerical values are desired, then one uses a compiler directive of the form

```
'timescale t_unit/t_precision
```

in the listing. In this expression, `t_unit` and `t_precision` can have values of 1, 10 or 100 followed by a time-scaling unit of s, ms, us, ns, ps, or fs for second, millisecond, microsecond, nanosecond, picosecond, or femtosecond, respectively. The `t_unit` gives the time-scale, while `t_precision` gives the resolution of the time-scale; obviously `t_unit` > `t_precision`. For example,

```
`timescale 1 ns/100 ps
```

gives a time-scale of 1 ns per unit, and a resolution of the time-scale as 100 ps. If a gate instance is written as

```
xor #(10) (out, A_0, A_1);
```

is used, the absolute delay through the gate is 10 × t-unit = 10 ns. If we change the time-scale to

```
`timescale 10 ns/1ns
```

the absolute delay is 10 × 10 ns = 100 ns. The value of t-precision = 1 ns determines the resolution; for example, if one specifies a time delay of 10.748 ns, the value would be rounded to 11 ns.

Gate delays allow us to monitor the response of a network in a dynamic environment. Let us simulate the module shown in Figure 15.6 for the inputs *a*, *b*, and *c*. The Verilog listing below introduces the concept of stimulus module that provides the signals.

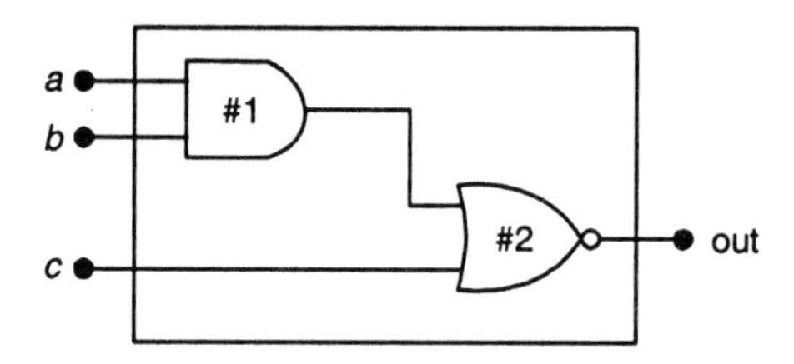

Figure 15.6 Gate delay example.

This module has gate delay

```
module Delay Ex (out, a,b,c);
input a, b, c;
output out;
wire w1;
and #1 (w1, a, b);
or # 2 (out, w1, c);
endmodule
```

The stimulus module provides the input signals

```
module stimulus;
reg A,B,C;
wire OUT; // This is a driven output values
```

The circuit instantiation is next

```
Delay Ex G1 (OUT, A, B, C);
initial
begin
$monitor ($time, "A=%b, B=%b, C=%b, OUT=%b, A, B, C, Out);
A = 1; B=0; C = 0
#1 B=1; C=1;
#2 A=0;
#1 B=0;
#1 C=0;
#3 $finish;
end
endmodule
```

15.4 SWITCH-LEVEL MODELLING

Verilog allows switch-level modelling that is based on the behaviour of MOSFETs. Although circuit-level simulators (such as SPICE) are much more accurate for performing critical electrical calculations. Verilog coding is useful for verifying logic flow through networks that consist of both transistors and logic gates. The ability to construct Verilog descriptions of complex system-level designs all the way down to basic CMOS circuits demonstrates the power of hierarchical design.

The switch primitives are named nmos and pmos, and behave in the same manner as the transistors with the same names. Figure 15.7 summarizes the behaviour of both. Verilog syntax for these primitives is in the form

```
nmos name (out, data, ctrl);
pmos name (out, data, ctrl);
```

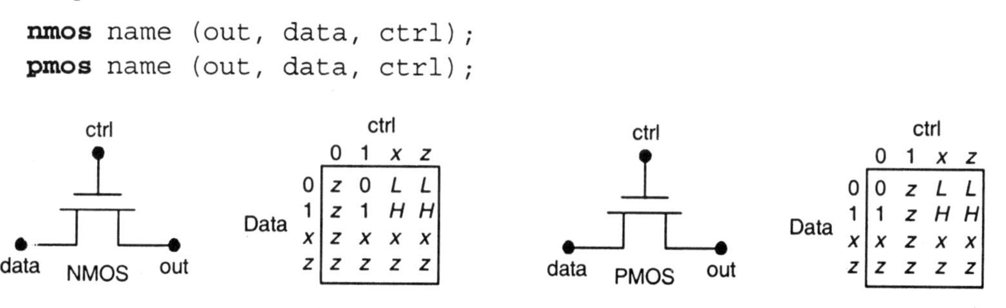

Figure 15.7 Switch level primitives.

where name is the optional instance identifier. For `ctrl` values (applied to the gate) that are 0 and 1, the behaviour is identical to FETs. The nmos switches are open for ctrl = 0 and closed for ctrl = 1, while pmos switches are closed for ctrl = 0 and open for ctrl = 1. An open switch induces a high impedance state with out = *z*. The tables also list two new entries, L and H, for the value of out when ctrl is *x* or *z*. The (low) symbol L stands for 0 or *z*, while the (high) symbol H represents 1 or *z*. The basis of this ambiguity is non-trivial. It is related to the physical concept that the output node can store charge, so that out may be related to an earlier value.

MOS switches can be used to describe CMOS logic gates. The simple NOT circuit in Figure 15.8 has the Verilog description.

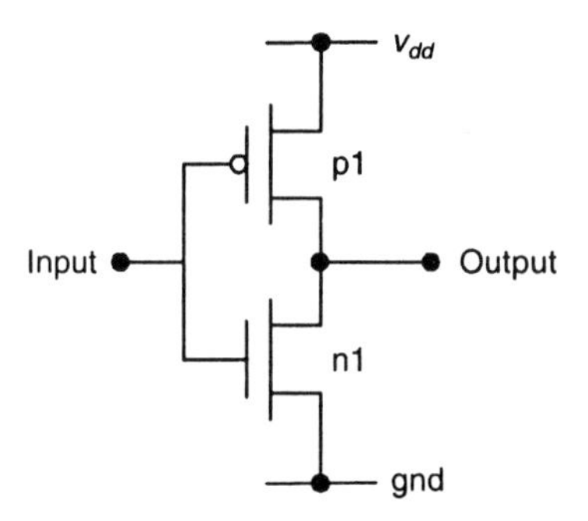

Figure 15.8 CMOS inverter using Verilog switches.

CMOS inverter switch network

```
module fet_not (out, in);
input input;
output output;
supply1 vdd;
supply0 gnd;
pmos p1 (vdd,output,input)
nmos n1 (gnd,output,input)
endmodule
```

The circuit and listing has been used to introduce two new verilog keywords **supply1** and **supply0** that define the power supply v_{dd} and ground gnd connections. These represent the strongest logic 1 and logic 0 drivers, respectively. The Verilog module treats these as the data into the FETs, while the gate input is the switch `ctrl`.

The same constructs can be used to model arbitrary CMOS logic gates.
The NAND2 and NOR2 switching networks in Figures 15.9(a) and (b) are described by the module

```
module fet_nand2 (out, in_a, in_b);
input in_a, in_b;
output out;
wire wn;//This fwire connects the series nmos switches
supply1 vdd;
supply0 gnd;
pmos p1 (vdd, out, in_a);
pmos p2 (vdd, out, in_b);
nmos n1 (gnd, wn, in_a);
nmos n2 (wn, out, iln_b);
endmodule
```

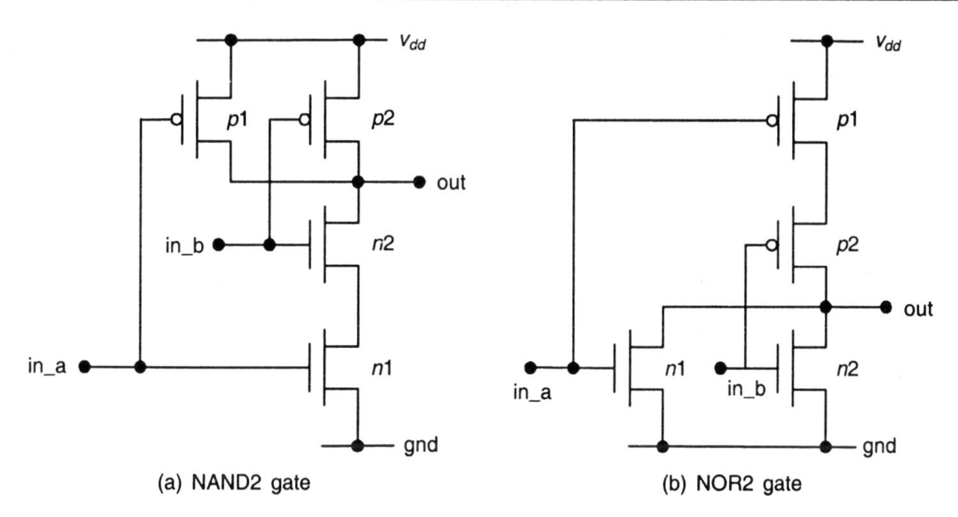

Figure 15.9 Logic gate construction.

for the NAND gate, and

```
module fet_nor2 (out, in_a, in_b);
input in_a, in_b;
output out;
wire wp;// this connects the series pmos switches
supply1 vdd;
supply0 gnd;
pmos p1 (vdd, wp, in_a);
pmos p2 (wp, out, in_b);
nmos n1 (gnd, out, in_a);
nmos n2 (gnd, out, in_b);
endmodule
```

for the NOR gate.
These can be verified using a line-by-line comparison.

Another useful set of primitives includes pull-up and pull-down components that have the keywords **pullup** and **pulldown**. These can be modelled as resistors that are connected to **supply1** and **supply0** as shown in Figure 15.9(a) and are described by

```
pullup(out_1); // This gives a high output
pulldown(out_0); // This gives a low output
```

In a Verilog listing, the output strengths are called `pull1` and `pull0`, and are weaker than the `supply1` and `supply0` levels as shown in Figure 15.10(a). `pull` primitives are used in various ways to model circuits. For example, a **pullup** can be used as a load device as in the NMOS NOR3 gate drawn in Figure 15.10(b). The Verilog description is

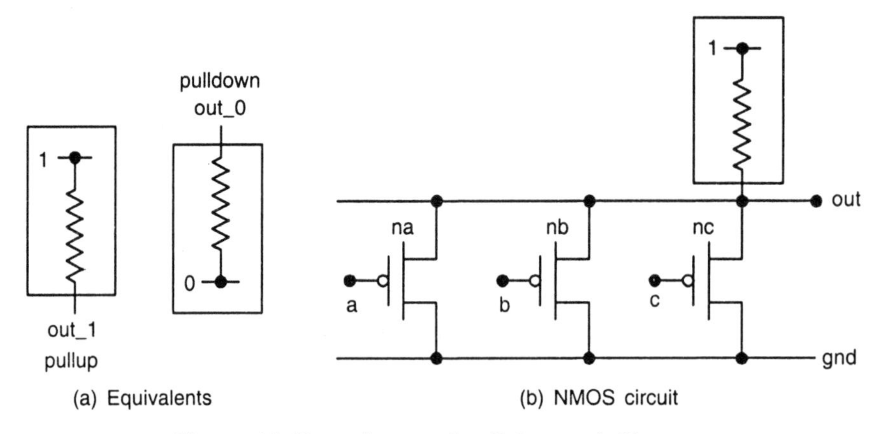

Figure 15.10 pull-up and pull-down primitives.

```
module fet_nor2 (out,in_a,in_b,in-c);
  input in_a, in_b;
  output out;
  supply0 gnd;
  nmos na (gnd,out,in_a),
    nb (gnd,out,in_b),
    nc (gnd,out,in_c);
  pullup (out);
endmodule
```

Note that `pullup` and `pulldown` require only one identifier. This is because only a single wire is provided out of each 'device' equivalent.

Resistive (rmos) Switches

Realistic MOSFETs have drain-source resistance that can modify the signal strength passing through them. Some of the effects can be included by using resistive MOS switches which are gate-controlled in the same manner as regular switches, but the devices alter the output strength. The FET equivalent primitives are **rnmos**, **rpmos**, and **rcmos**. The instancing syntax is the same as for non-resistive (ideal) switches. For example,

rnmos #(1, 2, 2) fet_1 (output, input, gate_ctrl);

specifies a resistive nFET. The main difference is that input-output strength relations are defined by the list in Figure 15.11. This is useful for including physical effects such as threshold voltage losses through nFET pass transistors. While a SPICE simulation at the electronics level is much more accurate, these are useful for modeling the switching behavior in non-critical paths.

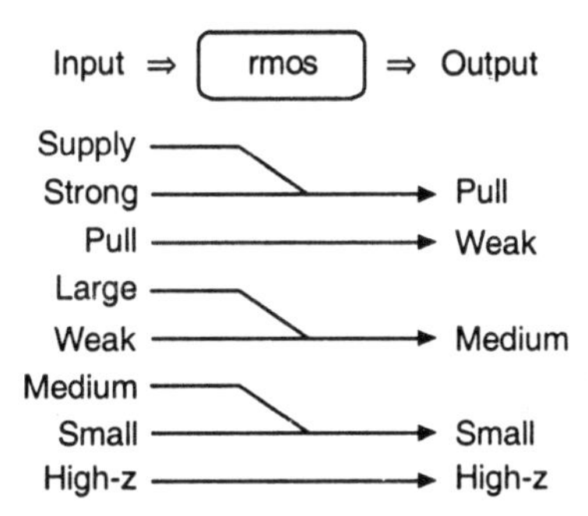

Figure 15.11 Resistive (**rmos**) input–output strength map.

15.5 DESIGN HIERARCHIES

The concept of primitives, modules and instancing provides the basis for hierarchical design in Verilog. Up to this point, we have learned how to write verilog code at the gate-level and the switch-level. These two levels can be used separately, or intermixed within a single module. We will use these two modelling levels as vehicles for learning the fundamentals of hierarchical design.

Let us construct a Verilog module for the gate by instancing the switch-level modules.

```
module fet_and2 (out,a,b,c,d);
input a,b,c,d;
output out;
wire out_not, out_nand1, out_nand2;
// Gate instance
  fet_nand2 g1 (out_nand1,a,b),
    g2 (out_nand2,c,d);
  fet_nor2 g3 (out,out_nand1,out_nand2);
endmodule
  /*. The nand and nor module litings must be
  included in the complete code to insure that they
  are defined for instancing*/
```

Figure 15.12 illustrates the instancing procedure, where it is assumed that the modules `fet_nand2` and `fet_nor2` have been defined using the previously written modules. Now suppose that we want to build a more complex network using the `fet_and4` module. The new module, which we will call `group_1`, can be constructed using any entries that have been defined. Figure 15.13 illustrates how the cell can be built using instances of `switch_level` modules and the `fet_and4` module, combined with the Verilog primitive XOR gate. The basic features of the module are summarized by the general form

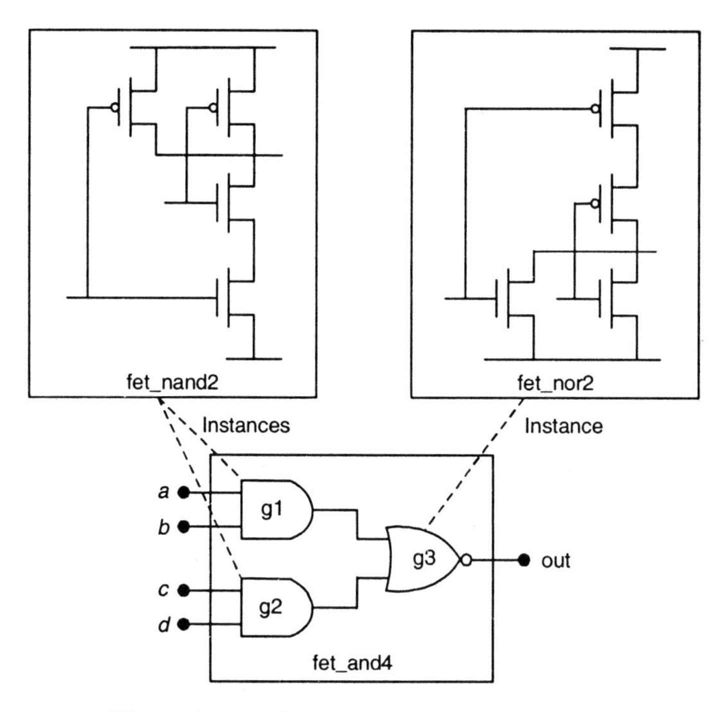

Figure 15.12 Creating an AND4 gate module.

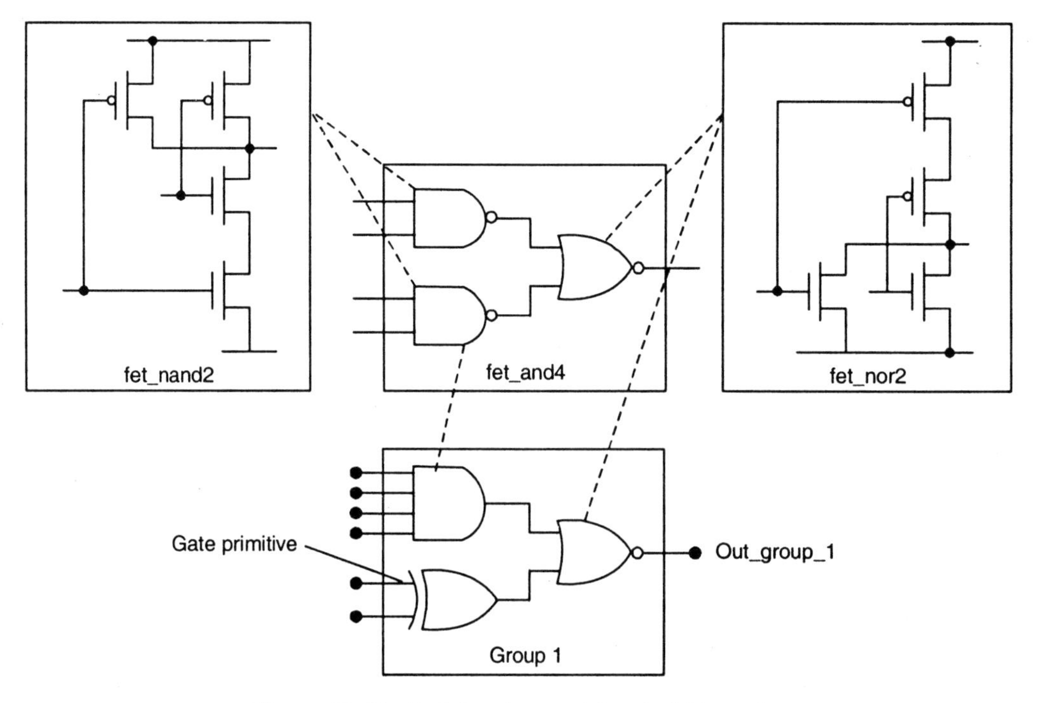

Figure 15.13 Building the next level of hierarchy.

```
module group_1(out_group_1,...);
  ...// input an wire declarations
  output out_group_1;
    // Gate instances
  fet_and4 (. . .);
  fet_nor2 (. . .);
  xor (. . .);
endmodule
```

SUMMARY

Hardware description languages (HDLs) are ideal vehicles for hierarchical design. A system can be specified from the highest abstract architectural level, down to primitive logic gates and switches. In this chapter, the VLSI design flow was discussed.

Verilog basic concepts were also discussed with simple examples. A number of examples were given for the Structural Gate-Level Modelling and Switch-Level Modelling. Finally design hierarchies were also analysed.

REVIEW QUESTIONS

1. Explain structural gate level modelling with a suitable example.
2. Write a Verilog coding for inverter, NAND and NOR gate program using gate level structure.
3. Construct the Verilog module for AND4 gate by instancing the switch level modules.

SHORT ANSWER QUESTIONS

1. ____________ are ideal vehicles for hierarchical design.

   Ans. Hardware Description Languages
2. What are identifiers in Verilog?

   Ans. Identifiers are the names of modules, variables and other objects.
3. Verilog language is ____________

   Ans. case sensitive
4. Write the syntax for delay in Verilog.

   Ans. timescale 1 ns/100 ps.

16 Behavioural Modelling

16.1 BEHAVIOURAL AND RTL MODELLING

The basis for behavioural modelling is the construction of **procedural blocks**. As implied by its name, a procedural block is a listing of statements that describe how a set of operations are performed. Many of these resemble constructs in the C programming language, and they introduce a new level of abstraction to the design process. Procedural blocks contain assignment statements, high-level constructs such as loops and conditional statements, and timing controls. There are two types of block that start with the keywords `initial` and `always`. An initial block executes once in the simulation and is used to set up initial conditions and step-by-step dataflow. An `always` block executes in a loop and repeats during the simulation. Block statements are used to group two or more statements together. Sequential statements are inserted between the keywords `begin` and `end`. It is also possible to write concurrently executed statements using the `fork` and `join` keywords.

Let us start by writing a module for a clock variable `clk`. We will assume a clock period of 10 time units so that the variable must change every 5 time units as illustrated in Figure 16.1.

```
module clock;
reg clk;
// The next statement starts the clock with a value of 0 at t=0
  initial
    clk=1'b0;
// When there is only one statement in the block, no grouping
is required
always
  #5 clk=~clk;
initial
  #500 $finish;// End of the simulation
endmodule
```

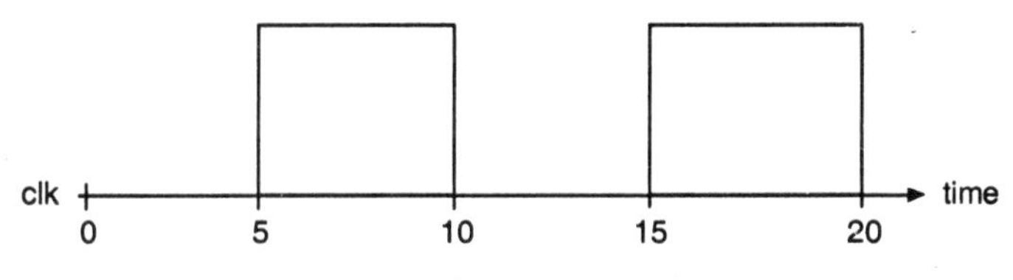

Figure 16.1 Clocking waveform clk.

The cyclic action is obtained using the NOT operator ~ in the statement

```
#5 clk=~clk;
```

Since this falls within an **always** statement the command is executed in a loop until the simulation ends at 500 time units.

Operators. The Verilog operators such as ~ are summarized in Figure 16.2 for future reference. Note that some symbols such as & are used differently depending upon the context. We will study a few to understand how they work.

Arithmetic	
+	addition
-	subtraction
*	multiplication
/	division
%	modulus

Bit-wise	
~	not
&	and
~&	nand
1	or
~1	nor
^	xor
~^	xnor
^~	xnor

Shift Operations	
>>	shift right
<<	shift left

Relational and Logical	
>	greater than
>=	greater than or equal to
<	less than
<=	less than or equal to
!	negation
&&	logical and
11	logical or
==	identity
!=	logical inequality
===	case equality
!==	case inequality

Figure 16.2 Verilog operators.

Consider first the behaviour of the reduction or unary operators (i.e., operations on a single number.) Suppose we assign the binary values `a=1101` and `b=0000`. Bit-wise negation gives

```
~a=0010
~b=1111
```

as it operates on each bit independently. A logical negation evaluates to

```
!a=0
!b=1
```

the logical operator `!a` gives the logical inverse of `a`. If A contains all 0s, then it is false (0). If it is `non_zero`, then it is true (1); `!a` gives the inverse of the value of `a`. Reduction operators operate on each bit of the number and results in a single bit true (1) or false (0) value. For example, with `a` and `b` defined as previously stated,

```
&a=0
&b=0
la=1
lb=0
^a=1
^b=0
```

The symbol 'l' used for the OR is called a pipe.

The next group contains the binary operators that have two operands. These are used in both bit-wise and logical contexts. With `a=1010` and `b=0011`, the bit-wise application of the operators acts in a bit-by-bit manner:

```
a&b=0010
alb=1011
a^b=1001
```

In a logical context, the answer is a single true (non-zero) or false (all 0s) number.

```
a&&b=1
allb=1
a&&c=0
```

where `c=0000`.

The equality operators are =, ==, and ===. The assignment operator = is used to copy the value from the right side of an expression to the left side as in

```
a=4'b1010
```

The equality operator == is used in

```
a==b
```

To express `a` is equal to `b`. Identity is written as

```
c==b
```

Which says that `c` is identical to `d`.

Timing controls. Timing controls statements dictate the times when actions take place. There are three types of timing controls that are used in a procedural block. A simple delay is specified using `# <time>` as the clock example. An edge-triggered control is of the form `@ (signal)`. In the statement

```
@(posedge clk) reg_1=reg2;
```

the **posedge** keyword is used to induce the assignment when the clock clk is rising from a 0 to a 1, or from *x* or *z* to 1. The positive edge of the clock is shown in Figure 16.3. Similarly, a negative-edge triggered event can be described by a statement of the form

```
@(negedge clk) output=a_in;
```

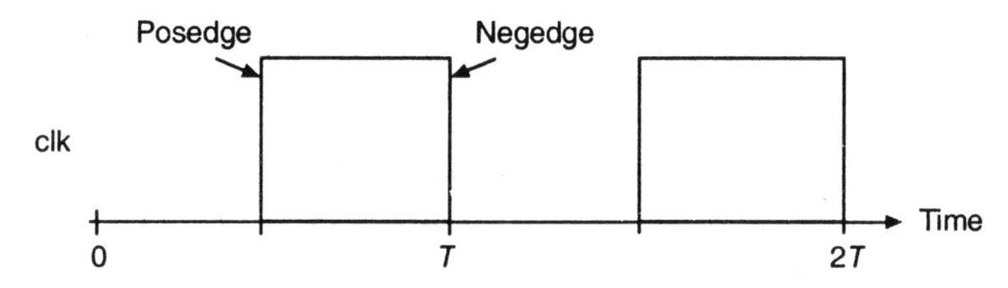

Figure 16.3 Clock edges.

A negative edge transition is from a 1 to a 0, or from *x* or *z* to 0. Edge-triggered statements can include the possibility of several signals changing by using the **or** keyword. Level-triggered events are modelled using the **wait** keyword.

```
wait(clk) q_out=d_in;
```

effects the transfer when clk is a 1. In general, the **wait** directive executes when the expression is logically TRUE (i.e., non-zero).

Procedural assignments. A procedural assignment is used to change or update the values of `reg` and other variables. They are usually divided into **blocking** and **non-blocking** assignments.

Blocking assignments are executed in the order that they are listed, and allow for straightforward sequential and parallel blocks. The assignment operator = is used in these statements. Consider the simple code listing

```
reg a,b,c, reg_1, reg_2;
initial
begin
  a=1;
  b=0;
  c=1;
  # 10 reg_1=1' bo;
  # 5 c= reg_1;
  # 5 reg_2= b;
end
```

The sequence of the statements is important. The assignments for a,b, and c are all performed at time 0. After 10 time units, `reg_1` is assigned the value of binary 1. Then 5 time units after this event (at 15 time units) `c` is assigned the value of `reg_1`. Finally, at 20 time units, `reg_2=b` is executed. The events are summarized in Figure 16.4.

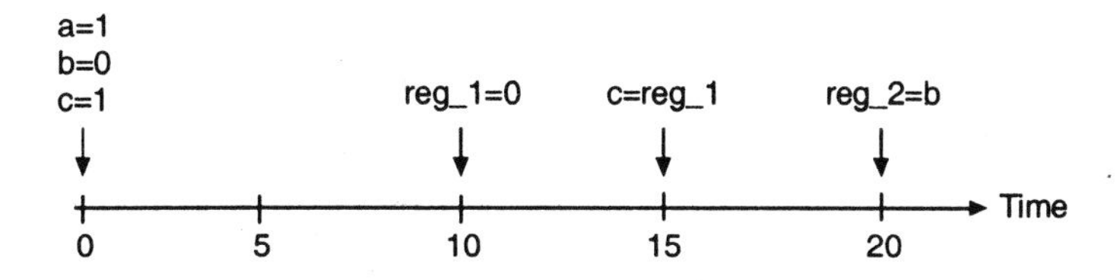

Figure 16.4 Timing for non-blocking assignment example.

Conditional statements. The **if/else if** constructs allow different outcomes depending upon current conditions. Let us examine the module listing

```
module if_else_example (ctrl,alu_op,clk);
input [1:0] alu_op;
input clk;
output ctrl_a;
reg ctrl_a;
always @ (posedge clk)
begin
if (alu_op ==0)
  ctrl=0;
else if (alu_op == 1)
  ctrl = 1;
else
  $display ("Signal ctrl is greater than 1");
end
endmodule
```

A case statement is another powerful construct. It has the form

```
case(condition)
```

and stipulates outcomes that depend on the value of condition. This is seen in the simple 2:1 multiplexer description below.

```
module simple_mux(mux_out, p0, p1,select);
input p0,p1;
input select;
output mux_out;
always @ (select or p0 or p1)
  case (select)
  1' b0 : mux_out=p0;
  1' b1: mux_out=p1;
  endcase
endmodule
```

Another type of coding is obtained using looping statements. The repeat loop executes a set of statements a specific number of times. For example, suppose that the variable counter has a value of 10. Then,

```
repeat(counter)
begin
 ...
end
```

performs the procedures listed in between the `begin/end` statements in a loop manner a total of counter = 10 times. A related keyword is `while` with the syntax

```
while (condition)
  begin
    ...
  end
```

This executes the `begin/end` block so long as `condition` is true (non-zero). If `condition` is initially false (zero), then the entire block is ignored. Verilog also has the `for` construct with the syntax

```
for (condition)...
```

That allows the sensing of `condition` expressions to execute the statement block.

A **forever** loop does exactly what its name implies: it executes for the entire length of the simulation. The clock generation module below illustrates this construct.

```
module clk_1;
reg clk;
initial
  begin
    clk=0
    forever
      begin
      #5 clk=1;
      #5 clk=0;
      end
  end
endmodule
```

A few other conditional constructs are available, but these illustrate the most-used coding styles.

16.1.1 Dataflow Modelling and RTL

Dataflow modelling describes a system by how the data moves and is processed. As with general behavioural modelling, a dataflow description is a high-level abstraction that does not provide structural details. Although the definition tends to vary, register-transfer level (RTL) modelling is usually interpreted as a combination of dataflow and behavioural coding styles. It uses high-level constructs that can be used as an input into a synthesis too, which is then used to generate a gate netlist.

We will be content with introducing the `assign` keyword. Continuous assignments define relationships and values. For example, statements of the form

```
assign a=~b&c;
assign out_1=(a l b) & (c l d);
```

can be used to define combinational logic operations. A useful conditional statement is

```
assign output=(something)?<true condition>:<false condition>
```

In this case, something represents a variable or a statement. The value of output depends if something is true or false. The description of a 2:1 MUX can be written as

```
module mux_2(out,p0,p1,select);
input p0,p1;
input select;
output mux_out;
assign out=(select) ? p1 :p0;
endmodule
```

16.2 GENERAL VLSI SYSTEM COMPONENTS

16.2.1 Multiplexers

Multiplexers are indispensable in modern digital design. A mux consists of n inputs and one output f.

The simplest example is a 2-to-1 multiplexers shown in Figure 16.5. There are several ways to describe the component. A behavioural description using the case statement was presented in the previous chapter, and is repeated here for reference. The input lines are designated as `p0` and `p1`, and the `select` bit is denoted by the identifier select.

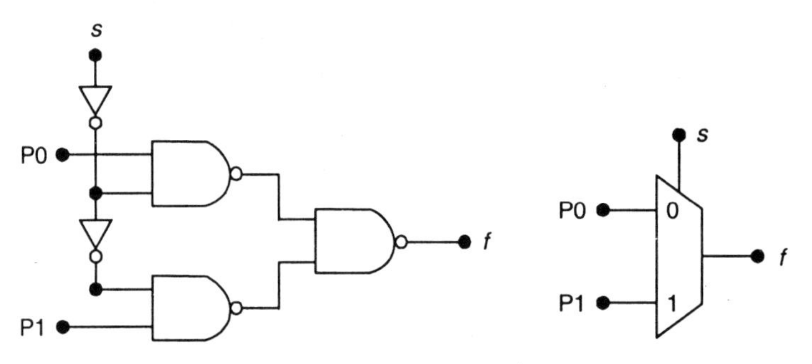

Figure 16.5 Gate-level NAND 2:1 multiplexer.

```
module simple_mux(mux_out,p0,p1,select);
input p0,p1;
input select;
output mux_out;
always @ (select)
  case (select)
  1' b0:mux_out=p0;
  1' b1:mux_out=p1;
  endcase
endmodule
```

The transmission-gate circuit in Figure 16.6(a) could also be a candidate in a CMOS technology. This circuit uses four FETs for the path logic (two for each TG). If we include a buffering NOT pair for the select bit, then the total FET count is increased to 8. The main

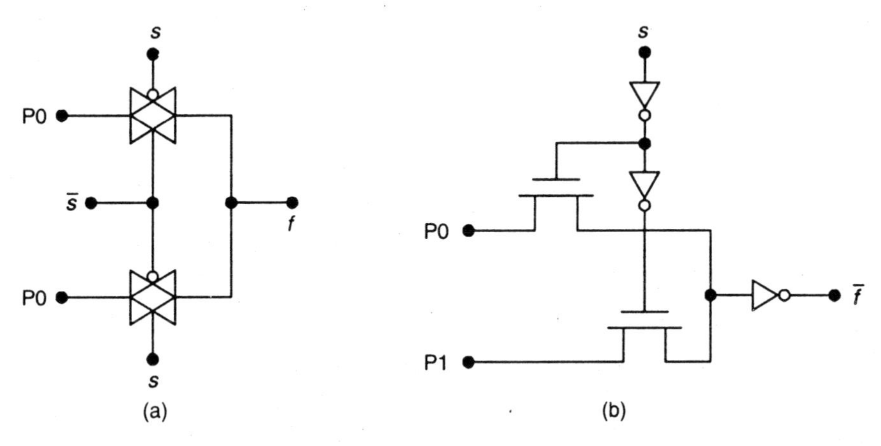

Figure 16.6 Multiplexer using switch logic.

problem with this circuit is that the TGs have parasitic resistance and capacitance that slow down the response. Figure 16.6(b) shows the same configuration using only NFET switches.

Larger multiplexers can be designed using primitive gates or by instancing 2:1 devices. Consider a 4:1 MUX described by

```
module bigger_mux(out_4,p0,p1,p2,p3,s0,s1);
input p0,p1,p2,p3;
input s0,s1;
output out_4;
assign out_4= s1 ? (s0 ? p3:p2) (s0 ? p1:p0);
endmodule
```

Since this is a high-level abstraction, no details about the internal structure are given. However, the assign statement can be interpreted as three 2:1 separate multiplexers with the first ?: using $s1$, and the second and third occurrences based on $s0$. This implies the structure illustrated in Figure 16.7. The select bit $s0$ is used to select ($p0$, $p2$) or ($p1$, $p3$) in the first stage devices. The final selection is achieved with $s1$ determining the actual output f which is the same as `out_4` in the Verilog listing.

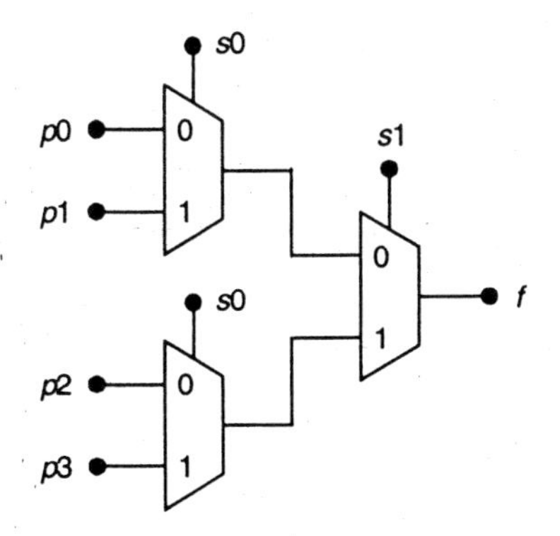

Figure 16.7 A 4:1 MUX using instanced 2:1 devices.

Another implementation is the gate-level construction as shown in Figures 16.8(a) and (b). Using this as a guide, the equivalent Verilog structural description would be

```
        module gate_mux_4(out_gate,p0,p1,p2,p3,s0,s1);
        input p0,p1,p2,p3;
        input s0,s1;
    wire w1,w2,w3,w3;
      output out_gate;
      nand(w1,p_0,~s1,~s0),
        (w2,p_1,~s1,s0),
        (w3,p_3,s1,s0),
        (out_gate,w1,w2,w3,w3);
    endmodule
```

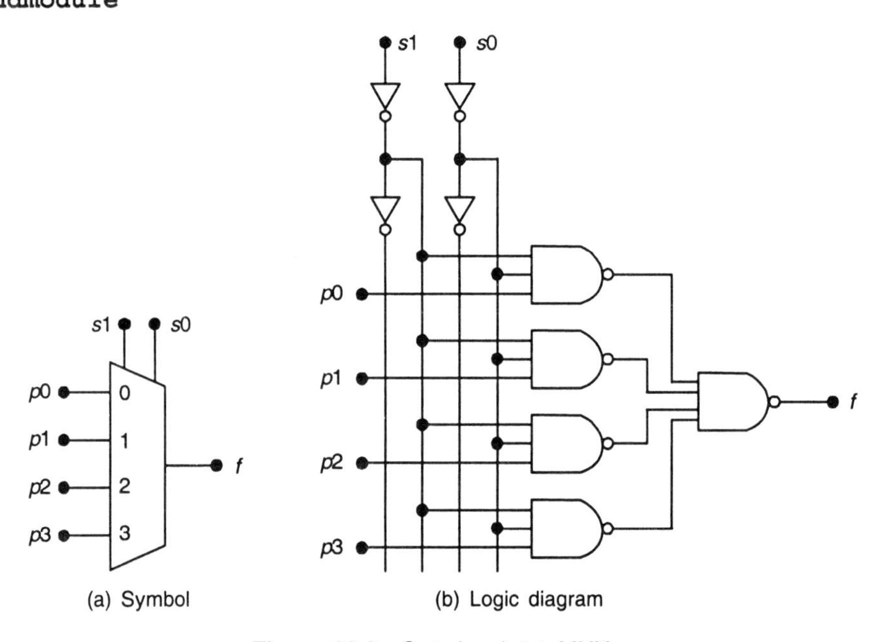

Figure 16.8 Gate-level 4:1 MUX.

The NOT gates have been modelled using ~ operators, but they could have been instanced with primitive NOT gate with the same result. In standard logic, this is equivalent to the SOP expression

$$f = p0 \cdot \overline{s}1 \cdot \overline{s}0 + p1 \cdot \overline{s}1 \cdot s0 + p2 \cdot s1 \cdot \overline{s}0 + p3 \cdot s1 \cdot s0$$

obtained from applying basic logic.

Yet another network is the pass-FET array shown in Figure 16.9. This uses the ANDing properties of NFETs to implement the logic expression directly. The structural description of this circuit is given by

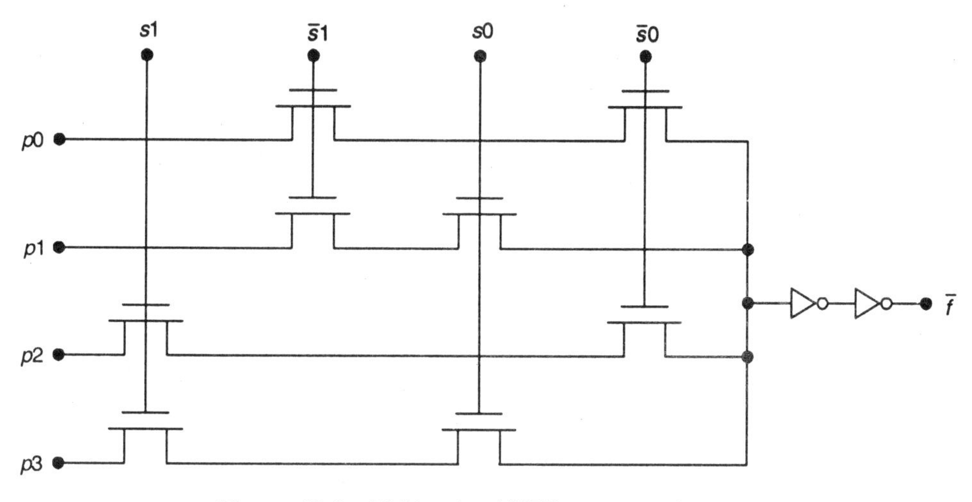

Figure 16.9 MUX using NFET pass transistors.

```
module tg_mux_4(f,p0,p1,p3,s0,s1);
input p0,p1,pf2,p3;
input s0,s1;
wire w0,w1,w2,w3,w_0,w_x;
output f;
nmos (p0,w0,~s1),(w0,w_0,~s0);
nmos (p1,w1,~s1),(w1,w_0,s0);
nmos(p2,w2,s1),(w2,2_0,~s0);
not(w_x,w_0),(f,w_x);
endmodule
```

16.2.2 Binary Decoders

A binary *n*/*m* row decoder accepts an *n*-bit control word and activates one of the *m*-output lines while the other ($m - 1$) lines are not affected. An active-high decoder sets a 1 on the selected line and keeps the others at 0. An active-low decoder is just the opposite, with the selected line reset to 0 while the remaining lines are at 1.

A 2/4 active-high decoder symbol and function table is shown in Figure 16.10(a). The 2-bit select word s1s0 activates the line corresponding to its specified decimal value 0,1,2,3. The function table yields the equations

$$\begin{aligned} d0 &= \bar{s}1 \cdot \bar{s}0 = \overline{s1 + s0} \\ d1 &= \bar{s}1 \cdot s0 = \overline{s1 + \bar{s}0} \\ d2 &= s1 \cdot \bar{s}0 = \overline{\bar{s}1 + s0} \\ d3 &= s1 \cdot s0 = \overline{\bar{s}1 + \bar{s}0} \end{aligned} \tag{16.1}$$

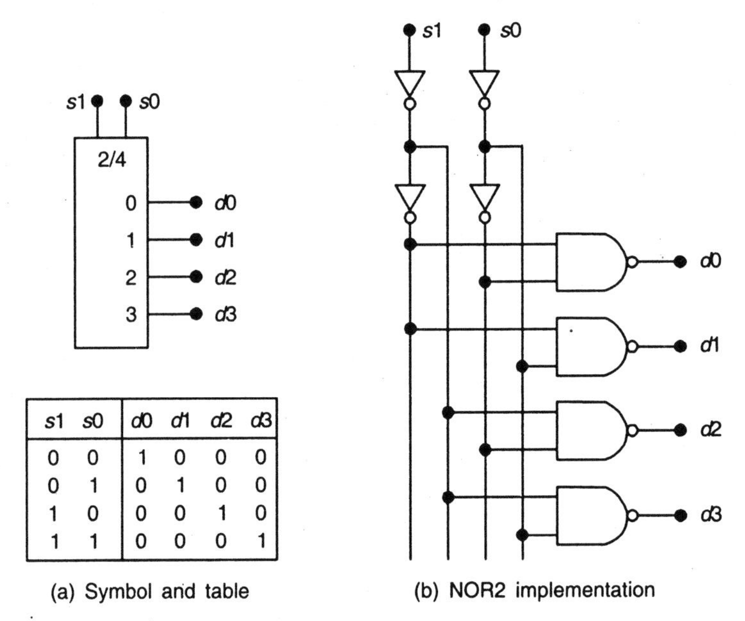

(a) Symbol and table

(b) NOR2 implementation

Figure 16.10 An active-high 2/4 decoder.

A straightforward NOR-gate implementation is shown in Figure 16.10(b). This gives the basis for the structural description

```
module decode_4 (d0,d1,d2,d3,s0,s1);
input s0,s1;
output d0,d1,d2,d3;
nor (d3,~s0,~s1),
     (d2,~s0,s1),
     (d1,s0,~s1),
     (d0,s0,s1);
endmodule
```

where we have absorbed the NOT drivers into the notation using the ~ operator.

An equivalent architectural description using **case** keywords can be written as

```
module dec_4 (d0,d1,d2,d3,sel);
input[1:0] sel;
output d0,d1,d2,d3;
case(sel)
  0:d0=1, d1=0, d2=0, d3=0;
  1:d0=0, d1=1, d2=0, d3=0;
  2:d0=0, d1=0, d2=0, d3=0;
  3:d0=0, d1=0, d2=1, d3=0;
endmodule
```

Which explicitly lists each possibility depending on the decimal value of `sel`. Another approach would be to use `assign` procedures. This represents an abstract high-level description of the operation that contains no structural information. While one can understand the operation of the unit, it must be translated into a lower level description before it can be built. This provides another example of equivalent hierarchical views.

An active-low decoder is shown in Figure 16.11. In this case, the selected output is driven low while the others remain at logic 1 value. The design is achieved by simply replacing the

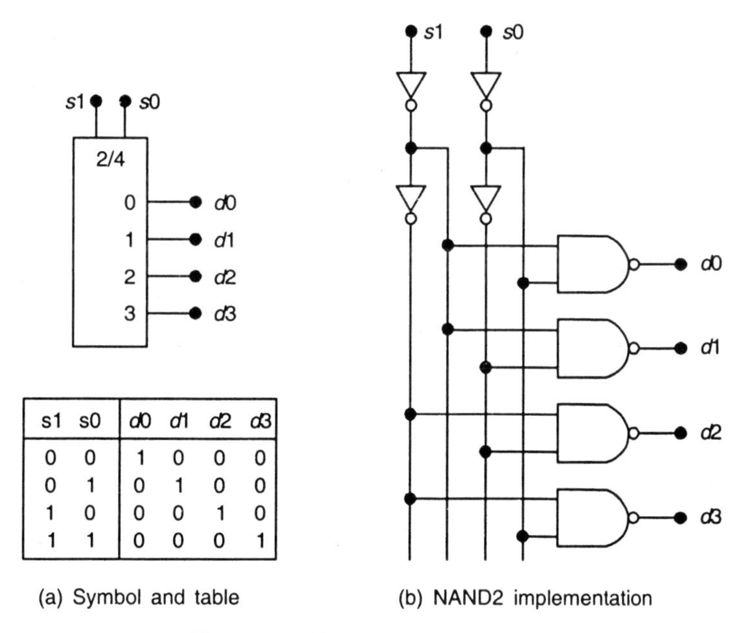

s1	s0	d0	d1	d2	d3
0	0	1	0	0	0
0	1	0	1	0	0
1	0	0	0	1	0
1	1	0	0	0	1

(a) Symbol and table

(b) NAND2 implementation

Figure 16.11 Active-low decoder.

NOR2 gates with NAND2 gates and complementing the inputs to each gate. The HDL code can be written by modifying the active-high listings by changing the logic. The gate-level structural Verilog description of the network can be constructed in the form

```
module dec_io(d0,d1,d2,d3,s0,s1);
input s0,s1;
output d0,d1,d2,d3;
nand(d0, ~s0, ~s1),
  (d1, ~s0,s1),
    (d2,s0, ~s1),
    (d3,s0,s1);
    endmodule
```

by inspection. These simple examples clearly show how large components can be described at various levels. The one that is most used in practice depends upon the problem and its level in the design hierarchy. In general, no single solution will be optimal for all situations.

16.2.3 Equality Detectors and Comparators

An equality detector compares two n-bit words and produces an output that is 1 if the inputs are equal on a bit-by-bit basis. A simple 4-bit circuit is shown in Figure 16.12. This uses the equality (XNOR) relation

$$\overline{a_i \oplus b_i} = 1 \tag{16.2}$$

if $a_i = b_i$ as a means to compare the inputs. If every XNOR produces a 1, then the output of AND gate gives Equal = 1; otherwise, Equal = 0.

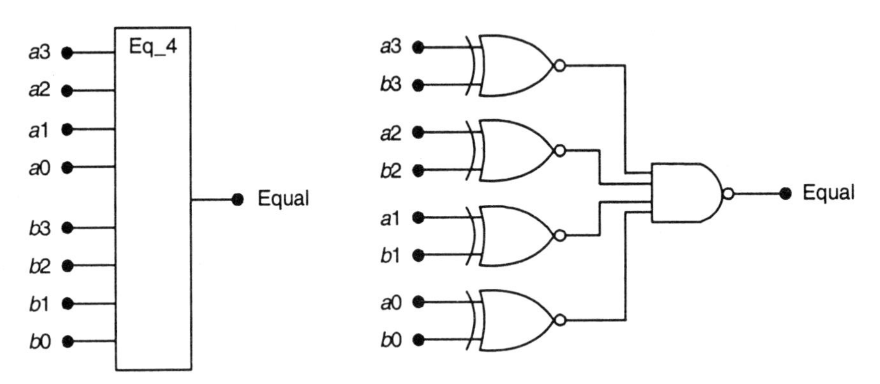

Figure 16.12 A 4-bit equality detector.

A Verilog listing for the operation is

```
module equality (Equal,a,b);
input [3:0] a,b;
output Equal;
always @ (a or b)
    begin
      if (a==b)
        Equal=1;
      else
        Equal=0;
    end
  endmodule
```

The internal structure of the circuitry is hidden in the logical equality condition a==b. The extension to an arbitrary word size is easily accomplished at both the circuit and HDL level. An example is the 8-bit version shown in Figure 16.13. This uses two 4-bit circuits with ANDed outputs to produce the final result.

Magnitude comparator circuits are used to compare two words a and b and determine if $a > b$ or $a < b$ is true; the equality condition $a = b$ may also be detected by the logic. The logic for a 4-bit magnitude comparator is shown in Figure 16.14. The input words are used on a bit-wise basis to produce two outputs, GT and LT, with the results summarized in Figure 16.15. The logic equations are a bit tedious to derive, but the signal paths can be traced from the diagram.

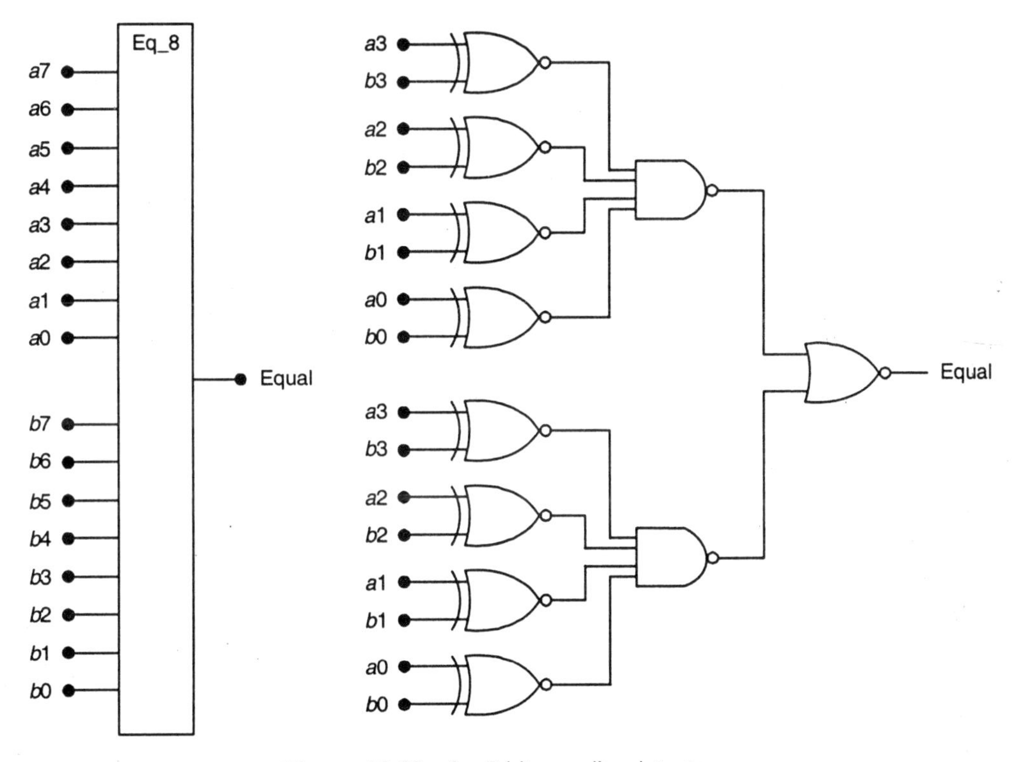

Figure 16.13 An 8-bit equality detector.

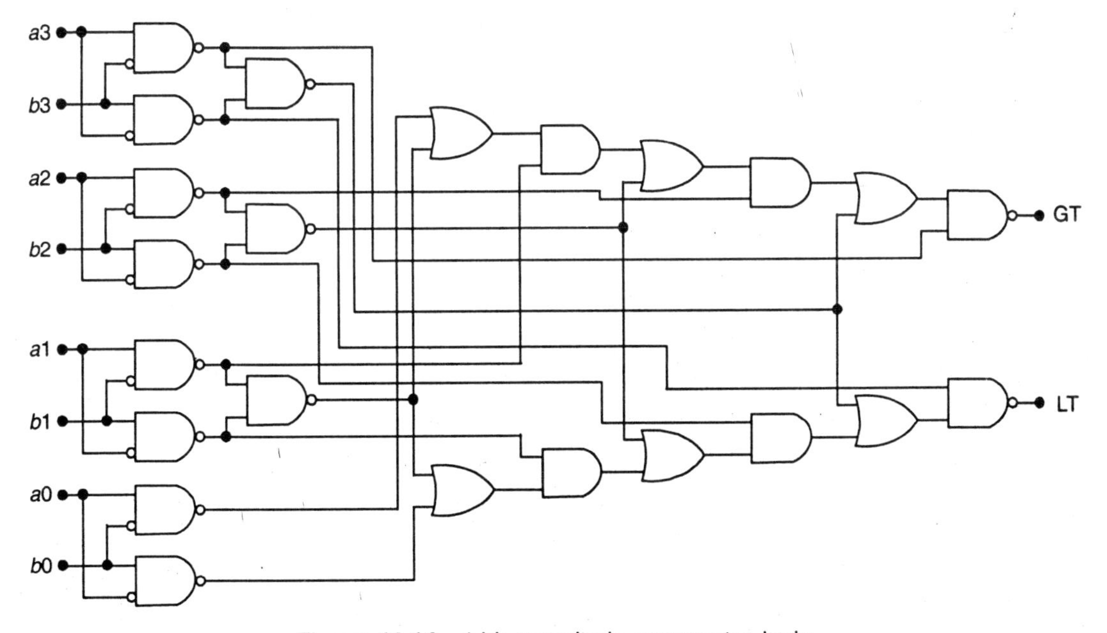

Figure 16.14 4-bit magnitude comparator logic.

Condition	GT	LT
$a > b$	1	0
$a < b$	0	1
$a = b$	0	0

Figure 16.15 Comparator output summary.

The symmetry of the upper and lower logic chains is the basis for producing a GT or LT result. Optional features of an equality detection output and an enable control can be added by cascading the circuit into the logic network shown in Figure 16.16.

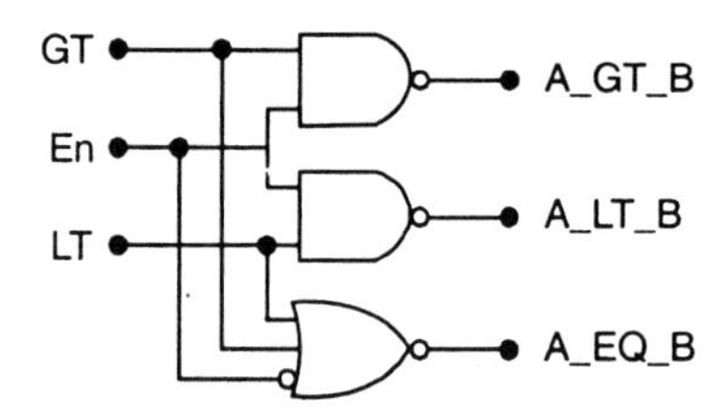

Figure 16.16 Additional logic for A_EQ_B and enable features.

A Verilog listing for the 4-bit comparator can be constructed as follows.

```
module comp_4(GT,LT,a,b);
input [3:0] a,b;
output GT,LT;
always @ (a or b)
begin
  if (a>b)
  GT=1, LT=0;
elseif (a<b)
  GT=0, LT=1;
else
  GT=0,LT=1;
end
endmodule
```

The high-level description masks the internal structure completely making it appropriate for architectural simulations. However, the logic and circuit implementations can be quite complicated.

The hierarchical design technique allows us to build an 8-bit comparator using two 4-bit circuits [Comp 4] and an interfacing network. The main circuit is shown in Figure 16.17. The lower Comp 4 block accepts the lower 4 bits of each word, while the upper block uses bits 4–7.

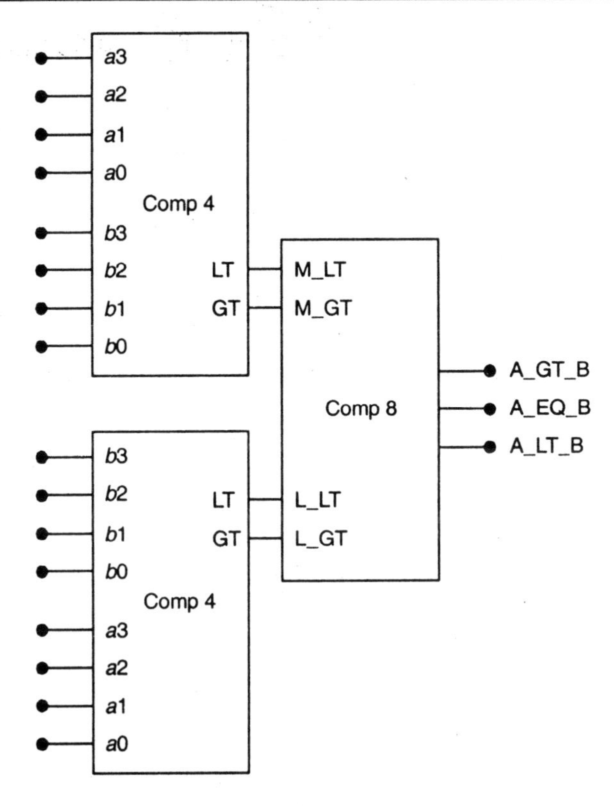

Figure 16.17 8-bit comparator system.

The interface block labelled Comp 8 includes an Enable input. The logic diagram for the interfacing network is shown in Figure 16.18. The upper inputs are the GT Comp 4 outputs, while the lower inputs are LT values from the 4-bit comparison circuits. These are then compared to produce the outputs. Including the AND gate and the NAND–NOT cascades (at the A_GT_B and A_LT_B outputs) allow us to generate the equality signal A_EQ_B that will be 1 if the words are equal. Note that A_GT_B and A_LT_B are both 0 when A_EQ_B = 1.

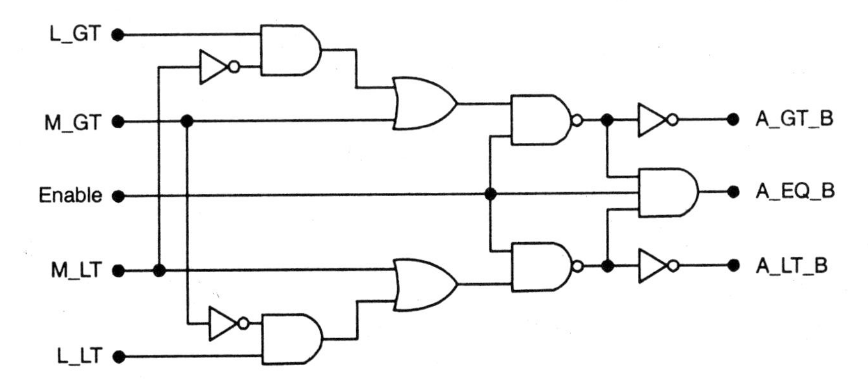

Figure 16.18 Comp 8 logic diagram.

16.2.4 Priority Encoder

A priority encoder examines the input bits of an n-bit word and produces an output that indicates the position of the highest priority logic 1 bit. Consider an 8-bit word

$$d = d7\, d6\, d5\, d4\, d3\, d2\, d1\, d0 \tag{16.3}$$

and let us assign the highest priority to $d7$, the second highest priority to $d6$ and so on. The operation of a priority encoder is to detect the presence of 1s in d; if two or more bits are at a logic 1 value, then the input with the highest priority takes precedence. If we use d as the input to an 8-bit priority encoder, then the output word

$$Q = Q1\, Q2\, Q3\, Q4 \tag{16.4}$$

is coded to indicate the highest priority bit. A function table for this scheme is provided in Figure 16.19. The bit $Q3$ is equal to 1 if any input bit is a 1. The 3-bit word $Q2Q1Q0$ is encoded to indicate the highest priority input bit. There is no formal logic symbol, so we will use the simple box shown in Figure 16.20 when the device is used in a system design.

d7	d6	d5	d4	d3	d2	d1	d0	Q3	Q2	Q1	Q0
0	0	0	0	0	0	0	1	1	0	0	0
0	0	0	0	0	0	1	–	1	0	0	1
0	0	0	0	0	1	–	–	1	0	1	0
0	0	0	0	1	–	–	–	1	0	1	1
0	0	0	1	–	–	–	–	1	1	0	0
0	0	1	–	–	–	–	–	1	1	0	1
0	1	–	–	–	–	–	–	1	1	1	0
1	–	–	–	–	–	–	–	1	1	1	1
0	0	0	0	0	0	0	0	0	0	0	0

d7 has the highest priority
d0 has the lowest priority

Q3 = 1 when di = 1
for any i = 0, ..., 7

Figure 16.19 Function table for an 8-bit priority encoder.

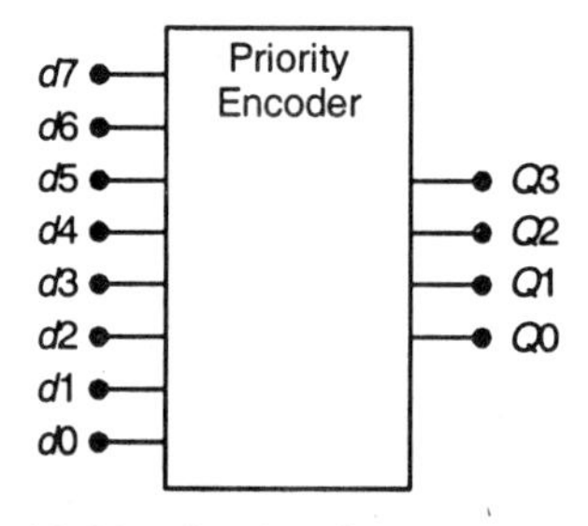

Figure 16.20 Symbol for priority encoder.

The logic for the network is drawn in two parts. The first section in Figure 16.21 shows the input buffers and complements generators for each bit. The output logic for the $Q2$ and $Q3$ is simple and is given by the expressions

$$Q2 = (d_0 + d_1 + d_2 + d_3) \cdot \overline{(d_4 + d_5 + d_6 + d_7)} \tag{16.5}$$

as can be

$$Q3 = (d_0 + d_1 + d_2 + d_3) + \overline{(d_4 + d_5 + d_6 + d_7)} \tag{16.6}$$

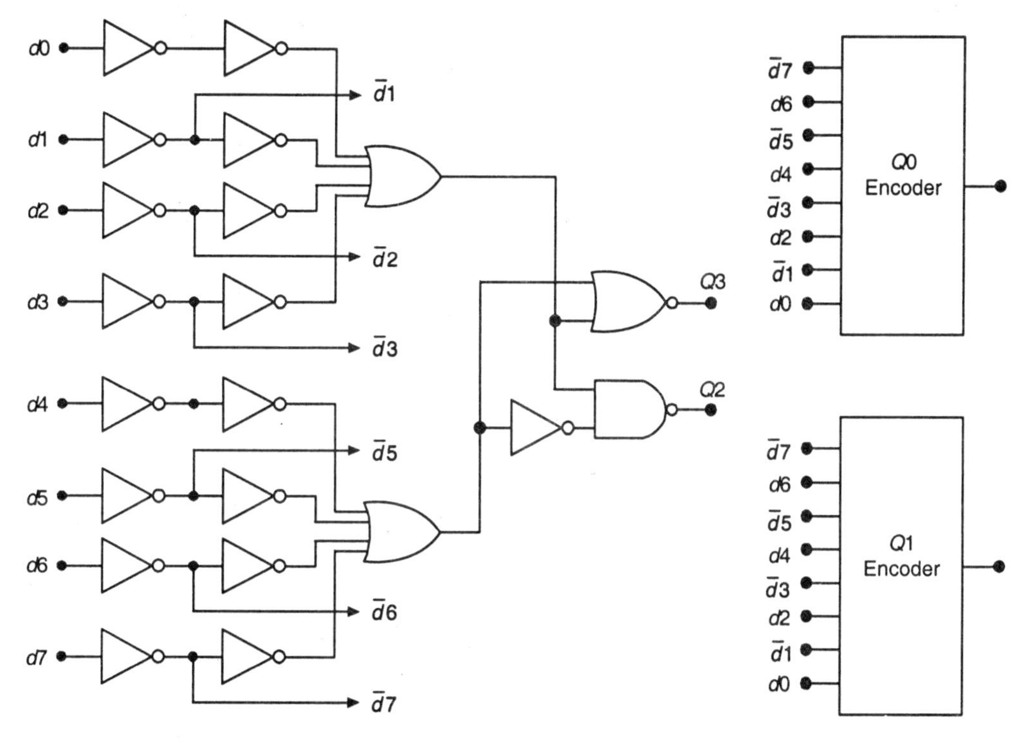

Figure 16.21 Logic diagram for the priority encoder.

verified from the schematic. The $Q0$ and $Q1$ encoders uses the buffered and complemented input as shown in Figures 16.22(a), (b). The logic equation for Q0 circuit is

$$Q0 = \bar{d}_7 \cdot [d_6 + d_5 \cdot (d_4 + \bar{d}_3[d_2 + \bar{d}_1 \cdot d_0])] \tag{16.7}$$

While

$$Q1 = \bar{d}7 \cdot \bar{d}6 \cdot \lfloor d5 + \bar{d}4 + \bar{d}3 \cdot \bar{d}2 \cdot (d1 + d0) \rfloor \tag{16.8}$$

gives the $Q1$ bit.

Even though the internal details of the circuit are complicated, the behavioural description is concerned only with the overall functional behaviour. One implementation for the module is

```
module priority_8 (Q,Q3,d);
  input [7:0] d;
  output Q3;
  always @ (d)
    begin
        Q3=1;
        if (A[7])Q=7;
      elseif (A[6]) Q=6;
    elseif (A[5]) Q=5;
    elseif (A[4]) Q=4;
    elseif (A[3]) Q=3;
      elseif (A[2]) Q=2;
    elseif (A[1]) Q=1;
    elseif (A[0]) Q=0;
```

```
    else
      begin
        Q3=0
        Q=3'b000;
      end
    end
endmodule
```

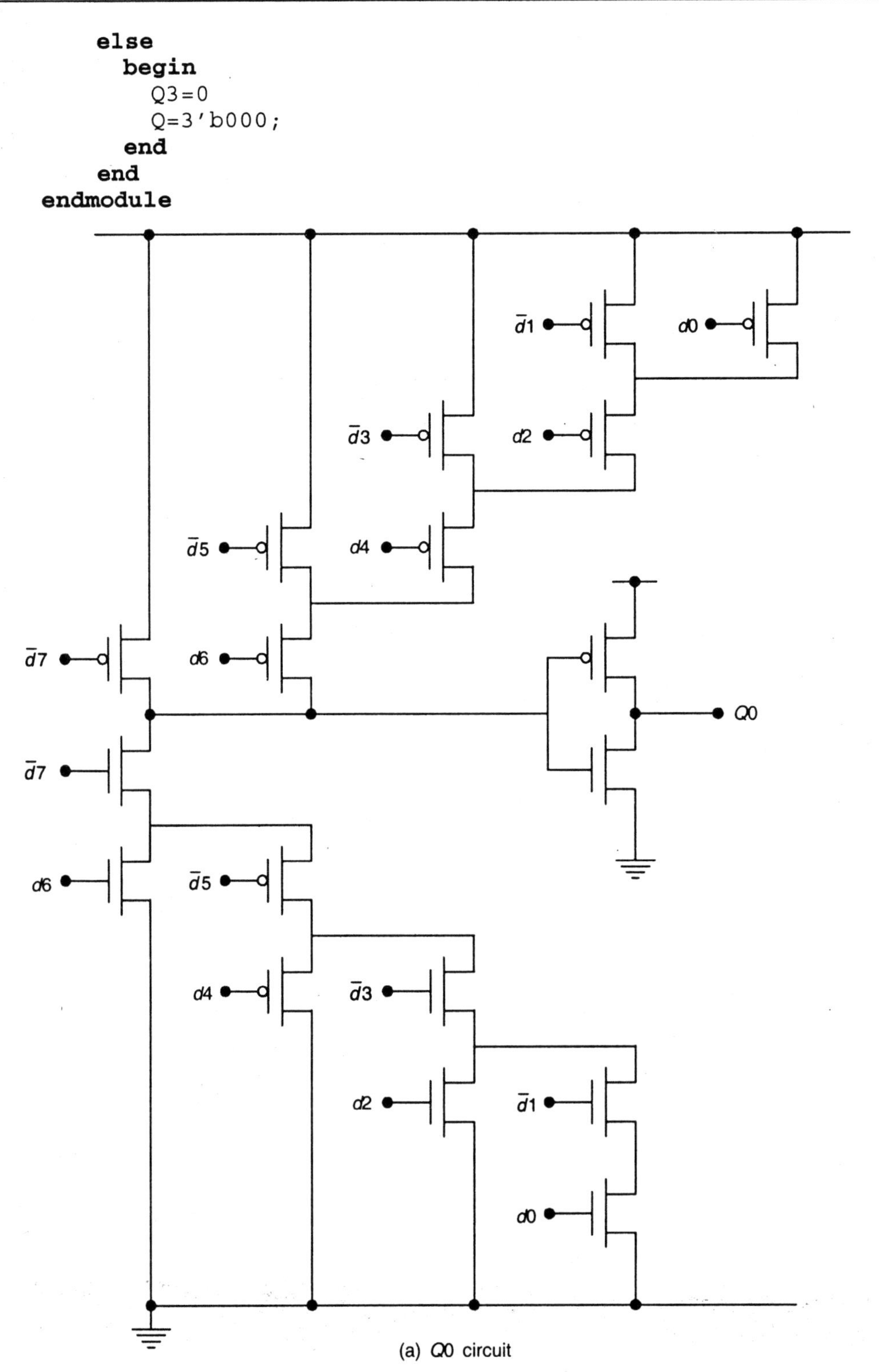

(a) Q0 circuit

Figure 16.22 (contd.).

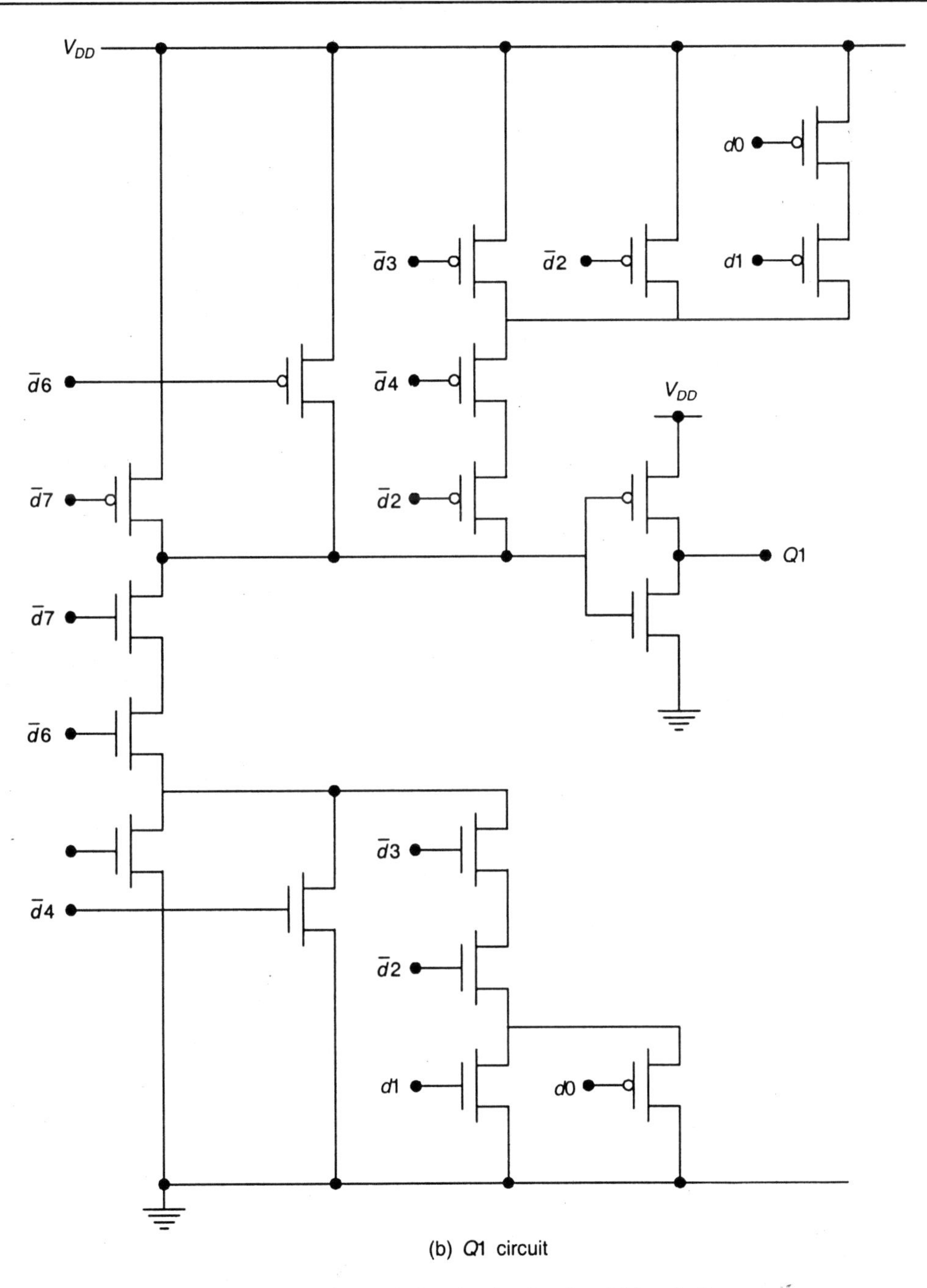

(b) Q1 circuit

Figure 16.22 Q0 and Q1 circuits for the 8-bit priority encoder.

We have defined $Q3$ as a scalar and Q as a 3-bit vector that is assigned a value corresponding to the decimal equivalent of $Q3Q1Q0$ listed in the function table. This example is particularly good at illustrating the separation of a high-level versus a low-level description. The translation of the HDL to the circuits and logic algorithms can be constructed, each with different area and switching properties.

16.2.5 Latches

A latch is a device that can receive and hold an input bit. A simple D-latch forms the basis for many designs. The symbol for the D-latch is shown in Figure 16.23(a), and a logic diagram is provided in Figure 16.23(b). By inspection we see that the circuit is formed using a NOR-based SR latch with complemented inputs. The latch is transparent in that a change in D is seen at the outputs Q and $\bar{Q}$ after a circuit delay time.

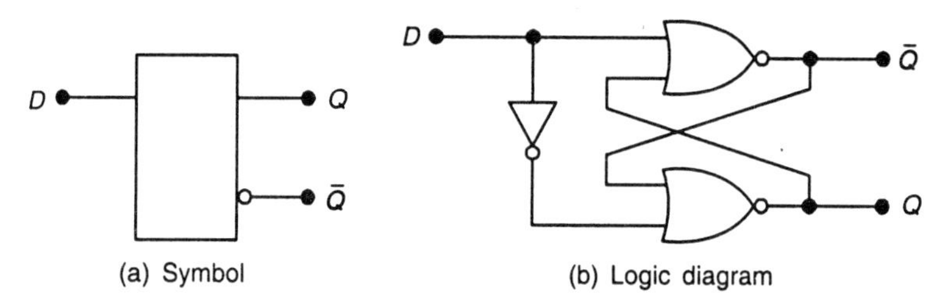

(a) Symbol (b) Logic diagram

Figure 16.23 D_latch.

A behavioural description for the device is

```
module d_latch (q,q_bar,d);
input d;
output q, q_bar;
reg q,q_bar;
  always @ (d)
    begin
      # (t_d) q=d;
      # (t_d) q_bar=~d;
    end
endmodule
```

We have used t_d for the time delay. The declaration neither models the action of the cross-coupled NOR circuit that can hold the input state. The equivalent structural description is

```
module d_latch_gates (q, q_bar,d);
input d;
output q,q_bar;
wire not_d;
not (not_d,d);
nor # (t_nor) g1 (q_bar,q,d),
  #(t_nor) g2 (q,q_bar,not_d);
endmodule
```

This provides a gate-by-gate guide for the device at the circuit and physical design level.

16.3 COMBINATIONAL LOGIC DESIGNS

The CMOS circuit can be constructed using either the logic diagram or the structural description. A direct translation is shown in Figure 16.24. At the physical level this can be created by instancing two NOR2 cells and one NOT cell, and then adding the interconnect wiring. Alternately, a custom layout would probably consume less area.

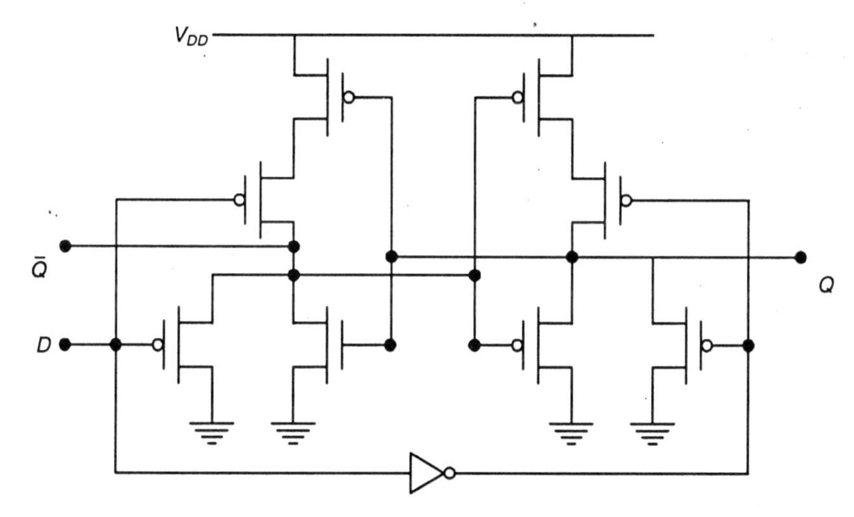

Figure 16.24 CMOS circuit for a D-latch.

A enable control En can be added to the basic D-latch by routing the inputs through AND gates as illustrated in Figures 16.25(a) and (b). An enable bit of En = 0 blocks the inputs by forcing 0s to the AND outputs, which places the SR latch into a hold state. If En = 1, then the values of D and $\bar{D}$ are

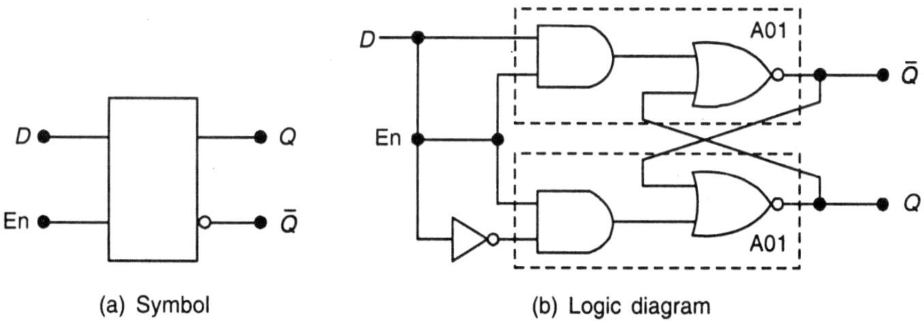

Figure 16.25 Gated D-latch with Enable control.

Admitted to the NOR circuits. To include this control in the behavioural description, we rewrite the code as

```
module d_latch(q,q_bar,d,enable);
input d, enable;
output q,q_bar;
```

```
reg q,q_bar;
always @ (d and enable)
  begin
    # (t_d) q=d;
    #(t_d) q_bar=~d;
  end
endmodule
```

A condition of En = 1 is required to change the state.

The structural description could be based on a brute-force listing of the circuit. However, since CMOS allows for complex logic gates as primitives, it makes more sense to describe the complex gate and then instance it in the listing.

```
// First define the AOI module
module aoi_2_1 (out,a,b,c);
input a,b,c,;
output out;
wire w1;
and (w1,a,b);
nor (out,w1,c);
endmodule
// Now use this to build the latch
module d_latch_aoi (q,q_bar,d,enable);
input d, enable;
output q, q_bar;
wire d_bar;
not (d_bar,d);
aoi_2_1 (q_bar,d,enable,q);
aoi_2_1(q,enable,d_bar,q_bar);
endmodule
```

The module name `aoi_2_i` is interpreted as an AOI gate with 2 inputs to the AND, and I input to the OR. Although the notation is not standard, it is widely used in practice. Note that the order of the inputs must be preserved with the module is instanced. The CMOS circuit diagram for this implementation is shown in Figure 16.26, and consists of one NOT gate and two AOI circuits.

16.4 CMOS VLSI LATCH

Many static D-latches CMOS VLSI are constructed from inverters and TGs or pass FETs. This design is based on the characteristics of the simplest static storage configuration called **bistable circuit**. A bistable circuit is one that can store (or, hold) either a logic 0 or a logic 1 indefinitely (or, at least as long as power is applied).

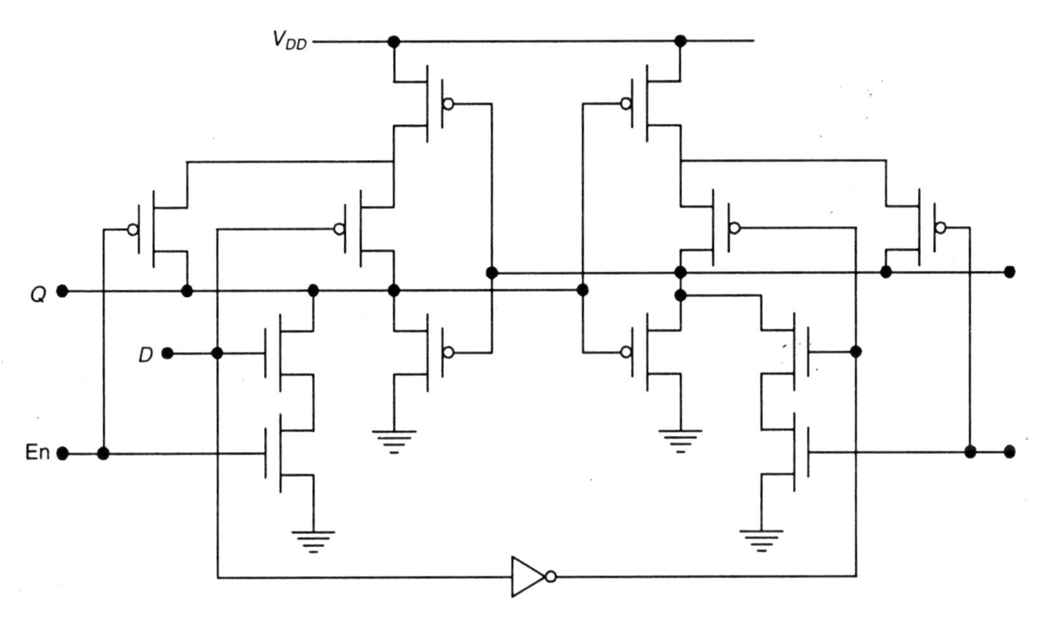

Figure 16.26 AOI CMOS gate for D-latch with enable.

A basic bistable circuit consists of two inverters as shown in Figure 16.27(a). The gates are wired such that the output of one is the input to the other, forming a closed loop. Any closed loop with an even number of inverters gives a bistable circuit. If we use three inverters as in Figure 16.27(b) the resulting circuit is unstable and cannot hold a bit value. A closed loop with an odd number of inverters is often called a **ring oscillator** as the signal at any point oscillates in time.

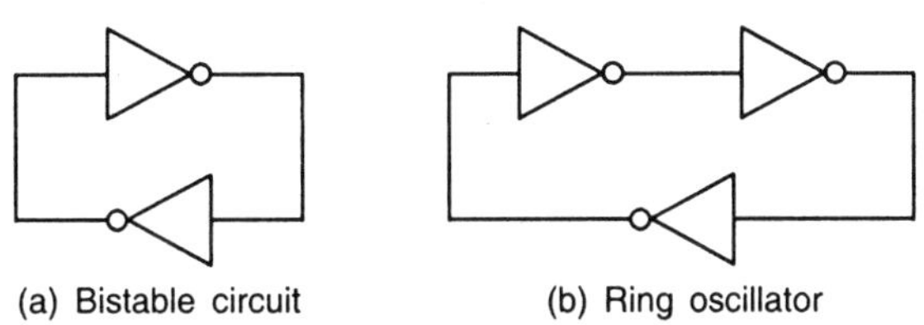

(a) Bistable circuit (b) Ring oscillator

Figure 16.27 Closed-loop inverter configurations.

To create a D-latch, we must provide an entry node for the input bit. One simple idea is the receiver circuit shown in Figure 16.28(a). The value of *D* is held by the bistable circuit formed by the inverter pair. The circuit helps the line resist changes in *D*, making it useful as an input stage to a receiver module that is driven by an external line. The presence of the feedback loop from the output of Inverter 2 to the input of Inverter 1 provides the desired latching, but complicates the design. Inverter 1 needs to detect changes, but Inverter 2 cannot be so strong that it prohibits a change in state. In general, Inverter 1 can use relatively large FETs, but Inverter 2 is purposefully made weaker by using small transistors.

Adding a transmission gate at the input as shown in Figure 16.28(b) gives us the ability to control the loading. When $C = 0$ (so $\overline{C} = 1$), the TG acts as an open switch and the circuit holds

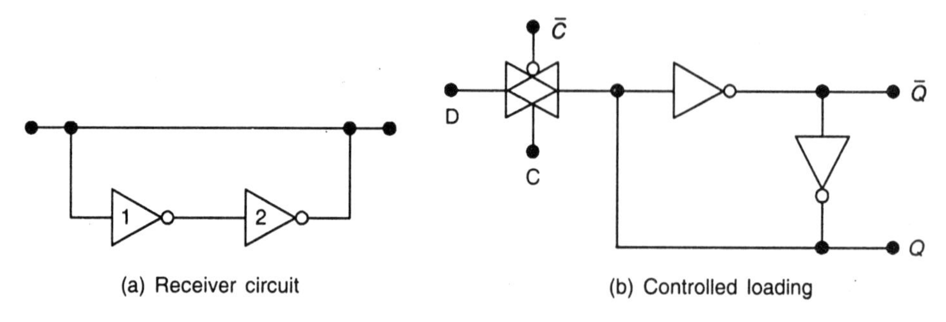

(a) Receiver circuit

(b) Controlled loading

Figure 16.28 Adding an input node to the bistable circuit.

the values of Q and $\bar{Q}$ at the output. When the control bit is set to $C = 1$ ($\bar{C} = 0$), the TG conducts and allows the input bit D to be transferred to the latching circuit. During this time the latch is transparent and the outputs go to $Q = D$ and $\bar{Q} = \bar{D}$. If C is reset to 0, then the state is held. The control bit C is thus equivalent to the enable signal En used previously. The inverter design constraints still apply.

16.4.1 D Flip-Flop

A flip-flop differs from a latch in that it is non-transparent. The D-type flip-flop (DFF) is the most commonly used flip-flop in CMOS circuits. The basic DFF design is a master–slave configuration obtained by cascading two oppositely phased D-latches as in Figure 16.29. The clock signal ϕ controls the operation and provides synchronization. The master latch allows an

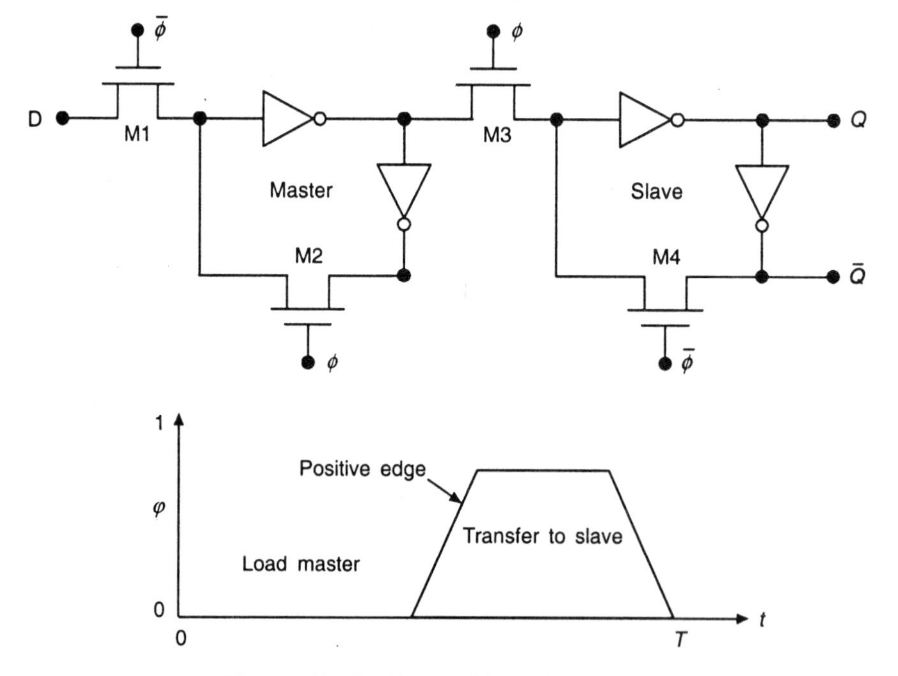

Figure 16.29 Master–Slave D-type flip-flop.

input of D when $\phi = 0$ and M1 acts as a closed switch. During this time, NFETs M2 and M3 are open-circuits. When the clock makes a transition to $\phi \to 1$, switches M2 and M3 close and effect the transfer of the bit to the slave. The input to the master is blocked since M1 is open with $\phi = 1$. The master–slave circuit acts as a positive edge-triggered device since the value of D during the positive clock edge defines the value of the bit that is transferred to the slave and available as Q. Note that once the bit is latched into the slave at time t_1, it does not appear at the output until

$$t_1 = t_{FET} + t_{NOT} \tag{16.9}$$

where t_{label} is the rise or fall time of the specified element. The symbol for a positive edge-triggered DFF is shown in Figure 16.30(a). The 'triangle' denotes an edge-triggering input. Adding a bubble produces the symbol for a negative edge-triggered DFF in Figure 16.30(b). At the circuit level, all that needs to be done is to interchange the ϕ and $\overline{\phi}$ signals.

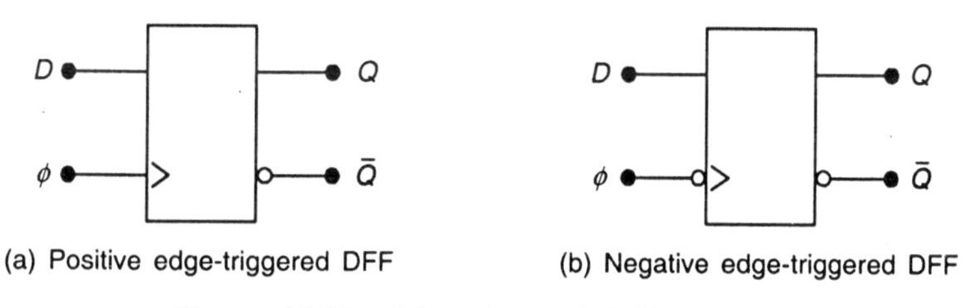

(a) Positive edge-triggered DFF (b) Negative edge-triggered DFF

Figure 16.30 Edge-triggered DFF symbols.

A Verilog behavioural description of a positive edge-triggered DFF can be written in the following manner.

```
module positive_dff (q,q_bar, d,clk);
input d,clk;
output q, q_bar;
reg q, q_bar;
always @ (posedge clk)
  begin
    q=d;
    q_bar=~d;
  end
endmodule
```

In a realistic application, a set of delay times would be needed. A negative edge-triggered module is obtained by modifying the always statement to

```
always @ (negdge clk)
```

An important point to reiterate there is that the design of the CMOS circuitry determines the delays through the DFF. Consider the alternate circuitry shown in Figure 16.31. This is logically equivalent to the circuit drawn in Figure 16.29, but the datapath from the input D to the output Q is through four inverters instead of two. Since every logic gate introduces additional signal delay, this circuit will be slower than the original design. We thus see that the circuit topology

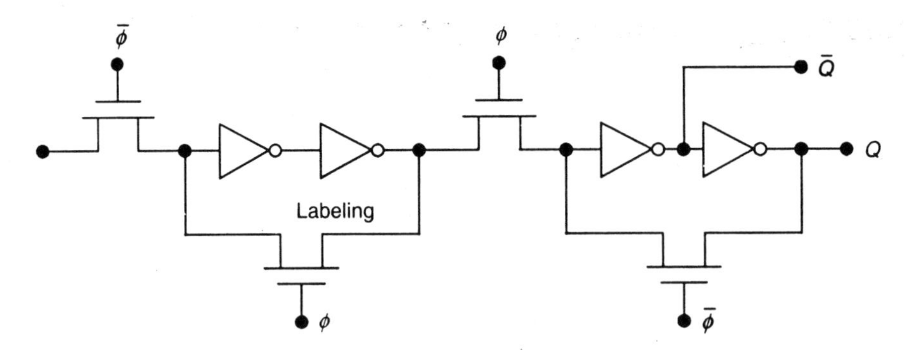

Figure 16.31 Alternate circuitry for the master–slave DFF.

and the resulting physical design directly affect the speed of the high-level construct described by the HDL listing. This type of consideration is one of the factors that distinguishes high-speed VLSI from other digital systems designs.

It is possible to add direct clear and set capabilities to the circuit by changing the gate functions. One approach is to use NAND2 logic. Consider the case in Figure 16.32 where one input is a control bit s and the other is a data value in. When s = 0, the output is 1 regardless

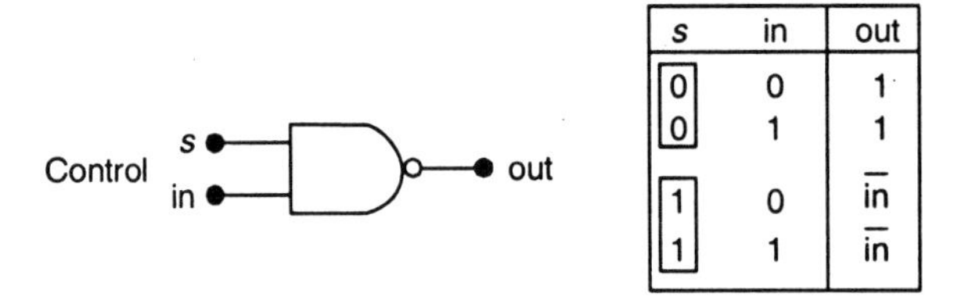

s	in	out
0	0	1
0	1	1
1	0	$\overline{in}$
1	1	$\overline{in}$

Figure 16.32 NAND2 used as a control element.

of the value of in. If s = 1, then out = $\overline{in}$ as shown. Replacing selected inverters with NAND2 gates yields DFFs with assert-low clear or set inputs, or both. Figure 16.33(a) provides the ability to clear (to 0) the contents of the latch using the clear control. With clear = 1, the NAND gates act as inverters and the circuit behaves as a normal DFF. When Clear = 0, the NAND in the slave forces an output of Q = 0. The output of the master is forced to a logic 1, which inverts to an output Q = 0. A Verilog listing that describes this device is

```
module dff_clear (q,d,clear,clk);
input d, clk, clear;
output q;
reg q;
always @ (posedge clk)
  q=d;
always @ (clear)
  if (clear)
    assign q=0;
  else
    deassign q;
endmodule
```

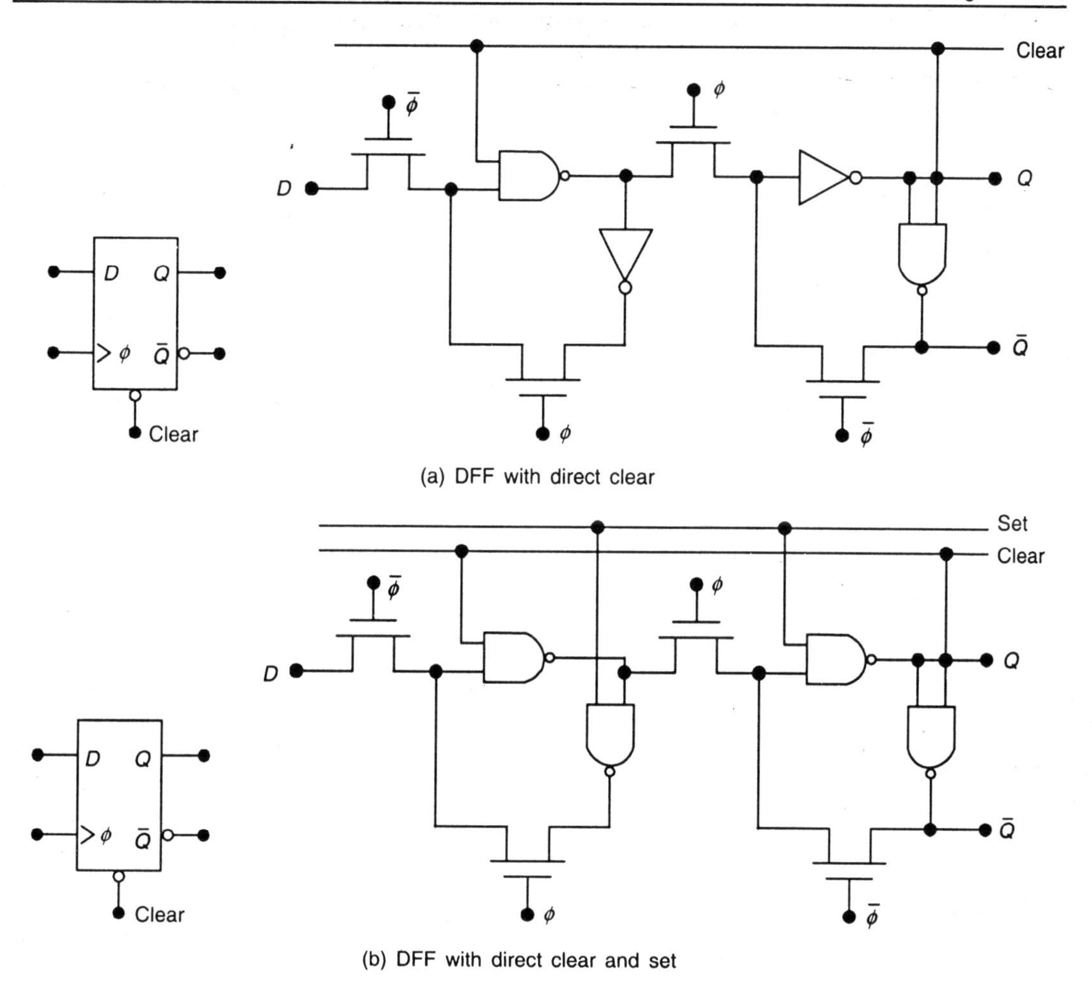

(a) DFF with direct clear

(b) DFF with direct clear and set

Figure 16.33 DFF circuits with assert-low clear and clear/set controls.

This uses the deassign statement that is executed if 'clear' is 0. This returns the value of q to its value established in the $q = d$ line.

A DFF with both clear and set controls is shown in Figure 16.33(b). The set capabilities are achieved by substituting NAND2 gates for the other two inverters. A condition of Set = 1 gives normal operation, while Set = 0 forces the slave to an output of $Q = 1$. This is reinforced by the output of the master. Note that Clear and Set cannot both be 0 at the same time as this forces $Q = \bar{Q}$. It is common to find several variations of DFFs in a cell library including features such as input buffers, clock buffers, and inputs with combinational logic gates. For example, a toggle flip-flop (TFF) that changes states on every rising clock edge can be created by adding a feedback loop from $\bar{Q}$ to D as shown in Figure 16.34. The assert-low set logic modification allows the initial value to be established as a 1.

A basic DFF loads a new data bit on every clock edge. Storage over an arbitrary number of clock cycles can be obtained by adding a control signal and associated logic to the circuit.

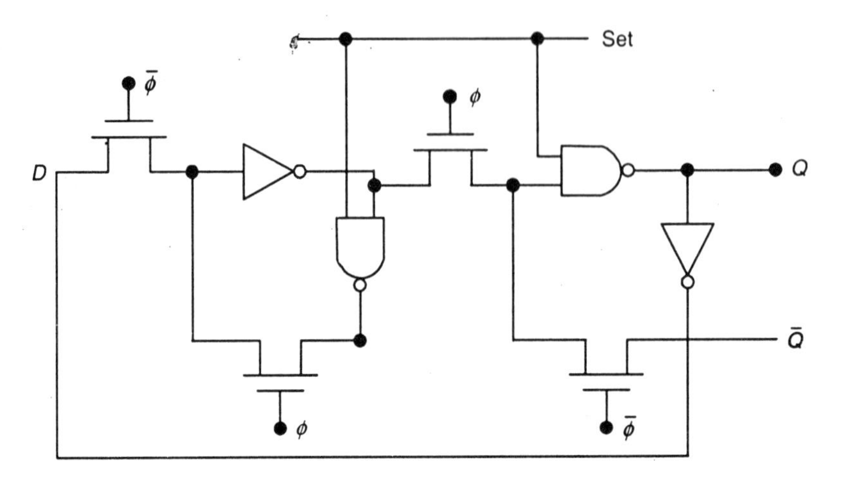

Figure 16.34 DFF modified to a TFF circuit using feedback.

A simple classical solution is shown in Figure 16.35(a). The control signal load controls a 2:1 MUX. If the control signal is asserted with Load = 1, the MUX allows a new data value D to enter the DFF. A control bit with a value Load = 0 takes the output Q and redirects it back to the input. This type of circuit is usually integrated as a single element with the simplified symbol shown in Figure 16.35(b). A simple Verilog description that includes the Load control is

```
module dff_load (q,q_bar,d,load,clk);
input d,clk,load;
output q,q_bar;
reg q,q_bar;
always @ (posedge clk)
  begin
    if (load) q=d;
    q_bar=~d;
  end
endmodule
```

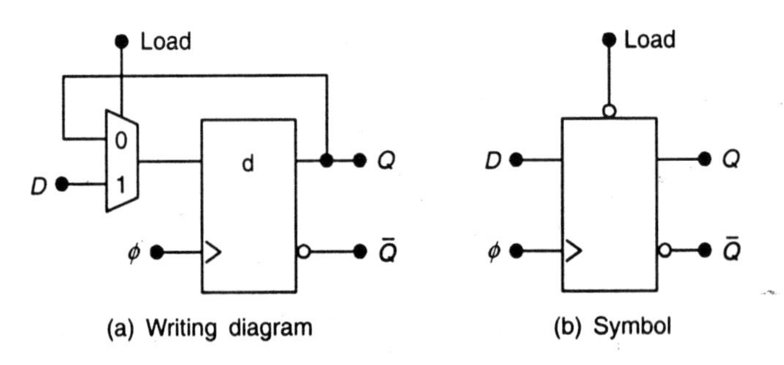

(a) Writing diagram (b) Symbol

Figure 16.35 D-type flip-flop with load control.

A CMOS master-slave FF with load control is shown in Figure 16.36.

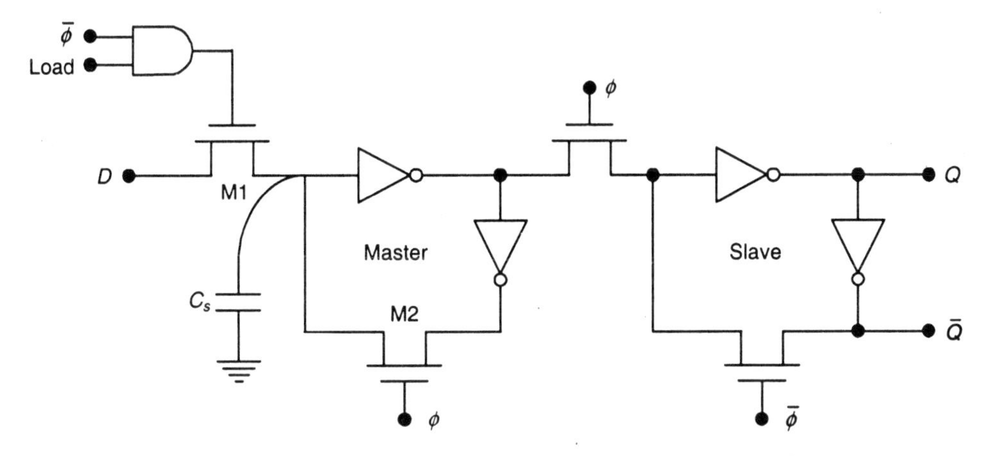

Figure 16.36 CMOS master–slave FF with load control.

SUMMARY

The basis for behavioural modelling is the construction of procedural blocks, which were introduced in this chapter. Examples are given for timing controls, procedural assignments, conditional statements. It was also discussed for general VLSI system components like multiplexers. Binary decoders, Equality detectors and comparators, priority encoder and latches. Discussion on combinational logic designs with examples like D flip-flop is also there.

REVIEW QUESTIONS

1. Explain the different types of modelling in Verilog and write the coding for decoder statement using dataflow modelling.
2. Draw the block diagram for 4x1 MUX using NFET pass transistors for the same and write the Verilog coding that.
3. Explain 8-bit equality detector and write a suitable source code.
4. Sketch a logic diagram for the priority encoder and draw $Q0$ and $Q1$ circuits for the 8-bit priority encoder.
5. Draw AOI CMOS gate for D-latch with enable, and write a suitable code for D-latch.

SHORT ANSWER QUESTIONS

1. Concurrently executed statements use the ____________ and ____________ keywords.
   Ans. fork, join
2. Level-triggered events are modelled using which keyword.
   Ans. wait. *Ex:* wait(clk) q_out=d_in;

3. A procedural assignment is usually divided into the ____________ and ____________ assignments.

   Ans. blocking; non-blocking

4. Explain equality detectors and comparators.

   Ans. An equality detector compares two *n*-bit words and produces an output that is 1 if the inputs are equal on a bit-by-bit basis.

5. How the cyclic action is obtained using the NOT operator?

   Ans. #5 clk=~clk;

6. How an edge-triggered control is specified?

   Ans. @(posedge clk) reg_1=reg2;

7. How the level-triggered events are modelled using the wait keyword.

   Ans. wait(clk) q_out=d_in;

8. What is the function of the **repeat** loop?

   Ans. **Repeat** loop executes a set of statements, a specific number of times. Ex: repeat(counter)

9. What is the form of assign statements?

   Ans. Assign statements are of the form *assign a=~b&c*;

10. What are the uses of the deassign statement in the d-ff?

    Ans. q=d; if clear is 0;

17 Arithmetic Circuits in CMOS VLSI

17.1 INTRODUCTION

Arithmetic functions such as addition and multiplication have a special significance in VLSI designs. Many application require these basic operations, but good silicon implementations have been a challenge since the early days of digital chip building. In this chapter, we will examine binary adders in detail, and extend the discussion to include multipliers.

17.2 BIT ADDER CIRCUITS

Consider two binary digits x and y. The binary sum is denoted by $x + y$ such that

$$\begin{aligned} 0 + 0 &= 0 \\ 0 + 1 &= 1 \\ 1 + 0 &= 1 \\ 1 + 1 &= 10 \end{aligned} \tag{17.1}$$

where the result in the last line is a binary 10 (i.e., 2 in base-10). This simple example illustrates the problem with addition. If we take two base-r numbers with digits 0, 1, ..., $(r - 1)$, then the sum of two numbers can be out of the range of the digit set itself. This, of course, is the origin of the concept of a carry-out. In the binary sum 1 + 1, the result 10 is viewed as a 0 with a 1 shifted to the left to give a "carry-out of 1."

A half-adder circuit has 2 inputs (x and y) and 2 outputs (the sum s and the carry-out c) and is described by the table provided in Figure 17.1. The outputs are given by the basic equations

$$s = x \oplus y \tag{17.2}$$

$$c = x \cdot y$$

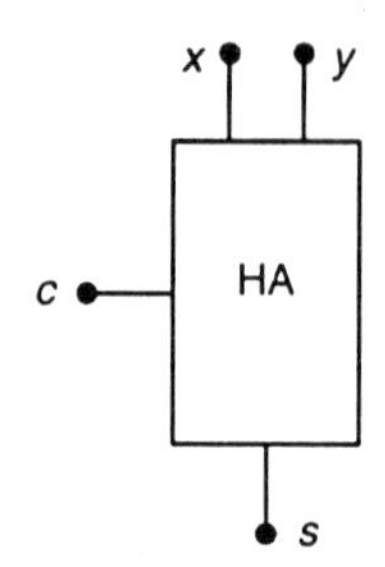

Figure 17.1 Half-adder symbol and operation.

x	y	s	c
0	0	0	0
0	1	1	0
1	0	1	0
1	0	0	1

Which are taken directly from the table. A high-level Verilog behavioural description of the cell can be written as

```
module half_add_gate(sum,c_out,x,y)
input x,y;
output sum, c_out;
assign {c_out,sum}=x+y;
endmode
```

This defines x and y as single-bit quantities, and then uses the concatenation operator { } to obtain a 2-bit result. This operator 'connects' binary segments in the order listed to create a single result. Alternatively, we may construct the gate-level network shown in Figure 17.2. This is described by the structural model

```
module half_add_gate(sum,c_out,x,y);
input x,y;
output sum, c_out;
and (c_out,x,y);
xor(sum,x,6);
endmodule
```

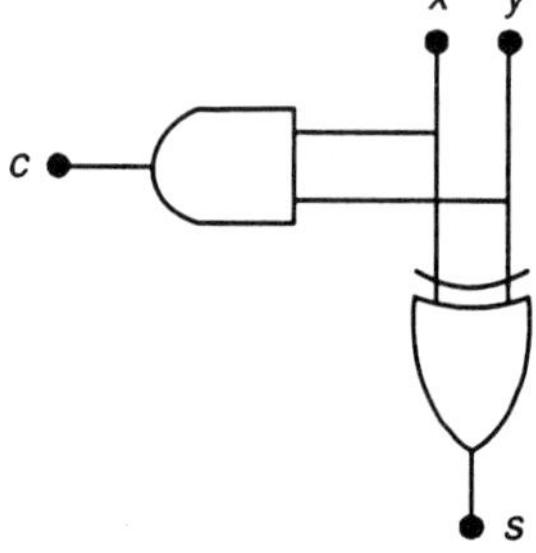

Figure 17.2 Half-adder logic diagram.

using primitive gate instances. Two more possibilities are shown in Figure 17.3. The half-adder in Figure 17.3(a) uses NAND2 gates, while the alternative in Figure 17.3(b) is a NOR-based design. Preference might be given to the NAND design since it avoids series PFET chains, but an half-adder is simple enough so that the difference is not a major factor.

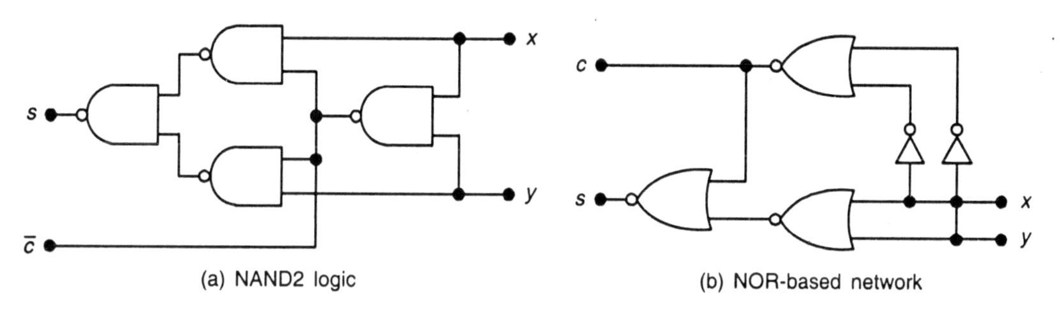

Figure 17.3 Alternative half-adder logic network.

A more complicated problem is adding n-bit binary words. Consider two 4-bit numbers $a = a_3a_2a_1a_0$ and $b = b_3b_2b_1b_0$. Adding gives

$$\begin{array}{r} a_3a_2a_1a_0 \\ +\, b_3b_2b_1b_0 \\ \hline c_4s_3s_2s_1s_0 \end{array} \tag{17.3}$$

where $s = s_3s_2s_1s_0$ is the 4-bit result and c_4 is the carry-out bit. To design an adder for the binary words, we break the problem down to bit level adders on a column-by-column basis. In the standard carry algorithm, each of the i-th columns ($i = 0, 1, 2, 3$) operates according to the **full-adder** equation

$$\begin{array}{r} c_i \\ a_i \\ +\, b_i \\ \hline c_{i+1}s_i \end{array} \tag{17.4}$$

where c_i is the carry-in bit from the $(i - 1)$-st column, and c_{i+1} is the carryout bit for the column. The operation is described by the full-adder table given in Figure 17.4 along with a simple schematic symbol. The most common expressions for the network are

$$s_i = a_i \oplus b_i \oplus c_i \tag{17.5}$$

$$c_i + 1 = a_i \cdot b_i + c_i \cdot (a_i \oplus b_i)$$

which can be derived directly from an SOP analysis of the function table. The carry-out bit can be written in alternate form

$$c_{i+1} = a_i \cdot b_i + (a_i + b_i) \tag{17.6}$$

if desired.

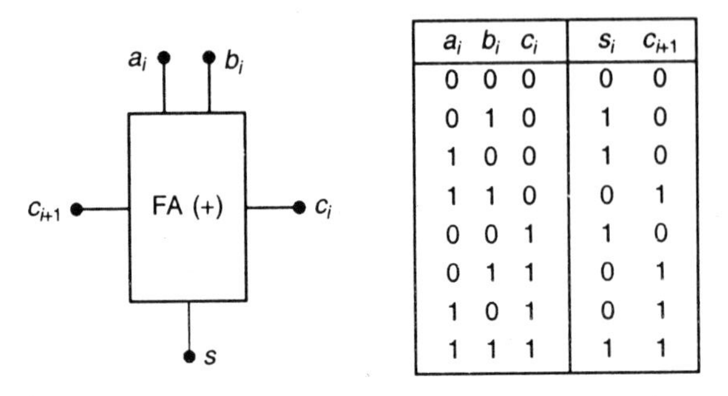

a_i	b_i	c_i	s_i	c_{i+1}
0	0	0	0	0
0	1	0	1	0
1	0	0	1	0
1	1	0	0	1
0	0	1	1	0
0	1	1	0	1
1	0	1	0	1
1	1	1	1	1

Figure 17.4 Full-adder symbol and function table.

A particularly compact circuit implementation is obtained using dual-rail complementary pass-transistor logic (CPL). The basic building block building block is the CPL 2-input array that has the generic form illustrated in Figure 17.5(a). The sum circuit is shown in Figure 17.5(b); the output of the first XOR/XNOR gate is the pair

$$a_i \oplus b_i \quad \text{and} \quad \overline{(a_i \oplus b_i)} \tag{17.7}$$

The second gate produces the sum by

$$s_n = \overline{(a_i \oplus b_i)} \cdot c_i + (a_i \oplus b_i) \cdot \overline{c}_i \tag{17.8}$$

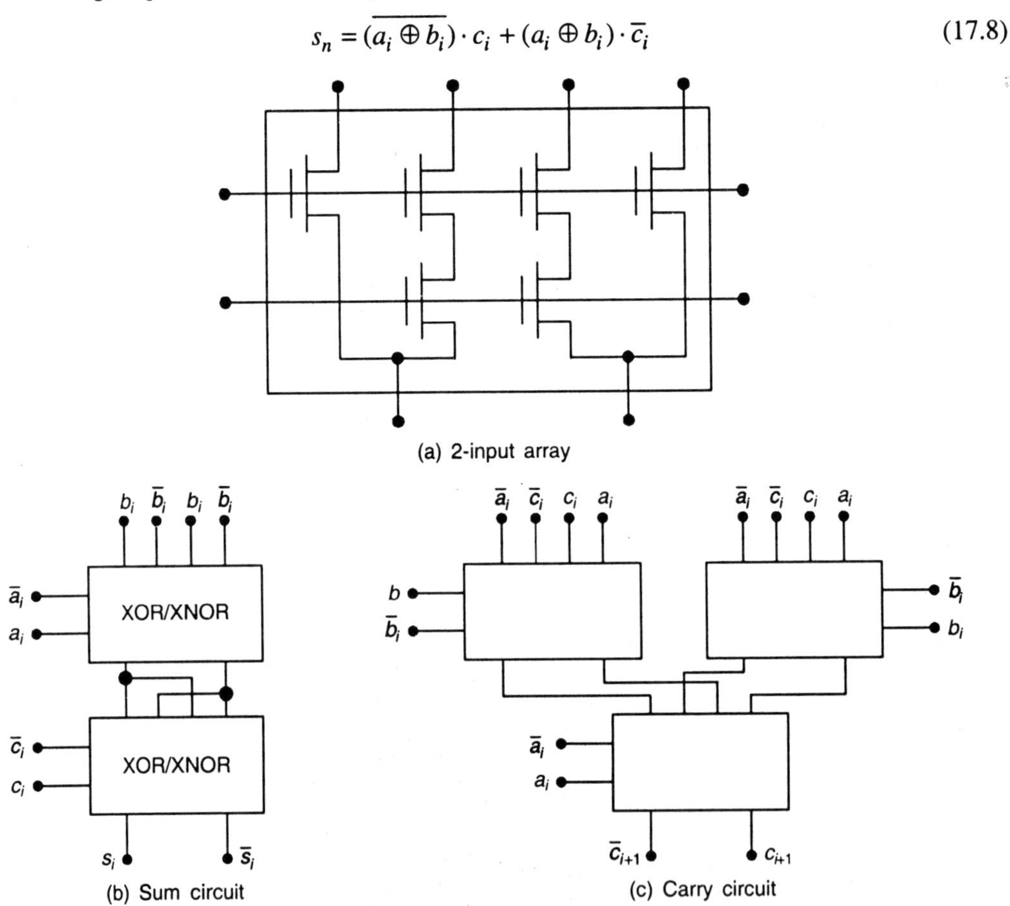

Figure 17.5 CPL full-adder design.

The carry circuit in Figure 17.5(c) employs the 2-input array as a combinational logic element. For example, the top array on the left-hand side produces outputs of

$$\overline{a}_i \cdot b_i + \overline{b}_i \cdot \overline{c}_i \tag{17.9}$$
$$\overline{b}_i \cdot c_i + a_i \cdot b_i$$

while the upper right circuit gives.

$$\overline{a}_i \cdot \overline{b}_i + b_i \cdot \overline{c}_i \tag{17.10}$$
$$b_i \cdot c_i + a_i \cdot \overline{b}_i$$

The last gate uses these to produce c_{i+1} and $\overline{c}_{i+1}$. Although this is a simple-looking solution, it must be remembered that CPL is a dual-rail technique that requires complementary variable paired such as $(a_i, \overline{a}_i)$ at every state. Also, because the threshold voltage loss drops the value of a logic 1 voltage as it passes through an NFET, restoring buffers or latching circuits are needed at the output. CPL is thus a somewhat specialized solution to implementing the CMOS full-adder. We note in passing that a CPL half-adder is easy to construct since it requires only the XOR/XNOR and AND/NAND functions.

A behavioural description of the full-adder is obtained by a simple modification of the half-adder model to the form

```
module full_adder (sum,c_out,a,b,c_in);
input a,b,c_in;
output sum, c_out;
assign {c_out,sum}=a+b+c_in;
endmodule
```

All variables are scalars (single bits) and the concatenation operator creates the two outputs. Structural modelling can be based on the gate-level network shown in Figure 17.6(a). This is a straightforward one-to-one translation of the equation set. At this level, the module takes the form

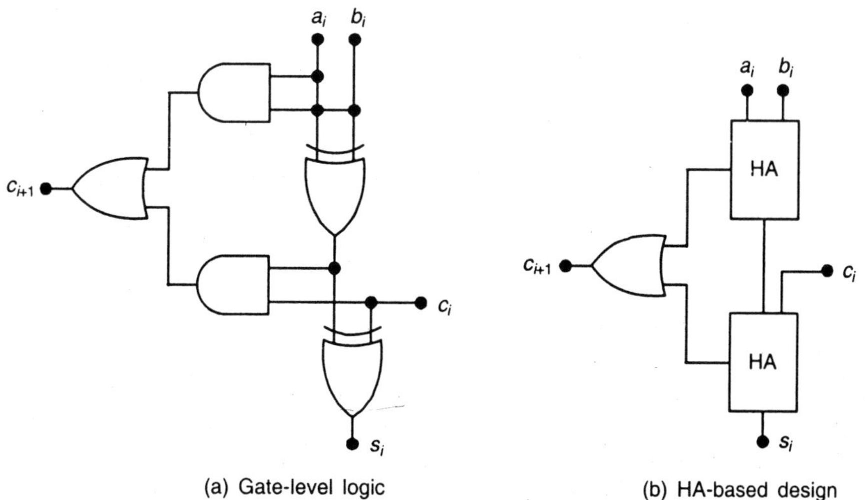

Figure 17.6 HA modules.

```
module full_adder_gate(sum,c_out,a,b,c_in);
input a,b,c_in;
output sum, c_out;
wire w1, w2, w3;
xor(w1,a,b),
  (sum,w1,w2,w3);
and (w2,a,b),
  (w3,w1,c_in);
or (c_out,w2,w3);
endmodule
```

where we have used slightly different variable identifiers to make the code more readable. A full-adder can also be built from two HA modules as shown in Figure 17.6(b). Using instance of the module defined by the listing

```
module half_add_gate(sum,c_out,x,y);
...
```

gives

```
module full_adder (sum,c_out,a,b,c_in);
input a,b,c_in;
output sum, c_out;
wire wa,wb,wc;
half_adder_gate(wa,wb,a,b);
half-adder_gate(sum,wc,wa,c_in);
or (c_out,wb,wc);
endmodule
```

as the description.

Owing to the importance of the full-adder, several implementations have been developed over the years. An AOI algorithm for static CMOS logic circuits can be obtained by writing the carry-out bit using Eq. (17.6). This allows us to write

$$\overline{s}_i = (a_i + b_i + \cdot c_i) \cdot \overline{c}_{i+1} + (a_i \cdot b_i \cdot c_i) \tag{17.11}$$

so that both c_{i+1} and s_i are in SOP form. Moreover, $\overline{s}_i$ uses $\overline{c}_{i+1}$ so that we can design an AOI gate for $\overline{c}_{i+1}$ and use the output to feed another AOI gate for $\overline{s}_i$. Figure 17.7 shows the construction of the two OAOI networks.

The upper circuit produces $\overline{c}_{i+1}$ and the lower one gives s_i after inversion. It is a straightforward exercise to design both OAOI circuits using series–parallel CMOS gates. Note, however, that Eq. (17.10) contains four OR operations, which indicates that the bottom AOI gate will have 4 series-connected PFETs. This may induce an unacceptably long delay in a word adder arrangement.

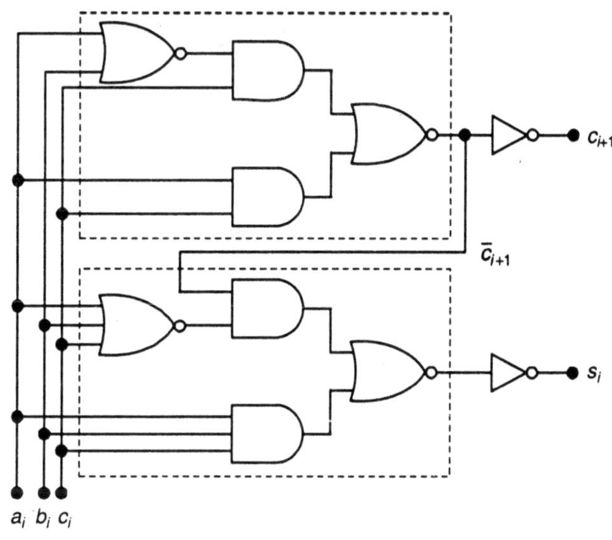

Figure 17.7 AOI full-adder logic.

To find a more efficient circuit, consider the NFET array for the carry-out circuit as implied by the AOI logic diagram. Using standard construction gives the NFET circuit in Figure 17.8(a). We see that there are two main pull-down paths corresponding to the terms

$$a_i \cdot b_i \qquad (17.12)$$

$$c_i \cdot (a_i + b_i)$$

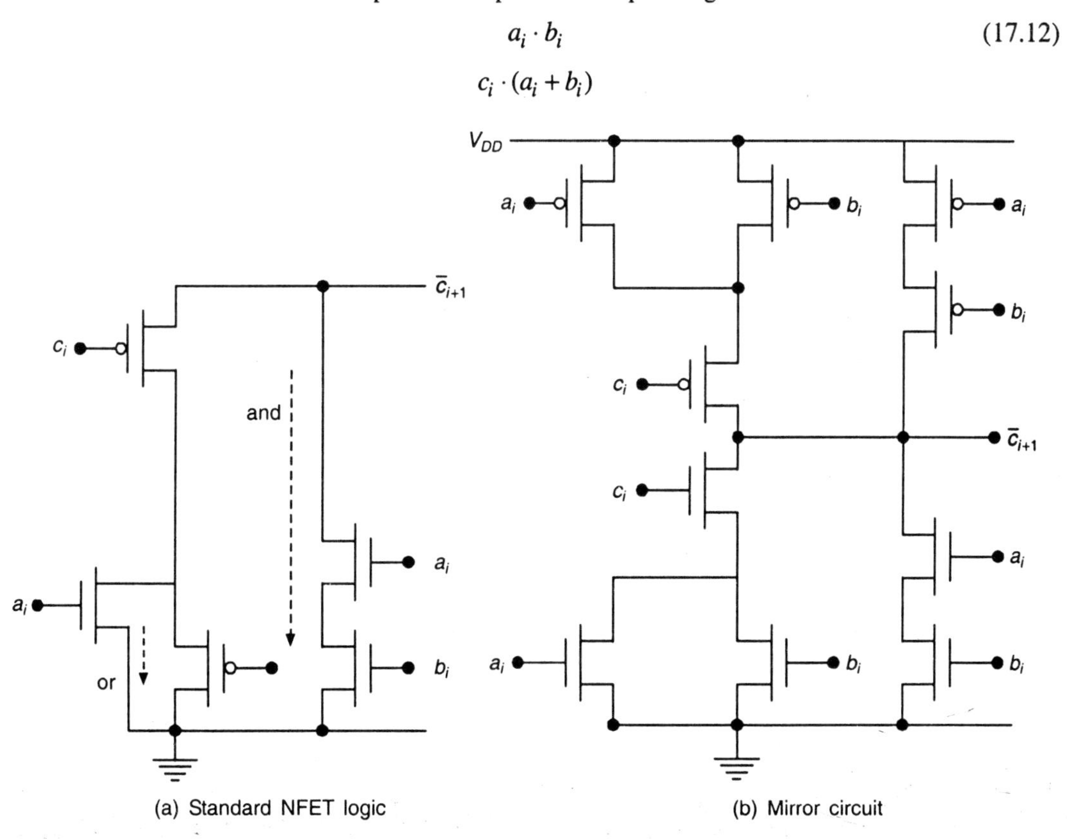

Figure 17.8 Evolution of carry-out circuit.

If either of these evaluates to 1, then the output is pulled to 0 (ground). The AND term is important when a_i and b_i are both 1; the OR term gives a pull-down if either $a_i = 1$ or $b_i = 1$ while $c_i = 1$. This leads us to construct a PFET mirror circuit to yield the total gate shown in Figure 17.8(b). The series PFETs give a pull up to 1 (V_{DD}) if $a_i = b_i = 0$, which is the opposite of the NFET pull-down condition. If only one of the outputs is 0, then the output is determined by the value of c_i.

To complete the building of a mirror CMOS full-adder, let us write the sum bit $\overline{s}_i$ as a simple OR gate such that

$$\overline{s}_i = A + B \tag{17.13}$$

where

$$A = (a_i + b_i + c_i) \cdot \overline{c}_{i+1} \tag{17.14}$$

$$B = (a_i \cdot b_i \cdot c_i)$$

This has the same characteristics as the carry-out circuit, and allows us to construct the complete full-adder shown in Figure 17.9. This is faster that a series–parallel realization, and facilitates the layout because of its mirrored FET arrays.

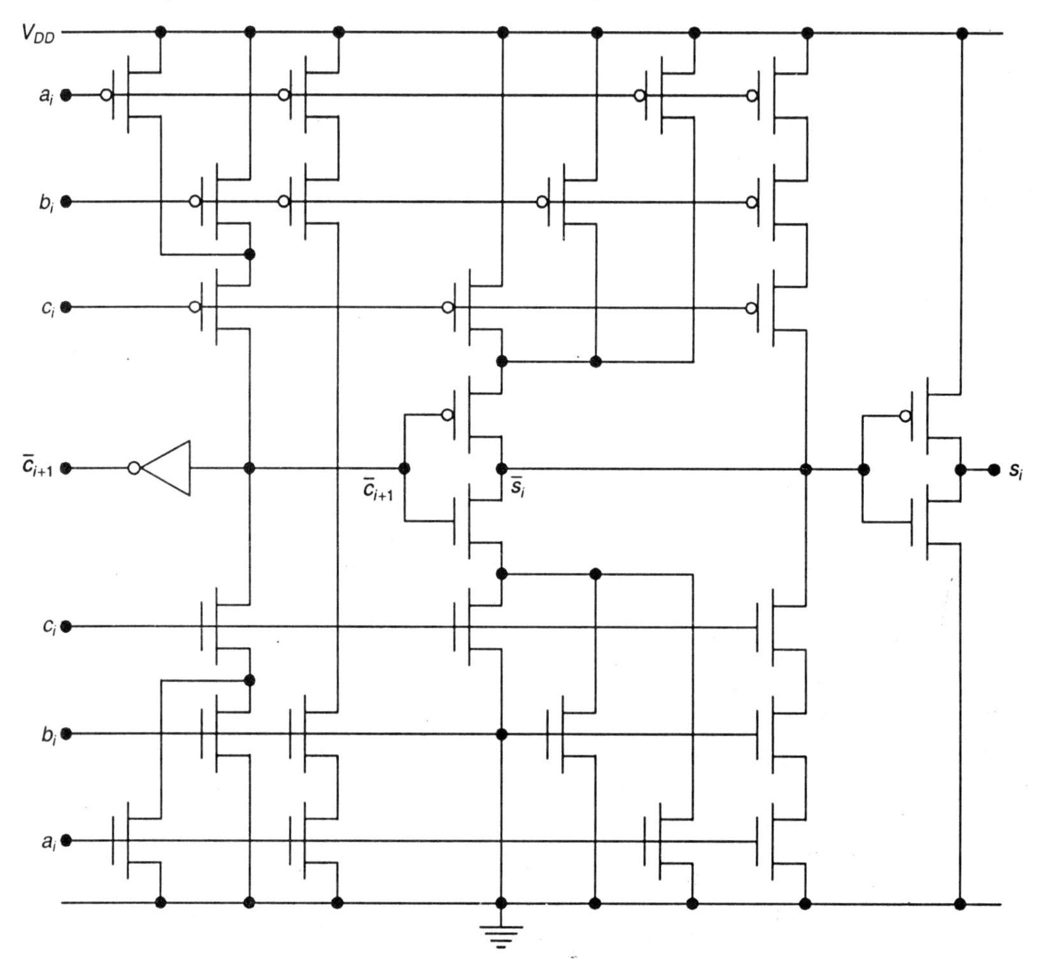

Figure 17.9 Mirror AOI CMOS full-adder.

A full-adder based on transmission gates (TGs) is shown in Figure 17.10. The input circuits provide the COR and XNOR operations which are then used by the output array of TGs to produce the sum and carry bits. This circuit has the characteristic that the delays for s_i and c_{i+1} are about the same as can be seen by tracing the logic flow path through the upper and lower sections. If the input bits are applied simultaneously, then both the sum and carry-out bits will be valid at about the same time. This is distinctly different from the AOI circuit in which the carry-out bit is produced first and then used in calculating the sum.

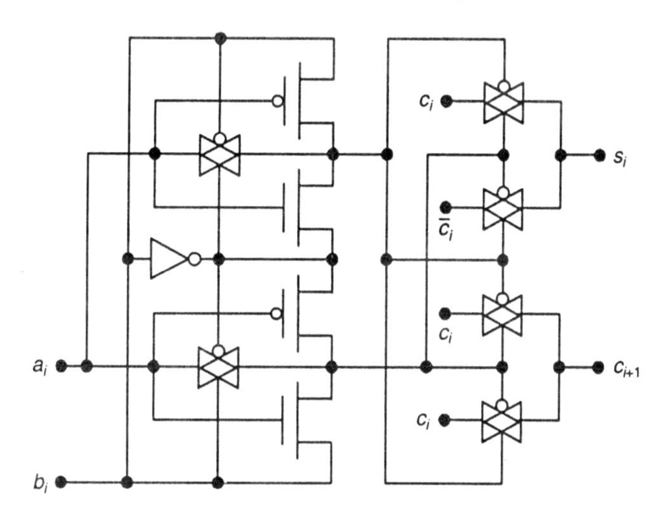

Figure 17.10 Transmission-gate full-adder circuit.

17.3 RIPPLE-CARRY ADDERS

Now that we have a basis for adding single bits, let us extend the problem to adding binary words. In general, adding two n-bit words yields an n-bit sum and a carry-out bit c_n that can either be used as the carry-in to another higher order adder, or act as an overflow flag. A general symbol is shown in Figure 17.11. We will use $n = 4$ in our initial discussions.

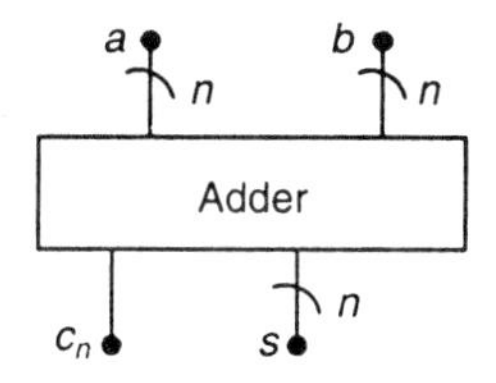

Figure 17.11 An n-bit adder.

Ripple-carry adders are based on the addition equation

$$\begin{array}{r} c_3c_2c_1c_0 \\ +\ a_3a_2a_1a_0 \\ +\ b_3b_2b_1b_0 \\ \hline c_4s_3s_2s_1s_0 \end{array} \tag{17.15}$$

where c_i represents the carry-in bit from the previous column. We will keep the 0-th carry-in bit c_0 for generality. Note that by including the carry-in word this is really adding three binary words. An n-bit ripple-carry adder requires n full-adders with the carry-out bit c_{i+1} used as in the carry-in bit to the next column. This is shown in Figure 17.12 for the case of 4-bit words.

A high-level model can be constructed using Verilog vectors as illustrated by the following

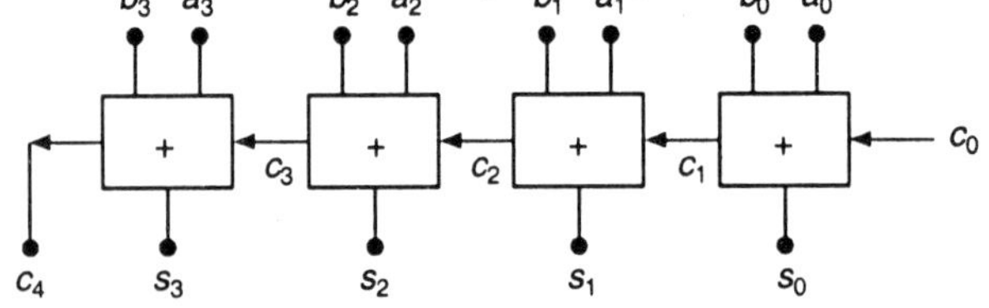

Figure 17.12 A 4-bit ripple-carry adder.

code.

```
module four_bit_adder (sum,c_4,a,b,c_0);
input [3:0]a,b;
input c_0;
output [3:0] sum;
output c_4;
assign {c_4,sum}=a+b+c_0;
endmode
```

This uses concatenation to create a 5-bit output that contains both sum and the carry-out bit c-4. Another approach to modelling this is to use four full-adder modules that are wired together as shown in the drawing:

```
module FA_modules (sum,c_4,a,b,c_0);
input [3:0]a,b;
input c_0;
output [3:0] sum;
output c_4;
wire c_1,c_2,c_3;
/* The single-bit FA modules instanced below have the
  syntax full-adder (sum,c_out,a,b,c_in)*/
full_adder fa0 (sum[0],c_1,a[0],b[0],c_0);
full_adder fa1 (sum[1], c_2,a[1],b[0],c_1);
full_adder fa2 (sum[2], c_3,a[2],b[2],c_2);
full_adder fa3 (sum[3], c_4,a[3],b[3],c_3);
endmodule
```

This example uses the notation

```
sum[i], a[i], and b[i]
```

to define the i-th bit of a vector. This is straightforward to understand. If we define a quantity such as

```
input [7:0] Q;
```

with Q = 10001110, then Q[0] = 0, Q[1] = 1, Q[2] = 1, and so on. As always, it is assumed that instanced modules such as

```
full-adder (sum,c_out,a,b,c_in);
```

are defined elsewhere in the listing. They may be written at any level from a behavioural description down to a gate-level structural listing.

The ripple-carry adder construction provides for easy connections of neighbouring circuits. It is this feature, however, that makes the design slow. Since the output of any full-adder is not valid until the incoming carry bit is valid, the left-most circuit is the last to react. The word result is not valid until this occurs.

The overall delay depends on the characteristics of the full-adder circuits. Different CMOS implementations will produce different worst-case delay paths. For our purposes, let us assume that the AOI mirror CMOS full-adder in Figure 17.9 is used in the 4-bit network. Since the carry-out is required to calculate the sum, the carry delay from c_i to c_{i+1} is minimized. Figure 17.13 shows the longest delay path for the adder where the carry bits are transferred through every stage; it is assumed that all inputs are valid at the same time. Let us start by summing the individual delays to get the total delay t_{4b} as

$$t_{4b} = t_{d3} + t_{d2} + t_{d1} + t_{d0} \tag{17.16}$$

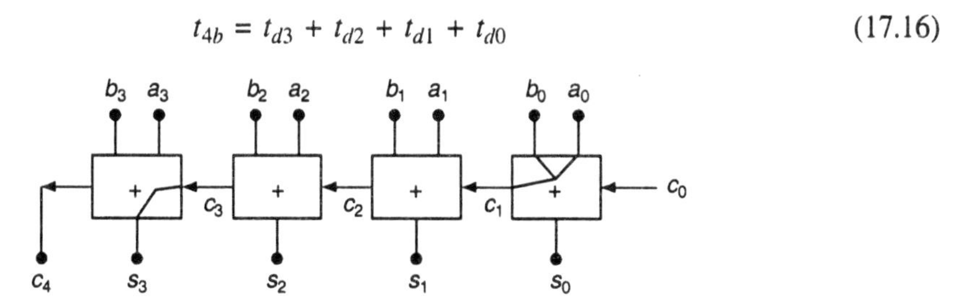

Figure 17.13 Worst-case delay through the 4-bit ripple adder.

where t_{di} is the worst-case delay through the i-th stage. The contributions for each stage can be evaluated. For the 0-th bit, $t_{d0} = t_d(a_0 \cdot b_0 \rightarrow c_1)$, which is the time for the inputs to produce the carry-out bit. The delay through sections 1 and 2 is the same, and is from the carry-in to the carry-out: $t_{d1} = t_{d2} = t_d(c_{in} \rightarrow c_{out})$. Finally, the delay in the last stage 3 in this design is the time needed to produce the output sum bit s_3, which we write as $t_{d3} = t_d(c_{in} \rightarrow s_3)$. Thus, the total delay is

$$t_{4b} = t_d(c_{in} \rightarrow s_3) + 2t_d(c_{in} \rightarrow c_{out}) + t_d\,(a_0 \cdot b_0 \rightarrow c_1) \tag{17.17}$$

If we extend this to an n-bit ripple-carry adder, then the worst-case delay is

$$t_{n\text{-bit}} = t_d(c_{in} \rightarrow s_{n-1}) + (n-2)\, t_d(c_{in} \rightarrow c_{out}) + t_d(a_0 \cdot b_0 \rightarrow c_1) \tag{17.18}$$

which shows that the delay is of order n. Symbolically, we express this as

$$\text{delay} \sim 0(\text{n}) \tag{17.19}$$

The ripple structure is therefore not a good choice for large word sizes.

Before progressing into more advanced adder designs, let us recall that a 2s complement subtractor can be built by adding XOR gates and an *add_sub* control bit as shown in

Figure 17.14. When *add_sub* = 0, the XORs pass the b_i bits and the output is the sum $(a + b)$. A control bit of *add_sub* = 1, changes the XORs into inverters, and the complemented values $\overline{b_i}$ enter the full adders; *add_sub* = 1 also acts as a carry-in of $c_0 = 1$. These operations combine to give the 2s complement algorithm for the difference $(a - b)$. This technique is also applicable in a limited manner to other adder networks.

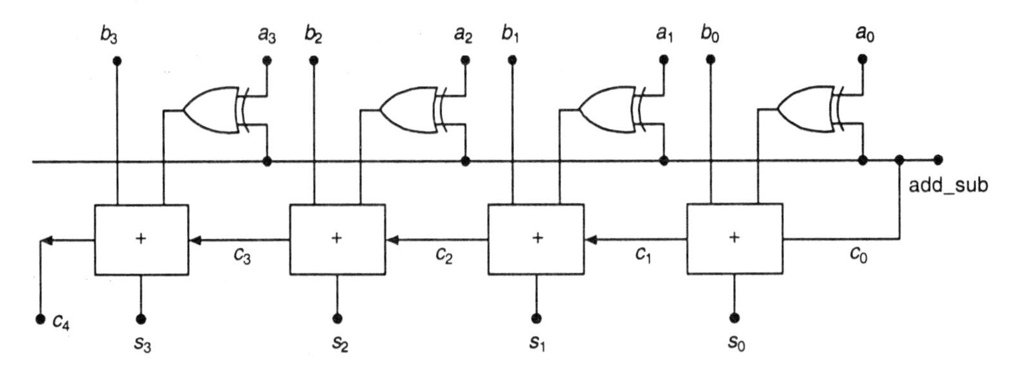

Figure 17.14 A 4-bit adder-subtractor circuit.

17.4 CARRY LOOK-AHEAD ADDERS

Carry look-ahead (CLA) adders are designed to overcome the latency introduced by the rippling effect of the carry bits. The CLA algorithm is based on the origin of the carry-out bit in the equation

$$c_{i+1} = a_i \cdot b_i + c_i \cdot (a_i \oplus b_i) \tag{17.20}$$

for the cases that give $c_{i+1} = 1$. Since either term may cause this output, we treat the two separately. First, if $a_i \cdot b_i = 1$, then $c_{i+1} = 1$. We call

$$g_i = a_i \cdot b_i \tag{17.21}$$

the **generate** term, since the inputs are viewed as "generating" the carry-out bit. Note that if $g_i = 1$, then few must have $a_i = b_i = 1$. The second term represents the case where an input carry $c_i = 1$ may be 'propagated' through the full-adder. This will happen if the propagate term

$$p_i = a_i \oplus b_i \tag{17.22}$$

is equal to 1; if $p_i = 1$ then $g_i = 0$ since the XOR operation produces a 1 iff the inputs are not equal. Figure 17.15 shows the behaviour of the generate and propagate terms. With these definitions, the equation for the carry-out bit is

$$c_{i+1} = g_i + p_i \cdot c_i \tag{17.23}$$

The main idea of the CLA is to firsts calculate the values of p_i and g_i for every bit, then use them to find the carry bits c_{i+1}. Once these are found, the sum bits are given by

$$s_i = p_i \oplus c_i \tag{17.24}$$

for every *i*. This avoids the need to ripple the carry bits serially down the chain.

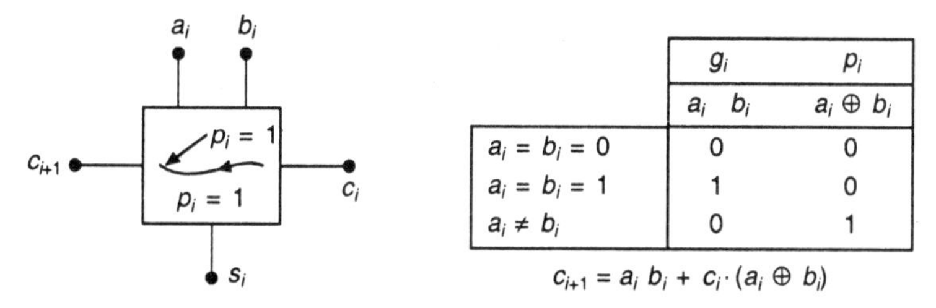

	g_i	p_i
	$a_i\ b_i$	$a_i \oplus b_i$
$a_i = b_i = 0$	0	0
$a_i = b_i = 1$	1	0
$a_i \neq b_i$	0	1

$c_{i+1} = a_i\ b_i + c_i \cdot (a_i \oplus b_i)$

Figure 17.15 Basis of the carry look-ahead algorithm.

Let us analyze the 4-bit CLA equations. With c_0 assumed known, we have

$$c_i = g_0 + p_0 \cdot c_0 \tag{17.25}$$

The expressions for c_2, c_3 and c_4 have the same form with

$$c_2 = g_1 + p_1 \cdot c_1 \tag{17.26}$$

$$c_3 = g_2 + p_2 \cdot c_2$$

$$c_4 = g_3 + p_3 \cdot c_3$$

These can be expressed using primitive generate and propagate terms by noting that c_i can be substituted into c_{i+1} in successions. The first reduction is obtained by substituting c_i into the c_2 equation to arrive at

$$c_2 = g_1 + p_1 \cdot (g_0 + p_0 \cdot c_0) \tag{17.27}$$

Expanding,

$$c_2 = g_1 + p_1 \cdot g_0 + p_1 \cdot p_0 \cdot c_0 \tag{17.28}$$

Similarly, substituting c_2 into c_3 gives

$$\begin{aligned} c_3 &= g_2 + p_2 \cdot (g_1 + p_1 \cdot g_0 + p_1 \cdot p_0 \cdot c_0) \\ &= g_2 + p_2 \cdot g_1 + p_2 \cdot p_1 \cdot g_0 + p_2 \cdot p_1 \cdot p_0 \cdot c_0 \end{aligned} \tag{17.28}$$

Finally, the carry-out bit is

$$\begin{aligned} c_4 &= g_2 + p_3 \cdot (g_2 + p_2 \cdot g_1 + p_2 \cdot p_1 \cdot g_0 + p_2 \cdot p_1 \cdot p_0 \cdot c_0) \\ &= g_3 + p_3 \cdot g_2 + p_3 \cdot p_2 \cdot g_1 + p_3 \cdot p_2 \cdot p_1 \cdot g_0 + p_3 \cdot p_2 \cdot p_1 \cdot p_0 \cdot c_0 \end{aligned} \tag{17.30}$$

These equation show that every carry bit can be found from the generate and propagate terms. Moreover, the algorithm yields nested SOP expressions. The logic diagram for the 4-bit network is shown in Figure 17.16 using the expanded expressions. Note the structure nature of the gate arrangement. Once the carry-out bits have been calculated, the sums are found using the simple XOR in Eq. (17.23). The complete adder circuit is shown in Figure 17.17 where the 'CLA Network' box represents the carry bit logic in Figure 17.16. This illustrates a marked departure from the ripple-carry design.

The high-level abstract Verilog description of a 4-bit adder can be used to describe any adder, including the CLA-based design. However, we can rewrite the behavioural code to better illustrate the internal algorithm in an explicit manner. The **assign**-based RTL module below illustrates this idea.

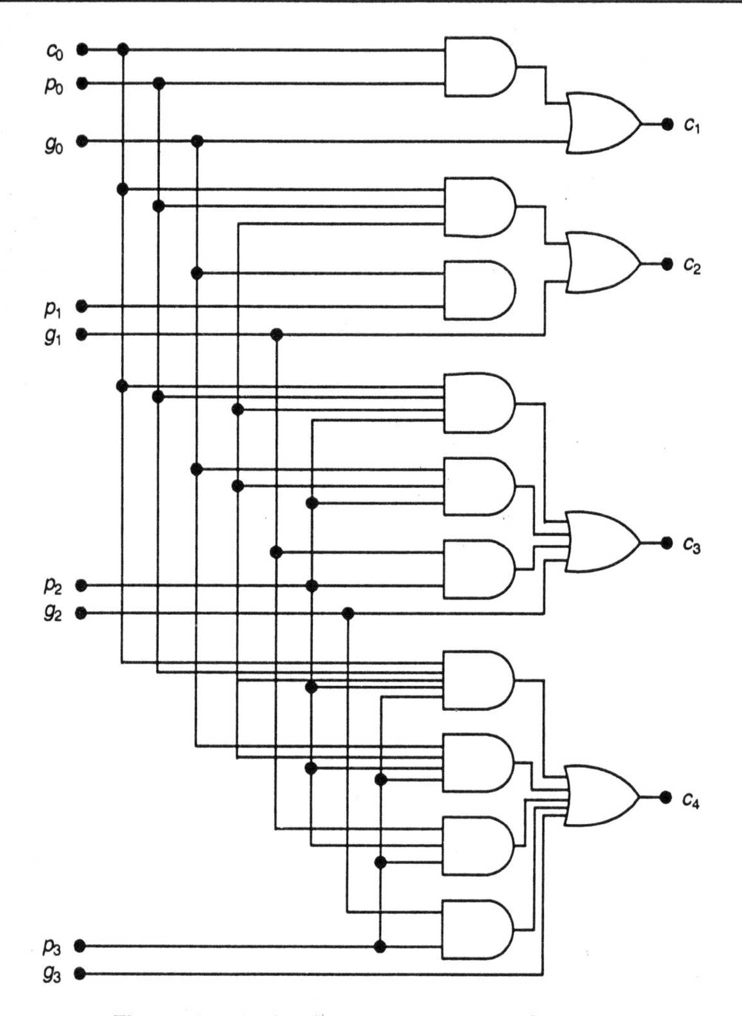

Figure 17.16 Logic network for 4-bit CLA carry bits.

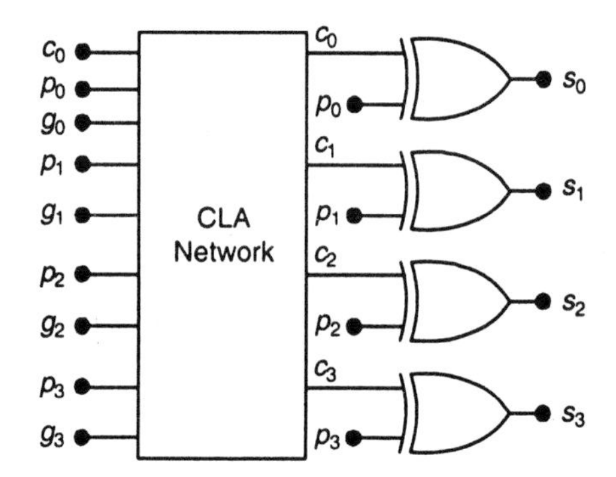

Figure 17.17 Sum calculation using the CLA network.

```
module CLA_4b (sum,c_4,a,b,c_0);
input [3:0]a,b;
input c_0;
output [3:0] sum;
output c_4;
wire p0,p1,p2,p3,g0,g1,g2,g3;
wire c1,c2,c3,cc4;
assign
  p0=a[0]^b[0],
  p1=a[1]^b[1],
  p2=a[2]^b[2],
  p3=a[3]^b[3],
  g0=a[0]&b[0],
  g1=a[1]&b[1],
  g2=a[2]&b[2],
  g3=a[3]&b[3],
assign
  c1=g0 l (p0 & c_0),
  c2=g1 l (p1 & g0) l (p1 & p0 & c_0),
  c3=g2l (p2 & g1) l (p2 & p1 & g0) l (p2 & p1 & p0 & c_0),
  c4=g3l (p3 & g2) l (p3 & p2 & g1) l (p3 & p2 & p1 & g0),
  l (p3 & p2 & p1 & p0 & c_0);
assign
  sum [0] = p0^c_0,
  sum [1] = p1^c1,
  sum [2] = p2^c2,
  sum [3] = p3^c3,
  c_4 =c4;
endmodule
```

Adding delay times on each statement, provides the final information needed for a simulation. The repetitive nature of the CLA equations can be implemented in a more efficient coding style by using the Verilog for procedure.

To translate the CLA algorithms into circuits, we use the logic construction techniques to create the NFET arrays shown in Figure 17.18. Note that teach carry-out circuit $\overline{c}_i$ forms the basis for the next higher term $\overline{c}_{i+1}$. This is due to the nesting property of the algorithm for the next higher term $\overline{c}_{i+1}$.

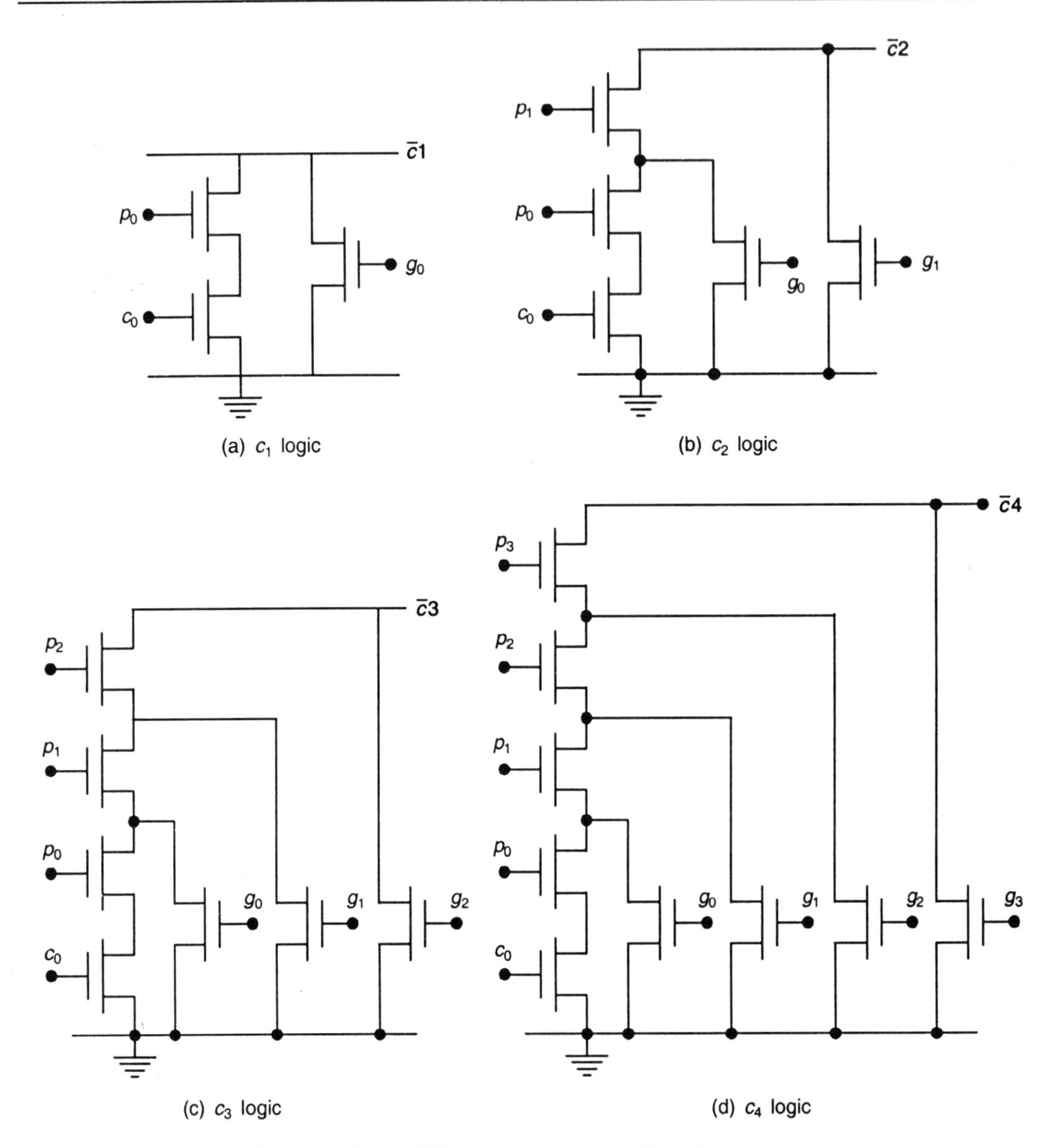

Figure 17.18 NFET logic arrays for the CLA terms.

SUMMARY

Many applications require arithmetic operations like addition and multiplication. In this chapter, we have examined binary adders (bit adders, ripple adders and carry lookahead adders) in detail. However, as we can design the multiplier with the adders, the multipliers are not discussed in this chapter.

REVIEW QUESTIONS

1. Explain the complementry pass transistor logic with a suitable example.
2. Explain mirror AOI CMOS full-adder circuit.
3. Explain ripple carry adder.
4. Derive the equation for carry look-ahead adders.

SHORT ANSWER QUESTIONS

1. What are the basic equations for half adder.
   Ans. $s = x \oplus y$; $c = x \cdot y$.
2. Write the verilog program for half adder using assign statement.
   Ans. assign {c-out, sum} = $a + b + c\_in$.

18 VHDL

18.1 INTRODUCTION TO VHDL

VHDL is an acronym for **VHSIC** Hardware Description Language (VHSIC is an acronym for Very High Speed Integrated Circuits). It is a hardware description language that can be used to model a digital system at many levels of abstraction, ranging from the algorithmic level to the gate level. The complexity of the digital system being modelled could vary from that of a simple gate to a complete digital electronic system, or anything in between. The digital system can also be described hierarchically. Timing can also be explicitly modelled in the same description.

The VHDL language can be regarded as an integrated amalgamation of the following languages:

Sequential language +
Concurrent language +
Net-list language +
Timing specifications +
Waveform generation language ⇒ VHDL

Therefore, the language has constructs that enable you to express the concurrent or sequential behaviour of a digital system with or without timing. It also allows you to model the system as an interconnection of components. Test waveforms are also generated by using the same constructs. All the above constructs, may be combined to provide a comprehensive description of the system in a single model.

The language not only defines the syntax but also defines very clear simulation semantics for each language construct. Therefore, models written in this language can be verified using a VHDL simulator. It is a strongly typed language and is often verbose to write. It inherits many of its features, especially the sequential language part, from the Ada programming language. Because VHDL provides an extensive range of modelling capabilities, it is often difficult to understand. Fortunately, it is possible to quickly assimilate a core subset of the language that is both easy and simple to understand without learning the more complex features. This subset is usually sufficient to model most applications. The complete language, however, has sufficient power to capture the descriptions of the most complex chips to a complete electronic system.

18.2 HARDWARE ABSTRACTION

VHDL is used to describe a model for a hardware device. This model specifies the external view of the device and one or more internal views. The internal view of the device specifies the functionality or structure, while the external view specifies the interface of the device through which it communicates with the other models in its environment.

The device-to-device model mapping is strictly one-to-many. That is, a hardware device model. For example, a device modelled at a high level of abstraction may not have a clock as one of its inputs, since the clock may not have been used in the description. Also, the data transfer at the interface may be treated in terms of, say, integer values. In VHDL, each device model is treated as a distinct representation of a unique device, called an entity. The entity is a hardware abstraction of the actual hardware device. Each entity is described using one model, which contains one external view and one or more internal views. At the same time, a hardware device may be represented by one or more entities.

18.2.1 Basic Terminology

VHDL is a hardware description language that can be used to model a digital system. The digital system can be as simple as a logic gate or as complex as a complete electronic system. A hardware abstraction of this digital system is called an entity. An entity X, when used in another entity Y, becomes a component for the entity Y. Therefore, a component is also an entity, depending on the level at which you are trying to model.

To describe an entity, VHDL provides five different types of primary constructs, called design units. They are:

1. Entity declaration
2. Architecture body
3. Configuration declaration
4. Package declaration
5. Package body

An entity is modelled using an entity declaration and at least one architecture body. The entity declaration describes the external view of the entity; for example, the input and output signal names. The architecture body contains the internal description of the entity; for example, as a set of interconnected components that represents the structure of the entity, or as a set of concurrent or sequential statements that represents the behaviour of the entity. Each style of representation can be specified in a different architecture body or mixed within a single architecture body.

A configuration declaration is used to create a configuration for an entity. It specifies the binding of one architecture body from the many architecture bodies that may be associated with the entity. It may also specify the bindings of components used in the selected body to other entities. An entity may have any number of different configurations.

A package declaration encapsulates a set of related declarations, such as type declarations, subtype declarations, and subprogram declarations, which can be shared across two or more design units. A package body contains the definitions of subprograms declared in package declaration.

Once an entity has been modelled, it needs to be validated by a VHDL system. A typical VHDL system consists of an analyzer and a simulator. The analyzer reads in one or more design units contained in a single file and compiles them into a design library after validating the syntax and performing some static semantic checks. The design library is a place in the host environment (that is, the environment that supports the VHDL system) where compiled design units are stored.

The simulator simulates an entity, represented by an entity-architecture pair or by a configuration, by reading in its compiled description from the design library and then performing the following steps:

1. Elaboration, 2. Initialization, 3. Simulation

A note on the language syntax: The language is case-insensitive; that is, lower-case and upper-case characters are treated alike (except in extended identifiers, string literals and character literals). For example, CARRY, carry or carry all refer to the same name. The language is also free-format, very much like in Ada and Pascal. Comments are specified in the language by preceding the text with two consecutive dashes (--) and all the text between the two dashes and the end of that line is treated as a comment.

18.2.2 Entity Declaration

The entity declaration specifies the name of the entity being modelled and lists the set of interface ports. Ports are signals through which the entity communicates with the other models and its external environment.

IN: The port IN specifies input port.

OUT: The port OUT specifies output port.

bit: `std_logic` is a predefined type of the language; it is an enumeration type containing the character literals ‘0’ and ‘1’. The port types for this entity have been specified to be of type BIT, which means that the ports can take the values ‘0’ or ‘1’.

The entity declaration does not specify anything about the internals of the entity. It only specifies the name of the entity and the interface ports.

18.2.3 Architecture Body

The internal details of an entity are specified by an architecture body using any of the following modelling styles:

1. As a set of interconnected (to represent structure)
2. As a set of concurrent assignment statements (to represent dataflow)
3. As a set of sequential assignment statements (to represent behaviour)
4. As any combination of the above three

18.2.4 Structural Style of Modelling

In the structural style of modelling, an entity is described as a set of interconnected components. The architecture body is composed of two parts:

The declarative part (before the keyword begin)

The statement part (after the keyword begin)

Two component declarations are present in the declarative part of the architecture body. These declarations specify the interface of components that are used in the architecture body. The declared components are instantiated in the statement part of the architecture body using component instantiation statements. In addition to the two component declarations, the architecture body contains a signal declaration that declares two signals.

18.2.5 Dataflow Style of Modelling

In this style of modelling, the flow of data through the entity is expressed primarily using concurrent signal assignment statements. The structure of the entity is not explicitly specified in this modelling style, but it can be implicitly deduced.

Delta delay: The delay of 0 ns is known as delta delay, and it represents an infinitesimally small delay.

18.2.6 Behavioural Style of Modelling

In contrast to the style of modelling, the behavioural style of modelling specifies the behaviour of an entity as a user of statements that are executed sequentially in the specified order. This set of sequential statements, which are specified inside a process statement, do not explicitly specify the structure of the entity but merely its functionality. A process statement is a concurrent statement that can appear within an architecture body. A process statement also has a declarative part and statement part.

Process statement: A process statement also has a declarative part and statement part.

Variable declaration: The variable declaration starts with the keyword variable. A variable is different from a signal in that it is always assigned a value instantaneously, and the assignment operator used is := compound symbol. Variables declared outside of a subprogram are called shared variables. These variables can be updated and read by more than one process.

Signal declaration: Signal declaration starts with the keyword signal. The assignment operator used is <= compound symbol. Signals cannot be declared within a process. Signal assignment statements appearing within the process are called sequential signal assignment statements.

18.2.7 Mixed Style of Modelling

It is possible to mix the three modelling styles. That is, within an architecture body, component instantiation statement (that represent structure), concurrent statements (that represent dataflow) and process statements (that represent behaviour) can be used.

Configuration declaration: A configuration declaration is used to select one of the possibly many architecture bodies that an entity may have, and to bind components, used to represent structure in that architecture body, to entities represented by an entity–architecture pair or by a configuration, which resides in a design library.

Package declaration: A package declaration is used to store a set of common declarations, such as components, types, procedures and functions. These declarations can then be imported into other design units using a use clause. The package STD_LOGIC_1164 which is an IEEE standard.

Package body: A package body is used to store the definitions of functions and procedures that were declared in the corresponding package declaration, and also the complete constant declarations for any deferred constants that appear in the package declaration. Therefore, a package body is always associated with a package declaration.

18.3 VHDL IS LIKE A PROGRAMMING LANGUAGE

The behaviour of a module may be described in programming language form. This chapter describes the facilities in VHDL which are drawn from the familiar programming language repertoire. If you are familiar with the Ada programming language, you will notice the similarity with that language. This is both a convenience and a nuisance. The convenience is that you don't have much to learn to use these VHDL facilities. The problem is that the facilities are not as comprehensive as those of Ada, though they are certainly adequate for most modelling purposes.

18.3.1 Lexical Elements

18.3.1.1 Comments

Comments in VHDL start with two adjacent hyphens ('--') and extend to the end of the line. They have no part in the meaning of a VHDL description.

18.3.1.2 Identifiers

Identifiers in VHDL are used as reserved words and as programmer defined names. They must conform to the rule:

```
identifier ::= letter {[underline] letter_or_digit}
```

Note that case of letters is not considered significant, so the identifiers cat and Cat are the same. Underlined characters in identifiers are significant, so `This_Name` and `ThisName` are different identifiers.

18.3.1.3 Numbers

Literal numbers may be expressed either in decimal or in a base between two and sixteen. If the literal includes a point, it represents a real number, otherwise it represents an integer. Decimal literals are defined by:

```
decimal_literal ::= integer [ . integer ] [ exponent ]
integer ::= digit { [ underline ] digit }
exponent ::= E [ + ] integer | E - integer
```

Examples:

```
0 1 123_456_789 987E6 -- integer literals
0.0 0.5 2.718_28 12.4E-9 -- real literals
```

18.3.1.4 Characters

Literal characters are formed by enclosing an ASCII character in single-quote marks.

Example:

```
'A', '*', '$', ' '.
```

18.3.1.5 Strings

Literal strings of characters are formed by enclosing the characters in double-quote marks. To include a double-quote mark itself in a string, a pair of double-quote marks must be put together. A string can be used as a value for an object which is an array of characters.

Examples of strings:

```
"A string"
"" empty string
```

18.3.1.6 Bit Strings (Std_logic)

VHDL provides a convenient way of specifying literal values for arrays of type bits (0s and 1s), the syntax is:

```
bit_value ::= extended_digit {[underline] extended_digit}
```

Base specifier B stands for binary, O for octal and X for hexadecimal.

Examples:

```
B"1010110" -- length is 7
O"126" -- length is 9, equivalent to B"001_010_110"
X"56" -- length is 8, equivalent to B"0101_0110"
```

18.3.2 Data Types and Objects

VHDL provides a number of basic, or *scalar* types, and a means of forming *composite* types. The scalar types include numbers, physical quantities, and enumerations (including enumerations of characters), and there are a number of standard predefined basic types. The composite types provided are arrays and records. VHDL also provides *access* types (pointers) and *files*, although these will not be fully described in this booklet. A data type can be defined by a type declaration. Examples of different kinds of type declarations are given in the following sections.

18.3.2.1 *Integer Types*

An integer type is a range of integer values within a specified range.
The syntax for specifying integer types is:

```
integer_type_definition ::= range_constraint
range_constraint ::= range range
range ::= simple_expression direction simple_expression
direction ::= to | downto
```

The expressions that specify the range must, of course, evaluate to integer numbers. Types declared with the keyword `to` are called *ascending* ranges, and those declared with the keyword `downto` are called *descending* ranges. The VHDL standard allows an implementation to restrict the range, but requires that it must at least allow the range –2147483647 to +2147483647.
Some examples of integer type declarations:

```
type byte_int is range 0 to 255;
type signed_word_int is range -32768 to 32767;
type bit_index is range 31 downto 0;
```

There is a predefined integer type called integer. The range of this type is implementation defined, though it is guaranteed to include –2147483647 to +2147483647.

18.3.2.2 *Arrays*

An array in VHDL is an indexed collection of elements all of the same type. Arrays may be one-dimensional (with one index) or multidimensional (with a number of indices). In addition, an array type may be constrained, in which the bounds for an index are established when the type is defined, or unconstrained, in which the bounds are established subsequently.
Some examples of constrained array type declarations:

```
type word is array (31 downto 0) of std_logic;
type memory is array (address) of word;
type transform is array (1 to 4, 1 to 4) of real;
type register_bank is array (byte range 0 to 132) of integer;
```

An example of an unconstrained array type declaration:

```
type vector is array (integer range <>) of real;
```

The symbol '<>' (called a box) can be thought of as a place-holder for the index range, which will be filled in later when the array type is used. For example, an object might be declared to be a vector of 20 elements by giving its type as:

vector (1 **to** 20)

There are two predefined array types, both of which are unconstrained.

They are defined as:

```
type string is array (positive range <>) of character;
type std_logic_vector is array (natural range <>) of std_logic;
```

The types positive and natural are subtypes of integer.

18.3.2.3 *Records*

VHDL provides basic facilities for records, which are collections of named elements of possibly different types. The syntax for declaring record types is:

```
record_type_definition ::=
record
element_declaration
{element_declaration}
end record
```

An example record type declaration:

```
type instruction is
record
op_code : processor_op;
address_mode : mode;
operand1, operand2: integer range 0 to 15;
end record;
```

18.3.2.4 *Object Declarations*

An object is a named item in a VHDL description which has a value of a specified type. There are three classes of objects: constants, variables and signals. Only the first two will be discussed in this section; signals will be covered in Section 3.2.1. Declaration and use of constants and variables is very much like their use in programming languages. A constant is an object which is initialised to a specified value when it is created, and which may not be subsequently modified.
The syntax of a constant declaration is:

```
constant_declaration ::=
constant identifier_list: subtype_indication [:= expression ];
```

Constant declarations with the initializing expression missing are called deferred constants, and may only appear in package declarations. The initial value must be given in the corresponding package body.

Examples:

```
constant e : real := 2.71828;
constant delay : Time := 5 ns;
constant max_size : natural;
```

18.4 VHDL PROGRAM

18.4.1 Basic Logic Gates

Every VHDL design description consists of at least one *entity/architecture* pair, or one entity with multiple architectures. The entity section of the HDL design is used to declare the *I/O ports* of the circuit, while the description code resides within architecture portion. Standardized design libraries are typically used and are included prior to the entity declaration. This is accomplished by including the code `library ieee;` and `use ieee.std_logic_1164.all;`.

OR gate

```
library ieee;
use ieee.std_logic_1164.all;

entity OR_ent is
port( x: in std_logic;
      y: in std_logic;
      F: out std_logic
);
end OR_ent;

architecture OR_arch of OR_ent is
begin

F<= x or y;
end OR_arch;
```

AND gate

```
library ieee;
use ieee.std_logic_1164.all;

entity AND_ent is
port( x: in std_logic;
      y: in std_logic;
      F: out std_logic
);
end AND_ent;

architecture behav1 of AND_ent is
begin

  begin
  F<= x and y;

end behav1;
```

XOR Gate

```
library ieee;
use ieee.std_logic_1164.all;

entity XOR_ent is
port( x: in std_logic;
      y: in std_logic;
      F: out std_logic
);
end XOR_ent;


architecture behv1 of XOR_ent is
begin

  begin
F<= x or y;
end behv1;
```

18.4.2 Combinational Logic Design

We use *port map statement* to achieve the *structural model* (components instantiations). The following example shows how to write the program to incorporate multiple components in the design of a more complex circuit. In order to simulate the design, a simple *test bench* code must be written to apply a sequence of inputs (stimulators) to the circuit under test (CUT). The output of the test bench and CUT interaction can be observed in the simulation waveform window.

18.4.2.1 *If Statement*

The `if` statement allows selection of statements to execute depending on one or more conditions.
The syntax is:

```
if_statement ::=
if condition then
sequence_of_statements
{elsif condition then
sequence_of_statements}
[else
sequence_of_statements]
end if;
```

Tristate buffer using `if` statement

```
library ieee;
use ieee.std_logic_1164.all;

entity tristate_dr is
port( d_in: in std_logic_vector(7 downto 0);
      en: in std_logic;
      d_out:  out std_logic_vector(7 downto 0)
);
end tristate_dr;

architecture behavior of tristate_dr is
begin

process(d_in, en)
begin
  if en= '1' then
    d_out <= d_in;
  else
    d_out <= "ZZZZZZZZ";
  end if;
end process;

end behavior;
```

18.4.2.2 *Signal vs. Variable*

Signals are used to connect the design components and must carry the information between current statements of the design. On the other hand, *variables* are used within process to compute certain values.

As in other programming languages, a variable and signal are given a new value using an assignment statement. The syntax is:

```
variable_assignment_statement ::= target := expression;
signal_assignment_statement ::= target := expression;
```

The following example shows their difference:

Signal vs. Variable

```
library ieee;
use ieee.std_logic_1164.all;
entity sig_var is
port( d1, d2, d3: in std_logic;
      res1, res2: out std_logic);
end sig_var;
```

```
architecture behv of sig_var is
signal sig_s1: std_logic;
begin
  proc1: process(d1,d2,d3)
  variable var_s1: std_logic;
  begin
    var_s1 := d1 and d2;
    res1 <= var_s1 xor d3;
  end process;
  proc2: process(d1,d2,d3)
  begin
    sig_s1 <= d1 and d2;
    res2 <= sig_s1 xor d3;
  end process;

end behv;
```

18.4.3 Typical Combinational Components

The following behaviour style codes demonstrate the concurrent and sequential capabilities of VHDL. The *concurrent statements* are written within the body of an architecture. They include *concurrent signal assignment, concurrent process* and *component instantiations (port map statement). Sequential statements* are written within a *process* statement, *function* or *procedure.* Sequential statements include *case statement*, *if-then-else* statement and *loop statement*.

18.4.3.1 Case Statement

The case statement allows selection of statements to execute depending on the value of a selection expression. The syntax is:

```
case_statement ::=
case expression is
case_statement_alternative
{case_statement_alternative}
end case;
case_statement_alternative ::=
when choices =>
sequence_of_statements
choices ::= choice {|choice}
choice ::=
simple_expression
| discrete_range
| element_simple_name
| others
```

The selection expression must result in either a discrete type, or a one-dimensional array of characters. The alternative whose choice list includes the value of the expression is selected and the statement list executed. Note that all the choices must be distinct, that is, no value may be duplicated. Furthermore, all values must be represented in the choice lists, or the special choice **others** must be included as the last alternative. If no choice list includes the value of the expression, the others alternative is selected.

Multiplexer using case statement

```
library ieee;
use ieee.std_logic_1164.all;

entity Mux is
port( I3: in std_logic_vector(2 downto 0);
      I2: in std_logic_vector(2 downto 0);
      I1: in std_logic_vector(2 downto 0);
      I0: in std_logic_vector(2 downto 0);
      S:  in std_logic_vector(1 downto 0);
      O:  out std_logic_vector(2 downto 0)
);
end Mux;

architecture behv1 of Mux is
begin
  process(I3,I2,I1,I0,S)
  begin

    case S is
      when "00" =>  O <= I0;
      when "01" =>  O <= I1;
      when "10" =>  O <= I2;
      when "11" =>  O <= I3;
      when others =>  O <= "ZZZ";
    end case;

  end process;
end behv1;
```

18.4.3.2 Null Statement

The null statement has no effect. It may be used to explicitly show that no action is required in certain cases. It is most often used in case statements, where all possible values of the selection expression must be listed as choices, but for some choices, no action is required.

Decoder using null statement

```
library ieee;
use ieee.std_logic_1164.all;

entity DECODER is
port( I:  in std_logic_vector(1 downto 0);
      O:  out std_logic_vector(3 downto 0)
);
end DECODER;

architecture behv of DECODER is
begin

  process (I)
  begin

    case I is
      when "00" => O <= "0001";
      when "01" => O <= "0010";
      when "10" => O <= "0100";
      when "11" => O <= "1000";
      when others => null;
    end case;

  end process;

end behv;
```

18.4.3.3 Loop Statements

VHDL has a basic loop statement, which can be augmented to form the usual while and for loops seen in other programming languages. The syntax of the loop statement is:

```
loop_statement ::=
[loop_label:]
[iteration_scheme] loop
sequence_of_statements
end loop [loop_label];
```

The `for` iteration scheme allows a specified number of iterations. The `loop` parameter specification declares an object which takes on successive values from the given range for each iteration of the loop. Within the statements enclosed in the loop, the object is treated as a constant, and so may not be assigned to. The object does not exist beyond execution of the loop statement.

An example:

```
for item in 1 to last_item loop
table(item) := 0;
end loop;
```

XOR-ing multiple bit

```
library ieee;
use ieee.std_logic_1164.all;
use ieee.std_logic_arith.all;

entity loop1 is
port(a,b:in std_logic_vector(7 downto 0);
c:out std_logic_vector(7 downto 0));
end loop1;
architecture loop1 of loop1 is
begin
process(a,b)
begin
for I in 7 downto 0 loop
c(i):=a(i) xor b(i);
end loop;
end process;
end loop1;
```

Adder

```
library ieee;
use ieee.std_logic_1164.all;
use ieee.std_logic_arith.all;
use ieee.std_logic_unsigned.all;


entity ADDER is

generic(n: natural :=2);
port( A:  in std_logic_vector(n-1 downto 0);
      B:  in std_logic_vector(n-1 downto 0);
      carry:  out std_logic;
      sum:  out std_logic_vector(n-1 downto 0)
);

end ADDER;
```

```
architecture behv of ADDER is

signal result: std_logic_vector(n downto 0);

begin

  -- the 3rd bit should be carry

  result <= ('0' & A)+('0' & B);
  sum <= result(n-1 downto 0);
  carry <= result(n);

end behv;
```

Comparator

```
library ieee;
use ieee.std_logic_1164.all;

entity Comparator is
generic(n: natural :=2);
port( A: in std_logic_vector(n-1 downto 0);
      B: in std_logic_vector(n-1 downto 0);
      less: out std_logic;
      equal: out std_logic;
      greater: out std_logic
);
end Comparator;

architecture behv of Comparator is

begin

  process(A,B)
  begin
    if (A<B) then
      less <= '1';
      equal <= '0';
      greater <= '0';
    elsif (A=B) then
      less <= '0';
      equal <= '1';
      greater <= '0';
```

```
    else
      less <= '0';
      equal <= '0';
      greater <= '1';
    end if;
  end process;

end behv;
```

ALU

```
library ieee;
use ieee.std_logic_1164.all;
use ieee.std_logic_unsigned.all;
use ieee.std_logic_arith.all;

entity ALU is

port( A:  in std_logic_vector(1 downto 0);
      B:  in std_logic_vector(1 downto 0);
      Sel:  in std_logic_vector(1 downto 0);
      Res:  out std_logic_vector(1 downto 0)
);

end ALU;

-----------------------------------------------------

architecture behv of ALU is
begin
  process(A,B,Sel)
  begin

    case Sel is
      when "00" =>
        Res <= A + B;
      when "01" =>
        Res <= A + (not B) + 1;
      when "10" =>
        Res <= A and B;
      when "11" =>
        Res <= A or B;
      when others =>
        Res <= "XX";
    end case;

  end process;
end behv;
```

18.4.4 Typical Sequential Components

18.4.4.1 Latch and Flip-Flops

Besides, from the circuit input and output signals, there are normally two other important signals, *reset* and *clock*, in the sequential circuit. The reset signal is either *active-high* or *active-low* status and the circuit status transition can occur at either clock *rising-edge* or *falling-edge*. Flip-Flop is a basic component of the sequential circuits.

D_latch

```
library ieee ;
use ieee.std_logic_1164.all;

entity D_latch is
port( data_in:  in std_logic;
      enable:   in std_logic;
      data_out: out std_logic
);
end D_latch;

architecture behv of D_latch is
begin

  process(data_in, enable)
  begin
    if (enable= '1') then
      data_out <= data_in;
    end if;
  end process;

end behv;
```

D_flip-flop

```
library ieee;
use ieee.std_logic_1164.all;
use work.all;

entity dff is
port( data_in: in std_logic;
      clock: in std_logic;
    data_out: out std_logic
);
```

```
end dff;

architecture behv of dff is
begin

  process(data_in, clock)
  begin

    -- clock rising edge

      if (clock= '1' and clock'event) then
        data_out <= data_in;
      end if;

  end process;

end behv;
```

JK flip-flop

The description of JK flip-flop is based on functional truth table.
The concurrent statement and signal assignment are used in this example.

```
--------------------------------------------------------------------
library ieee;
use ieee.std_logic_1164.all;

entity JK_FF is
port (  clock: in std_logic;
        J, K: in std_logic;
        reset: in std_logic;
        Q, Qbar: out std_logic
);
end JK_FF;

architecture behv of JK_FF is

    signal state: std_logic;
    signal input: std_logic_vector(1 downto 0);

begin

  input <= J & K;

  p: process(clock, reset) is
```

```
  begin

    if (reset= '1') then
      state <= '0';
    elsif (rising_edge(clock)) then

    -- compare to the truth table
      case (input) is
        when "11" =>
          state <= not state;
        when "10" =>
          state <= '1';
        when "01" =>
          state <= '0';
        when others =>
          null;
        end case;
      end if;

  end process;

  -- concurrent statements
  Q <= state;
  Qbar <= not state;

end behv;
```

Register

```
library ieee ;
use ieee.std_logic_1164.all;
use ieee.std_logic_unsigned.all;

entity reg is

generic(n: natural :=2);
port( I: in std_logic_vector(n-1 downto 0);
      clock: in std_logic;
      load: in std_logic;
      clear: in std_logic;
      Q: out std_logic_vector(n-1 downto 0)
);
end reg;
```

```
architecture behv of reg is

  signal Q_tmp: std_logic_vector(n-1 downto 0);

begin

  process(I, clock, load, clear)
  begin

    if clear = '0' then
    Q_tmp <= (Q_tmp'range => '0');
    elsif (clock= '1' and clock'event) then
      if load = '1' then
        Q_tmp <= I;
      end if;
    end if;

  end process;

  -- concurrent statement
  Q <= Q_tmp;

end behv;
```

Shift Register

```
library ieee ;
use ieee.std_logic_1164.all;

entity shift_reg is
port( I: in std_logic;
      clock: in std_logic;
      shift: in std_logic;
      Q: out std_logic
);
end shift_reg;

architecture behv of shift_reg is

  signal S: std_logic_vector(2 downto 0):="111";

begin

  process(I, clock, shift, S)
```

```
  begin

    -- everything happens upon the clock changing
    if clock'event and clock= '1' then
      if shift = '1' then
        S <= I & S(2 downto 1);
      end if;
    end if;

  end process;

  Q <= S(0);

end behv;
```

N_bit counter

```
library ieee;
use ieee.std_logic_1164.all;
use ieee.std_logic_unsigned.all;

entity counter is

generic(n: natural :=2);
port( clock: in std_logic;
      clear: in std_logic;
      count: in std_logic;
      Q: out std_logic_vector(n-1 downto 0)
);
end counter;


architecture behv of counter is

  signal Pre_Q: std_logic_vector(n-1 downto 0);

begin

  process(clock, count, clear)
  begin
    if clear = '1' then
      Pre_Q <= Pre_Q - Pre_Q;
    elsif (clock= '1' and clock'event) then
```

```
      if count = '1' then
        Pre_Q <= Pre_Q + 1;
      end if;
    end if;
  end process;

  -- concurrent assignment statement
  Q <= Pre_Q;

end behv;
```

18.5 FINITE STATE MACHINE

The most important description model presented here may be the Finite State Machine (FSM). A general model of an FSM consists of both the combinational logic and sequential components such as state registers, which record the states of circuit and are updated synchronously on the rising-edge of the clock signal. The output function computes the various outputs according to different states. Another type of sequential model is the memory module, which usually takes a long time to be synthesized due to the number of design cells.

```
State_reg and comb_logic
```

We use case statement to describe the state transition. All the inputs and signals are put into the process sensitive list.

```
library ieee ;
use ieee.std_logic_1164.all;

entity seq_design is
port( a: in std_logic;
      clock: in std_logic;
      reset: in std_logic;
      x: out std_logic
);
end seq_design;

architecture FSM of seq_design is

  type state_type is (S0, S1, S2, S3);
  signal next_state, current_state: state_type;

begin

  -- concurrent process#1: state registers
  state_reg: process(clock, reset)
```

```
    begin
      if (reset= '1') then
        current_state <= S0;
      elsif (clock'event and clock= '1') then
        current_state <= next_state;
      end if;
    end process;
    comb_logic: process(current_state, a)
    begin
        case current_state is
      when S0 =>  x <= '0';
        if a= '0' then
          next_state <= S0;
        elsif a = '1' then
          next_state <= S1;
        end if;
      when S1 =>  x <= '0';
        if a= '0' then
          next_state <= S1;
        elsif a= '1' then
          next_state <= S2;
        end if;
      when S2 =>  x <= '0';
        if a= '0' then
          next_state <= S2;
        elsif a= '1' then
          next_state <= S3;
        end if;
      when S3 =>  x <= '1';
        if a= '0' then
          next_state <= S3;
        elsif a= '1' then
          next_state <= S0;
        end if;
      when others =>
        x <= '0';
        next_state <= S0;
      end case;
    end process;
  end FSM;
```

18.6 MEMORIES

A simple 4*4 RAM module

```
library ieee;
use ieee.std_logic_1164.all;
use ieee.std_logic_arith.all;
use ieee.std_logic_unsigned.all;

entity SRAM is
generic( width: integer:=4;
         depth: integer:=4;
         addr: integer:=2);
port( Clock: in std_logic;
      Enable: in std_logic;
      Read: in std_logic;
      Write: in std_logic;
      Read_Addr: in std_logic_vector(addr-1 downto 0);
      Write_Addr: in std_logic_vector(addr-1 downto 0);
      Data_in: in std_logic_vector(width-1 downto 0);
      Data_out: out std_logic_vector(width-1 downto 0)
);
end SRAM;

architecture behav of SRAM is

-- use array to define the bunch of internal temporary signals

type ram_type is array (0 to depth-1) of
  std_logic_vector(width-1 downto 0);
signal tmp_ram: ram_type;

begin

  -- Read Functional Section
process(Clock, Read)
begin
  if (Clock'event and Clock= '1') then
    if Enable= '1' then
      if Read= '1' then

        Data_out <= tmp_ram(conv_integer(Read_Addr));
      else
```

```
          Data_out <= (Data_out'range => 'Z');
        end if;
      end if;
    end if;
  end process;



    --Write Functional Section

  process(Clock, Write)
    begin
      if (Clock'event and Clock= '1') then
        if Enable= '1' then
          if Write= '1' then
            tmp_ram(conv_integer(Write_Addr)) <= Data_in;
          end if;
        end if;
      end if;
    end process;
  end behav;
```

18.6.1 wait Statement

The primary unit of behavioural description in VHDL is the *process*. A process is a sequential body of code which can be activated in response to changes in state. A process is activated initially during the initialisation phase of simulation. It executes all of the sequential statements, and then repeats, starting again with the first statement. A process may suspend itself by executing a `wait` statement. This is of the form:

```
wait_statement ::=
wait [sensitivity_clause] [condition_clause] [timeout_clause];
sensitivity_clause ::= on sensitivity_list
sensitivity_list ::= signal_name {, signal_name }
condition_clause ::= until condition
timeout_clause ::= for time_expression
```

The sensitivity list of the `wait` statement specifies a set of signals to which the process is sensitive while it is suspended. When an event occurs on any of these signals (that is, the value of the signal changes), the process resumes and evaluates the `condition`. If it is true or if the condition is omitted, execution proceeds with the next statement, otherwise the process suspends. If the sensitivity clause is omitted, then the process is sensitive to all of the signals mentioned in the `condition` expression. The `timeout` expression must evaluate to a positive duration, and indicates the maximum time for which the process will wait. If it is omitted, the process may wait indefinitely. If a sensitivity list is included in the header of a `process` statement, then the

process is assumed to have an implicit wait statement at the end of its statement part. The sensitivity list of this implicit wait statement is the same as that in the process header. In this case, the process may not contain any explicit wait statements.
An example of a process statement with a sensitivity list:

```
process (reset, clock)
variable state : bit := false;
begin
if reset then
state := false;
elsif clock = true then
state := not state;
end if;
q <= state after prop_delay;
-- implicit wait on reset, clock
end process;
```

During the initialization phase of simulation, the process is activated and assigns the initial value of state to the signal *q*. It then suspends at the implicit wait statement indicated in the comment. When either reset or clock change value, the process is resumed, and execution repeats from the beginning. The next example describes the behaviour of a synchronization device called a Muller-C element used to construct asynchronous logic. The output of the device starts at the value '0', and stays at this value until both the inputs are '1', at which time the output changes to '1'. The output then stays '1' until both the inputs are '0', at which time the output changes back to '0'.

```
muller_c_2 : process
begin
wait until a = '1' and b = '1';
q <= '1';
wait until a = '0' and b = '0';
q <= '0';
end process muller_c_2;
```

This process does not include a sensitivity list, so explicit wait statements are used to control the suspension and activation of the process. In both wait statements, the sensitivity list is the set of signals a and b, determined from the condition expression.

---

wait on

---

wait on statement is always followed by the input signal.

```
library ieee;
use ieee.std_logic_1164.all;
entity w is
  port(reset,clk,d:in std_logic;
```

```
    q:out std_logic);
  end w;
  architecture w of w is
  begin
    process
    begin
      if reset= '1' then
        q<='0';
        elsif clk'event and clk= '1' then
          q<=d;
        end if;
        wait on reset,clk;
      end process;
    end w;
```

This process statement contains a `wait on` statement that causes the process to halt the execution until an event occurs on either `reset` or `clk`.

wait until

`wait until` statement is always followed by a condition clause or a Boolean value. This `wait until` statement suspends the process until the condition is met or the Boolean expression returns a true value.

```
  library ieee;
  use ieee.std_logic_1164.all;
  entity w is
    port(reset,clk,d:in std_logic;
    q:out std_logic);
  end w;
  architecture w of w is
  begin
    process
    begin
  wait until clk= '1' and clk'event;
  if reset= '1' then
    q<='0;;
  else
    q<=d;
  end process;
  end w;
```

wait for

`wait for` statement is always followed by a time clause

```
library ieee;
use ieee.std_logic_1164.all;
entity arr is
  port(a,b:in std_logic;
  c:out std_logic);
  end arr;
architecture arr of arr is
begin
process
begin
wait for 25 ns;
c<=a and b;
end arr;
```

This program suspends the execution of and operation for 25 ns, after the time specified in the time expression has elapsed execution continues on the statement following the `wait` statement.

18.6.2 Concurrent Signal Assignment Statements

Often a process describing a driver for a signal contains only one signal assignment statement. VHDL provides a convenient short-hand notation, called a `concurrent_signal_assignment_statement`, for expressing such processes. The syntax is:

```
concurrent_signal_assignment_statement ::=
[label:] conditional_signal_assignment
|[label:] selected_signal_assignment
For each kind of concurrent signal assignment, there is a
corresponding process statement with the same meaning.
```

18.6.2.1 Conditional Signal Assignment

A `conditional_signal_assignment_statement` is a shorthand for a process containing signal assignments in an `if` statement.
The syntax is:

```
conditional_signal_assignment ::= target <= options
conditional_waveforms;
options ::= [guarded] [transport]
conditional_waveforms ::=
{waveform when condition else} waveform
```

Use of the word **guarded** is not covered in this booklet. If the word **transport** is included, then the signal assignments in the equivalent process use transport delay.
Suppose we have a `conditional_signal_assignment`:

```
s <= waveform_1 when condition_1 else
waveform_2 when condition_2 else
...
waveform_n;
```

2x1 mux

```
library ieee;
use ieee.std_logic_1164.all;
entity arr is
port(a,b,sel:in std_logic;
c:out std_logic);
end arr;
architecture nn of arr is
begin
c<=a when sel='0' else
  b;
end nn;
```

Equivalent process

```
process
if condition_1 then
s <= waveform_1;
elsif condition_2 then
s <= waveform_2;
elsif ...
else
s <= waveform_n;
wait [ sensitivity_clause ];
end process;
```

If none of the waveform value expressions or conditions contains a reference to a signal, then the `wait` statement at the end of the equivalent process has no sensitivity clause. This means that after the assignment is made, the process suspends indefinitely. For example, the conditional assignment:

```
reset <= '1', '0' after 10 ns when short_pulse_required else
'1', '0' after 50 ns;
```

schedules two transactions on the signal `reset`, then suspends for the rest of the simulation. On the other hand, if there are references to signals in the waveform value expressions or conditions, then the `wait` statement has a sensitivity list consisting of all of the signals referenced. So the conditional assignment:

```
mux_out <= 'Z' after Tpd when en = '0' else
in_0 after Tpd when sel = '0' else
in_1 after Tpd;
```

is sensitive to the signals en and sel. The process is activated during the initialization phase, and thereafter whenever either of en or sel changes value. The degenerate case of a conditional signal assignment, containing no conditional parts, is equivalent to a process containing just a signal assignment statement. So:

```
s <= waveform;
is equivalent to:
process
s <= waveform;
wait [sensitivity_clause];
end process;
```

18.6.2.2 selected_signal_assignment

A selected_signal_assignment statement is a shorthand for a process containing signal assignments in a case statement.
The syntax is:

```
selected_signal_assignment ::=
with expression select
target <= options selected_waveforms ;
selected_waveforms ::=
{waveform when choices,}
waveform when choices
choices ::= choice {|choice}
```

The options part is the same as for a conditional_signal_assignment. So if the word **transport** is included, then the signal assignments in the equivalent process use transport delay. Suppose we have a selected signal assignment:

```
with expression select
s <= waveform_1 when choice_list_1,
waveform_2 when choice_list_2,
...
waveform_n when choice_list_n;
Then the equivalent process is:
process
case expression is
when choice_list_1=>
s <= waveform_1;
when choice_list_2=>
s <= waveform_2;
...
```

```
when choice_list_n=>
s <= waveform_n;
end case;
wait [sensitivity_clause];
end process;
```

The sensitivity list for the `wait` statement is determined in the same way as for a `conditional_signal_assignment`. That is, if no signals are referenced in the selected signal assignment expression or waveforms, the `wait` statement has no sensitivity clause. Otherwise the sensitivity clause contains all the signals referenced in the expression and waveforms. An example of a selected signal assignment statement:

```
with alu_function select
alu_result <= op1 + op2 when alu_add | alu_incr,
op1 - op2 when alu_subtract,
op1 and op2 when alu_and,
op1 or op2 when alu_or,
op1 and not op2 when alu_mask;
```

In this example, the value of the signal `alu_function` is used to select which signal assignment to `alu_result` to execute. The statement is sensitive to the signals `alu_function`, `op1` and `op2`, so whenever any of these change value, the selected signal assignment is resumed.

2x1 mux

```
library ieee;
use ieee.std_logic_1164.all;
entity arr is
port(a,b,sel:in std_logic;
c:out std_logic);
end arr;
architecture nn of arr is
begin
with sel select
c<=a when '0',
  b when others;
end nn;
```

18.7 SUBPROGRAMS

Subprograms consist of `function` and `procedure` used to perform common operations. `function` can return just one value whereas procedure can return more than one argument. In `function` all parameters are input values, whereas `procedure` can contain input, inout, and output parameters. All statements inside the subprogram are sequential.

18.7.1 Function

Syntax:

```
function function_name (parameters) return type;
function function_name (parameters) return type is
  begin
    sequential statements
    end function function_name;
```

Find the maximum of two numbers

```
package att is
  function max(a,b:integer)
return integer;
end;
package body att is
function max(a,b:integer)
return integer is
begin
if a> b then
return a;
else
return b;
end if;
end max;
end att;

  ---function calling program

use work.att.all;
library ieee;
use ieee.std_logic_1164.all;
entity arr is
port(a,b:in integer;
c:out integer);
end arr;
architecture arr of arr is
begin
  c<=max(a,b);
end arr;

---------------------------------------------------------------
```

18.7.2 Procedure

syntax:

```
procedure procedure_name (formal_parameter_list)
procedure procedure_name (formal_parameter_list) is
  procedure_declarations
  begin
    sequential statements
    end procedure procedure_name;
```

2x1 mux

```
package pro is
procedure mux (signal a,b,sel:in std_logic;
signal c:out std_logic);
end;
package body pro is
procedure mux (signal a,b,sel:in std_logic;
signal c:out std_logic) is
begin
if sel='0' then
c<=a;
else
c<=b;
end if;
return;
end mux;
end pro;

--procedure calling program

library ieee;
use ieee.std_logic_1164.all;
entity arr is
port(a,b,sel:in std_logic;
c:out std_logic);
end arr;
architecture be of arr is
begin
mux(a,b,sel,c);
end be;
```

18.8 STRUCTURAL MODELLING

An entity is modelled as a set of components connected by signals as a netlist. The behaviour of the entity is not explicitly apparent from its model. The component instantiation statement is a primary mechanism used for describing such a model of an entity.

This structural model design can be implemented in the following method.

1. Design of individual component
   Program of individual component
2. Entity declaration
   The entity declaration specifies the name of the entity being modelled and lists the set of interface ports. Ports are signals through which the entity communicates with the other models in its external environment.
3. Component Declaration
   Syntax:
   Component *component_name* is
   [port (list of interface port);]
   end component [component name]
4. Component Instantiation

A component instantiation statement defines a sub-component of the entity in which it appears. It associates the signals in the entity with the ports of that sub-components. A format of that Component Instantiation statement is

Component label: component name [port map (association list)]

Example

1. Design of individual component

```
Library ieee;
Use ieee.std_logic_1164.all;
Entity and1 is
Port( a,b:in std_logic;
C:out std_logic);
End and1;
Architecture and1 of and1 is
Begin
C<= a and b;
End and1;
```

```
Library ieee;
Use ieee.std_logic_1164.all;
Entity not1 is
Port( c:in std_logic;
d:out std_logic);
End not1;
```

```
Architecture not1 of not1 is
Begin
D<=not c;
End not1;
```

2. Entity Declaration

```
Library ieee;
Use ieee.std_logic_1164.all;
Entity overall is
Port( a,b:in std_logic;
d:out std_logic);
End overall;
```

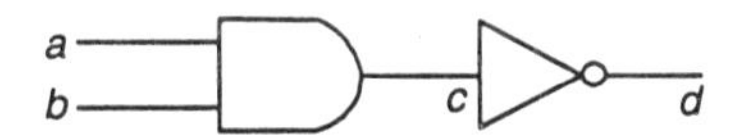

3. Component Declaration

```
Component and1 is
Port( a,b:in std_logic;
C:out std_logic);
End component;
Component not1 is
Port( c:in std_logic;
d:out std_logic);
End not1;
```

4. Component Instantiation

```
u1: and1 port map (a,b,c);
u2: not1 port map (c,d);
end overall;
```

18.9 TEST BENCHES

The test bench is used to verify the functionality of the design. The test bench allows the design to verify the functionality of the design at each step in the HDL synthesis-based methodology. When the designer makes a small change to fix an error, the change can be tested to make sure that it didn't affect the other part of the design.

As mentioned in the block diagram, a test bench is at the highest level in the hierarchy of the design. The test bench instantiates the design under test (DUT). The test bench provides the necessary input stimulus to the DUT and examines the output from the DUT.

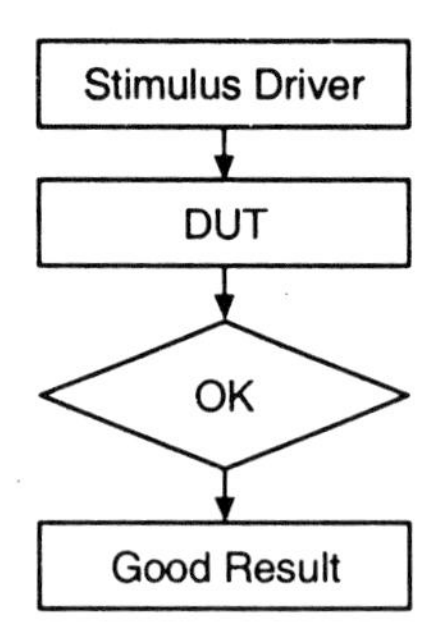

Figure 18.1 Test bench block diagram.

Test bench for counter

```
Library ieee;
Use ieee.std_logic_1164.all;
use ieee.std_logic_unsigned.all;
use ieee.std_logic_arith.all;
Entity counter is
Port( reset,clk:in std_logic;
count:out integer range 0 to 10);
End counter;
Architecture counter of counter is
Begin
process(reset,clk)
variable counting:integer range 0 to 10;
begin
if reset='1' then
counting:=0;
elsif
clk'event and clk='1' then
if counting=10 then counting:=0;
else
counting:=counting+1;
end if;
end if;
count<=counting;
end process;
end counter;
```

Test bench creation

```
Library ieee;
Use ieee.std_logic_1164.all;
Entity testbench is
End testbench;
Architecture testbench of testbench is
```

```
Signal reset,clk:std_logic;
Signal count:std_logic_vector(0 to 3);
COMPONENT counter
  PORT(
      reset: IN std_logic;
      clk: IN std_logic;
      count: OUT std_logic_vector(0 to 3)
      );
  END COMPONENT;

  Inst_counter: counter PORT MAP(
    reset => reset,
    clk => clk,
    count => count
  );
process
begin
wait for 50 ns;
reset<='1',' 0 after 50ns;
clk<='1',' 0 after 50ns;
end process;
end;
```

18.10 VHDL vs. VERILOG

There are now two industry standard hardware description languages, VHDL and Verilog. It is important that a designer knows both of them although we are using only VHDL in class. Verilog is easier to understand and use. For several years, it has been the language of choice for industrial applications that required both simulation and synthesis. It lacks, however, constructs needed for system level specifications. VHDL is more complex, thus difficult to learn and use. However, it offers a lot more flexibility of the coding styles and is suitable for handling very complex designs.

SUMMARY

In this chapter:

We have had a basic introduction to VHDL and how it can be used to model the behaviour of devices and designs.

We also discussed how `types` can be used by three different types of objects: the signal variable and constant.

How signals are the main mechanism for the connection of entities and how signals are used to pass information between entities.

How sub-program consists of function and procedure—functions have only one parameters and a single return value; procedure can have any number of in out, inout parameters.

How `component` can be used to specify which entity to use for each component.

REVIEW QUESTIONS

1. Explain the difference between the signal and variable, and give an example for each statement.
2. Write the syntax for case statement and write the coding for 4x1 multiplexer using case and null statements.
3. Give an example for asynchronous and synchronous sequential statements.
4. Write the importance of FSM and coding for traffic light controller.
5. Explain the different between Verilog and VHDL.

SHORT ANSWER QUESTIONS

1. Expand the acronym VHDL?

   Ans. VHDL is an acronym for VHSIC Hardware Description Language (VHSIC is an acronym for Very High Speed Integrated Circuits).

2. What are the different types of modelling VHDL?

   Ans. (a) Structural modelling, (b) Data flow modelling, (c) Behavioural modelling, and (d) Mixed type modelling.

3. Give the syntax of an unconstrained array type declaration:

   Ans. Type vector is array (integer range <>) of real.

4. Explain the behaviour style codes to demonstrate the concurrent capabilities of VHDL.

   Ans. The concurrent statements are written within the body of architecture.

5. For what purpose null statement is used?

   Ans. Null statement may be used to explicitly show that no action is required in certain cases.

6. What are the three (suspension) wait statements?

   Ans. (a) wait on, (b) wait until, and (c) wait for

7. What are concurrent_signal_assignment statements?

   Ans. Conditional_signal_assignment and selected_signal_assignment

8. Two subprograms are...

   Ans. (a) Function, and (b) procedure

9. Test bench is used for what purpose?

   Ans. The test bench is used to verify the functionality of the design.

10. Which language is suitable for handling very complex designs?

    Ans. VHDL

Index

2-/4- input NAND gate, 161–162
3-input tally circuit, 118
3–2 compressor, 223

ACTEL, 283
  logic cell, 281
  I/O pad, 282
Adders, 200
  AOI mirror CMOS full-adder, 407
  binary adder, 200
  Brent–Kung adder, 217
  carry lookahead adder, 213
  carry select adder, 208–210
  carry-bypass adder, 207
  equal-length adder, 210
  four-bit adder, 207
  full-adder, 202
  Kogge–Stone logarithmic adder, 215–216
  linear adder, 215
  logarithmic lookahead adder, 214
  mirror adder, 204
  ripple carry adder, 201–202
  square-root carry select adder, 210
  state CMOS adder, 202
  transmission gate based adder, 205
Algotronix FPGA cell/chip, 253–254
Altera Max 7000, 241
ALU/Arithmetic Building Blocks, 198
AND-OR-INVERT, 116, 120
AOI
  logic, 119
  mirror CMOS full-adder, 407
Area pick-up point, 292
Array, 420
  array organization, 221
  Ball Grid Array (BGM), 243
  base array, 267
  channelled/channelless gate array, 267
  OR arrays, 233
  Pin Grid Array (PGA), 243
  regular logic arrays, 343
  reprogrammable Gate arrays, 262
  Sea-Of-Gates (SOG) array, 268
  Weinberger array, 232
  XILINX programmable array, 283
Arrival time profile, 224
ASICs/Application Specific Integrated Circuits, 262
  ASIC design flow, 256
  cell-based ASIC (CBIC), 264
  Gate Array Based ASICs, 262
  programmable ASICs, 262
  semicustom ASICs, 262
  standard cell based, 262
Automatic Test Pattern Generation (ATPG), 322
Average Failure Rate (AFR), 313

Barrel shifter, 127–128
Base cell/primitive cell, 267
Bathtub curve, 311
BiCMOS, 137
  2-input NAND gate, 145
  2-input NOR gate, 146
  fabrication/technology, 32, 91
  inverters/gates, 144
Bit-of-nails inverter, 344
Body effect, 37
Bonding pads, 137
Booth's recording, 219
  modified Booth's recording, 219–220
Bottom-up/top-down approach, 289
Break even graph/volume, 273
BSIM model, 61

Built-In Logic Block Observation (BILBO), 341
Bulk threshold parameter, 57
Buried contact, 99
Buried subcollector, 32
Bus-oriented test techniques, 331
Butting contacts/layers, 99–100

C^2MOS (Clocked CMOS), 189
CAD (Computer Aided Design), 111
Canonical product-of-sums form (OR–AND), 119
Canonical sum-of-products form (AND–OR), 119
Capacitance
 bit capacitance, 26
 channel capacitance, 44
 coupling capacitance, 302
 density, 58–59
 distributed capacitance, 138
 gate/channel capacitance, 43
 of bulk-drain/bulk-source/bulk-channel junctions, 59
Capacitive coupling, 170
Capacitive loads, 147
Capture-IR mode, 347
Carry-generation circuitry, 204
Carry-inverting gate, 204
Carry-lookahead equation, 213
Carry-propagate, 222
Carry-propagation, 214
Carry-save structure, 222–223
Cascaded
 complementary inverters, 85
 dynamic gates/inverters, 171
Channel
 length modulation parameter, 42, 57
 length modulation, 41–42
 length/width, 57, 68
Channel stop-implant, 35
Channelled/channelless gate arrays, 267
Charge leakage, 168
Charge sharing, 169
Charge-steering (pass-transistor) logic, 118
Chemical Vapour Deposition (CVD), 13
 Atmospheric CVD (ATCVD), 23, 102
Circuit extraction, 111
Circuit level design extraction, 259
Circuit simulator, 60
Clock overlap/skew, 186
 distribution network, 190
 feedthrough, 170
CMOS (Complementary MOS)
 ASICs, 263–265
 DC characteristics of CMOS inverter, 65
 design rules, 102
 Ex-OR, 122
 function blocks, 155
 implementation of Schimtt trigger, 194
 inverter state characteristics, 71
 inverter, 63, 75–79, 106
 latch-up condition, 65
 multiplexers, 125–127
 NAND gate, 107
 NOR gate, 108
 PLA design, 230
 process/fabrication sequence, 110, 194
 p-well CMOS inverter, 16
 super buffer, 143
 switching characteristics of CMOS inverter, 65, 110
 transmission gate, 131
Colour coding, 91
Combinational 0 or 1 controllability, 325
Combinational logic circuit, 158
Combinational measures/observability, 325
Complete Feedback Shift Register (CFSR), 340
Computative Distribution Function (CDF), 312
Configurable Logic Blocks (CLB), 250
Contamination delay, 180
Controllability, 330
CPL/Complementary Pass-transistor Logic, 165–166
CPLDs/Complex Programmable Devices, 226, 237–240
Critical path, 210–211
Cross-coupled latch, 339
Current starved inverter, 195–196

D-algorithm (DALG), 324
Datapath logic/library, 267
Datapath, 199
Defect density, 276
Delay element, 195
Delay time, 80
Delta delay, 417
Depletion mode NMOS device, 228
Depletion region, 8, 36
Depletion-Mode Device (DMD), 6
Design
 Design Rule Checkers (DRCs), 111
 flow/entry, 256
 for test, 274
 rules, 90, 98, 102
Design pass/turn/spin, 275
Die cost, 277
Die size, 276
Dies per wafer, 276
Differential Cascade Voltage Switch Logic (DCVSL), 164
Differential transistor-pass logic, 165
Diffusion mask, 14, 17
Diffusion process/region, 99–100
Diffusion wires, 17
Dogleg router, 301
Domino logic, 171

Dopants/Doping concentrations, 13
D-propagation phase, 324
Dual edge registers, 190
Dual-rail complementary pass-transistor logic (CPL), 400
DUT board, 310
Dynamic,
  Dynamic Logic Arrays (DLAs), 233
  hazards, 167
  implementation, 205
  inverter circuit, 51
  latch, 132
  logic, 166
  positive edge-triggered register, 188
  ratioless inverter, 146

Electrical Rule Checker (ERC), 111
Electrically
  Alterable/Erasable ROM (EAROM/EEROM), 27
  Erasable Programmable ROM (EEPROM), 278–279
Electronic Design Automation (EDA) system, 273
Elmore
  constact, 288
  delay model, 162
Embedded gate array, 269
End around shifter (barrel shifter), 127
Enhancement mode, 137, 148
  devices, 153
  PMOS devices, 230
  pull-down devices, 228
Epitaxial layer, 29
Equal-sized gate strategy, 135
Erasable PLD (EPLD), 271
Evaluation, 168, 171
Exit-IR-mode, 347

Failure In Time (FIT), 312
Fan-In/Fan-Out, 134
Fault grading/simulation, 326
Fault sampling, 329
FET switches, 128
Field Programmable Gate Arrays (FPGAs), 226, 242, 263
Field Programmable Logic Array (FPLA), 232
Finite State Machine, 235, 436
  importance, 237
  modularity, 237
Finite-difference equation, 139
Flat cell, 112
Foreground memory, 181
Fouler–Nordheim tunnelling, 27
Full custom ICs/ASICs, 262, 264
Functional Standard Blocks (FSBs), 264
Functionality tests, 309

Gate
  array, 267
  density, 267
  four-input NAND gate, 161
  GaAs gates, 31
  Gate Under Test (GUT), 323
  gate utilization, 276
  primitives, 355
Gate-array based ASICs, 267
Generic Array Logic (GAL), 226
Glitching, 163, 167
Global
  clock lines, 255
  routing, 285, 289

Hardware and software cost, 273
Hardware Description Language (HDL), 256
Hazard rate, 313
Hierarchical design, 109
High frequency MOSFET model, 57
High impedance CMOS transistors, 32
High noise margin, 30, 65
High-to-low state, 79, 194
Hightower algorithm, 302
Hold mode, 182
Hold time, 132, 180
Horizontal and Vertical routing channels, 250
Horizontal constraint graph, 300
Hot-electron phenomenon, 47
HVH routing, 303
Hysteresis, 193–194

I/O level test, 309–310
IDD test, 309
IDDQ testing, 342
Initialization vector, 320
Instantaneous Failure Rate (hazard rate), 313
In-System Programming (ISP) technique, 240, 246
Interconnect, 265
  delay, 286
  matrix, 226
Inverse binary tree, 217
Inverter,
  cascaded complementary inverters, 85
  pair delay, 85
  ratio, 150
  switching threshold, 71, 74

JTAG port, 240

Kink effect, 21
Kogge–Stone logarithmic adder, 215–216

Lambda-based design rules, 97
Latch up, 50
Latches, 131
Latency, 198
Layout editor, 111
Leakage current, 168
Left-Edge Alogorithm (LEA), 299
Level Array Block (LAB), 241
Library cells, 110
Linear Feedback Shift Register (LFSR), 339
Line-search/line-probe algorithm, 302
Local density, 298
Logic,
  cell, 265
  design, 257
  synthesis, 256
Loop-counter, 328
Low-to-High state, 79, 194
Lumped RC transmission line, 138
Lumped-delay model, 288

Macrocell, 232
Macros, 267
Magacells, 264
Manhattan routing, 295
Manhattan-carry chain, 205, 208
Manufacturing (CMOS) tests, 308
Masked ROM/Masked PLD, 271
Mead and Conway, 91, 95, 98
Mealy machine, 236
Mean Time To Failure (MTTF), 305, 312
Micron design rules, 97
Modularity, 261
Monochrome encoding, 91
Moore machine, 236
Moore's law, 276
MOS,
  layers, 91
  structure, 43
  transistor, 60, 91
Multichip Modules (MCMs), 344
Multiplexer-based
  latches, 183
  registers, 186
Multiplexers,
  by-pass multiplexers, 209
  CMOS multiplexers, 125–127
  NMOS (NOR/NAND), 124–125
Multiplicand, 219
Multiplier, 217
  array multiplier, 218, 220
  carry-save multiplier, 222
  Wallace tree multipler, 223
  tree multiplier, 222
Multivibrator
  astable, 193, 195
  bistable element, 181, 193
  monostable, 193–194
  non-bistable sequential circuits, 193

NAND DLA, 234
NAND PLA/NAND–NAND/NOR–NOR, 227
n-channel devices, 21, 65
n-channel pass gates, 251
Negative edge triggered register, 182, 186
NFETs,
NMOS,
  multiplexers, 124
  NMOSFETs, 107
  pass-transistor, 130, 150
  PLA, 227
  precharging circuitry, 147
  super buffers, 139–141
  switch, 159
  tansistor, 42, 64, 72, 256
  tristate super buffer, 142
Noise margin/immunity, 77, 189
Non Return to Zero (NRZ), 310
Non-inverting Boolean function, 160
Non-inverting super buffer, 141, 147
Non-overlapping clocks, 186–187
Non-Recurring Engineering (NRE), 274
n-substrate, 63
n-tree gates, 172
n-well process, 15

Observability, 321, 330
Off-grid/on-grid, 297
One-controllability/one-counter, 328
One-level sensitization probability, 329
One-shot circuits, 193–195
OR–AND, 119
OR–AND–INVERT, 172
Output buffer/inverter, 228
Overlap period, 186
Overpowering, 183

Package body/declaration, 418
Pad drivers, 142
Parallel scan, 338
Parallel simulation, 327
Parasitic
  capacitors, 58, 60
  transistors, 63–65
Partial product, 218–219
Partial scan, 336–337

Pass-transistor, 149, 150, 154
Pause-IR state, 347
Personality matrix, 228–229
Pick-up point, 292
Pipelining, 224
PLCC package, 239
PLDs, 235
PLICE, 280
PMOS,
  PMOSFETs, 107
  transistor, 34–35
PODEM/PODEM-X algorithms, 324
Polysilicon/refractory metal interconnect, 22, 26
Postlayout simulation, 257
Power cells, 266
Power dissipation, 86
  dynamic dissipation, 87
  power economy, 88
  short-circuit dissipation, 87
  static dissipation, 86–87
  total power dissipation, 87
Predefused array, 465
Prelayout simulation, 256
Primitive cell, 267
Priority encoder, 382
Procedural blocks, 366
Product life time, 275
Production test, 274
Productivity, 274
Profit margin, 277
Profit model/flow, 275
Programmable Array Logic (PAL), 233
Programmable Logic Array (PLA), 116–117, 227–228
Programmable,
  Programmable Array Logic (PAL), 226, 271
  programmable ASICs, 262
  programmable interconnect points, 251
  Programmable Logic Array (PLA) devices, 226, 271
  Programmable Logic Devices (PLDs), 226, 263
  Programmable Read-Only Memory (PROM), 226, 246
Programming costs, 274
PROM, 232
Propagate-generate, 205
Propagation,
  delay, 162, 180, 209, 211, 217, 221, 242, 148
  time, 139
Pseudo-NMOS gate, 163, 166
Pseudoterminals, 297
p-tree gates, 171–172
Pull-down/Pull-up Network (PDN/PUN), 159
Pull-up/Pull-down, 361
p-well process, 15

Quad Flat Pack (QFP), 240

Race conditions, 192
Rail-to-tail swing, 161, 192
RAM cells, 251
Random logic, 343–344
Ratioed logic, 163, 140
Regenerative circuits, 193
Register, 183, 189
  edge-triggered register, 182–183
  edge-triggered storage element, 185
  Instruction Register (IR), 347
  pulse registers, 192
  register files, 198
  Register Transfer Level (RTL), 259
Regular logic arrays, 343
Reliability, 311–312
Ring oscillator, 195
Rippling effect, 213
Robustness, 159
ROM (Read Only Memory), 116, 270
Routing
  HVH/VHV routing, 303
  multilevel, 303
Routing procedures, 205
  3/2.5 layer routing, 303
  clock routing, 304
  global routing, 285, 289
  HVH/VHV routing, 303
  Manhattan-routing, 303
  unreserved layer routing, 303
  reserved layer routing, 303
  routing steps, 304
Row-end calls, 266
RTL (Register Transfer Level), 352–353
Run test/idle mode, 347

SA0/SA1 approach, 317
Schimitt trigger, 193–194
SCOAP algorithm, 326
Sea-Of-Gates (SOG) array, 268
Select-TR-scan/Select–DR-scan, 347
Self-test techniques, 339
Semicustom ASIC, 262
Sensitization-counter, 328
Sequential
  0/1 controllability, 325
  modules, 325
  routing, 289
Serial scan, 336
Shared expanders, 242
Side-wall junction, 45
Signal declaration, 417
Signature analysis and BILBO, 339–340
Silicon on insulator/sapphire process, 15
Sneak path, 150–151
Spacer cells, 266
SPICE models, 60

SRAM, 256
Stacked gate structure, 27
Stacked via, 295
Stage ratio, 86
Standard cells/standard cell library, 264
State RAM, 252
Static CMOS gate, 159
Static memory, 181
Statistical Fault Analysis, 328
Steiner tree, 293
Stick diagrams, 91
Structured design, 260
Structured gate array, 269
Stuck at/on/closed/open, 314–317
Subthreshold region, 49
Super buffers, 139
  intermediate-sized super buffer, 148
Surround By Zero, 310
System on a Chip (SOC), 262
System partitioning, 256
System-Level Macros (SLM), 264

Tally circuit, 117
Tap controller/pins, 346
Test program, 310, 321
Testability measures (TMs), 325
Test-Date Registers (TDRs), 348
Testing, 308
  ad hoc testing, 330
  multiplexer-based testing, 332
Timing controls, 368
Timing metrics/parameters, 180–181
Top-down/Bottom-up approach, 289
Tracks/track spacing, 297
Transconductance, 42
  parameter, 57
  ratio, 72
Transistor event, 194
Transmission gate, 11, 184
  multiplexer, 183
  register, 186
Transparent mode/high/low, 182
True Single-Phased Clocked Register (TSPCR), 190
TSPC circuit, 191
Turnaround time, 267
Twin-tub process, 15, 19
Two-/Three-level MOS technology, 265
Two-input LUT, 245

Universal logic module, 152
Update-IR mode, 347
UV-erasable EPROM, 278
UVPROM, 270

Value set, 355
Variable declaration, 417
Vector-merging adder/operation, 222–223
Verilog constraint graph, 299
Verilog HDL, 353
Verilog, 256
Vertical ion bombardment, 23
VHDL, 256, 414
VHV routing, 303
VIAs, 23
Voltage controlled current source/resister, 41
Voltage Controlled Oscillator (VCO), 195

Wafer size/cost, 276
Wafer testing procedure, 310
Waffle Via, 294
Wallace tree multiplier, 223–224
Worst-case arrival times, 211
Worst-case prapagation delay, 181
Wrap around, 239, 127

Y-chart, 259

ZALG, 324
Zero-controllability, 329
Zero-counter, 328
Zero-threshold devices, 142